Oluwatosin Ademola Ijabadeniyi, Christiana Eleojo Aruwa, Titilayo Adenike Ajayeoba (Eds.)

Food Biotechnology

Also of interest

Food Science and Technology.
Fundamentals and Innovation
Oluwatosin Ademola Ijabadeniyi (Ed.), 2nd Edition 2023
ISBN 978-3-11-101309-1, e-ISBN (PDF) 978-3-11-101310-7,
e-ISBN (EPUB) 978-3-11-101348-0

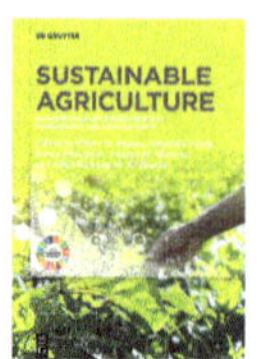

Sustainable Agriculture.
Nanotechnology, Biotechnology, Management and Food Security
Vishnu D. Rajput, Abhishek Singh, Karen Ghazaryan, Tatiana M. Minkina,
Abdel Rahman M. Al-Tawaha (Eds.), 2025
ISBN 978-3-11-136197-0, e-ISBN (PDF) 978-3-11-136975-4,
e-ISBN (EPUB) 978-3-11-136994-5

Industrial Biotechnology
Mark Anthony Benvenuto, 2nd Edition 2024
ISBN 978-3-11-132993-2, e-ISBN (PDF) 978-3-11-133025-9,
e-ISBN (EPUB) 978-3-11-133082-2

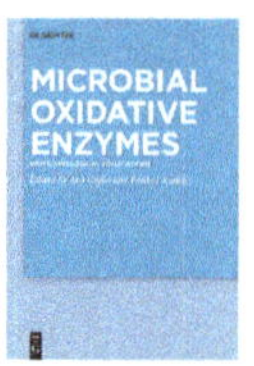

Microbial Oxidative Enzymes.
Biotechnological Applications
Arti Gupta, Pankaj Kumar (Eds.), 2024
ISBN 978-3-11-106172-6, e-ISBN (PDF) 978-3-11-106223-5,
e-ISBN (EPUB) 978-3-11-106271-6

Engineered Nanoparticles in Agriculture.
From Laboratory to Field
Vasileios Fotopoulos, Gholamreza Gohari (Eds.), 2023
ISBN 978-1-5015-2315-1, e-ISBN (PDF) 978-1-5015-2322-9,
e-ISBN (EPUB) 978-1-5015-1552-1

Food Biotechnology

Food Processing, Gene Editing and Safety

Edited by
Oluwatosin Ademola Ijabadeniyi, Christiana Eleojo Aruwa
and Titilayo Adenike Ajayeoba

DE GRUYTER

Editors
Prof. Oluwatosin Ademola Ijabadeniyi
Durban University of Technology
Department of Biotechnology and Food Science
Steve Biko Road, Berea 16
4001 Durban
South Africa
oluwatosini@dut.ac.za

Dr. Christiana Eleojo Aruwa
Durban University of Technology
Department of Biotechnology/Food Technology
Steve Biko Campus, Berea
PO Box 1334
19 Steve Biko
4000 Durban
South Africa
christianaa@dut.ac.za

Dr. Titilayo Adenike Ajayeoba
Adeleke University
Food science and Nutrition program
Department of Microbiology
Faculty of Science
P.M.B 250
Lougun-Ogberin Road
232103 Ede
Nigeria
ajayeoba.titilayo@adelekeuniversity.edu.ng

ISBN 978-3-11-144121-4
e-ISBN (PDF) 978-3-11-144123-8
e-ISBN (EPUB) 978-3-11-144157-3

Library of Congress Control Number: 2025935637

Bibliographic information published by the Deutsche Nationalbibliothek
The Deutsche Nationalbibliothek lists this publication in the Deutsche Nationalbibliografie; detailed bibliographic data are available on the Internet at http://dnb.dnb.de.

Cover image: metamorworks/iStock/Getty Images Plus
Typesetting: Integra Software Services Pvt. Ltd.

www.degruyter.com
Questions about General Product Safety Regulation:
productsafety@degruyterbrill.com

Foreword

Professor Oluwatosin Ijabadeniyi has over 15 years of experience in the food science industry and in academia in the fields of food development and safety. His interests have also included the world of blockchain and these elements have allowed him to be an expert in some of the leading fields that will shape the food economy. His cumulative knowledge in the field of food science has produced this book along with a curated choice of authors that who are specifically selected for their knowledge in their respective fields.

This collection of experts under his editorship has produced a definitive book in the food industry covering multiple areas of emerging interest including food genetics, microbial engineering, enzyme technology, human microbiome studies, functional foods, and artificial intelligence for food production. In each chapter, experts have unpacked concepts for food scientists that will be relevant to the current and upcoming academic market of both established academics and students who wish to explore some or all of these concepts. This is a comprehensive collection of topics, encompassing a large body of knowledge in the food product optimiszation sector, and it will serve as a reference for all who are interested in the food sciences.

Prof. Feroz Mahomed Swalaha
Head of Department
Biotechnology and Food Science
Durban University of Technology
PO Box 1334, Durban, 4000, South Africa

https://doi.org/10.1515/9783111441238-202

Acknowledgement

Oluwatosin Ademola Ijabadeniyi, PhD, MBA, sincerely acknowledges the Association of Public Health Laboratories (APHL) and the Centers for Disease Control and Prevention (CDC) for their invaluable support through the APHL-CDC Public Health Laboratory Fellowship during the period of writing this book.

I am especially grateful to Dr. Sinisa Urban for graciously hosting me and serving as a mentor at the Maryland Department of Health, Laboratories Administration, in Baltimore, Maryland, USA.

https://doi.org/10.1515/9783111441238-203

Contents

List of contributing authors

Benard Odhiambo Oloo
Department of Dairy and Food Science
and Technology
Egerton University

Martha Bosibori Ombonga
Department of Dairy and Food Science
and Technology
Egerton University
Kenya

Jumbale Mwarome
Department of Food Science
Nutrition and Technology
Faculty of Agriculture
University of Nairobi-Kenya
Kenya

George Ooko Abong
Department of Food Science
Nutrition and Technology
Faculty of Agriculture
University of Nairobi-
Kenya

Duke Omayio Gekonge
Department of Food Science
Nutrition and Technology
Faculty of Agriculture
University of Nairobi-
Kenya

Abisoye Solomon Fiyinfoluwa
Department of Biological Sciences
Olusegun Agagu University
of Science and Technology
Okitipupa, Ondo State
Nigeria

Soji Fakoya
Department of Biological Sciences
Olusegun Agagu University
of Science and Technology
Okitipupa, Ondo State
Nigeria

Omotola Folake Olagunju
Department of Science
School of Health and Life Sciences
Teesside University
Middlesbrough TS1 3BX
North Yorkshire
England,
United Kingdom

Blessing Nwokocha
Department of Science
School of Health and Life Sciences
Teesside University
Middlesbrough
England,
United Kingdom

Abiola Folakemi Olaniran
Landmark University SDG 12
(Responsible Consumption
and Production research Group)
Department of Food Science and Nutrition
Department of Microbiology
College of Pure and Applied Sciences
P.M.B. 1001
Landmark University
Omu-Aran, Kwara State
Nigeria

Yusuf B. O.
Department of Biotechnology
and Food Science
Faculty of Applied Sciences
Durban University of Technology
Durban
South Africa

Saheed Sabiu
Department of Biotechnology
and Food Science
Faculty of Applied Science
Durban University of Technology
P. O. Box 1334
Durban 4000
South Africa

https://doi.org/10.1515/9783111441238-205

Oladunni Mary Ayodele
Department of Biotechnology and Food Science
Faculty of Applied Sciences
Durban University of Technology
P.O. Box
South Africa

Melania Dandadzia
Department of Food Science and Technology
Chinhoyi University of Technology
P. Bag 7724, Chinhoyi
Zimbabwe

Tinotenda Nhovorob
Department of Livestock
Wildlife and Fisheries
Gary Magadzire School of Agriculture
and Engineering
Great Zimbabwe University
PO Box 1235, Masvingo
Zimbabwe

Morleen Muteveric
Department of Physics
Geography and Environmental Science
School of Natural Sciences
Great Zimbabwe University
PO Box 1235, Masvingo
Zimbabwe

Faith Matiza Ruzengweb
Department of Livestock
Wildlife and Fisheries
Gary Magadzire School of Agriculture
and Engineering
Great Zimbabwe University
PO Box 1235, Masvingo
Zimbabwe

Temitope Ruth Olopade
Department of Microbiology
College of Basic Medical and Allied Sciences
Salem University
Lokoja, Kogi State
Nigeria

Oluwatobi Victoria Obayomi
Department of Food Science and Nutrition
College of Pure and Applied Sciences
P.M.B. 1001
Landmark University
Omu-Aran, Kwara State
Nigeria

and
Department of Microbiology
College of Pure and Applied Sciences
P.M.B. 1001
Landmark University
Omu-Aran, Kwara State
Nigeria

Ajayeoba Titilayo Adenike
Food Science and Nutrition Program
Department of Microbiology
Faculty of Science
Adeleke University
Ede, Osun State
Nigeria

Lala Opeyemi Titilayo
Food Science and Nutrition Program
Department of Microbiology
Faculty of Science
Adeleke University
Ede, Osun State
Nigeria

Opeyemi Christianah Ogunbiyi
Department of Basic Sciences
Faculty of Science and Technology
Babcock University
Ilishan-Remo, Ogun State
Nigeria

Oluwayomi Christianah Olaoye
Department of Animal Nutrition
College of Animal and Livestock Production
Federal University of Agriculture
Abeokuta
Nigeria

Adetinuke Aina
Department of Works and Physical Planning
University of Lagos
Nigeria

Awoyemi O. Blessing
Department of Microbiology
School of Life Sciences
Federal University of Technology
Akure PMB 704
Ondo State
Nigeria

Ogundolie Frank Abimbola
Faculty of Computing and Applied Sciences
Department of Biotechnology
Baze University
Abuja
Nigeria

Charlene Pillay
Department of Biotechnology and Food Science
Faculty of Applied Sciences
Durban University of Technology
Durban 4001
South Africa

Christiana Eleojo Aruwa
Department of Biotechnology and Food Science
Faculty of Applied Sciences
Durban University of Technology
Durban 4001
South Africa

Adeoluwa Iyiade Adetunji
Labworld/Philafrica Foods (Pty) Ltd
11 Quality Road
Isando, Kempton Park
South Africa

Rose Oluwaseun Adetunji
JBS Innovation Lab
Johannesburg Business School
University of Johannesburg
JBS Park, 69 Kingsway Avenue
Auckland Park
South Africa

Ifeoluwa Komolafe
School of Health and Life Sciences
Teesside University
Middlesbrough
United Kingdom

Oluwaseun Julius
School of Health and Life Sciences
Teesside University
Middlesbrough
United Kingdom

Rodney Owusu-Darko
School of Health and Life Sciences
Teesside University
Middlesbrough
United Kingdom

Jokodola Temilola Tolulope
Department of Food Technology
University of Ibadan
Nigeria

Idowu-Mogaji
Department of Wildlife
and Ecotourism Management
University of Ibadan
Nigeria

Grace Oluwatoyin
Department of Wildlife
and Ecotourism Management
University of Ibadan
Nigeria

Abiona Stella Olusola
Food Science and Technology Department
Federal Polytechnic Ado Ekiti
Nigeria

Victor Ntulia
Department of Food Science and Technology
Faculty of Science, Engineering and Agriculture
University of Venda
P. Bag X5050
Thohoyandou 0950
Limpopo
South Africa

Bono Nethathea
Department of Food Science and Technology
Faculty of Science, Engineering and Agriculture
University of Venda
P. Bag X5050
Thohoyandou 0950
Limpopo
South Africa

James A. Elegbeleyeb
Phytochemical Food Network Research Group
Department of Crop Sciences
Tshwane University of Technology
Pretoria
South Africa

Josphat N. Gichurec
Department of Consumer and Food Sciences
Faculty of Natural and Agricultural Sciences
University of Pretoria
Private bBag X20
Hatfield 0028
Pretoria
South Africa

Elna M. Buysc
Department of Consumer and Food Sciences
Faculty of Natural and Agricultural Sciences
University of Pretoria
Private bBag X20
Hatfield 0028
Pretoria
South Africa

Vivian Chiamaka Nwokorogu
Department of Biotechnology and Food Science
Faculty of Applied Science
Durban University of Technology
P. O. Box 1334
Durban 4000
South Africa

Titilayo Ibironke Ologunagba
Department of Medical Biochemistry
School of Basic Medical Sciences
The Federal University of Technology
Akure
Nigeria

Taofeeq Garuba
Department of Plant Biology
Faculty of Life Sciences
University of Ilorin
P.M.B. 1515
Ilorin, Kwara State
Nigeria

Buka Magwaza
Department of Biotechnology and Food Science
Faculty of Applied Sciences
Durban University of Technology
PO Box 1334
Durban 4000
South Africa

Grace Abel
Department of Biotechnology and Food Science
Faculty of Applied Sciences
Durban University of Technology
PO Box 1334
Durban 4000
South Africa

Hassan T. Abdulameed
Department of Toxicology
Advanced Medical and Dental Institute
Universiti Sains Malaysia
Kepala Batas
13200 Penang
Malaysia
And
Department of Biochemistry
Kwara State University
Malete
Nigeria

Chinmay Hazare
Department of Biotechnology and Food Science
Faculty of Applied Sciences
Durban University of Technology
PO Box 1334
Durban 4000
South Africa

Santhosh Pillai
Department of Biotechnology and Food Science
Faculty of Applied Sciences
Durban University of Technology
PO Box 1334
Durban 4000
South Africa

Fimanekeni Ndaitavela Shivute
Multidisciplinary Research Services
Centre for Research Services
University of Namibia
Windhoek 10026
Namibia

Natascha Cheikhyoussef
Ministry of Higher Education
Technology, and Innovation (MHETI)
Windhoek
Namibia

Ahmad Cheikhyoussef
Multidisciplinary Research Services
Centre for Research Services
University of Namibia
Windhoek 10026
Namibia

Oluwatosin Ademola Ijabadeniyi, Christiana Eleojo Aruwa, and Titilayo Adenike Ajayeoba

Introduction

Food biotechnology: transforming food systems through innovation

Food biotechnology is a rapidly advancing field that addresses pressing global challenges such as population growth, resource scarcity, climate change, and evolving dietary needs. Biotechnology offers innovative solutions to create more resilient, sustainable, and health-focused food systems. It is reshaping how food is produced, processed, and consumed, with significant implications for public health, the environment, and the global economy by integrating innovative technologies.

Biotechnology has transformed traditional food systems by increasing efficiency, improving nutritional value, and minimizing environmental impact [1]. Through innovations like genetic engineering, fermentation, and molecular biology, food production has evolved to create nutrient-rich crops, biofortified foods, and sustainable agricultural practices [2]. These biotechnological advancements are essential in combating global food insecurity while adapting food systems to meet the challenges of climate change and population growth. For instance, genetic engineering has enabled the development of crops with enhanced nutritional profiles, such as vitamin- and mineral-enriched varieties that help combat malnutrition in developing countries [3]. Additionally, biotechnology extends the shelf life of perishable goods and increases crop yields by producing pest-resistant and climate-resilient crops. This minimizes resource consumption and reduces environmental impact, making food production more sustainable.

A notable development in food biotechnology is personalized nutrition, which customizes dietary plans based on genetic makeup. Nutrigenomics and metabolomics are emerging technologies tailored dietary recommendations to optimize health and prevent diseases [4]. Personalized nutrition is gaining popularity for its potential to improve dietary outcomes and lower the risk of chronic conditions such as diabetes and cardiovascular diseases. By empowering individuals to make informed food choices based on their unique genetic profiles, personalized nutrition enhances overall well-being. As it progresses, it could become a key tool in addressing public health challenges worldwide.

Oluwatosin Ademola Ijabadeniyi, Department of Biotechnology and Food Science, Durban University of Technology, Steve Biko Road, Berea 16, Durban, 4000, South Africa, e-mail: oluwatosini@dut.ac.za
Christiana Eleojo Aruwa, Department of Biotechnology and Food Science, Faculty of Applied Sciences, Durban University of Technology, Durban 4000, South Africa, e-mail: christianaa@dut.ac.za
Titilayo Adenike Ajayeoba, Food Science and Nutrition Programme, Microbiology Department, Adeleke University, Ede, Osun

https://doi.org/10.1515/9783111441238-001

Biotechnology is also making strides in the production of hypoallergenic foods. Through precision techniques such as gene editing, scientists can remove or reduce allergenic compounds in foods, making them safer for individuals with food allergies [5]. This advancement not only improves consumer safety but also creates opportunities for allergen-free food products, meeting the rising demand for safe food alternatives. As food allergies continue to increase globally, biotechnology offers solutions that can improve the quality of life for millions of consumers by reducing allergic reactions in commonly allergenic foods like peanuts, wheat, and dairy.

Gene-edited foods, produced using technologies like CRISPR, also offer promising improvements in crop yields, nutritional quality, and resistance to pests and diseases [6]. However, the commercialization of these foods faces regulatory challenges and public safety concerns that should address widespread market adoption. Transparent regulatory frameworks and rigorous safety assessments are essential to establish public trust [7]. Collaboration between regulatory bodies, scientists, and industry leaders is mandatory for these foods to meet safety standards while fostering innovation in food biotechnology.

Functional foods and nutraceuticals, which provide health benefits beyond basic nutrition, are also at the forefront of biotechnological advancements. These products, designed using molecular and biochemical approaches, are formulated to support specific health outcomes such as improved cardiovascular function, immune response, and cognitive performance [8]. Research into bioactive compounds, probiotics, and other functional ingredients underscores biotechnology's potential to shape the future of nutrition and health. As health-conscious consumers seek foods that offer additional benefits, functional foods will play a pivotal role in preventive healthcare, reducing the prevalence of chronic diseases [9].

Recent studies have emphasized the importance of gut microbiota in overall health, including digestive and metabolic functions [10, 11]. Nutrigenomics and microbiome research are uncovering complex interactions between diet, gut bacteria, and health outcomes. These findings are driving the creation of microbiome-targeted foods that improve digestive health and metabolic processes [11] to have a profound impact on future dietary recommendations, offering personalized solutions to enhance gut health and prevent metabolic disorders.

Finally, blockchain technology is transforming food supply chains by improving traceability and transparency [12]. By tracking each stage of food production, blockchain enhances consumer confidence in food safety and supports sustainable, ethical sourcing. This technology not only reduces food fraud and waste but also promotes accountability across the supply chain, building trust in food systems. As consumers increasingly prioritize sustainability and safety, blockchain will play a critical role in shaping the future of transparent and reliable food systems.

Figure 1 explores how interconnected biotechnological advancements reshape modern food systems. Key areas include emerging technologies, functional foods, gene-nutrient interactions, enzyme kinetics, food microbiomes, microbial engineering, food informatics, blockchain traceability, and predictive modeling. These innovations enhance food production, personalized nutrition, and safety, offering sustainable solutions for global food challenges. By addressing health, transparency, and safety, especially with gene-edited foods, biotechnology paves the way for a more resilient, efficient, and health-focused food ecosystem.

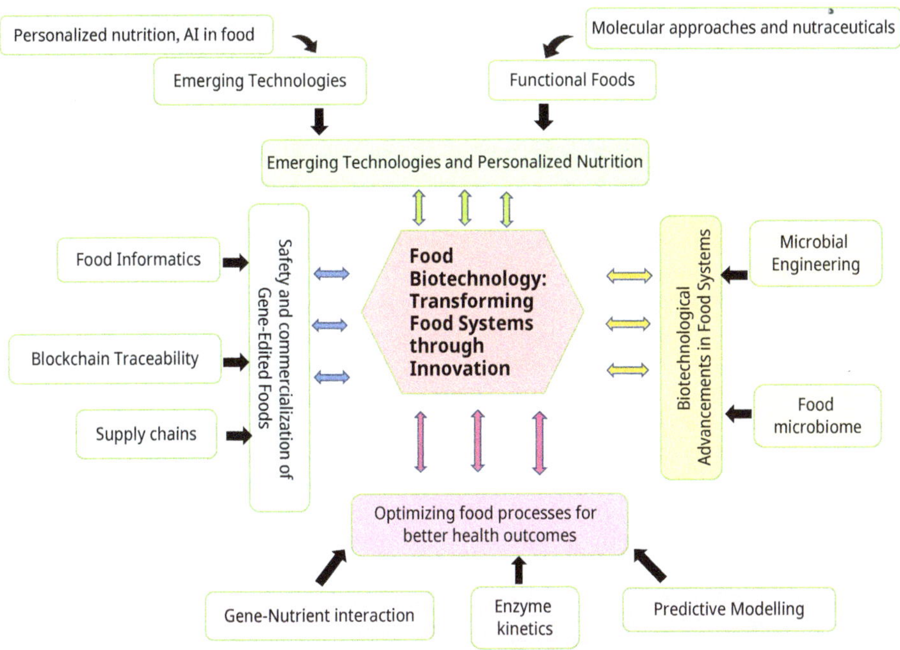

Figure 1: Key innovations in food biotechnology: transforming modern food systems.

Global food systems are under increasing pressure due to population growth, climate change, and resource limitations. Traditional agricultural practices struggle to meet these challenges, leading to food insecurity, environmental degradation, and public health concerns [13]. As consumer needs evolve, there is a growing demand for sustainable, nutritious, and safe food solutions that can address malnutrition, improve crop resilience, and reduce waste.

Food biotechnology offers transformative potential by integrating advanced techniques like genetic engineering, fermentation, and precision gene editing. These innovations are key to enhancing food production efficiency, improving nutritional value, and minimizing environmental impact. Through sustainable practices, biotechnology provides solutions for global food challenges, from personalized nutrition to allergen-free foods. This approach is essential for ensuring food security, promoting public health, and fostering sustainable agricultural practices. By leveraging biotechnology, modern food systems can meet the demands of the future.

This book is essential for students, industry professionals, and academics as it sets the stage for understanding the transformative role biotechnology plays in modern food systems. For students, it provides a foundation, highlighting key concepts like gene editing, nutrigenomics, and sustainable practices that are shaping the future of food. Industry professionals gain insight into emerging technologies that offer competitive advantages in food production, safety, and market expansion, particularly in areas like hypoallergenic foods and blockchain traceability. For academia, the introduction frames the current landscape of food biotechnology, emphasizing the intersection of research, innovation, and public health, and encouraging further exploration into critical areas such as functional foods and gut health. Overall, the introduction connects the broader societal challenges with technological solutions, making it a valuable resource for all three groups.

References

[1] Khan, N., Ray, R. L., Kassem, H. S., Hussain, S., Zhang, S., Khayyam, M., et al. Potential role of technology innovation in transformation of sustainable food systems: A review. Agriculture 2021, 11(10), 984.

[2] Wikandari, R., Manikharda,, Baldermann, S., Ningrum, A., & Taherzadeh, M. J. Application of cell culture technology and genetic engineering for production of future foods and crop improvement to strengthen food security. Bioengineered 2021, 12(2), 11305–11330.

[3] Harsonowati, W., Utami, D. W., Prathama, M., Soetopo, D., Ahmad, H. M., Oranab, S., et al. Use of biotechnological techniques for augmenting micro-and macronutrients use efficiency in plant system. Essential Minerals in Plant-Soil Systems: Elsevier 2024, 405–421.

[4] Lagoumintzis, G., & Patrinos, G. P. Triangulating nutrigenomics, metabolomics and microbiomics toward personalized nutrition and healthy living. Human Genomics 2023, 17(1), 109.

[5] Hammond, B. G., & Fuchs, R. L. Safety Evaluation for New Varieties of Food Crops Developed through Biotechnology. Biotechnology and Safety Assessment, CRC Press, 2019, 61–79.

[6] Guo, Y., Zhao, G., Gao, X., Zhang, L., Zhang, Y., Cai, X., et al. CRISPR/Cas9 gene editing technology: a precise and efficient tool for crop quality improvement. Planta 2023, 258(2), 36.

[7] Chartres, N., Sass, J. B., Gee, D., Bălan, S. A., Birnbaum, L., Cogliano, V. J., et al. Conducting evaluations of evidence that are transparent, timely and can lead to health-protective actions. Environmental Health 2022, 21(1), 123.

[8] Handa, A. K., Fatima, T., & Mattoo, A. K. Polyamines: bio-molecules with diverse functions in plant and human health and disease. Frontiers in Chemistry 2018, 6, 10.

[9] Baker, M. T., Lu, P., Parrella, J. A., & Leggette, H. R. Consumer acceptance toward functional foods: A scoping review. International Journal of Environmental Research and Public Health 2022, 19(3), 1217.

[10] Gomaa, E. Z. Human gut microbiota/microbiome in health and diseases: a review. Antonie Van Leeuwenhoek 2020, 113(12), 2019–2040.

[11] Fan, Y., & Pedersen, O. Gut microbiota in human metabolic health and disease. Nature Reviews Microbiology 2021, 19(1), 55–71.

[12] Menon, S., & Jain, K. Blockchain technology for transparency in agri-food supply chain: Use cases, limitations, and future directions. IEEE Transactions on Engineering Management 2021, 71, 106–120.

[13] Sasson, A. Food security for Africa: an urgent global challenge. Agriculture & Food Security 2012, 1, 1–16.

Benard Odhiambo Oloo* and Martha Bosibori Ombonga

Chapter 1
Biotechnological advancements in food systems

Abstract: Biotechnology has developed at a relatively faster pace and has made a positive impact in the global food systems with regard to food security, nutritional quality, and sustainability. This chapter focuses on the advanced biotechnology solutions implemented in the production of food to meet the demands of a growing global population in the face of challenges such as climate change. Genetic engineering, precision fermentation, and synthetic biology have changed the nature of agriculture by enhancing the production of crops, resistance to diseases and pests, and renouncing on chemical fertilizers. Notably, the use biotechnology in food processing and preservation has further helped in the production of functional foods, new protein sources, and bio fortification, thereby improving the well-being of human beings.

Biotechnology also has a major role to play in minimizing environmental effects that are there in conventional agricultural practices. GMOs and gene editing technologies like CRISPR-Cas9 provide the means through which emissions control, water conservation, and land degradation can be achieved. Also, cellular agriculture such as cultured meat and fermentation using microorganisms offer more sustainable protein sources cheaper and with less environmental impact than animal farming.

However, biotechnology practice in food systems has some challenges which include regulation, ethical issues, and consumer acceptance barriers. Such issues as the safety of genetically modified foods and ethical issues associated with lab-grown products make it important for there to be clear guidelines and good practices concerning such labeling. Sustainability considerations and public concerns need to also be integrated by policymakers, scientists, and industry players so as to enhance biotechnological advancements.

The remainder of this chapter briefly discusses the current state and future prospects of biotechnology in food systems. It underlines the necessity of cross-sectoral collaboration in facilitating innovation for safe and affordable food production that can meet the standards of sustainability and profitability. Although the application of biotechnology in food systems is still in its infancy, it provides a ray of hope that the food problems of the world can be solved, if there is careful regulation and participa-

***Corresponding author: Benard Odhiambo Oloo**, Department of Dairy and Food Science and Technology, Egerton University, Njoro, Nakuru, Kenya, e-mail: olooo.odhiambo@gmail.com
Martha Bosibori Ombonga, Department of Dairy and Food Science and Technology, Egerton University

https://doi.org/10.1515/9783111441238-002

tion by society. Finally, the future of biotechnology use in agriculture will be a key component in establishing sustainable and socially optimal food supplies that meet future generations' needs.

Keywords: Biotechnology, food security, genetic engineering, CRISPR-Cas9, GMOs, synthetic biology, precision fermentation, cellular agriculture, lab-grown meat, bio fortification, alternative proteins, sustainable agriculture, food innovation, environmental sustainability

1.1 Introduction

Sustainable Development Goal (SDG) 2 aims to ensure food security, nutrition and a sustainable food supply. This is challenging because of the constantly growing population and strained natural resources, among other factors. It is predicted that in 2050, the global population will reach 10 billion [1–3]. In addition, an increase of up to 60% in global agricultural food production is needed annually to meet food demand in 2050 [4, 5]. African countries are most affected in regard to food security and nutrition. Africa accounts for 20% of the global population, and most countries have recorded the lowest gross domestic product GDP [1]. In spite of abundant natural resources, African countries are struggling to achieve food security. An increase of 77% in food production is required annually to meet the demand for food by 2050. On the other hand, developed countries need a 24% food production increase to meet their food demand by 2050 [5]. One of the reasons for strained food production is the limited agricultural technology, such as the application of advanced biotechnology in food systems.

Apart from Africa, globally, there is a need to find alternative food sources that will be sustainable to feed the rising population. One of the ways of achieving this is by applying biotechnology to food systems. Traditional agriculture has faced many setbacks, thus creating a need for improvement. Climate change is one of the factors affecting traditional farming. Destruction of the ozone layer has led to regional climatic changes characterized by extreme weather events such as heat waves caused by the warming up of the earth (rise in temperature) due to increased greenhouse gases such as methane. Heat waves are detrimental to crops, animals and the entire agricultural sector, including aquatic life. High temperatures cause snowpacks and glaciers in the mountains to melt at a high rate even in winter; thus, water is not as reserved as before. Consequently, this leads to the unavailability of enough fresh water needed by fish and other aquatic animals throughout the year [6]. In addition, extreme temperatures increase survival of pests and diseases even in winter, and as a result, crops are destroyed [7]. On the other hand, high temperatures cause an increase in evaporation, leading to heavy rainfall that results in flooding in some areas. Approximately 80% of the total amount of food crops depends on rainfall. For this reason, changing weather patterns affect food production as most crops are prone to destruction in extreme drought periods and floods [6, 7]. Global food production, especially cereals such as maize and wheat, is projected to decline by 3.8% and 5.5%, respectively, due to the adverse effects of climate change [5].

One of the ways of mitigating the effects of climate change and ensuring food security despite the increasing population is to apply biotechnology in food systems. In farming, biotechnology has been used to come up with crops that have superior attributes compared to their predecessors. Moreover, biotechnology has advanced over the years with the aim of continuously improving food production. For instance, genetic engineering (GE), the current biotechnology advancement, has been used to create genetically modified (GM) crops that are resistant to abiotic stresses (salinity, heat stress, drought, and cold stress) that are currently advancing because of climate change [5]. *Bacillus thuringiensis* crops have proven to have desirable traits such as resistance to pests and diseases, high yield and little dependence on herbicides, pesticides and other chemicals [8, 9]. Tissue culture has been used to come up with superior planting materials that are free of diseases and high quality [10]. Biofortification has improved the nutritional value of foods by incorporating essential nutrients needed by the body [11]. Biotechnology has been applied in the production of alternative proteins, such as mycoproteins and lab-grown meat, which have the potential to ensure food security in the future [12].

1.2 Applications of biotechnology in food and crop production

1.2.1 Genetically modified organisms

GM organisms (GMOs) refer to plants, animals, or microorganisms whose genetic material has been altered using recombinant deoxyribonucleic acid (rDNA) to confer specific benefits. GE techniques are used to cut and insert genes from a particular organism of a species and transfer them to an organism of a different species [13, 14]. For instance, scientists are able to insert genes into crops from a soil bacterium known as *Bacillus thuringiensis* (Bt), enabling them to be insect resistant. When Bt genes are inserted in the desired crop and an insect attack, toxic protein crystal (Cry) and cytolytic (Cyt) are produced by the digestive system of the insect and proceed to solubilize in the midgut. They are then activated by proteases and bind to specific receptors in the insect cell membrane, leading to cell disruption, and the insect dies [15, 16]. Such transgenic crops are known as Bt crops, and they are resistant to pests and insects from Lepidoptera, Coleopteran, and Diptera. Plant lectins have also been used against such insects and pests. Lectins work by inhibiting nutrient absorption and disrupting digestive cells by stimulating endocytosis and producing toxin metabolites, leading to death [17].

Traditional crops are easily attacked by pests and insects, resulting in adverse negative effects. Insect pests endure plant growth by sucking sap and chewing plants, along with acting as vectors for transmitting viruses, fungal and bacterial diseases [18]. This leads to low crop yields, and farmers lose the economic value of their crops. Approximately 37% of crops are lost due to insect pests and diseases, with 13% from insects globally [17]. Generally, crop loss to pests accounts for 20–40% of global crop loss annually [13]. Farmers opt for insecticides to control pests and diseases, and this has been linked to negative impacts. The health of farm workers who spray pesticides is compromised, and approximately 1–3% suffer from pesticide poisoning [17]. Besides, such chemicals also cause environmental degradation, and they are expensive.

Insect-resistant crops have proven to be more reliable than conventional crops due to their numerous advantages. In the USA, approximately 90% of corn, cotton, and soybeans are GM [14]. The adoption of such crops was influenced by the noticeable benefits of GM crops. For instance, in Arizona, US, Bt cotton led to increased yields by 5% and US$25–65 were saved per acre due to low pesticide use [19, 20]. In India, Bt cotton led to an approximately 50% increase in profits and a 2% increase in yields compared to conventional cotton [13]. Reduced pesticide use helps conserve beneficial soil insects, reduce water contamination, and prevent soil erosion due to reduced tillage. In Bangladesh, field trials for Bt eggplant turned out successful. There was a 5% increase in yields for farmers who plant Bt brinjal (Bt eggplant). Reduced pesticide use lowered costs meant for their purchase and application by 37.5%, leading to low input costs. Net revenues increased by 128% because of less pesticide use and high yields [20]. In another trial, Bt eggplant farmers saved 22% in human labor, 29% in chemical fertilizer, 20% in mechanical power and 61% from low pesticide application. Besides, farmers got yields higher by 13% and 83% higher income compared to farmers who planted conventional eggplant [21]. Bt cotton has been a game-changer for cotton production in Africa. In Kenya, before Bt cotton, they could produce only 20,000 bales a year while their demand was 140,000. That meant relying on imports to fill the gap. However, since adopting Bt cotton in 2012, Kenya's cotton production has soared, potentially reaching self-sufficiency as Bt cotton yields can be up to 260,000 bales [22].

Herbicide resistance can also be achieved through genetic modification. In Australia, Roundup Ready GM variety of canola was introduced. The variety contained EPSP (5-enol pyruvylshikimate-3-phosphate) synthase alterations combined with a glyphosate oxidoreductase gene that made it glyphosate-resistant. The variety led to more canola being cultivated because it was tolerant to blackleg diseases, and weeds made traditional varieties produce a lot of anti-nutritional compounds that reduced the quality of canola oil. In 2014, GM canola planting increased to 21% from zero in 2009 [13].

GM crops can be made to tolerate abiotic stresses such as drought, salinity, extreme heat or cold, and heavy metals, thus improving agricultural productivity [14]. For example, gene modification by increasing the activity of a specific gene (AtNHX1) that removes sodium from cells has improved salt tolerance in various crops like Arabidopsis, tomato, and soybean. In rice, introducing genes like OsGrx_C7 from other plants helps to boost the activity of other genes involved in salt tolerance [18]. Extremely cold temperatures are detrimental to plants. For this reason, making them resistant to frosty temperatures is necessary. A gene from cold-water fish transferred to plants such as potatoes and tobacco allows them to withstand cold conditions [23]. To mitigate the effects of drought in sub-Saharan Africa, a public–private partnership led by the African Agricultural Technology Foundation (AATF) developed Bt MON810, a drought-tolerant maize variety. This GM maize is cultivated in South Africa, and field trials continue in Kenya, Ethiopia, Nigeria, Mozambique, Uganda, and Tanzania, with the goal of commercialization.

The ongoing debate surrounding GM foods includes safety concerns and ethical considerations, among others. There is concern about the safety of viral promoters used during genetic modification that might cause infections in people who consume GM plants. For instance, the cauliflower mosaic virus (CaMV) promoter is commonly used to activate genes in engineered corn, cotton, and canola. Research suggests these weakened viruses can pick up missing pieces from nearby genes and become infectious again [24]. GMOs may have a negative impact on the environment by killing a lot of species of insect larvae. For instance, monarch butterfly caterpillars died after eating milkweed crops containing fallen pollen from Bt corn [24].

Many African countries have diverse agricultural traditions. This strong cultural connection to agriculture can lead to skepticism about GM crops, with some communities fearing disruption to their traditional practices. Public awareness is necessary to enlighten them on both the potential benefits and drawbacks of GM crops, facilitating a productive conversation about this technology [25]. A common misconception is that only GE technology alters our food, while conventional breeding does not yet; both methods rely on genetic changes to achieve desired traits in crops. However, GE techniques offer a more precise approach than conventional breeding. This precision allows scientists to identify the specific mutations made to the DNA. In contrast, conventional breeding methods like radiation mutagenesis can cause numerous, and often untracked, changes to the plant's DNA [26]. Besides Africa, other developed countries raise concerns about the impact of GMOs on biodiversity. Strong evidence suggests that genes from GMO plants can be transferred to other plants or organisms through pollen dispersal, particularly among closely related species. This could lead to a competitive disadvantage for natural wild varieties, potentially causing their decline or even disappearance. The other concern is that some weed species might develop resistance to herbicides [27].

1.2.2 Gene editing

Traditionally, plant breeders improve crops by crossing different varieties and selecting the best offspring. This works well for most crops, but it is difficult with wild species that rarely or never cross with domesticated ones. This crossing is often done to introduce genes for stress tolerance (like salt resistance) from wild relatives into cultivated crops. Scientists employ mutagenesis, where high-dose radiation randomly damages a plant's DNA, forcing it to repair itself, leading to random mutations. This can lead to beneficial mutations, but it also creates many unwanted changes throughout the genome. Sorting through these changes to find the desired one is a lengthy process, often involving back-crossing the mutated plant with its original variety for several generations [28, 29]. Therefore, the process is time-consuming because it can take years to isolate the desired mutation from the unwanted mutation.

The ideal solution to the major challenge of conventional breeding would be a way to make specific, targeted changes to a plant's genetic material. Genome editing offers a much more precise and efficient way to alter specific genes at defined locations. It can either suppress unwanted genes or boost the activity of beneficial ones, all with greater control and speed compared to traditional mutagenesis [28]. Generally, gene editing refers to the alteration of the genetic material of an organism at specific locations by deletion, insertion, or substitution to achieve desired attributes such as resistance to abiotic stress [30, 31]. Gene editing is different from GMOs because it does not involve the introduction of foreign genetic material [32]. This is done with the help of gene editing tools such as clustered regularly interspaced short palindromic repeats (CRISPR/Cas9), transcription activator-like (TAL) effector nucleases (TALENs), and zinc-finger nucleases (ZFNs) [33]. The CRISPR/Cas9 technique is the most commonly used tool for gene editing because it offers precision alteration with the ability to create specific, heritable genetic changes in a cost-effective manner, and the crops produced are similar to their conventional types. Furthermore, CRISPR/Cas9 boasts additional advantages like its ability to edit multiple genes simultaneously (multiplexing), precisely insert or replace small pieces of DNA (prime editing), and directly modify single DNA bases (base editing) [31].

Gene editing has tremendous benefits in the food chain because of its ability to confer desirable traits in crops that make them resistant to abiotic stress that is being accelerated by climate change. For instance, the CRISPR/Cas9 gene-editing tool has been used to identify the role of two genes, SAPK1 and SAPK2, in plant response to salt stress. These genes are activated by osmotic stress and a plant hormone called ABA [18]. Gene-edited crops range from tomatoes, rice, wheat, and maize, among others. Some of the gene-edited crops have been made to improve nutrition compared to their traditional counterparts. In Asia, a company called Sanatech Seed, in collaboration with the University of Tsukuba, launched a new type of gene-edited tomato, Sicilian Rouge High gamma-aminobutyric acid (GABA), on September 15, 2021. The edited tomato was produced by applying CRISPR-Cas9 technology to elevate the levels of GABA that is known to promote relaxation and lower blood pressure [34]. In the USA and Argentina, field trials of genetic edited high oleic soybean oil were done [32]. Scientists have employed the CRISPR/Cas9 gene editing technique to target the mildew-resistance locus O (TaMLO) gene in wheat. This gene controls resistance to powdery mildew disease caused by the fungus *Blumeria graminis* f. sp. *tritici* (Btg). By knocking out TaMLO using CRISPR/Cas9, researchers have successfully created wheat mutants that display resistance to the disease [35, 36].

Genome editing has led to the creation of improved food products that have been made available in markets like the USA and Japan. These include soybeans with a high oleic acid content, corn with a waxy texture, faster-growing red sea bream fish, mustard greens without pungency, and even a type of pufferfish free of its natural toxins and tomatoes rich in GABA (an amino acid) [37]. In developing countries like Kenya, CRISPR/Cas9 gene editing has led to the development of safer cassava varieties. This was achieved by targeting and deleting a specific gene called cytochrome P450 (CYP79D1), known to be involved in cassava's production of cyanide. By deleting this gene, reduced cyanide levels in cassava by sevenfold were recorded [38].

Gene-edited foods face the following challenges: The public's lack of awareness of the differences between GMOs and gene-edited foods makes it difficult for quick adoption. Gene editing is now limited to a number of crop species, and initial setup costs for specialized laboratories are expensive. Regulatory systems that treat gene editing like GMOs limit the potential positive impact of such foods [31].

1.2.3 Tissue culture

Conventional plant breeding posed a challenge to scientists, who had to seek other ways of solving the problems. Normally, plants reproduce by either sexual or asexual reproduction. In sexual reproduction, seeds are formed where the female and male DNA are fused, resulting in new plants that might be genetically different from the parent plants [39]. Scientists could take years to breed crops with desirable characteristics. On the other hand, asexual reproduction involves vegetative propagation, where new plants grow from a vegetative part of the parent plant, such as a stem, a leaf, or roots. This results in new plants that have genes similar to those of the parent. This means that the new plants could be prone to pests, diseases, or any other negative characteristics that were in the parent plants. Tissue culture has been adopted to mitigate such challenges posed by traditional breeding.

Tissue culture involves growing plant cells, organs, and tissues on a culture media containing growth factors (nutrient media). Tissue culture was first predicted by Theodor Schwann in 1832, who realized that cells can grow outside their original host if similar growth conditions are provided. In 1907, Ross Granville Harrison obtained nerve cells from a frog and was able to culture them in solidified lymph, and he became the father of tissue culture [40]. In 1980, there was mass propagation of plants using tissue culture [41].

Tissue culture is one of the biotechnological tools, and it has proved to be reliable in regard to the production of quality and quantity planting materials. Plants from tissue culture have superior features compared to their conventional plants [42]. For instance, they can be high-yielding, free of pests and diseases, have a short maturity time, and have the ability to grow on infertile soils. Examples of crops that can be produced through tissue culture include bananas, dates, sweet potatoes, palms, and pineapples, among others. Micropropagation, one of the forms of tissue culture, allows the production of numerous high-quality planting materials [43]. This has led to the mass production of copies of plants in a short period of time, besides being of high quality. Tissue culture has also provided an avenue for the application of advanced modern biotechnology in plants. For instance, tissue culture and recombinant DNA are required to produce transgenic crops [44, 45].

Farmers around the globe have benefited from this technique, especially those who rely on planting materials for reproduction. For instance, in Thailand, tissue culture accelerated the growth rate of orchids, enabling the country to produce over 50 million plantlets annually and becoming one of the main exporters of orchids globally [46, 47]. The Philippines and Indians were able to prevent banana top virus and banana bract mosaic virus from invading their banana through tissue culture [48]. In developing countries like Kenya, production of bananas has started to decline because of soil infertility and invasion of pests and diseases mostly transmitted through propagation using infected suckers. Tissue culture was adopted, and immense benefits were recorded. Tissue culture reduced maturity from 2–3 years to 12–16 months, banana bunch weights were increased from 10–15 kg to 30–45 kg, and annual yields increased per unit of land [34, 49].

Tissue culture, like any other biotechnology technique, faces challenges. Sometimes, there is uncertainty about whether the crops cultured in the laboratory environment will be able to handle stress when planted out in the fields. If contamination occurs, resources and time are wasted. Finally, there are no clear and established protocols to be followed for all plant species [44].

1.3 Biotechnology in food processing and preservation

1.3.1 Precision fermentation

With the world population expected to hit over 9 billion people by the year 2050 [50], there have been concerns about whether the current food system would be able to keep up with the food demands. Statistics show that in order to meet global food demands and healthy dietary needs, production must be at least doubled [50]. Doubling food production would be problematic environmentally because more production means more greenhouse gas emissions, which would be detrimental to environmental sustainability efforts. Precision fermentation has been fronted as the possible solution to facilitate the availability of sufficient and nutritious food to people. Precision fermentations uses GM microbes such as yeast and bacteria to produce desired ingredients such as proteins [51]. The process of precision fermentation starts through the identification of desired molecules. Thereafter, the best host is selected, as each molecule has a preferred host. Some molecules can do well in yeast, and others in fungi. After the identification of the host, the genes are inserted into the genome of the microbes and then later transferred and grown in fermentation tanks that have nutrient broth under controlled conditions [51]. After maturity, the molecules are separated from the broth through filtration.

Precision fermentation is the future in the face of limited land and unpredictable weather patterns. These patterns, characterized by extreme weather events such as long periods of drought and heavy rainfall, undermine farming in developing countries that heavily rely on weather for their agricultural activities [52]. As noted earlier, the form of fermentation occurs in bioreactors. These fermenters make it easy to control temperatures and other conditions, such as pH acidity, to meet particular needs [51]. This overcomes the naturally fluctuating weather conditions that have become unpredictable and destabilizing farming activities across the globe. Therefore, with precision fermentation, the world will stand to overcome hurdles that have hampered humankind for ages, thus ensuring that the cultured cells turn into food that people can use with external environmental interventions. If these technologies are adopted, production cycles will remain uninterrupted, resulting in higher yields, unlike in the past. Moreover, this kind of technology would ensure a shorter production cycle, meaning more yields in a relatively short period of time than traditional ways.

1.3.2 Mycoproteins

Mycoproteins, produced through precision fermentation with the help of a fungus called *Fusarium venenatum*, serve as a substitute for animal meat. Experts prefer to use this type of fungi because research has shown that it leads to end products that are rich in protein in a shorter production cycle. Further research studies affirm that mycoproteins are nutritious – they contain low fat and high fiber. Since they are grown in a controlled environment, sustainability is achieved, which will be instrumental in meeting our environmental obligations [20]. Notably, these forms of proteins could play a huge role in healthcare in a number of ways, as essential amino acids comprise 41% of the total proteins [53]. They have the ability to maintain healthy skeletal muscle through maintenance and reconditioning. Also, mycoprotein maintains cholesterol in the blood. In addition to that, it controls glucose and insulin. Such usage points to a bright future if precision fermentation is called up to allow the production of products such as mycoproteins that are highly nutritious and sustainable. Sarcopenia, a healthcare issue, would be a long-term solution because of the affected demographic of older adults who require high protein content to maintain their muscle protein synthesis for good health. The promising future of mycoproteins has been cemented by recent health research on their effect on health. The findings show that mycoproteins cannot be categorized as allergic because the possibility of triggering allergic reactions in the human body is low, thus making them safe for us [53].

Despite its promising future in producing proteins, precision fermentation suffers from scaling up the production due to a number of problems. The problems emanate first from the unavailability of capacity to support the production process. The production process itself is tedious to manage to bring about optimum growth because it involves the control of various aspects, such as ensuring that there are appropriate substrate, temperature, and pH, which are hard to achieve as the scaling up results in heat transfer, among other production challenges that will overall impact on the production output [54]. Other research studies points to the possibility of lack of know-how to go about scaling up in precision fermentation. However, preliminary research studies claim that solid-state precision fermentation (SSPF) could be the solution to the problem [55]. In addition to SSPF, other processing governing procedures have been deemed fit to optimize the productions. Such include the process hazard analyses. These processes deploy manufacturing practices in a systematic way to identify issues and impacts for easy mitigation.

1.3.3 Biofortification

Biofortification has been hailed as the most efficient and cost-effective way to combat the macronutrient malnutrition problem that affects many people in the world. Biofortification entails the production of crops through crop breeding that have sufficient bioavailable nutrients in their edible portions. Conventionally, biofortification has been achieved through crop breeding of staple foods, leading to improved levels of minerals and vitamins. However, sometimes conventional crop breeding experiences are practically impossible if there are no sufficient genetic variations of the desired trait, such as protein, in the staple crop populations. This has made it hard to improve certain micronutrients in some staple crops; for instance, the enhancement of Vitamin A using traditional crop breeding in rice became problematic due to the fact there are no rice types rich or have a precursor of vitamin A [56]. Fortunately, this limitation was overcome using biotechnology, especially through GE, which has led to the production of golden rice.

1.3.4 Golden rice

In a fight against vitamin A deficiency that causes irreversible blindness and makes one prone to diseases, research has been conducted resulting in Golden Rice2 which is rich in beta-carotene. Notably, beta-carotene is unavailable in the edible part of the rice seeds – endosperm. Thus, two genes were introduced that encoded phytoene synthase (PSY) and phytoene desaturase CRTI. For example, the golden rice is obtained by replacing PSY in golden rice with the maize gene [57]. This led to increased levels of beta-carotene by 23 times, which is a significant increase that would be key in addressing challenges associated with deficiency of vitamin A [34]. Such demonstrations show that biotechnology stands to solve the macronutrient malnutrition that ails our societies, especially the regions of the global south that depend on staple foods that lack critical dietary nutrients that could keep them safe.

1.3.5 Iron-bottled beans and provitamin A maize

Other research studies also indicate that biotechnology can be used to produce beans with higher protein bioavailability. Such research study conducted with the aim of biofortifying crops and fighting anemia that causes death among children and adults affirmed that biofortified red mottled colored beans contain more bioavailable iron than standard red beans [58]. Other research has uncovered that biotechnology could be deployed in biofortifying maize with vitamin A. Through transgenic incorporation, maize can be enriched, leading to the production of provitamin A hybrids that contain key ingredients instrumental in fighting the hidden hunger that occurs through eating food deficient in vitamin A [11]. The above research, among many others, gives a sneak view of the potential of biotechnology. This confirms that biotechnology in the biofortification process is a viable and long-term solution to the macronutrient solution in that once crops have been biofortified, the desired traits remain in the crops' seeds. In that regard, it is evident that although biotechnology may be resource-intensive and time-consuming through long periods of research, their outcomes are worth the pursuit because the desired traits let beta carotene in golden rice remain in the edible parts of the seeds, which can be used for future planting.

1.3.6 Lab-grown meat

The key drivers of lab-grown meat are food security, environmental conservation, human health, quality food, and animal welfare, among others. The increase in population has led to an increase in food demand. For instance, the consumption of meat is projected to increase by 80% by 2050 [3]. Lab-grown meat will play a significant role in ensuring food security. Traditional meat production is dependent on rearing livestock. However, livestock keeping requires the utilization of depleting natural resources such as land. Currently, land under livestock rearing is approximately 75% of the total land under agriculture. In addition, animal rearing has led to the emission of greenhouse gases. Approximately 8–15% of greenhouse gas emissions are a result of livestock [5, 59, 60]. Methane gas results from the burning of crop residues, enteric fermentation, and manure management. Carbon IV oxide stems from the grazing of animals' fossil fuels when using machines for feed processing. Nitrous oxide from crop residues, drained organic soils, manure management, synthetic fertilizers, and manure left on pasture and applied to soils [61–63]. Besides, greenhouse gas emissions, pesticides, and all chemicals used to grow feeds may be leached and end up polluting water sources, thus posing a risk to aquatic life [60]. Consumption of meat has been linked to cardiovascular diseases such as obesity and arteriosclerosis due to high levels of saturated fat in meat and meat products such as sausages. In addition, traditional meat could pose a risk to consumers in regard to diseases. Sickness might strike meat consumers if the meat contains pathogens such as *Salmonella, Listeria,* and *Escherichia coli*, among others, if the meat is contaminated. On the other hand, the use of excess antibiotics to ensure animal health may lead to antibiotic resistance, which could be projected onto consumers of the meat. Approximately 80% of antibiotics are used in veterinary drugs in the United States [59]. Slaughtering of animals for meat is considered torture and cruelty. Therefore, lab-grown meat is an ideal alternative to producing meat without necessarily rearing lots of animals for the same.

The first meat made in the laboratory dates back to 2013, when Mark Post cultured bovine skeletal muscle cells and eventually, meat burgers were made and sensory evaluated in August 2013 in London. In 2020, Singapore became the first country to approve the sale of cultured meat. This achievement came from a US startup, Eat Just. The company aimed at producing 12,000 tons of pork annually, equivalent to eliminating the need to slaughter 170,000 pigs. The meat was not sourced from slaughtered animals but rather grown in a bioreactor using a plant-based nutrient supply. While the initial process involved fetal bovine serum, it was removed before the meat reached consumers [64]. Cultured meat will be beneficial by meeting the high demand for meat, reducing greenhouse gas emissions, preserving natural resources such as land and reducing the risk of contracting zoonotic diseases brought about by rearing livestock [59].

Cultured meat is made by first getting cells from live animals that will multiply indefinitely and also mature into functional skeletal muscle tissue, mimicking what happens in a living animal. Secondly, a three-dimensional scaffold is needed to provide a supportive structure for the growing muscle cells. This scaffold should allow for efficient nutrient delivery and waste removal, mimicking the role of a circulatory system. Finally, the cells need to be cultured in a bioreactor that provides the ideal environment for their growth and maturation. An essential component of any cell culture system, including in vitro meat production, is an affordable nutrient medium. This medium must provide all the necessary nutrients that the growing cells would normally obtain through digestion. Muscle fibers are then processed into meat products such as meat burgers [12].

1.4 The future of food systems: challenges and opportunities

Significant literature affirms that biotechnology is going to be revolutionary in the long term if it is widely adopted in solving micronutrient malnutrition, a health issue that affects almost half of the world's population [65]. It will ensure that the availability of sufficient and nutritious food in the world in a cost-effective manner. Such affirmations are due to the production of crops with desired concentrations of bioavailable nutrients and other desirable attributes, such as resistance to abiotic stress. Such progress in agriculture will have a tremendous effect on people's lives and in many countries, especially those in the global south that are struggling with food security-related problems. Golden rice, an improved variant of rice that is a staple food to numerous millions across the globe, has enhanced levels of vitamin A, thus preventing blindness and other ailments that are vitamin A deficient [66]. Other biotechnology crops are available that have been engineered to contain bioavailable nutrients such as zinc, iron, and vitamin A, these include maize and beans, and many others have been produced in various countries to solve region-specific lack of critical macronutrient issues [67]. Such advancements reveal that solutions to micronutrient malnutrition, a global problem for children and pregnant women among other demographics of world populations, are available. Application of gene editing and genetic modification leads to crops that are adaptable to climate change, are resistant to abiotic stress and are able to withstand pests, insects, and diseases. Notably, despite this, biotechnology products are not widely accepted due to information or misinformation that is misleading, which has led to partial acceptance due to fear that they are not safe for use through the use of negative propaganda that has led to frustrating legal bottlenecks that stifle this responsible innovation.

Biotechnology products that involve the alteration of genes of plants and animals face mounting criticism from GMO critics who cite that the research behind them might be malicious and unsafe. For instance, Greenpeace is one organization that has been at the forefront of criticizing all GM products. Unfortunately, its stance on biotechnology and GM crops is exaggerated and based on misinterpretations and misunderstanding of science, thus leading to unsubstantiated claims that have no basis in biology [68]. As a result, such organizations have led to wide speculation that GMOs is unsafe, and this narrative has been taken up by the masses, leading to slow acceptance in the market by the consumers [69]. Importantly, the misinformation thrives on inappropriate risk assignment of the biotechnology products and to some, it is due to long-held moral belief that the alteration of genes does not align with God's will [70, 71]. Despite this backlash, biotechnology has continued to gain traction across the world. Notably, the criticism has not been overly destructive, but it has been informative. There have been legitimate concerns that have shaped and instituted policies to ensure GM products are produced in legal and ethical ways to safeguard them. The dark sides of regulations that have been strict are becoming legal bottlenecks that stifle the adoption of biotechnology.

Looking through the propaganda or misinformation around biotechnology products, it is quite apparent that there is a knowledge gap among the general public/consumers in regard to biotechnology that the industry experts and academia need to bridge so that people understand the facts and logic behind it [72, 73]. Therefore, more knowledge dissemination is required to inform the public of potential benefits so that they can understand what biotechnology aims to achieve to avoid being misled [70, 71]. This necessitates deliberate efforts from academia, industry experts, and responsible authorities to educate the masses on biotechnology and GMOs and demystify myths, misinterpretations, and misinformation surrounding these breakthroughs. They would be through seminars, workshops and other outreach programs that would create awareness of what biotechnology aims to address and show how the science involved does not threaten human life. In doing so, people would gain insights and become accepted, thus making the fight against micronutrient malnutrition a reality. That is not to say that there are opposing views that, with reasonable legitimacy, should be brought forth for discussion and further appropriate action.

1.5 Conclusion and future directions

Biotechnology presents a unique set of tools that will continue to provide ardent solutions for the multifaceted challenges facing food systems across the world. It is particularly important because of its versatility in application and the huge untapped potential it presents. Not only is biotechnology poised to address the matters of food sufficiency and hidden hunger but also the issues of over nutrition and attendant problems such as obesity. In addition, biotechnology is currently tackling some of the wicked problems such as climate change, droughts, and pest and diseases outbreaks through production of drought-tolerant, water-efficient, and pest- and disease-resistant varieties of crops, particularly gaining prominence in the whole area of cell agriculture, tackling the problem of diminishing protein sources, and biohazards. Overall, this development in biotechnology comes with the need for improved stewardship and research transparency which can be enhanced through open access research.

References

[1] Adeyeye, S. A. O., Ashaolu, T. J., Bolaji, O. T., Abegunde, T. A., & Omoyajowo, A. O. Africa and the Nexus of poverty, malnutrition and diseases. Critical Reviews in Food Science and Nutrition 2023, 63(5), 641–656.

[2] Dalki, E., & DeSutter, A. (2021). Feeding 10 Billion: Cultured Meat as a Sustainable Protein Source for a Growing Population.

[3] Nadathur, S., Wanasundara, J. P., Marinangeli, C. P. F., & Scanlin, L. Proteins in our diet: Challenges in feeding the global population. In: Sustainable Protein Sources, Academic Press, 2024, 1–29.

[4] Falcon, W. P., Naylor, R. L., & Shankar, N. D. Rethinking global food demand for 2050. Population and Development Review 2022, 48(4), 921–957.

[5] Malhi, G. S., Kaur, M., & Kaushik, P. Impact of climate change on agriculture and its mitigation strategies: A review. Sustainability 2021, 13(3), 1318.

[6] Prakash, S. Impact of climate change on aquatic ecosystem and its biodiversity: An overview. International Journal of Biological Innovations 2021, 3(2).

[7] Skendžić, S., Zovko, M., Živković, I. P., Lešić, V., & Lemić, D. The impact of climate change on agricultural insect pests. Insects 2021, 12(5), 440.

[8] Ibrahim, R. A., & Shawer, D. M. Transgenic Bt-plants and the future of crop protection (an overview). International Journal of Agricultural and Food Research 2014, 3(1).

[9] Sadek, H. E., Ebadah, I. M., & Mahmoud, Y. A. Importance of biotechnology in controlling insect pests. Journal of Modern Agriculture and Biotechnology 2023, 2(1), 5.

[10] Chadipiralla, K., Gayathri, P., Rajani, V., & Reddy, P. V. B. Plant tissue culture and crop improvement. Sustainable Agriculture in the Era of Climate Change 2020, 391–412.

[11] Maqbool, M. A., Aslam, M., Beshir, A., & Khan, M. S. Breeding for provitamin A biofortification of maize (Zea mays L.). Plant Breeding 2018, 137(4), 451–469.

[12] Mengistie, D. Lab-growing meat production from stem cell. Journal of Nutrition & Food Sciences 2020, 3(1), 100015.

[13] Raman, R. The impact of Genetically modified (GM) crops in modern agriculture: A review. GM Crops & Food 2017, 8(4), 195–208.

[14] Yali, W. Application of genetically modified organism (GMO) crop technology and its implications in modern agriculture. International Journal of Applied Agricultural Sciences 2022, 8, 14–20.

[15] Gbashi, S., Adebo, O., Adebiyi, J. A., Targuma, S., Tebele, S., Areo, O. M., . . . & Njobeh, P. Food safety, food security and genetically modified organisms in Africa: A current perspective. Biotechnology and Genetic Engineering Reviews 2021, 37(1), 30–63.

[16] Sheikh, A. A., Wani, M. A., Bano, P., Un, S., Nabi, T. A. B., Bhat, M. A., & Dar, M. S. An overview on resistance of insect pests against Bt crops. Journal of Entomology and Zoology Studies 2017, 5(1), 941–948.

[17] Bakhsh, A., Khabbazi, S. D., Baloch, F. S., Demirel, U., Çalişkan, M. E., Hatipoğlu, R., . . . & Özkan, H. Insect-resistant transgenic crops: Retrospect and challenges. Turkish Journal of Agriculture and Forestry 2015, 39(4), 531–548.

[18] Wikandari, R., Manikharda,, Baldermann, S., Ningrum, A., & Taherzadeh, M. J. Application of cell culture technology and genetic engineering for production of future foods and crop improvement to strengthen food security. Bioengineered 2021, 12(2), 11305–11330.

[19] Souza Filho, P. F., Andersson, D., Ferreira, J. A., & Taherzadeh, M. J. Mycoprotein: Environmental impact and health aspects. World Journal of Microbiology and Biotechnology 2019, 35(10), 147.

[20] Ahmed, A. U., Hoddinott, J., Abedin, N., & Hossain, N. The impacts of GM foods: Results from a randomized controlled trial of Bt eggplant in Bangladesh. American Journal of Agricultural Economics 2021, 103(4), 1186–1206.

[21] Rashid, M. A., Hasan, M. K., & Matin, M. A. (2018). Socio-economic performance of Bt eggplant cultivation in Bangladesh.
[22] Vitalis, N. E., & Sun, Y. (2023). A comparative study of transgenic cotton development, impacts, challenges and prospects with respect to China and Africa.
[23] Maryam, B. M., Datsugwai, M. S. S., & Shehu, I. The role of biotechnology in food production and processing. Industrial Engineering 2017, 1(1), 24–35.
[24] Tiruneh, A. Review on debates over genetically modified crops in the context of agriculture. Advances in Crop Science and Technology 2023, 11, 615.
[25] Sadikiel Mmbando, G. The adoption of genetically modified crops in Africa: The public's current perception, the regulatory obstacles, and ethical challenges. GM Crops & Food 2024, 15(1), 1–15.
[26] Sax, J. K. The GMO/GE debate. Texas A&M Law Review 2016, 4, 345.
[27] Karalis, D. T., Karalis, T., Karalis, S., & Kleisiari, A. S. Genetically modified products, perspectives and challenges. Cureus 2020, 12(3).
[28] Georges, F., & Ray, H. Genome editing of crops: A renewed opportunity for food security. GM Crops & Food 2017, 8(1), 1–12.
[29] Lamichhane, S., & Thapa, S. Advances from conventional to modern plant breeding methodologies. Plant Breeding and Biotechnology 2022, 10(1), 1–14.
[30] Majid, A., Parray, G. A., Wani, S. H., Kordostami, M., Sofi, N. R., Waza, S. A., . . . & Gulzar, S. Genome editing and its necessity in agriculture. International Journal of Current Microbiology and Applied Sciences 2017, 6(2017), 5435–5443.
[31] Salgotra, R. K., Johar, P., Sood, M., & Chauhan, B. S. Dynamics of genome edited products in the world market for future food security: A review. Plant Archives 2024, 24(1).
[32] Menz, J., Modrzejewski, D., Hartung, F., Wilhelm, R., & Sprink, T. Genome edited crops touch the market: A view on the global development and regulatory environment. Frontiers in Plant Science 2020, 11, 586027.
[33] Ahmar, S., Saeed, S., Khan, M. H. U., Ullah Khan, S., Mora-Poblete, F., Kamran, M., & Jung, K. H. A revolution toward gene-editing technology and its application to crop improvement. International Journal of Molecular Sciences 2020, 21(16), 5665.
[34] ISAAA Inc. (n.d.). Pocket K No. 27: Biotechnology and Biofortification https://www.isaaa.org/resources/publications/pocketk/27/default.asp
[35] Fiaz, S., Ahmar, S., Saeed, S., Riaz, A., Mora-Poblete, F., & Jung, K. H. Evolution and application of genome editing techniques for achieving food and nutritional security. International Journal of Molecular Sciences 2021, 22(11), 5585.
[36] Roychowdhury, R., Choudhury, S., Hasanuzzaman, M., & Srivastava, S., Eds. Sustainable Agriculture in the Era of Climate Change, 2020.
[37] Molla, K. A., Chakravorty, N., & Bansal, K. C. Genome editing for food, nutrition, and health. The Nucleus 2024, 1–4.
[38] Juma, B. S. (2022). Generation of Cassava with Reduced Levels of Cyanogens Through CRISPR/Cas9 Targeted Mutagenesis of the Cytochrome P450 (CYP79D1) Gene (Doctoral dissertation, JKUAT-IBR).
[39] Sawada, H., Inoue, N., & Iwano, M. Sexual Reproduction in Animals and Plants, Springer Nature, 2014.
[40] Evans, D. H. Pioneers in Cell Physiology: The Story of Warren and Margaret Lewis, Springer Nature, 2022.
[41] Arora, K., Rai, M. K., & Sharma, A. K. Tissue culture mediated biotechnological interventions in medicinal trees: Recent progress. Plant Cell, Tissue and Organ Culture (PCTOC) 2022, 150(2), 267–287.
[42] Tefera, A. A. Review on application of plant tissue culture in plant breeding. Journal of Natural Sciences Research 2019, 9(3), 20–25.

[43] Gupta, N., Jain, V., Joseph, M. R., & Devi, S. A review on micropropagation culture method. Asian Journal of Pharmaceutical Research and Development 2020, 8(1), 86–93.
[44] Chandran, H., Meena, M., Barupal, T., & Sharma, K. Plant tissue culture as a perpetual source for production of industrially important bioactive compounds. Biotechnology Reports 2020, 26, e00450.
[45] Naik, R., Bhushan, A., Gupta, R. K., Walia, A., & Gaur, A. Low cost tissue culture technologies in vegetables: A review. International Journal of Biochemistry Research & Review 2020, 29(9), 66–78.
[46] Thammasiri, K. Current status of orchid production in Thailand. II International Orchid Symposium 1078 2014, February, 25–33.
[47] Thammasiri, K., Jitsopakul, N., & Prasongsom, S. Micropropagation of some orchids and the use of cryopreservation. Orchids Phytochemistry, Biology and Horticulture: Fundamentals and Applications 2022, 225–260.
[48] Navik, P., Kumar, D., Raghavendra, S., Padmavati, A., Shamim, M. D., & Srivastava, D. Tissue culture technology intervention in Red Dacca and Cavendish banana for nutritive value enhancement under organic farming in India. In: Transforming Organic Agri-Produce into Processed Food Products, Apple Academic Press, 2023, 395–424.
[49] Omari, E. N., Mucheru-Muna, M., Mburu, B. K., & Odawa, A. A. Assessing the advantages of tissue culture bananas technology production of banana farmers in Kisii county, Kenya. Environmental Challenges 2024, 14, 100843.
[50] Hubert, B., Rosegrant, M., Van Boekel, M. A., & Ortiz, R. The future of food: Scenarios for 2050. Crop Science 2010, 50, S–33.
[51] Augustin, M. A., Hartley, C. J., Maloney, G., & Tyndall, S. Innovation in precision fermentation for food ingredients. Critical Reviews in Food Science and Nutrition 2023, 1–21.
[52] Hilgendorf, K., Wang, Y., Miller, M. J., & Jin, Y. S. Precision fermentation for improving the quality, flavor, safety, and sustainability of foods. Current Opinion in Biotechnology 2024, 86, 103084.
[53] Finnigan, T. J., Wall, B. T., Wilde, P. J., Stephens, F. B., Taylor, S. L., & Freedman, M. R. Mycoprotein: The future of nutritious nonmeat protein, a symposium review. Current Developments in Nutrition 2019, 3(6), nzz021.
[54] Knychala, M. M., Boing, L. A., Ienczak, J. L., Trichez, D., & Stambuk, B. U. Precision fermentation as an alternative to animal protein, a review. Fermentation 2024, 10(6), 315.
[55] Boukid, F., Ganeshan, S., Wang, Y., Tülbek, M. Ç., & Nickerson, M. T. Bioengineered enzymes and precision fermentation in the food industry. International Journal of Molecular Sciences 2023, 24(12), 10156.
[56] Khush, G. S., Lee, S., Cho, J. I., & Jeon, J. S. Biofortification of crops for reducing malnutrition. Plant Biotechnology Reports 2012, 6, 195–202.
[57] Shahbaz, U., Yu, X. B., Akhtar, W., Ndagijimana, R., & Rauf, H. Golden rice to eradicate the vitamin A deficiency in the developing countries. European Journal of Nutrition & Food Safety 2020, 12(1), 53–63.
[58] Tako, E., Blair, M. W., & Glahn, R. P. Biofortified red mottled beans (Phaseolus vulgaris L.) in a maize and bean diet provide more bioavailable iron than standard red mottled beans: Studies in poultry (Gallus gallus) and an in vitro digestion/Caco-2 model. Nutrition Journal 2011, 10, 1–10.
[59] Ching, X. L., Zainal, N. A. A. B., Luang-In, V., & Ma, N. L. Lab-based meat the future food. Environmental Advances 2022, 10, 100315.
[60] George, A. S. The development of lab-grown meat which will lead to the next farming revolution. Proteus Journal 2020, 11(7), 1–25.
[61] Liu, Z., & Liu, Y. Mitigation of greenhouse gas emissions from animal production. Greenhouse Gases: Science and Technology 2018, 8(4), 627–638.
[62] Lynch, J., & Pierrehumbert, R. Climate impacts of cultured meat and beef cattle. Frontiers in Sustainable Food Systems 2019, 3, 421491.

[63] Tubiello, F. N., Rosenzweig, C., Conchedda, G., Karl, K., Gütschow, J., Xueyao, P., . . . & Sandalow, D. Greenhouse gas emissions from food systems: Building the evidence base. Environmental Research Letters 2021, 16(6), 065007.

[64] Kumar, P., Sharma, N., Sharma, S., Mehta, N., Verma, A. K., Chemmalar, S., & Sazili, A. Q. In-vitro meat: A promising solution for sustainability of meat sector. Journal of Animal Science and Technology 2021, 63(4), 693.

[65] Sasmita, P., Nugraha, Y., & Lestari, P. Nutritionally enhanced food crops for their potential health benefits in Indonesia. AIP Conference Proceedings 2022, November, 2513(1). AIP Publishing.

[66] Amna,, Qamar, S., Tantray, A. Y., Bashir, S. S., Zaid, A., & Wani, S. H. Golden rice: Genetic engineering, promises, present status and future prospects. Rice Research for Quality Improvement: Genomics and Genetic Engineering: Volume 2: Nutrient Biofortification and Herbicide and Biotic Stress Resistance in Rice 2020, 581–604.

[67] Siwela, M., Pillay, K., Govender, L., Lottering, S., Mudau, F. N., Modi, A. T., & Mabhaudhi, T. Biofortified crops for combating hidden hunger in South Africa: Availability, acceptability, micronutrient retention and bioavailability. Foods 2020, 9(6), 815.

[68] Holmes, M. Perspectives on biotechnology: Public and corporate narratives in the GM archives. Plants, People, Planet 2022, 4(5), 476–484.

[69] Bearth, A., Kaptan, G., & Kessler, S. H. Genome-edited versus genetically-modified tomatoes: An experiment on people's perceptions and acceptance of food biotechnology in the UK and Switzerland. Agriculture and Human Values 2022, 39(3), 1117–1131.

[70] Sax, J. K. Biotechnology and consumer decision-making. Seton Hall Law Review 2016, 47, 433.

[71] Zurek, M., Ingram, J., Sanderson Bellamy, A., Goold, C., Lyon, C., Alexander, P., . . . & Withers, P. J. Food system resilience: Concepts, issues, and challenges. Annual Review of Environment and Resources 2022, 47, 511–534.

[72] Nezhmetdinova, F. T., Guryleva, M. E., Sharypova, N. K., Zinurova, R. I., & Tuzikov, A. R. Risks of modern biotechnologies and legal aspects of their implementation in agriculture. In: BIO Web of Conferences, vol. 17, EDP Sciences, 2020, 00227.

[73] Obi-Egbedi, O., Oluwatayo, I. B., & Ogungbite, O. Genetically modified crops' technology and its awareness among smallholder farmers in Nigeria. Problems of World Agriculture/Problemy Rolnictwa Światowego 2020, 20(4), 58–67.

Jumbale Mwarome, George Ooko Abong*, and Duke Omayio Gekonge

Chapter 2 Emerging technologies in food biotechnology and personalized nutrition

Abstract: While food is primarily meant to perform dietary functions, food manufacturers have dedicated resources to produce processed foods that could offer further functional advantages to consumers' overall health apart from the primary dietary functions. This has further necessitated intrinsic regulations in different countries to promote effective monitoring of food quality and safety. Furthermore, new food safety and quality controls, smart food technologies, artificial intelligence, nutrigenomics and nutrigenetics, data analytics, microbiome research, and food engineering characterize modifications in the food industry. The new trends have had implications on personalized diet, health, and convenience. Effective application of biotechnology in food presents significant environmental, economic, and social potential, and with benefits of technology which range from enhanced production, processing, and quality. Moreover, the quality and safety of food is a public health concern with projected global population demand matching the exponentially increase in population by 2050. The technologies applied to food production and processing must be sustainable in the short and long run to make meaningful sense and promote nutritional outcomes. The current chapter, therefore, illustrates the current technological advances in food production, processing, and the relevant dietary implications of personalized nutritional approaches.

Keywords: Processing, diet, nutrition, technology, artificial intelligence, genetics, machine learning

2.1 Introduction

New consumer demands call for food producers to process food that can offer dietary functions to catter for their health [1]. Significant changes continue to be experienced in food biotechnology and personalized nutrition due to technological evolution in

***Corresponding author: George Ooko Abong**, Department of Food Science, Nutrition and Technology, Faculty of Agriculture, University of Nairobi-Kenya, Nairobi, Kenya, e-mails: ooko.george@uonbi.ac.ke, georkoyo@yahoo.com
Jumbale Mwarome, Duke Omayio Gekonge, Department of Food Science, Nutrition and Technology, Faculty of Agriculture, University of Nairobi-Kenya

https://doi.org/10.1515/9783111441238-003

the respective fields. As a result, consumer perceptions, purchasing behaviors, and food consumption patterns are developing digitally. New food safety and quality controls, smart food technologies, artificial intelligence (AI), nutrigenomics and nutrigenetics, data analytics, microbiome research, and food engineering characterize these modifications. Commonly referred to as the fourth industrial revolution [2], these new trends have had implications on personalized diet, health, and convenience. The current chapter will therefore discuss the current technological advances in food production, processing, and the relevant dietary implications of personalized nutritional approaches.

2.2 Consumer acceptance and ethics

The World Health Organization defines biomedic ethics as the systemic reflection of a moral constituent of life and its conflicts [3]. Effective application of biotechnology in food presents significant environmental, economic, and social potential. The benefits of this technology range from enhanced production, processing, and quality [4, 5]. However, the introduction of biotechnology in food is subject to public perception. Consumers' attitudes influence the rejection or acceptance of these technologies. Dating back to the 1990s, several factors including awareness of benefits, level of knowledge, confidence, and trust influenced their acceptability [6]. Nevertheless, taking the example of genetic engineering, research has shown that consumers do not reject its technology outrightly with their objections solely attributed to its application. Moreover, public reactions to genetic engineering are defined by the technology's association with negative constructs, including harm, ethical concerns, unnaturalness, danger, risk, and tempering with nature [5].

Similar trends are observed in personalized nutrition and biotechnology. In personalized nutrition, an individual's nutrition service, and products are tailored to meet an individual's needs based on their medical history, phenotype, and nutritional needs [7]. While people in modern societies have exceptional possibilities for healthy living, through better access to disease and health risk mitigation, these opportunities give more hope implications making people focus more on external well-being than internal dietary focus (physical exercise and healthy food) [8]. Based on the four main pillars of ethical analysis; beneficence, autonomy, nonmaleficence, and justice, food is considered an important aspect of well-being while health is a limited aspect of well-being. Consequently, modern society perceives well-being with external determinants leading to a creation of exaggerated health expectations. Nevertheless, previous studies have shown higher adverse health outcomes for patients without personalized nutrition aid compared to those who had the aid [9]. These variations might, however, vary subject to a person's genetic makeup which is key in customizing the individual's dietary plans. Whereas there has been an upward focus on disease prevention through individualized nutritional interventions, it is less evident how user data is utilized to provide tailored nutritional recommendations. There are gaps in identifying factors that influence consumer acceptance of tailored advice and optimal design strategies for crafting a successful personalized recommendation based on existing studies [10]. These breaches may be subject to ethical adherence and acceptance in the long run. Contemporary issues further demand for regulatory and policy enforcement in emerging food and nutrition technologies.

2.3 Food safety and quality control

The quality and safety of food is a public health concern. With a projected global population of 9 billion people by 2050, the demand for safe and quality food will similarly exponentially increase [11]. Advances in detecting contaminants and ensuring food safety have been taking shape worldwide. Developing trends including biotechnology applications in enhancing traceability and transparency have been under focus in developing and developed countries with most countries creating laws and regulations that address related challenges. More so, the emergence of AI has allowed its applications in food quality control and safety. On top of safeguarding the health of populations, these technologies further aim at maintaining nutritional quality. However, with globalized food production and distribution, anthropogenic and environmental risks expose the food value chain to extensive pollution. These pollutants range from drugs, chemical pollutants, microbial organisms, and physical hazards [12]. Of importance is the biochemical composition of food and the mechanisms by which they are processed, transported, and stored. Modernized safety and quality control technologies therefore put into consideration the aforementioned factors to enable consumers to make well-informed decisions.

With the complexity of modern food supply chains, transparent systems in the food value chain are important in enhancing the traceability and safety of food. This is not only beneficial to consumers but also advantageous to other stakeholders. Technologically, food safety detection is buttressed by three main pillars including smart indicators and sensors integrated into food packages, portable detection devices, and data-assisted whole-genome sequencing [11]. By investing in data management platforms, biotechnology has promoted high security and quality data while enhancing data collection within all supply chain procedures [13]. This helps to identify possible hazards, improves efficiency, and reduces wastage of resources.

2.4 Sustainable food technologies

Within food technologies, sustainability determinants are based on three main pillars of sustainable food systems: environmental, economic, and social [14]. As environmental sustainability is a prerequisite of planetary health consideration, it demands technological application and innovation for waste reduction. According to Kaushik and Sharma [15], 30–50% of food goes to waste. Food wastage varies within the food value chain in developed and developing countries. Most of the food wastage in developed countries happens at the end of the value chain while in developing countries, much of the wastage occurs in the initial stage of the chain [16]. As a result, there is a need to contextualize methodologies to identify and control food waste and loss.

Food adequacy is one of the main determinants of food security [17]. For stable and continuous access to safe, nutritious, and adequate food, sustainable food sources have to be available within food systems. As a result, modern agricultural production has been focusing on long-lasting production mechanisms of adequate food to offer sustainable remedies that could incorporate technology, innovation, and high mechanization to improve yields and assure the availability of alternative food sources [18, 19]. Transgenic crops currently in use are highly appropriate for agricultural producers and consumers in underdeveloped nations, as shown by two case studies from Kenya and Mexico [20]. Besides, since all of the technology is contained in the seed, integrating it into conventional smallholder farming methods is simple. Aside from its contribution to boosting crop yields, biotechnology through genetic engineering has been used to improve the quality of food and improve the economic status of populations [21].

2.5 Trends in personalized nutrition

Advanced technologies have been vital in modern nutritional interventions. AI not only establishes data-driven personalized nutrition recommendations but also plays a significant role in predicting dietary needs and health outcomes through machine learning [21]. In microbiome research, new technologies enable assessment of the impact of microbiota on nutrition and health. This allows quantification of prebiotics and probiotics tailored to individual microbiomes. The same is applicable in nutrigenetics and nutrigenomics where dietary recommendations are personalized from genetic variation outcomes [22].

2.6 Technologies in food processing

The food industry is highly competitive with consumers being cautious regarding what they take. Recently, consumers have sought food that is convenient, diverse, is self-stable, has low calories, and is less costly. Characteristics of food quality including, taste, texture, and nutritional value are strongly determined by how food is processed [23]. Technologies used in food processing produce nutrient-dense, fresh-tasting, and safe food items. These technologies include high-pressure processing (HPP), irradiation, extraction and separation, membrane technologies, heat processing, radiofrequency electric fields, pulsed electric fields (PEFs), ultrasound processing, cold plasma treatment, nanotechnology, advanced packaging, robotics, and automation [24, 25]. These technologies have attracted much attention in recent years because compared to their conventional counterparts, they allow significant reduction in processing times with savings in energy consumption and production costs while ensuring food safety [26]. In addition to energy saving, food processing technologies generally allow for the saving of water and energy, increase reliability, lower emissions, improve product quality, and enhance productivity [27]. However, the growing global population necessitates more sustainable and resilient food sources, making it even more difficult to meet. The food industry's top concerns now include customer awareness, safety, shelf life, quality, and nutritional qualities of foods, as well as the frequency of various food-related disease outbreaks.

2.6.1 Ultrasound processing

Sound waves at a frequency higher than the human hearing threshold (>16 kHz) cause a sequence of compression and rarefaction cycles on the molecules of the material they travel through. Cavitation bubbles are created when these mechanical waves pass through the food product's surface. This event causes several physical and chemical characteristics that can be used in food preparation, such as the disintegration of water molecules and the disruption of biological tissue's cell walls [28, 29]. The food processing industry utilizes ultrasound processing extensively for a variety of purposes, including homogenization, crystallization, drying, dispersion, defoaming, improving solubility, emulsification, and texture, changing viscosity, plant sanitation, and fermentations [29, 30]. While there are many benefits to ultrasonic processing, such as its suitability for commercial scale-up, research has shown that high-amplitude ultrasound treatment can degrade food qualities including color, flavor, and nutritional value [31, 32]. Thus, a deeper comprehension of ultrasound's intricate process and how it affects food qualities will promote the industry's use of this technology. To enable substantial industrial scale-up with considerable financial gains, it is also necessary to take into account notable advancements in high power process design, enhanced energy efficiency, simple installation, competitive energy consumption, and low maintenance costs [33, 34].

2.6.2 Cold plasma treatment

Cold plasma technology (CPT) is a nonthermal processing method that inactivates food-related microorganisms using energetic, reactive gases. An ionized gas called plasma is made up of several different charged species, including electrons, ions, photons, and free radicals. These charged carriers are created when energy is added to a neutral gas [35]. To guarantee that every component of the product is fully treated, plasma circulates around it. Additionally, there are numerous possible uses for the technique in terms of surface cleaning of food items and food packaging materials. Microorganisms are subjected to intensely bombarded charged species during surface cleaning, which causes surface lesions on the bacterial cell wall that rupture [36]. A new and quick food processing method that produces no harmful byproducts or exhaust fumes after processing is cold plasma treatment. However, one must take into account the nutritional value, color, texture, chemical alterations, and general quality of the meal [37, 38]. Although the food sector has not fully embraced CPT for large-scale commercial settings due to a lack of understanding about certain crucial characteristics, the technology is easily scalable and has the potential for extensive use.

2.6.3 High-pressure processing

The food industry is increasingly utilizing HPP, also known as high hydrostatic pressure and ultra-high pressure, as a thermal food processing technique. The technology uses a cold pasteurization process to denaturize proteins, inactivate pathogens, and extend shelf life [39]. In HPP, food is subjected to extremely high pressure, which is instantly and consistently distributed throughout the food product, irrespective of its dimensions. HPP technology is safe, works at room temperature, and is less time-consuming, energy-efficient, and waste-free. Additionally, the method satisfies the highest hygienic standards as the product can be handled after packaging and is independent of the size, shape, or content of the product. HPP produces a higher-quality product by preserving the product's flavor and freshness to a greater extent than traditional processing techniques [40]. HPP has been studied commercially in a variety of foods, such as fish and seafood, meat-based goods (raw and cooked sausages and dry ham), fruits, vegetables, juices and beverages, and ready-to-eat meals [39, 41].

2.6.4 Irradiation

Radiation is a nonthermal food preservation technique that minimizes or eliminates microorganisms without changing the food. Food's natural qualities are maintained due to the penetration by X-rays. Thereafter, the X-rays inactivate microorganisms, and generate no heat. There is no production of radioactive waste at the food processing hub, but the radiation dose given to food products is based on their makeup and potential to contain microorganisms. The food is exposed to radiation during processing for a specific amount of time, but it never makes contact with the radiation source. Food quality features such as color, flavor, vitamin content, and taste are minimally altered, and the procedure utilizes less energy to inactivate bacteria. Nevertheless, the type of radiation source, its dose level, and the raw materials employed are linked to this change in food quality [42, 43]. The food business may use irradiation methods in a variety of ways. In addition to reducing post-harvest loss and preserving the color of fresh meat, the irradiation technology also prevents sprouting in products like potatoes and controls post-packaging contamination in a variety of food items, including cereals, legumes, spices, poultry, fish, seafood, meat, fruits, vegetables, and tubers [43–46]. However, not all food kinds can be exposed to radiation; for example, milk and foods with high fat and vitamin contents cannot be exposed to radiation.

2.6.5 Pulsed electric field

Food preparation uses a nonthermal method called pulsed electric field (PEF) processing. Using a set of electrodes inside a PEF chamber, the product is exposed to brief bursts of strong electric fields that last anywhere from microseconds to milliseconds [47, 48]. Using this method lessens the undesirable alterations that heat, chemical, and enzymatic processing procedures might cause to biological materials [49, 50]. Hamilton and Sale conducted the first comprehensive study on the impact of PEF treatment on yeast and bacteria [51]. The scientists discovered that electric field strength, treatment duration, pulse number, and pulse width were the most crucial PEF parameters for the inactivation of microorganisms. Fruit juices, milk, yogurt, soups, cooked meats, liquid eggs, and other food products can all be processed and preserved using this technology, which works well for liquid or semi-solid food products [47]. PEF has been effectively used in the commercial sector for a range of fruit juices. It improves the shelf life, functional qualities, and textural qualities of juices while having little negative impact on their sensory and physical qualities, according to studies [52]. The potato sector has been one of PEF's largest markets over the years. In addition to increasing drying efficiency and sugar leaching, PEF can produce French fries of superior quality. It is also frequently used to lessen the amount of cutting force required when making French fries. Because the method inactivates bacteria while preserving the food's nutritional value and sensory appeal, it is seen to be superior to conventional thermal processing. The technique is inexpensive, waste-free, energy-efficient, and simple to integrate into the current processing lines [53]. Implementing PEF treatment in the food business has been reported to present some problems, such as the difficulty of processing items with high electrical conductivity and the high initial capital cost of establishing PEF treatment systems.

2.6.6 Radiofrequency electric fields

In this thermal process, a dielectric substance is heated by a high-frequency electric field produced by an RF generator. No matter the food product's density, size, or thermal conductivity, the RF technology's innate characteristic feature allows heat to be produced instantly, precisely, uniformly, and selectively at its center [54]. The creation of heat through dielectric energy caused by molecular friction in water molecules indicates that the moisture content of the food affects RF heating. RF uses less energy than traditional heat processing and penetrates food particles of any size deeply, quickly, and uniformly [54–56]. RF heating can be used commercially in a variety of food processing applications. The post-baking drying of biscuits, crackers, and breakfast cereals is the first and most well-known use of radiofrequency technology in the food sector. STALAM [57], an Italian maker of RF equipment, reports that by dispersing the right amount of energy over a dough matrix, electromagnetic waves may be used to complete a baking process (without the need for traditional heating) [58]. When comparing the use of radiofrequency electric fields to traditional thermal methods for processing fruit juices (orange and peach), superior bacteriological and organoleptic properties were found. Additionally, RF heating can be used to sterilize packaged solid or viscous liquid food goods, bake bread, defrost food products, and disinfect and sanitize dry food commodities like grains, seeds, legumes, and dry fruits [56, 59–61].

2.7 Challenges in the implementation of personalized nutrition and food processing technologies

Even though effective application of biotechnology in food presents significant environmental, economic, and social potential, their introduction is subject to public perception. Sadly, some are rejected based on consumers' attitudes that influence the rejection or acceptance of these technologies. There is need to create awareness of benefits, and build trust. Moreover, in personalized nutrition, an individual's nutrition service, and products are tailored to meet an individual's needs and people focus more on external well-being than internal dietary focus (physical exercise and healthy food).

Despite the fact that developing trends including biotechnology applications in enhancing traceability and transparency have made significant gains, globalized food production and distribution, anthropogenic and environmental risks continue to expose the food value chain to extensive pollution. The need to enhance modernized safety and quality control technologies to enable consumers make well-informed decisions cannot be overemphasized. Sustainability of the technologies remains a key ingredient to their success. Since environmental sustainability is a prerequisite of planetary health consideration that demands technological application and innovation for waste reduction, there is a need to contextualize methodologies to identify and control food waste and loss while minimizing use of resources.

2.8 Conclusion

Modern technologies enhance food security by promoting intensive production of food as well as promoting nutritional outcomes. However, these innovations, including genetic modification of crops, active packaging, smart food packaging, and the use of AI in dietary recommendations, are subject to consumer perspective and ethical acceptability. For effective monitoring of the safety and quality of food, segregation of data in the food value chain is essential for enhanced traceability using the advanced technologies.

References

[1] Granato, D., Barba, F. J., Bursać Kovačević, D., Lorenzo, J. M., Cruz, A. G., & Putnik, P. Functional foods: Product development, technological trends, efficacy testing, and safety. Annual Reviews of Food Science and Technology 2020, 11, 93–118. https://doi.org/10.1146/annurev-food-032519-051708.

[2] Boland, M., Alam, F., & Bronlund, J. Modern technologies for personalized nutrition. Trends in Personalized Nutrition 2019, 195–222. https://doi.org/10.1016/B978-0-12-816403-7.00006-4.

[3] WHO. WHO Code of Ethics 2023.

[4] Powell, L. F., Lawes, J. R., Clifton-Hadley, F. A., Rodgers, J., Harris, K., Evans, S. J., et al. The prevalence of Campylobacter spp. in broiler flocks and on broiler carcases, and the risks associated with highly contaminated carcases. Epidemiology and Infection 2012, 140, 2233–2246. https://doi.org/10.1017/S0950268812000040.

[5] Blaine, K., Kamaldeen, S., & Powell, K. Public perception of biotechnology. Basic Biotechnology Third Edition 2006, 67, 3–24. https://doi.org/10.1017/CBO9780511802409.003.

[6] Hoban, T. J. Food industry innovation: Efficient consumer response. Agribusiness 1998, 14, 235–245. https://doi.org/10.1002/(SICI)1520-6297(199805/06)14:3<235::AID-AGR6>3.0.CO;2-1.

[7] Palmnäs, M., Brunius, C., Shi, L., Rostgaard-Hansen, A., Torres, N. E., González-Domínguez, R., et al. Perspective: Metabotyping – A potential personalized nutrition strategy for precision prevention of

cardiometabolic disease. Advances in Nutrition 2020, 11, 524–532. https://doi.org/10.1093/advances/nmz121.

[8] Görman, U. Some ethical issues raised by personalized nutrition. Genes Nutrition 2007, 2, 55–58. https://doi.org/10.1007/s12263-007-0013-x.

[9] Kolodziejczyk, A. A., Zheng, D., & Elinav, E. Diet–microbiota interactions and personalized nutrition. Nature Reviews Microbiology 2019, 17, 742–753. https://doi.org/10.1038/s41579-019-0256-8.

[10] Reinders, M. J., Starke, A. D., Fischer, A. R. H., Verain, M. C. D., Doets, E. L., & Van Loo, E. J. Determinants of consumer acceptance and use of personalized dietary advice: A systematic review. Trends in Food Science and Technology 2023, 131, 277–294. https://doi.org/10.1016/j.tifs.2022.12.008.

[11] Yu, Z., Jung, D., Park, S., Hu, Y., Huang, K., Rasco, B. A., et al. Smart traceability for food safety. Critical Reviews in Food Science and Nutrition 2022, 62, 905–916. https://doi.org/10.1080/10408398.2020.1830262.

[12] Li, S., Tian, Y., Jiang, P., Lin, Y., Liu, X., & Yang, H. Recent advances in the application of metabolomics for food safety control and food quality analyses. Critical Reviews in Food Science and Nutrition 2021, 61, 1448–1469. https://doi.org/10.1080/10408398.2020.1761287.

[13] Astill, J., Dara, R. A., Campbell, M., Farber, J. M., Fraser, E. D. G., Sharif, S., et al. Transparency in food supply chains: A review of enabling technology solutions. Trends in Food Science and Technology 2019, 91, 240–247. https://doi.org/10.1016/j.tifs.2019.07.024.

[14] FAO. Sustainable food systems: Concept and framework. Food Engineering Innovations Across the Food Supply Chain 2021, 15–46. https://doi.org/10.1016/B978-0-12-821292-9.00015-7.

[15] Kaushik, M., & Sharma, D. Characterization and treatment of waste from food processing industries. 2021. https://doi.org/10.1007/978-981-15-8967-6_3.

[16] UNEP. Food Waste Index Report 2024 Appendix 1 Methodology for Level 1 2024.

[17] HLPE. Food Security and Nutrition: Building A Global Narrative Towards 2030. A Report by the High Level Panel of Experts on Food Security and Nutrition of the Committee on World Food Security, Rome, Italy, 2020. https://doi.org/10.1109/ccgrid57682.2023.00010.

[18] Mbow, C., Rosenzweig, C., Barioni, L. G., Benton, T. G., Herrero, M., Krishnapillai, M., Liwenga, E., Pradhan, P., Rivera-Ferre, M. G., Sapkota, T., Tubiello, F. N., & Xu, Y. City Food Security, vol. 2020, 2019. https://doi.org/10.4324/9780203593066-20.

[19] Abah, J., Ishaq, M. N., & Wada, A. C. The role of biotechnology in ensuring food security and sustainable agriculture. African Journal of Biotechnology 2010, 9, 8896–8900.

[20] Moloney, M. The role of biotechnology and new breeding technologies for global food security. 2016. https://doi.org/10.1603/ice.2016.106552.

[21] Areche, F. O., Gondal, A. H., Sumarriva-Bustinza, L. A., Zela-Payi, N. O., Sumarriva-Bustinza, J. M., Oscanoa-León, R. B., et al. Role of biotechnology in food security: A Review. SABRAO Journal of Breeding and Genetics 2023, 55, 1496–1509. https://doi.org/10.54910/sabrao2023.55.5.5.

[22] De Moraes Lopes, M. H. B., Ferreira, A. C. B. H., Ferreira, D. D., Da Silva, G. R., Caetano, A. S., & Braz, V. N. Use of artificial intelligence in precision nutrition and fitness. 2020. https://doi.org/10.1016/B978-0-12-817133-2.00020-3.

[23] Capitanio, F., Coppola, A., & Pascucci, S. Product and process innovation in the Italian food industry. Agribusiness 2010, 26, 503–518. https://doi.org/10.1002/agr.20239.

[24] Priyadarshini, A., Rajauria, G., O'Donnell, C. P., & Tiwari, B. K. Emerging food processing technologies and factors impacting their industrial adoption. Critical Reviews in Food Science and Nutrition 2019, 59, 3082–3101. https://doi.org/10.1080/10408398.2018.1483890.

[25] Hassoun, A., Jagtap, S., Trollman, H., Garcia-Garcia, G., Abdullah, N. A., Goksen, G., et al. Food processing 4.0: Current and future developments spurred by the fourth industrial revolution. Food Control 2023, 145, 109507. https://doi.org/https://doi.org/10.1016/j.foodcont.2022.109507.

[26] Misra, N. N., Koubaa, M., Roohinejad, S., Juliano, P., Alpas, H., Inácio, R. S., et al. Landmarks in the historical development of twenty first century food processing technologies. Food Research International 2017, 97, 318–339. https://doi.org/10.1016/j.foodres.2017.05.001.
[27] Chemat, F., Vian, M. A., & Cravotto, G. Green extraction of natural products: Concept and principles. International Journal of Molecular Sciences 2012, 13, 8615–8627. https://doi.org/10.3390/ijms13078615.
[28] Cheng, X., Zhang, M., Xu, B., Adhikari, B., & Sun, J. The principles of ultrasound and its application in freezing related processes of food materials: A review. Ultrasonics Sonochemistry 2015, 27, 576–585. https://doi.org/https://doi.org/10.1016/j.ultsonch.2015.04.015.
[29] Soria, A. C., & Villamiel, M. Effect of ultrasound on the technological properties and bioactivity of food: A review. Trends in Food Science and Technology 2010, 21, 323–331. https://doi.org/https://doi.org/10.1016/j.tifs.2010.04.003.
[30] Guamán-Balcázar, M. C., Setyaningsih, W., Palma, M., & Barroso, C. G. Ultrasound-assisted extraction of resveratrol from functional foods: Cookies and jams. Applied Acoustics 2016, 103, 207–213. https://doi.org/https://doi.org/10.1016/j.apacoust.2015.07.008.
[31] Farkas, J., & Mohácsi-Farkas, C. History and future of food irradiation. Trends in Food Science and Technology 2011, 22, 121–126. https://doi.org/https://doi.org/10.1016/j.tifs.2010.04.002.
[32] Harder, E., Damm, W., Maple, J., Wu, C., Reboul, M., Xiang, J. Y., et al. OPLS3: A force field providing broad coverage of drug-like small molecules and proteins. Journal of Chemical Theory and Computation 2016, 12, 281–296. https://doi.org/10.1021/acs.jctc.5b00864.
[33] Zinoviadou, K. G., Galanakis, C. M., Brnčić, M., Grimi, N., Boussetta, N., Mota, M. J., et al. Fruit juice sonication: Implications on food safety and physicochemical and nutritional properties. Food Research International 2015, 77, 743–752. https://doi.org/https://doi.org/10.1016/j.foodres.2015.05.032.
[34] Rentería, A., & García, I. Quality and sensory profile of ultrasound-treated beef. 2017, 29, 463–475.
[35] Misra, N. N., Tiwari, B. K., Raghavarao, K. S. M. S., & Cullen, P. J. Nonthermal plasma inactivation of food-borne pathogens. Food Engineering Reviews 2011, 3, 159–170. https://doi.org/10.1007/s12393-011-9041-9.
[36] Pankaj, S. K., Bueno-Ferrer, C., Misra, N. N., Milosavljević, V., O'Donnell, C. P., Bourke, P., et al. Applications of cold plasma technology in food packaging. Trends in Food Science and Technology 2014, 35, 5–17. https://doi.org/https://doi.org/10.1016/j.tifs.2013.10.009.
[37] Mason, T. J., Chemat, F., & Ashokkumar, M. 27 – Power ultrasonics for food processing. In: Gallego-Juárez, J. A., & Graff, K. F., editors. Power Ultrasonics, Oxford: Woodhead Publishing, 2015, 815–843. https://doi.org/https://doi.org/10.1016/B978-1-78242-028-6.00027-2.
[38] Korachi, M., Ozen, F., Aslan, N., Vannini, L., Guerzoni, M. E., Gottardi, D., et al. Biochemical changes to milk following treatment by a novel, cold atmospheric plasma system. International Dairy Journal 2015, 42, 64–69. https://doi.org/https://doi.org/10.1016/j.idairyj.2014.10.006.
[39] Tribst, A. A. L., de Castro Leite Júnior, B. R., de Oliveira, M. M., & Cristianini, M. High pressure processing of cocoyam, Peruvian carrot and sweet potato: Effect on oxidative enzymes and impact in the tuber color. Innovative Food Science and Emerging Technologies 2016, 34, 302–309. https://doi.org/https://doi.org/10.1016/j.ifset.2016.02.010.
[40] Tsevdou, M. S., Eleftheriou, E. G., & Taoukis, P. S. Transglutaminase treatment of thermally and high pressure processed milk: Effects on the properties and storage stability of set yoghurt. Innovative Food Science and Emerging Technologies 2013, 17, 144–152. https://doi.org/https://doi.org/10.1016/j.ifset.2012.11.004.
[41] Georget, E., Sevenich, R., Reineke, K., Mathys, A., Heinz, V., Callanan, M., et al. Inactivation of microorganisms by high isostatic pressure processing in complex matrices: A review. Innovative Food Science and Emerging Technologies 2015, 27, 1–14. https://doi.org/https://doi.org/10.1016/j.ifset.2014.10.015.

[42] Gautam, R. K., Nagar, V., & Shashidhar, R. Effect of radiation processing in elimination of Klebsiella pneumoniae from food. Radiation Physics and Chemistry 2015, 115, 107–111. https://doi.org/https://doi.org/10.1016/j.radphyschem.2015.06.016.
[43] Urbain, W. Food Irradiation, First. Elsevier, 2012.
[44] Rogers, E. M. Diffusion of Innovations, 4th Edition, New York: Simon and Schuster, 2010.
[45] Rawson, A., Patras, A., Tiwari, B. K., Noci, F., Koutchma, T., & Brunton, N. Effect of thermal and non thermal processing technologies on the bioactive content of exotic fruits and their products: Review of recent advances. Food Research International 2011, 44, 1875–1887. https://doi.org/https://doi.org/10.1016/j.foodres.2011.02.053.
[46] Kumar, S., Saxena, S., Verma, J., & Gautam, S. Development of ambient storable meal for calamity victims and other targets employing radiation processing and evaluation of its nutritional, organoleptic, and safety parameters. LWT – Food Science and Technology 2016, 69, 409–416. https://doi.org/https://doi.org/10.1016/j.lwt.2016.01.059.
[47] Toepfl, S., Heinz, V., Knorr, D., & Sun, D.-W. Overview of pulsed electric field processing for food. Emerging Technologies for Food Processing 2005, 69–97. https://doi.org/10.1016/B978-012676757-5/50006-2.
[48] Mohammed, M., & Eissa, A. H. A. Pulsed Electric Fields for Food Processing Technology, 2012.
[49] Donsì, F., Ferrari, G., & Pataro, G. Applications of pulsed electric field treatments for the enhancement of mass transfer from vegetable tissue. Food Engineering Reviews 2010, 2, 109–130. https://doi.org/10.1007/s12393-010-9015-3.
[50] Vorobiev, E., & Lebovka, N. ChemInform abstract: Enhanced extraction from solid foods and biosuspensions by pulsed electrical energy. Food Engineering Reviews 2010, 2, 95–108. https://doi.org/10.1007/s12393-010-9021-5.
[51] Hamilton, W. A., & Sale, A. J. H. Effects of high electric fields on microorganisms: II. Mechanism of action of the lethal effect. Biochimica et Biophysica Acta – General Subjects 1967, 148, 789–800. https://doi.org/https://doi.org/10.1016/0304-4165(67)90053-0.
[52] Shakhova, N. B., Yurmazova, T. A., Hoang, T. T., & Anh, N. T. Pulsed electric discharge in active metallic grains for water purification processes. Procedia Chemistry 2015, 15, 292–300. https://doi.org/10.1016/j.proche.2015.10.047.
[53] Mittal, G. S., & Griffiths, M. W. 5 – Pulsed electric field processing of liquid foods and beverages. In: Sun, D.-W., editor. Emerging Technologies for Food Processing, London: Academic Press, 2005, 99–139. https://doi.org/https://doi.org/10.1016/B978-012676757-5/50007-4.
[54] Maloney, N., & Harrison, M. Chapter 8 – Advanced heating technologies for food processing. In: Leadley, C. E., editor. Innovation and Future Trends in Food Manufacturing and Supply Chain Technologies, Woodhead Publishing, 2016, 203–256. https://doi.org/https://doi.org/10.1016/B978-1-78242-447-5.00008-3.
[55] Zheng, A., Zhang, B., Zhou, L., & Wang, S. Application of radio frequency pasteurization to corn (Zea mays L.): Heating uniformity improvement and quality stability evaluation. Journal of Stored Products Research 2016, 68, 63–72. https://doi.org/https://doi.org/10.1016/j.jspr.2016.04.007.
[56] Jiao, Y., Tang, J., & Wang, S. A new strategy to improve heating uniformity of low moisture foods in radio frequency treatment for pathogen control. Journal of Food Engineering 2014, 141, 128–138. https://doi.org/10.1016/j.jfoodeng.2014.05.022.
[57] Stalam 2025. https://www.stalam.com/en/milestones/ (accessed January 5, 2025).
[58] Awuah, G. B., Ramaswamy, H. S., & Tang, J. Radio-Frequency Heating in Food Processing, CRC Press, 2014.
[59] Mishra, R. R., & Sharma, A. K. Microwave–material interaction phenomena: Heating mechanisms, challenges and opportunities in material processing. Composites: Part A Applied Science and Manufacturing 2016, 81, 78–97. https://doi.org/https://doi.org/10.1016/j.compositesa.2015.10.035.

[60] Zhou, L., & Wang, S. Verification of radio frequency heating uniformity and Sitophilus oryzae control in rough, brown, and milled rice. Journal of Stored Products Research 2016, 65, 40–47. https://doi.org/https://doi.org/10.1016/j.jspr.2015.12.003.
[61] Uyar, R., Bedane, T. F., Erdogdu, F., Koray Palazoglu, T., Farag, K. W., & Marra, F. Radio-frequency thawing of food products – A computational study. Journal of Food Engineering 2015, 146, 163–171. https://doi.org/https://doi.org/10.1016/j.jfoodeng.2014.08.018.

Abisoye Solomon Fiyinfoluwa* and Fakoya Soji*

Chapter 3
Processing of hypoallergenic foods with biotechnological approach

Abstract: Food allergies are a significant public health concern, necessitating innovative strategies for managing allergenicity in food products. This chapter provides an in-depth exploration of biotechnological approaches to developing hypoallergenic foods, focusing on the reduction or elimination of allergenic components. Beginning with a detailed background, the chapter highlights the significance of hypoallergenic food processing in ensuring dietary safety and improving the quality of life for individuals with food allergies. The fundamentals of hypoallergenic food development are discussed, including the definition, characteristics, and immunological basis of allergenic reactions, as well as common allergens in widely consumed foods. Biotechnological techniques such as genetic modification, enzymatic treatment, and nanotechnology are explored for their potential in reducing allergenic properties without compromising nutritional quality. Additionally, this chapter delves into analytical methods, molecular characterization of allergenic proteins, and bioinformatics tools for allergen prediction to enhance the identification and characterization of allergens. Case studies illustrate the successful application of these techniques, including genetically modified crops, enzyme-modified products, and nanotechnology-based solutions. The regulatory framework and safety assessment protocols surrounding hypoallergenic food development are thoroughly examined, highlighting international standards and safety testing protocols to ensure consumer protection. Challenges such as regulatory barriers, consumer acceptance, and the evolving nature of food allergies are discussed alongside future perspectives in the field. This chapter underscores the transformative role of biotechnology in addressing food allergies, paving the way for innovative solutions that align with public health priorities. By integrating scientific advancements with regulatory and consumer considerations, this chapter provides a holistic approach to hypoallergenic food processing, offering valuable insights for researchers, policymakers, and industry stakeholders.

Keywords: Hypoallergenic foods, food biotechnology, allergen reduction, biotechnological food processing, allergen safety assessment

***Corresponding author: Abisoye Solomon Fiyinfoluwa**, Department of Biological Sciences, Olusegun Agagu University of Science and Technology, Okitipupa, Ondo, Nigeria, e-mail: fiyinfoluwapraiz50@gmail.com

***Corresponding author: Fakoya Soji**, Department of Biological Sciences, Olusegun Agagu University of Science and Technology, Okitipupa, Ondo, Nigeria, e-mail: so.fakoya@oaustech.edu.ng

https://doi.org/10.1515/9783111441238-004

3.1 Introduction

3.2 Background and significance

Food allergies have become a growing concern worldwide, affecting individuals of all ages and demographics. The prevalence of food allergies has increased significantly in recent decades, with notable impacts on public health, quality of life, and healthcare costs [1]. Allergic reactions to common food items such as nuts, dairy, eggs, and shellfish can range from mild discomfort to life-threatening anaphylaxis, posing serious challenges for affected individuals and their caregivers. In response to this rising health issue, researchers and food industry professionals have intensified efforts to develop innovative strategies for producing hypoallergenic foods that minimize the risk of allergic reactions. The concept of hypoallergenic foods emerged as a promising solution to address the needs of individuals with food allergies. These foods are specifically designed and processed to reduce or eliminate allergenic components, while retaining nutritional value and palatability. The development of hypoallergenic food products holds significant potential to improve the quality of life for millions of people worldwide who are affected by food allergies. By providing safe and nutritious alternatives to allergen-containing foods, hypoallergenic products offer a sense of reassurance and empowerment to individuals with dietary restrictions [2].

The significance of hypoallergenic foods extends beyond the realm of individual health outcomes to broader societal and economic implications. Food allergies impose a substantial burden on healthcare systems, with costs associated with medical care, emergency interventions, and lost productivity [3]. Moreover, families of individuals with food allergies often experience heightened stress and anxiety due to concerns about accidental exposure and allergic reactions. Hypoallergenic food options alleviate these concerns by providing safer alternatives that enable individuals with food allergies to participate more fully in social activities and everyday dining experiences. Advances in biotechnological approaches have revolutionized the development and production of hypoallergenic foods, offering innovative solutions to longstanding challenges in food allergy management. Biotechnological techniques such as genetic modification, enzymatic treatment, and nanotechnology enable targeted modification or removal of allergenic proteins from food matrices, thereby reducing the risk of allergic reactions [4]. These approaches leverage scientific understanding of allergen structure and immunological mechanisms to engineer safer and more tolerable food products for individuals with food allergies.

In addition to enhancing food safety and accessibility, biotechnological innovations in hypoallergenic food processing contribute to sustainable agricultural practices and resource conservation. By reducing dependence on traditional allergen-containing crops and livestock, biotechnological approaches promote diversification of food sources and resilience in the face of environmental challenges [5]. Furthermore, advancements in biotechnology have the potential to optimize nutritional profiles, flavor profiles, and shelf stability of hypoallergenic foods, enhancing overall acceptability and marketability. Considering these considerations, this book chapter aims to provide a comprehensive overview of biotechnological approaches for the processing of hypoallergenic foods. By examining the scientific principles, technological advancements, regulatory considerations, and socioeconomic implications of hypoallergenic food development, this chapter seeks to inform researchers, industry professionals, policymakers, and consumers about the current state of the field and future directions for innovation. Through collaborative efforts and interdisciplinary approaches, the goal is to advance knowledge and promote the development of safe, nutritious, and accessible hypoallergenic food options for individuals with food allergies.

3.2.1 Overview of hypoallergenic foods

Hypoallergenic foods represent a specialized category of food products that are formulated and processed to minimize the risk of triggering allergic reactions in susceptible individuals. These foods are designed to be safe for consumption by individuals with known food allergies or sensitivities, providing a viable alternative to conventional allergen-containing foods. The concept of hypoallergenicity revolves around the reduction or elimination of allergenic proteins and other potential triggers from food formulations, while ensuring nutritional adequacy and palatability. Hypoallergenic foods encompass a wide range of products, including infant formulas, baked goods, snacks, beverages, and dietary supplements, tailored to meet the diverse dietary needs and preferences of consumers [6].

One of the primary goals of hypoallergenic food development is to address the needs of individuals with common food allergies, such as those to peanuts, tree nuts, eggs, milk, soy, wheat, fish, and shellfish. These allergens account for most food-related allergic reactions and pose significant challenges for affected individuals in terms of dietary restrictions and avoidance strategies [7]. Hypoallergenic foods offer a valuable solution by providing alternative sources of essential nutrients and dietary components without the risk of allergen exposure. By utilizing carefully selected ingredients and processing techniques, hypoallergenic food manufacturers aim to create products that are safe, reliable, and enjoyable for consumers with food allergies.

The formulation of hypoallergenic foods typically involves several key strategies for allergen reduction or elimination. One approach is the selection of hypoallergenic ingredients that are less likely to provoke allergic reactions in sensitive individuals. These may include alternative protein sources, such as plant-based proteins (e.g., pea protein and rice protein) or hydrolyzed proteins (e.g., hydrolyzed whey protein and hydrolyzed soy protein), which have undergone processing to break down allergenic epitopes and reduce allergenicity [8]. Additionally, manufacturers may employ technological methods such as heat treatment, enzymatic hydrolysis, or filtration to further degrade or remove allergenic proteins from food matrices, thereby enhancing safety and tolerability [9].

In recent years, advances in biotechnological approaches have revolutionized the development of hypoallergenic foods, offering innovative solutions for allergen mitigation and risk management. Genetic modification techniques allow for precise manipulation of crop plants to reduce the expression of allergenic proteins or introduce novel traits that confer hypoallergenic properties [10]. Enzymatic modification of allergenic proteins through targeted proteolysis or enzymatic cleavage offers another avenue for reducing allergenicity while preserving nutritional quality and functional properties [11]. Furthermore, nanotechnology applications enable the encapsulation or immobilization of allergenic components within food matrices, thereby preventing their interaction with immune cells and reducing the likelihood of allergic reactions [12].

Despite significant progress in hypoallergenic food development, challenges remain in ensuring the safety, efficacy, and acceptability of these products for consumers with food allergies. Issues such as cross-contamination, labeling accuracy, and consumer education pose ongoing concerns in the production and marketing of hypoallergenic foods [13]. Regulatory agencies also play a critical role in establishing standards and guidelines for allergen management, labeling requirements, and safety assessments to protect public health and ensure transparency in the food industry [14]. Moreover, ongoing research efforts are needed to advance scientific understanding of food allergens, immune mechanisms, and personalized approaches to allergy management, with the goal of improving quality of life for individuals with food allergies.

3.2.2 Importance of biotechnological approaches

Biotechnological approaches play a crucial role in the development and production of hypoallergenic foods, offering innovative solutions to challenges associated with allergen management and food safety. These advanced techniques leverage principles of molecular biology, genetics, and nanotechnology to engineer food products with reduced allergenicity and improved tolerability for individuals with food allergies. The significance of biotechnological approaches lies in their ability to target specific allergenic components within food matrices while preserving nutritional quality, sensory attributes, and shelf stability [15]. One of the key advantages of biotechnological approaches is their precision and specificity in allergen mitigation. Genetic modification techniques allow for the precise manipulation of plant genomes to silence or modify genes encoding allergenic proteins, thereby reducing their expression levels, or altering their structural characteristics [16]. By selectively targeting allergenic epitopes, genetic engineering enables the development of hypoallergenic crop varieties with reduced potential for triggering allergic reactions in sensitive individuals. This targeted approach minimizes unintended changes to the overall composition and functionality of the food product, ensuring that nutritional value and sensory properties are maintained [17].

Enzymatic modification represents another powerful biotechnological strategy for reducing allergenicity in food proteins. Enzymes such as proteases, peptidases, and transglutaminases can be utilized to hydrolyze allergenic proteins into smaller fragments or modify their conformational structure, thereby reducing their allergenic potential [18]. Enzymatic treatment offers a flexible and scalable approach to allergen reduction, allowing for precise control over processing conditions and enzymatic activity levels to achieve desired outcomes. Furthermore, enzymatic modification can be applied to a wide range of food matrices, including plant-based proteins, animal proteins, and processed foods, making it a versatile and adaptable technology for hypoallergenic food production [19]. Nanotechnology has also emerged as a promising frontier in the development of hypoallergenic foods, offering novel approaches for allergen encapsulation, delivery, and controlled release. Nanoparticle-based delivery systems enable the targeted delivery of bioactive compounds, including allergenic proteins, to specific sites within the gastrointestinal tract, thereby minimizing their interaction with immune cells and reducing the risk of allergic reactions [20]. Additionally, nanoscale carriers such as liposomes, nanoparticles, and nano emulsions can be used to encapsulate allergenic components within protective matrices, preventing their exposure to the external environment and preserving their stability during processing and storage [21].

The adoption of biotechnological approaches in hypoallergenic food production offers numerous benefits beyond allergen mitigation, including enhanced sustainability, efficiency, and scalability. By reducing dependence on conventional allergen-containing crops and livestock, biotechnological innovations contribute to resource conservation, land use optimization, and environmental sustainability [22]. Moreover, biotechnological techniques enable the development of tailored solutions to address specific allergenic challenges associated with different food matrices, processing methods, and consumer preferences, thereby facilitating personalized approaches to allergy management [23].

In conclusion, biotechnological approaches represent a cornerstone of innovation in the field of hypoallergenic food production, offering precision, versatility, and scalability in allergen mitigation. These advanced techniques hold promise for improving food safety, accessibility, and acceptability for individuals with food allergies, while also addressing broader societal and economic challenges associated with food allergen management. Continued research and development efforts in biotechnology are essential to advancing knowledge, expanding capabilities, and realizing the full potential of hypoallergenic foods in promoting health and well-being for all consumers.

3.3 Fundamentals of hypoallergenicity

3.3.1 Definition and characteristics of hypoallergenic foods

Hypoallergenic foods are a specialized category of food products designed and processed to minimize the risk of triggering allergic reactions in susceptible individuals. These foods are formulated with the intention of reducing or eliminating allergenic components that commonly provoke immune responses in individuals with food allergies or sensitivities. The term "hypoallergenic" implies that these foods are less likely to cause allergic reactions compared to conventional allergen-containing foods, making them suitable options for individuals with known food allergies or intolerances [24].

3.3.2 Characteristic of hypoallergenic foods

1. **Selection of hypoallergenic ingredients**: One characteristic of hypoallergenic foods is the careful selection of ingredients that are less likely to induce allergic reactions in sensitive individuals. Common allergens such as peanuts, tree nuts, milk, eggs, soy, wheat, fish, and shellfish are typically avoided or substituted with alternative ingredients that have a lower allergenic potential. These may include hypoallergenic protein sources such as pea protein, rice protein, or hydrolyzed proteins, as well as alternative grain flours, starches, and oils derived from non-allergenic sources [7].

2. **Processing techniques for allergen reduction**: Another characteristic of hypoallergenic foods is the use of specialized processing techniques aimed at reducing or eliminating allergenic proteins from food matrices. Techniques such as heat treatment, enzymatic hydrolysis, filtration, and nanotechnology-based encapsulation may be employed to degrade allergenic proteins or modify their structure to reduce their allergenicity [25]. These processing methods help to ensure that hypoallergenic foods are safe for consumption by individuals with food allergies while retaining nutritional quality and sensory attributes.

3. **Labeling and certification**: Proper labeling and certification are essential characteristics of hypoallergenic foods to ensure transparency and inform consumers about the allergen content and safety of the product. Manufacturers are required to accurately label hypoallergenic foods with information regarding the presence of allergenic ingredients, potential cross-contamination risks, and compliance with regulatory standards for allergen management [26]. Certification programs such as gluten-free certification, nut-free certification, and allergen-free certification provide additional assurance to consumers with food allergies and facilitate informed purchasing decisions.

4. **Clinical validation and testing**: Hypoallergenic foods may undergo clinical validation and testing to assess their safety and efficacy for individuals with food allergies.

Clinical trials and allergen challenge studies are conducted to evaluate the tolerance and immunological response of sensitive individuals to hypoallergenic products [27]. These studies help to establish the hypoallergenic properties of foods and provide scientific evidence to support their use in allergy management.

5. **Nutritional composition and fortification**: Despite the focus on allergen reduction, hypoallergenic foods are required to meet nutritional standards and provide essential nutrients for optimal health and growth. Manufacturers may fortify hypoallergenic products with vitamins, minerals, and other nutrients to compensate for any potential deficiencies resulting from allergen avoidance or processing [28]. Careful formulation and nutrient analysis ensure that hypoallergenic foods are nutritionally balanced and suitable for individuals with specific dietary requirements.

6. **Consumer acceptance and palatability**: Finally, a critical characteristic of hypoallergenic foods is consumer acceptance and palatability. While allergen avoidance is essential for managing food allergies, it is equally important for hypoallergenic foods to be appealing and enjoyable to consume. Factors such as taste, texture, aroma, appearance, and packaging play a significant role in determining consumer acceptance and satisfaction with hypoallergenic products [29]. Manufacturers may invest in sensory testing, consumer surveys, and product development efforts to optimize the sensory attributes and overall appeal of hypoallergenic foods.

3.4 Allergens in common food products

Understanding the allergens present in common food products is essential for the development of effective strategies to mitigate the risk of allergic reactions in sensitive individuals. Several foods are known to be major allergenic sources, accounting for most food-related allergic reactions worldwide. These allergens are typically proteins that can trigger immune responses in susceptible individuals, leading to a range of symptoms from mild to severe. Common food allergens include peanuts, tree nuts, milk, eggs, soy, wheat, fish, and shellfish, which are widely used in various culinary preparations and processed food products [28]:

a. **Peanuts and tree nuts**: Peanuts and tree nuts are among the most common food allergens, with peanuts being one of the leading causes of severe allergic reactions (anaphylaxis) in both children and adults. The proteins present in peanuts and tree nuts can elicit strong immune responses in sensitive individuals, leading to symptoms such as itching, hives, swelling, difficulty breathing, and gastrointestinal distress [30]. Cross-reactivity between different nut varieties poses additional challenges for individuals with nut allergies, necessitating strict avoidance of all nut-containing foods.

b. Milk and dairy products: Cow's milk and dairy products are another significant source of food allergens, particularly in infancy and early childhood. Proteins such as casein and whey found in milk can trigger allergic reactions in susceptible individuals, manifesting as skin rashes, gastrointestinal symptoms, respiratory distress, and in severe cases, anaphylaxis [31]. While cow milk allergy is common in young children, it may persist into adulthood in some cases, necessitating lifelong dietary restrictions and careful allergen avoidance.

c. Eggs: Eggs are a versatile ingredient used in various culinary applications and processed food products, making them a common allergenic source for individuals with egg allergies. The proteins found in egg whites and egg yolks, particularly ovalbumin and ovomucoid, are known allergens that can provoke immune responses in sensitive individuals [32]. Symptoms of egg allergy may include skin reactions, respiratory symptoms, gastrointestinal distress, and anaphylaxis, depending on the severity of the allergic reaction.

d. Soy: Soybeans and soy-based products are prevalent in many cuisines and processed food items due to their versatility and nutritional value. However, soy proteins such as glycinin and beta-conglycinin can trigger allergic reactions in susceptible individuals, leading to symptoms ranging from mild itching and swelling to severe respiratory distress and anaphylaxis [33]. Soy allergy is particularly common in infants and young children, often presenting as part of a broader pattern of food allergies in early childhood.

e. Wheat: Wheat is a staple food ingredient consumed worldwide in various forms, including bread, pasta, cereals, and baked goods. However, wheat proteins such as gluten and gliadin can induce allergic reactions in individuals with wheat allergies or celiac disease, an autoimmune condition triggered by gluten consumption [34]. Symptoms of wheat allergy may include skin rashes, gastrointestinal discomfort, respiratory symptoms, and, in severe cases, anaphylaxis, or gluten-induced enteropathy.

f. Fish and shellfish: Fish and shellfish are major allergenic sources, with proteins such as parvalbumin in fish and tropomyosin in shellfish known to provoke allergic reactions in sensitive individuals [35]. Fish allergy is more common in adults than in children and can manifest as skin reactions, gastrointestinal symptoms, respiratory distress, and anaphylaxis upon ingestion or inhalation of fish proteins. Shellfish allergy encompasses reactions to crustaceans (e.g., shrimp, crab, and lobster) and mollusks (e.g., clams, oysters, and mussels) and can cause similar symptoms to fish allergy, ranging from mild to severe.

g. Cross-contamination and hidden allergens: It is important to note that allergens present in common food products may also pose risks of cross-contamination and hidden allergen exposure in processed foods. Cross-contact can occur during food processing, handling, and preparation, leading to unintended allergen exposure in individuals with food allergies [36]. Additionally, hidden allergens may be present in foods as undeclared ingredients or as contaminants, posing challenges for individuals with food allergies in identifying and avoiding potential allergens.

The awareness of the allergens present in common food products is crucial for individuals with food allergies, caregivers, healthcare professionals, and food industry stakeholders. Strict allergen avoidance, accurate labeling, and effective communication strategies are essential for preventing allergic reactions and ensuring the safety of individuals with food allergies. Continued research and education efforts are needed to improve allergen management practices, enhance consumer awareness, and develop innovative solutions for allergen detection and mitigation in the food industry.

3.5 Immunological basis of food allergies

Food allergies are immune-mediated adverse reactions to specific proteins present in food items. The immunological basis of food allergies involves complex interactions between the immune system and allergenic proteins, resulting in the production of allergic antibodies and the release of inflammatory mediators. When an individual with a predisposition to food allergies ingests an allergenic food protein, the immune system recognizes it as foreign and mounts an inappropriate immune response, leading to the development of allergic symptoms [37]:

a. Sensitization phase: The first step in the immunological cascade of food allergies is sensitization, wherein the immune system becomes primed to recognize and respond to specific food proteins. During sensitization, exposure to allergenic proteins triggers the production of allergen-specific immunoglobulin E (IgE) antibodies by B lymphocytes [38]. These IgE antibodies bind to high-affinity receptors on the surface of mast cells and basophils, sensitizing these cells to subsequent exposure to the same allergenic protein.

b. Activation phase: Upon re-exposure to the allergenic protein, cross-linking of IgE antibodies on mast cells and basophils by the allergen leads to cell activation and degranulation, resulting in the release of inflammatory mediators such as histamine, leukotrienes, and cytokines [39]. These mediators induce a cascade of physiological changes, including vasodilation, increased vascular permeability, smooth muscle contraction, and recruitment of inflammatory cells, leading to the characteristic symptoms of allergic reactions.

c. Type I hypersensitivity reaction: Food allergies are classified as type I hypersensitivity reactions, also known as immediate hypersensitivity reactions, due to the rapid onset of symptoms following allergen exposure. Type I hypersensitivity reactions are mediated by IgE antibodies and involve the activation of mast cells and basophils, leading to the release of preformed and newly synthesized inflammatory mediators [40]. The clinical manifestations of type I hypersensitivity reactions can vary widely depending on the route of allergen exposure, the dose of allergen ingested, and the individual's immune reactivity.

d. Organ manifestations: Food allergies can affect various organ systems, including the skin, gastrointestinal tract, respiratory system, and cardiovascular system. Cutaneous symptoms such as itching, hives, and angioedema are common manifestations of allergic reactions and may occur alone or in combination with other symptoms [41]. Gastrointestinal symptoms such as nausea, vomiting, abdominal pain, and diarrhea are also prevalent in food-allergic individuals, reflecting the involvement of the gastrointestinal mucosa in the allergic response.

e. Risk factors and triggers: Several factors contribute to the development and severity of food allergies, including genetic predisposition, environmental exposures, and immunological dysregulation. Individuals with a family history of allergies, especially immediate relatives with food allergies, are at increased risk of developing food allergies themselves [42]. Early exposure to allergenic foods, alterations in gut microbiota composition, and environmental factors such as pollution and climate change may also influence the development and expression of food allergies.

The immunological basis of food allergies involves a complex interplay between allergenic proteins, IgE antibodies, mast cells, basophils, and inflammatory mediators, resulting in the development of allergic symptoms. Understanding the mechanisms underlying food allergies is crucial for the development of effective diagnostic tests, preventive measures, and therapeutic interventions to manage this growing public health concern. Continued research efforts are needed to elucidate the factors contributing to the rising prevalence of food allergies and to identify novel targets for allergy prevention and treatment.

3.6 Biotechnological techniques in food processing

3.6.1 Genetic modification for allergen reduction

Genetic modification represents a promising biotechnological approach for reducing allergenicity in food products by targeting specific genes associated with allergen expression or immunoreactivity. This technique involves the precise manipulation of an organism's genetic material to introduce desirable traits or suppress undesirable traits, such as allergen production [43]. By selectively modifying the expression of allergen genes in crop plants, researchers aim to develop hypoallergenic varieties that produce lower levels of allergenic proteins or lack specific allergenic epitopes altogether.

One strategy for genetic modification involves the use of gene silencing techniques to suppress the expression of allergen genes in crop plants. RNA interference (RNAi) is a commonly employed method for gene silencing, wherein double-stranded RNA molecules are introduced into plant cells to target and degrade messenger RNA (mRNA) transcripts corresponding to allergen genes [44]. By inhibiting the expression of allergen genes at the transcriptional level, RNAi-mediated gene silencing can effectively reduce the accumulation of allergenic proteins in plant tissues, resulting in hypoallergenic crops with enhanced safety profiles.

Another approach to genetic modification for allergen reduction involves the use of gene editing technologies such as CRISPR-Cas9 to precisely modify allergen genes in crop plants. CRISPR-Cas9 enables targeted modifications to the plant genome by introducing specific DNA edits at desired genomic loci [45]. By precisely editing allergen genes to disrupt key epitopes or regulatory sequences, researchers can engineer crop varieties that produce modified allergenic proteins with reduced immunoreactivity or functional impairment. This approach offers a precise and efficient means of generating hypoallergenic crops with tailored modifications to allergen genes.

In addition to gene silencing and gene editing, genetic modification techniques may also involve the introduction of novel genes or regulatory elements to alter the expression or functionality of allergen genes in crop plants. For example, researchers may introduce transcription factors or regulatory sequences that modulate the expression of allergen genes in response to environmental cues or developmental signals [46]. By controlling the timing and level of allergen expression, these regulatory elements can help to minimize allergen accumulation in plant tissues and enhance the safety of genetically modified (GM) crops.

One of the key advantages of genetic modification for allergen reduction is its specificity and precision in targeting allergen genes while minimizing unintended changes to the overall composition and functionality of the crop plant. Unlike conventional breeding methods, which rely on random genetic recombination and selection, genetic modification allows researchers to precisely manipulate specific genes of interest without introducing unwanted genetic variability [43]. This targeted approach enables the development of hypoallergenic crop varieties with consistent and predictable allergen profiles, facilitating regulatory approval and consumer acceptance.

Moreover, genetic modification offers the potential for rapid and scalable deployment of hypoallergenic crop varieties to address pressing food safety challenges and meet the growing demand for allergen-reduced food products. Once developed, GM crops can be propagated through conventional agricultural practices to produce large quantities of hypoallergenic raw materials for food processing and manufacturing [47]. This scalability and efficiency make genetic modification an attractive biotechnological approach for allergen reduction in food processing, with the potential to improve the accessibility and affordability of hypoallergenic foods for consumers worldwide.

3.6.2 Enzymatic modification of allergenic proteins

Enzymatic modification of allergenic proteins represents a promising biotechnological approach for reducing the allergenicity of food products while preserving their nutritional value and sensory attributes. Enzymes, biological catalysts that accelerate chemical reactions, are employed to modify the structure and properties of allergenic proteins in ways that render them less immunoreactive to sensitive individuals [18]:

a. Selection of enzymes: A critical step in enzymatic modification is the selection of appropriate enzymes that can selectively target and modify allergenic proteins without compromising the overall quality of the food product. Various enzymes, including proteases, peptidases, and transglutaminases, have been investigated for their ability to hydrolyze allergenic epitopes or modify protein structures to reduce allergenicity [18].

b. Proteolytic cleavage: Proteases are enzymes that catalyze the hydrolysis of peptide bonds within proteins, leading to the cleavage of protein chains into smaller peptides or amino acids. Proteolytic cleavage of allergenic proteins can disrupt the conformational structure of allergens and eliminate specific epitopes responsible for eliciting immune responses [48]. This process reduces the allergenic potential of proteins while preserving their nutritional quality and functional properties.

c. Selective hydrolysis: Enzymatic hydrolysis can be tailored to selectively target allergenic epitopes within protein molecules while leaving non-allergenic regions intact. By controlling reaction conditions such as enzyme concentration, reaction time, and pH, enzymatic hydrolysis can be optimized to achieve desired levels of allergen reduction without excessive protein degradation [49].

d. Modification of protein structure: In addition to proteolytic cleavage, enzymes such as transglutaminases can be utilized to modify the structure of allergenic proteins through cross-linking or polymerization reactions. Transglutaminases catalyze the formation of covalent bonds between protein molecules, resulting in the formation of protein aggregates or complexes [50]. This modification alters the antigenic properties of allergenic proteins, reducing their ability to bind to IgE antibodies and trigger allergic reactions.

e. Optimization of processing conditions: The effectiveness of enzymatic modification depends on various factors, including enzyme specificity, substrate specificity, reaction conditions, and protein composition. Optimization of processing conditions is crucial to maximize allergen reduction while minimizing undesirable effects such as changes in taste, texture, and nutritional value [51].

f. Integration into food processing: Enzymatic modification techniques can be integrated into existing food processing workflows to reduce allergenicity in a wide range of food products. Applications include the production of hypoallergenic infant formulas, baked goods, dairy products, meat analogs, and processed foods [52]. Enzymatic modification offers a versatile and scalable approach to allergen reduction that can be adapted to different food matrices and processing conditions.

g. Safety and regulatory considerations: The safety of enzymatically modified foods is a critical consideration, particularly regarding the potential for unintended allergen cross-reactivity or adverse reactions in sensitive individuals. Regulatory agencies such as the Food and Drug Administration (FDA) and the European Food Safety Authority (EFSA) require comprehensive safety assessments for novel food ingredients and processing techniques, including enzymatic modifications, to ensure consumer safety [53].

h. Consumer acceptance: Consumer acceptance of enzymatically modified foods is influenced by factors such as taste, texture, appearance, and perceived health benefits. Sensory testing and consumer surveys are essential for evaluating the palatability and marketability of hypoallergenic products and informing product development efforts [54]. Effective communication strategies are also important for building consumer trust and confidence in the safety and efficacy of enzymatically modified foods.

Enzymatic modification of allergenic proteins offers a promising approach for reducing allergenicity in food products while maintaining nutritional quality and sensory attributes. By selectively targeting allergenic epitopes and modifying protein structures, enzymes can mitigate the risk of allergic reactions in sensitive individuals and expand food options for consumers with food allergies. Continued research and development efforts are needed to optimize enzymatic modification techniques, address safety and regulatory concerns, and enhance consumer acceptance of hypoallergenic foods.

3.7 Nanotechnology applications in hypoallergenic food production

Nanotechnology has emerged as a promising frontier in the development of hypoallergenic foods, offering innovative solutions for allergen management and risk reduction. By harnessing the unique properties of nanoparticles, nanotechnology enables precise control over the delivery, encapsulation, and release of bioactive compounds, including allergenic proteins, within food matrices. The application of nanotechnology in hypoallergenic food production holds potential to enhance safety, stability, and efficacy while minimizing the risk of allergic reactions in sensitive individuals [55]:

a. Nanoparticle-based encapsulation: One key application of nanotechnology in hypoallergenic food production is the encapsulation of allergenic proteins within protective matrices to prevent their interaction with immune cells and reduce allergenicity. Nanoparticles such as liposomes, polymeric nanoparticles, and lipid nanoparticles can be used to encapsulate allergenic proteins, forming a physical barrier that shields the proteins from enzymatic degradation and immune recognition [56]. This encapsulation technology allows for controlled release of allergenic components in the gastrointestinal tract, minimizing the risk of allergic reactions while preserving nutritional quality and sensory attributes of the food product.

b. Targeted delivery systems: Nanotechnology enables the development of targeted delivery systems for hypoallergenic foods, facilitating the selective release of bioactive compounds at specific sites within the gastrointestinal tract. Surface modification of nanoparticles with ligands or peptides that target specific receptors or cell types allows for precise delivery of encapsulated allergenic proteins to immune cells or tissues involved in allergic responses [57]. This targeted delivery approach enhances the bioavailability and efficacy of hypoallergenic food components while minimizing off-target effects and systemic exposure.

c. Biopolymer nanocomposites: Biopolymer nanocomposites represent another innovative application of nanotechnology in hypoallergenic food production, offering enhanced stability, solubility, and bioactivity of encapsulated allergenic proteins. Nanocomposite materials composed of biopolymers such as chitosan, alginate, or gelatin, reinforced with nanoparticles such as silica, titanium dioxide, or montmorillonite, provide a biocompatible and biodegradable matrix for encapsulation and delivery of bioactive compounds [58]. These biopolymer nanocomposites offer superior mechanical properties, barrier properties, and controlled release characteristics compared to conventional polymer matrices, making them ideal candidates for hypoallergenic food applications.

d. Nanoscale sensing and detection: Nanotechnology offers novel approaches for the detection and quantification of allergenic proteins in food matrices, enabling rapid and sensitive analysis of allergen content for quality control and safety assurance purposes. Nanoscale sensing platforms based on techniques such as surface plasmon resonance (SPR), fluorescence resonance energy transfer, and nanoparticle-enhanced immunoassays allow for real-time detection of allergenic proteins at trace levels in complex food samples [59]. These nanoscale sensing technologies offer advantages such as high sensitivity, specificity, and multiplexing capabilities, facilitating reliable and efficient allergen detection in hypoallergenic food production.

e. Nanoparticle-based immunoassays: Nanoparticle-based immunoassays represent a powerful tool for allergen detection and quantification in hypoallergenic food matrices, offering enhanced sensitivity and specificity compared to traditional immunoassay methods. Nanoparticle labels such as gold nanoparticles, quantum dots, and magnetic nanoparticles can be conjugated with allergen-specific antibodies or aptamers to facilitate recognition and binding to target allergenic proteins [4]. This nanoparticle-based immunoassay approach allows for rapid, multiplexed, and quantitative analysis of allergen content in hypoallergenic foods, enabling rigorous quality control and allergen management throughout the production process.

f. Regulatory considerations: Despite the promising potential of nanotechnology in hypoallergenic food production, regulatory considerations regarding the safety and labeling of nanomaterials in food products remain a critical consideration. Regulatory agencies such as the FDA and the EFSA have established guidelines and criteria for the safety assessment and approval of nanotechnology-based food ingredients and additives [60]. Comprehensive risk assessments, toxicity studies, and characterization of nanomaterials are essential to ensure the safety and regulatory compliance of hypoallergenic foods incorporating nanotechnology-based approaches.

g. Consumer acceptance and perception: Consumer acceptance and perception of nanotechnology-based hypoallergenic foods are important factors influencing market adoption and commercial success. Public awareness, education, and communication regarding the safety, efficacy, and benefits of nanotechnology in food production play a crucial role in shaping consumer attitudes and preferences [61]. Transparent labeling, clear communication of nanotechnology applications, and engagement with stakeholders can help build trust and confidence in hypoallergenic foods incorporating nanotechnology-based approaches.

3.8 Identification and characterization of allergens

3.8.1 Analytical methods for allergen detection

Accurate and reliable detection of allergenic proteins in food matrices is essential for ensuring food safety and compliance with labeling regulations, particularly for individuals with food allergies or intolerances. Analytical methods for allergen detection encompass a diverse range of techniques, each with its unique principles, advantages, and limitations. These methods play a critical role in identifying allergenic ingredients, assessing cross-contamination risks, and verifying allergen-free claims in food products, thereby safeguarding the health and well-being of consumers [62]:

a. Immunological methods: Immunological methods such as enzyme-linked immunosorbent assays (ELISA), lateral flow devices (LFDs), and immunoblotting are widely used for allergen detection due to their high specificity, sensitivity, and ease of use. ELISA-based assays utilize allergen-specific antibodies for the quantification of target allergenic proteins in food samples, offering rapid and quantitative results [63].

LFDs provide rapid, on-site detection of allergens through antibody-antigen interactions on a lateral flow strip, making them suitable for field testing and point-of-care applications. Immunoblotting techniques such as Western blotting enable qualitative detection and characterization of allergenic proteins based on their molecular weight and antigenic properties, allowing for identification of specific allergens in complex food matrices [8].

b. DNA-based methods: DNA-based methods such as polymerase chain reaction (PCR), real-time PCR (qPCR), and multiplex PCR are employed for allergen detection by targeting DNA sequences encoding allergenic proteins. PCR-based assays offer high specificity and sensitivity for the detection of allergen-encoding genes in food samples, allowing for identification of allergenic ingredients even at trace levels [64]. Real-time PCR techniques enable quantitative analysis of allergen content in food samples, providing accurate and reliable results for allergen labeling and regulatory compliance [65]. Multiplex PCR assays allow for simultaneous detection of multiple allergens in a single reaction, offering efficiency and cost-effectiveness for comprehensive allergen analysis in complex food matrices.

c. Mass spectrometry (MS): MS techniques such as liquid chromatography-MS (LC-MS) and matrix-assisted laser desorption/ionization MS (MALDI-MS) are powerful tools for allergen detection and characterization based on protein profiling and peptide sequencing. LC-MS analysis enables identification and quantification of allergenic proteins in food samples through separation by LC followed by MS detection [66]. MALDI-MS techniques offer rapid and sensitive detection of allergens by analyzing protein or peptide profiles directly from food samples, facilitating rapid screening and confirmation of allergenic ingredients [67].

d. Biosensor technologies: Biosensor technologies such as SPR, quartz crystal microbalance (QCM), and electrochemical biosensors offer label-free, real-time detection of allergens in food samples based on specific molecular interactions. SPR biosensors utilize changes in refractive index upon antibody-antigen binding to detect allergenic proteins in food matrices with high sensitivity and specificity [68]. QCM sensors measure changes in mass upon allergen binding to a sensor surface, providing rapid and quantitative detection of allergenic ingredients in complex food samples [69]. Electrochemical biosensors offer portable, low-cost solutions for allergen detection, utilizing changes in electrical properties upon allergen binding for rapid and on-site analysis of food samples.

e. Proteomic approaches: Proteomic approaches such as two-dimensional gel electrophoresis (2DGE), LC-tandem MS (LC-MS/MS), and protein microarrays are utilized for comprehensive profiling and characterization of allergenic proteins in food matrices. 2DGE separates proteins based on their isoelectric point and molecular weight, allowing for visualization and identification of allergenic proteins in complex mixtures [69]. LC-MS/MS techniques enable identification and sequencing of allergenic peptides from food samples, providing valuable insights into the composition and structure of allergenic ingredients. Protein microarrays offer high-throughput screening of allergen-antibody interactions, facilitating multiplexed detection and quantification of allergenic proteins in food samples.

f. Validation and standardization: Validation and standardization of analytical methods for allergen detection are essential to ensure reliability, reproducibility, and accuracy of results across different laboratories and food matrices. International organizations such as the International Organization for Standardization (ISO) and the Codex Alimentarius Commission provide guidelines and protocols for method validation, performance criteria, and quality assurance in allergen analysis [70]. Interlaboratory proficiency testing programs and reference materials help to assess the proficiency and competency of analytical laboratories in allergen detection, driving continuous improvement and harmonization of methods within the food industry [71].

3.8.2 Molecular characterization of allergenic proteins

Molecular characterization of allergenic proteins is essential for understanding their structural properties, allergenic epitopes, and cross-reactivity patterns, providing valuable insights into allergen sensitization and immune responses in individuals with food allergies. Various techniques and approaches are employed for the molecular characterization of allergenic proteins, enabling identification, purification, sequencing, and structural analysis of allergenic components in food matrices [72]:

a. Protein purification and isolation: Protein purification and isolation techniques are utilized to obtain pure allergenic proteins from complex food matrices for subsequent molecular characterization. These techniques involve extraction of allergenic proteins from food sources using methods such as solvent extraction, aqueous extraction, or enzymatic digestion, followed by separation and purification steps such as chromatography, electrophoresis, or filtration [73]. Purified allergenic proteins serve as valuable substrates for molecular analysis, including sequencing, epitope mapping, and structural studies.

b. MS analysis: MS techniques such as - LC-MS and MS/MS are powerful tools for the molecular characterization of allergenic proteins based on their mass-to-charge ratios and fragmentation patterns. LC-MS analysis enables identification and quantification of allergenic proteins in food samples through separation by LC followed by mass spectrometry detection [74]. MS/MS techniques provide detailed sequence information and structural insights into allergenic proteins by analyzing peptide fragments generated through protein fragmentation and ionization.

c. Protein sequencing and peptide mapping: Protein sequencing and peptide mapping techniques are employed to determine the amino acid sequence and structural organization of allergenic proteins, facilitating the identification of allergen-specific epitopes and cross-reactivity patterns. Techniques such as Edman degradation, peptide mass fingerprinting, and MS/MS are utilized for sequencing allergenic proteins and mapping allergen-specific peptides [75]. These approaches enable the characterization of primary, secondary, and tertiary protein structures, providing insights into allergen structure–function relationships and immunological properties.

d. Epitope mapping: Epitope mapping techniques are employed to identify and characterize allergen-specific epitopes responsible for immune recognition and antibody binding in individuals with food allergies. Epitope mapping methods include epitope scanning, alanine scanning, and SPR analysis, which involve systematic analysis of allergen-derived peptides or mutated allergen variants for their ability to interact with allergen-specific antibodies or T cells [76]. Epitope mapping studies provide valuable information on the immunodominant regions of allergenic proteins and their role in allergen sensitization and allergic responses.

e. Structural biology approaches: Structural biology approaches such as X-ray crystallography, nuclear magnetic resonance (NMR) spectroscopy, and molecular modeling are employed to elucidate the three-dimensional structure of allergenic proteins and their interactions with allergen-specific antibodies or receptors. X-ray crystallography enables high-resolution structural determination of allergenic proteins by analyzing the diffraction patterns of X-rays scattered by protein crystals [77]. NMR spectroscopy provides insights into the dynamic properties and conformational flexibility of allergenic proteins in solution, complementing X-ray crystallography data [72]. Molecular modeling techniques such as homology modeling and molecular docking are utilized to predict protein structures and simulate protein–ligand interactions, facilitating the design of hypoallergenic variants and allergen-specific immunotherapies.

f. Bioinformatics analysis: Bioinformatics analysis plays a crucial role in the molecular characterization of allergenic proteins by facilitating sequence alignment, structural prediction, epitope mapping, and allergenicity assessment. Bioinformatics tools and databases such as Allergen Online, Allergome, and Structural Database of Allergenic Proteins (SDAP) provide curated allergen sequences, structural information, and computational resources for allergen characterization and analysis [78]. Sequence homology searches, phylogenetic analysis, and sequence motif identification are utilized to identify potential allergenic proteins, predict cross-reactivity, and assess allergen variability across different species or food sources.

g. Functional studies: Functional studies of allergenic proteins are conducted to evaluate their biological activities, immunomodulatory effects, and allergenic potency in vitro and in vivo. These studies involve assessing allergen specific IgE binding, T-cell activation, cytokine production, and basophil activation in allergic individuals or animal models [79]. Functional assays such as basophil activation test (BAT), T-cell proliferation assay, and cytokine release assay provide insights into the allergenic potential and immunogenicity of allergenic proteins, guiding risk assessment and allergen management strategies in food production and labeling.

h. Clinical relevance and translation: Molecular characterization of allergenic proteins contributes to our understanding of allergen sensitization, cross-reactivity, and immune responses in individuals with food allergies, informing clinical diagnosis, management, and treatment strategies. Molecular insights into allergen structure–function relationships, epitope specificity, and allergen variability guide the development of allergen-specific immunotherapies, diagnostic tests, and personalized treatment approaches for individuals with food allergies [80]. Translation of molecular findings into clinical practice facilitates improved allergen labeling, allergen risk assessment, and patient counseling, enhancing the safety and quality of life for individuals with food allergies.

3.8.3 Bioinformatics in allergen prediction

Bioinformatics plays a crucial role in the identification and characterization of allergenic proteins by providing computational tools and resources for allergen prediction, classification, and analysis. Bioinformatics approaches utilize sequence-based, structure-based, and functional-based methods to predict allergenicity, assess cross-reactivity, and identify potential allergenic proteins in food sources and other allergenic sources [81]:

a. Sequence homology searches: Sequence homology searches are commonly used in bioinformatics for allergen prediction by comparing amino acid sequences of known allergens with sequences of newly identified proteins. Tools such as BLAST (Basic Local Alignment Search Tool) and FASTA (Fast Alignment Search Tool) facilitate sequence similarity searches against allergen databases and protein sequence repositories, enabling the identification of homologous sequences and potential allergenic proteins [82].

b. Sequence motif identification: Sequence motif identification is employed to identify conserved amino acid motifs or signatures associated with allergenic proteins. Bioinformatics tools such as MEME (Multiple Em for Motif Elicitation) and PROSITE enable the detection of sequence motifs characteristic of allergens, such as cysteine residues, hydrophobic patches, or IgE-binding epitopes [83]. Identification of these motifs' aids in the prediction and classification of potential allergenic proteins based on their sequence features.

c. Machine learning algorithms: Machine learning algorithms are utilized in bioinformatics for allergen prediction by training computational models on large datasets of known allergens and non-allergenic proteins. Supervised learning algorithms such as support vector machines (SVMs), random forests, and artificial neural networks are trained to classify proteins as allergens or non-allergens based on sequence-derived features, physicochemical properties, and structural characteristics [84]. These machine learning approaches provide predictive models for allergen identification with high accuracy and efficiency.

d. Allergenicity assessment tools: Allergenicity assessment tools utilize bioinformatics algorithms and databases to predict the allergenic potential of proteins based on their sequence, structure, and physicochemical properties. Tools such as AllerCat, Allermatch, and AllerTool enable the assessment of allergenicity using sequence-based similarity, motif matching, and physicochemical properties of proteins [82]. These tools incorporate various parameters such as sequence identity, sequence motifs, and structural features to predict allergenicity and prioritize proteins for experimental validation.

e. Structural modeling and simulation: Structural modeling and simulation techniques are employed in bioinformatics to predict the three-dimensional structure of proteins and assess their allergenic potential. Homology modeling, threading, and ab initio modeling are utilized to generate protein models based on known protein structures and sequence similarities [85]. Molecular dynamics simulations and docking studies are performed to analyze protein–protein interactions, epitope binding, and structural stability, providing insights into allergen structure–function relationships and allergen-antibody interactions.

f. Genomic and proteomic databases: Genomic and proteomic databases serve as valuable resources for bioinformatics analysis and allergen prediction by providing comprehensive datasets of allergenic proteins, their sequences, and structural annotations. Databases such as the Allergome, UniProt, and the Structural Database of Allergenic Proteins (SDAP) compile curated allergen sequences, structural information, and experimental data for computational analysis [82]. These databases facilitate data mining, sequence comparison, and structural analysis of allergenic proteins, supporting allergen prediction and characterization efforts.

g. Cross-reactivity prediction: Bioinformatics approaches are employed to predict cross-reactivity between allergenic proteins based on sequence similarity, structural homology, and epitope conservation. Tools such as AllerCross utilize sequence alignment algorithms and structural comparison methods to identify cross-reactive allergens and assess potential cross-reactivity risks [86]. These predictive models aid in the assessment of allergen sensitization profiles, the design of hypoallergenic variants, and the development of allergen-specific immunotherapies.

h. Integration of multiple data sources: Integrating multiple data sources and bioinformatics tools enables comprehensive allergen prediction and characterization by incorporating sequence, structure, and functional information. Bioinformatics pipelines and workflows combine sequence analysis, structural modeling, and machine learning techniques to prioritize potential allergenic proteins, assess allergenicity, and identify cross-reactive epitopes [87]. Integration of diverse data sources enhances the accuracy and reliability of allergen prediction models, facilitating informed decision-making in allergen management and food safety assessment.

3.9 Hypoallergenic food development: case studies

3.9.1 Case study 1: genetically modified crops

GM crops have emerged as a promising approach in hypoallergenic food development by offering the potential to reduce allergenicity while maintaining nutritional quality and crop yield. In this case study, we examine the application of genetic modification techniques to develop hypoallergenic crops with reduced allergenic proteins and improved safety for individuals with food allergies [22]:

a. **Identification of target allergenic proteins**: The first step in developing hypoallergenic GM crops involves the identification of target allergenic proteins present in conventional crops. Through bioinformatics analysis, proteomic profiling, and clinical studies, researchers identify allergenic proteins responsible for allergic reactions in sensitive individuals [88]. Common allergenic proteins such as gliadins in wheat, storage proteins in soybeans, and profilins in fruits are among the targets for genetic modification to reduce allergenicity.

b. **Gene silencing techniques**: Gene silencing techniques such as RNA interference (RNAi) are utilized to downregulate the expression of target allergenic proteins in GM crops. RNAi technology allows for specific suppression of allergen-encoding genes by introducing double-stranded RNA molecules that target and degrade mRNA transcripts, resulting in reduced protein synthesis [89]. By silencing allergen genes, GM crops can be engineered to produce reduced levels of allergenic proteins, thereby mitigating the risk of allergic reactions in sensitive individuals.

c. **Allergen-free variants**: Genetic modification also enables the development of allergen-free variants of crops by disrupting or eliminating the expression of allergenic proteins altogether. Gene editing techniques such as CRISPR-Cas9 are employed to precisely edit the genomes of crops and create knockout mutations in allergen genes, resulting in the production of hypoallergenic varieties [90]. Allergen-free crops offer a safe alternative for individuals with severe food allergies, providing reassurance and peace of mind in food consumption.

d. Enhanced nutritional quality: In addition to reducing allergenicity, genetic modification can be used to enhance the nutritional quality of crops, thereby improving overall food safety and health outcomes. By introducing genes encoding beneficial nutrients such as vitamins, minerals, and antioxidants, GM crops can offer increased nutritional value and health benefits for consumers [91]. Through genetic engineering, hypoallergenic crops can be enriched with essential nutrients to address nutrient deficiencies and promote dietary diversity.

e. Field trials and safety assessment: Before commercialization, GM hypoallergenic crops undergo rigorous field trials and safety assessment to evaluate their performance, allergenicity, and environmental impact. Field trials assess agronomic traits, yield potential, and crop performance under various environmental conditions to ensure suitability for cultivation. Safety assessment includes allergenicity testing, compositional analysis, and environmental risk assessment to ensure the safety and regulatory compliance of GM crops [92].

f. Regulatory approval and labeling: Regulatory approval is obtained from government agencies such as the FDA and the EFSA before commercialization of GM hypoallergenic crops. Regulatory agencies evaluate the safety, efficacy, and environmental impact of GM crops based on comprehensive data from field trials, safety assessment studies, and allergenicity testing [93]. Upon approval, GM hypoallergenic crops are labeled to inform consumers of their GM status and hypoallergenic properties, ensuring transparency and informed decision-making.

g. Market adoption and consumer acceptance: The successful development and commercialization of GM hypoallergenic crops depends on market adoption and consumer acceptance. Education, communication, and engagement with stakeholders are essential to build trust and confidence in GM technologies and hypoallergenic foods [94]. Transparent labeling, clear communication of benefits, and adherence to regulatory guidelines are crucial for fostering consumer acceptance and uptake of GM hypoallergenic crops in the marketplace.

3.9.2 Case study 2: enzyme-modified products

Enzyme modification represents a promising approach in hypoallergenic food development by utilizing enzymatic treatments to alter the allergenic properties of food proteins while preserving their nutritional integrity and sensory attributes [9]:

a. Selection of suitable enzymes: The first step in enzyme modification is the selection of suitable enzymes capable of modifying allergenic proteins without compromising food quality or safety. Enzymes such as proteases, carbohydrases, and lipases are chosen based on their specificity, activity, and compatibility with the target food matrix [95]. Proteases such as papain and bromelain are commonly used to hydrolyze allergenic proteins into smaller peptides, reducing their allergenicity while maintaining nutritional value.

b. Enzymatic hydrolysis: Enzymatic hydrolysis is employed to cleave peptide bonds within allergenic proteins, resulting in the generation of smaller peptides with reduced allergenic potential. Proteases target specific amino acid residues within allergenic proteins, breaking them down into peptides of varying lengths [11]. This enzymatic process disrupts the structure of allergenic epitopes, thereby reducing their ability to elicit allergic reactions in sensitive individuals.

c. Optimization of processing conditions: Optimization of processing conditions such as enzyme concentration, reaction time, temperature, and pH are essential to maximize the efficiency of enzymatic modification while minimizing undesirable effects on food quality. Enzyme kinetics studies and response surface methodology are utilized to optimize processing parameters and achieve the desired level of allergen reduction without compromising sensory attributes [96]. Controlled enzymatic hydrolysis ensures uniform modification of allergenic proteins throughout the food matrix, enhancing safety and consistency.

d. Characterization of modified products: Characterization of enzyme-modified products involves assessing changes in allergenicity, protein composition, structural integrity, and functional properties. Analytical techniques such as ELISA, SDS-PAGE, and mass spectrometry are utilized to quantify residual allergenic proteins, identify peptide fragments, and evaluate structural modifications [72]. Physicochemical analyses assess changes in protein solubility, emulsifying properties, and rheological behavior, providing insights into the functional properties of modified products.

e. Allergenicity testing: Allergenicity testing is conducted to evaluate the allergenic potential of enzyme-modified products and ensure their safety for individuals with food allergies. In vitro assays such as IgE-binding assays and basophil activation tests assess the ability of modified proteins to elicit allergic reactions in sensitized individuals [53]. Clinical studies may also be conducted to evaluate the allergenicity of modified products in allergic individuals under controlled conditions.

f. Regulatory approval and labeling: Regulatory approval is obtained from government agencies such as the FDA and the EFSA before commercialization of enzyme-modified products. Regulatory agencies evaluate the safety, efficacy, and allergenicity of modified products based on comprehensive data from allergenicity testing, compositional analysis, and safety assessment studies [97]. Upon approval, modified products are labeled to inform consumers of their hypoallergenic properties and ensure transparency in food labeling.

g. Market adoption and consumer acceptance: Market adoption and consumer acceptance of enzyme-modified products depend on factors such as taste, texture, price, and perceived health benefits. Consumer awareness campaigns, educational initiatives, and sensory evaluation studies are conducted to increase public awareness and acceptance of modified products [98]. Transparent communication of benefits, safety assurances, and regulatory compliance are essential for building trust and confidence in enzyme-modified foods among consumers.

3.9.3 Case study 3: successful nanotechnology applications

Nanotechnology has emerged as a promising approach in hypoallergenic food development, offering innovative solutions for reducing allergenicity, improving food safety, and enhancing nutritional quality [99]:

a. **Nanoencapsulation of allergenic proteins**: Nanoencapsulation involves the encapsulation of allergenic proteins within nanoscale carriers such as liposomes, nanoparticles, or biopolymeric matrices to protect them from enzymatic degradation and immune recognition [25]. Through controlled release mechanisms, nanoencapsulation enables targeted delivery of allergenic proteins to specific sites in the gastrointestinal tract, reducing their exposure to immune cells and mitigating allergic reactions.

b. **Surface modification and functionalization**: Surface modification and functionalization of nanoparticles with biocompatible coatings such as polyethylene glycol (PEG) or chitosan enhance their stability, biocompatibility, and targeting efficiency [100]. Functionalized nanoparticles exhibit improved colloidal stability, prolonged circulation time, and enhanced cellular uptake, facilitating efficient delivery of allergen-targeted therapies and immunomodulatory agents.

c. **Immobilization of allergens**: Immobilization of allergens onto nanomaterials such as gold nanoparticles or silica nanoparticles reduces their allergenicity while preserving their structural integrity and antigenic properties [4]. Immobilized allergens exhibit decreased IgE-binding capacity and reduced ability to trigger allergic reactions in sensitized individuals, offering a safe and effective approach for allergen-specific immunotherapy.

d. **Nanoparticle-based allergen detection**: Nanoparticle-based assays and biosensors are employed for rapid and sensitive detection of allergenic proteins in food matrices, enabling accurate allergen labeling and quality control [101]. Functionalized nanoparticles conjugated with allergen-specific antibodies or aptamers selectively capture and detect target allergens, providing quantitative and qualitative analysis of allergenic contaminants in food products.

e. **Nanoparticle-mediated immunomodulation**: Nanoparticle-mediated immunomodulation involves the delivery of immunomodulatory agents such as allergen peptides, adjuvants, or tolerogenic nanoparticles to regulate immune responses and induce allergen-specific tolerance [102]. Tolerogenic nanoparticles coated with allergen-derived peptides or immune-regulatory molecules promote the generation of regulatory T cells and suppress allergic reactions, offering a potential therapeutic strategy for allergen desensitization [103].

f. **Regulatory approval and safety assessment**: Regulatory approval and safety assessment of nanotechnology-based hypoallergenic foods involve comprehensive evaluation of nanoparticle characteristics, stability, biocompatibility, and toxicity profiles [4]. Regulatory agencies assess the safety and efficacy of nanoparticle-based formulations through preclinical studies, toxicological assessments, and clinical trials to ensure their compliance with food safety standards and regulatory guidelines [104].

g. **Consumer acceptance and market adoption**: Consumer acceptance and market adoption of nanotechnology-based hypoallergenic foods depend on factors such as safety assurances, transparency, and perceived benefits. Effective communication, education, and engagement with stakeholders are essential to address consumer concerns, build trust, and promote the acceptance of nanotechnology-based food products in the marketplace [105].

3.9.4 Future directions and challenges

Advancing hypoallergenic food production requires addressing several technical, regulatory, and consumer-related challenges across various biotechnological approaches. Nanotechnology applications face hurdles such as scalability, cost-effectiveness, and regulatory compliance, necessitating advances in nanomaterial design and manufacturing processes to optimize performance [106]. Similarly, enzyme modification techniques must overcome issues related to enzyme specificity, scalability, and cost, with research needed to explore novel enzymes and optimize processing technologies. GM hypoallergenic crops continue to face public skepticism, regulatory hurdles, and technological limitations, which can be addressed through advancements in gene editing and effective communication strategies [107]. Analytical methods for allergen detection require improvements in sensitivity, specificity, and multiplexing capabilities to mitigate matrix interference and cross-reactivity issues, while meeting regulatory requirements for allergen labeling [108]. Interdisciplinary collaboration among scientists, engineers, regulators, industry stakeholders, and consumer advocacy groups is crucial for driving innovation, addressing regulatory concerns, and ensuring the safety and scalability of these technologies [109]. By integrating advancements across these domains, the future of hypoallergenic food development holds promises for enhancing food safety and meeting the dietary needs of individuals with food allergies.

3.10 Regulatory framework and safety assessment

3.10.1 International regulations for hypoallergenic foods

The development and commercialization of hypoallergenic foods are subject to a complex regulatory framework established by international organizations and government agencies to ensure food safety, labeling accuracy, and consumer protection [110]:

a. Codex Alimentarius Commission: The Codex Alimentarius Commission, established by the Food and Agriculture Organization (FAO) and the World Health Organization (WHO), sets international standards for food safety, quality, and labeling. Codex guidelines provide recommendations for the assessment and management of allergenic foods, including labeling requirements, risk assessment methodologies, and good manufacturing practices [110].

b. European Union regulations: In the European Union (EU), allergen labeling regulations require the mandatory labeling of major allergens in prepackaged foods, including cereals containing gluten, crustaceans, eggs, fish, peanuts, soybeans, milk, nuts, celery, mustard, sesame seeds, sulfur dioxide, lupin, and mollusks. Additionally, the EU Novel Foods Regulation governs the approval and labeling of novel foods, including GM hypoallergenic foods, ensuring their safety and regulatory compliance [110].

c. United States Food and Drug Administration (FDA): The FDA regulates hypoallergenic foods in the United States under the Food Allergen Labeling and Consumer Protection Act (FALCPA), which mandates the labeling of major food allergens in packaged foods. FALCPA requires food manufacturers to clearly list allergenic ingredients on food labels, providing consumers with accurate information to make informed choices and prevent allergic reactions [110].

d. Joint FAO/WHO expert consultation: The Joint FAO/WHO Expert Consultation on Allergenicity of Foods Derived from Biotechnology provides scientific guidance and recommendations for the assessment and safety evaluation of GM hypoallergenic foods. Expert consultations review scientific evidence, risk assessment methodologies, and allergenicity testing protocols to ensure the safety and regulatory compliance of hypoallergenic food products [110].

e. Risk assessment and safety evaluation: Regulatory agencies conduct risk assessment and safety evaluation of hypoallergenic foods to assess their allergenicity, toxicity, and potential health risks. Safety assessments include allergen testing, compositional analysis, toxicity studies, and allergenicity testing in sensitized individuals to evaluate the safety and efficacy of hypoallergenic food products [110].

3.10.2 Safety assessment protocols for biotechnologically processed foods

The safety assessment of biotechnologically processed foods involves comprehensive protocols and methodologies established by regulatory agencies to evaluate the safety, allergenicity, and nutritional integrity of GM and biotechnologically modified food products [111]:

a. **Composition analysis:** Composition analysis is conducted to compare the nutritional composition of biotechnologically processed foods with their conventional counterparts and assess any unintended changes resulting from genetic modification or biotechnological processing. Analytical techniques such as chromatography, spectroscopy, and mass spectrometry are utilized to quantify macronutrients, micronutrients, vitamins, minerals, and other nutritional components, ensuring the nutritional equivalence and safety of biotechnologically processed foods [112].

b. **Allergenicity assessment**: Allergenicity assessment is a critical component of safety evaluation for biotechnologically processed foods to assess the potential risk of allergenic reactions in sensitive individuals. In vitro assays such as ELISA, western blotting, and IgE-binding assays are employed to evaluate the allergenicity of novel proteins introduced through genetic modification or biotechnological processing [53]. Animal models and human clinical studies may also be conducted to assess allergenicity under simulated physiological conditions and evaluate the potential for allergic sensitization or elicitation of allergic reactions [113].

c. **Bioinformatics analysis**: Bioinformatics analysis plays a key role in safety assessment protocols for biotechnologically processed foods by providing computational tools and databases for allergen prediction, protein sequence analysis, and structural modeling. Bioinformatics approaches such as sequence homology searches, protein structure prediction, and epitope mapping enable the identification and characterization of potential allergens, cross-reactive epitopes, and structural similarities between novel proteins and known allergens [114].

d. **Toxicological studies:** Toxicological studies are conducted to assess the safety of biotechnologically processed foods by evaluating their potential to cause adverse effects on human health. Animal feeding studies, acute toxicity tests, sub chronic toxicity studies, and genotoxicity assessments are performed to evaluate the safety profile of novel proteins, metabolic byproducts, and unintended effects associated with genetic modification or biotechnological processing [115].

e. **Allergen-free variants**: In cases where genetic modification or biotechnological processing aims to reduce allergenicity, safety assessment protocols focus on confirming the absence or reduction of allergenic proteins in the final food product. Analytical techniques such as ELISA, PCR, and mass spectrometry are utilized to detect and quantify residual allergens, ensuring compliance with allergen labeling regulations and minimizing the risk of allergic reactions in sensitive individuals [116].

f. Comparative assessment: Comparative assessment is conducted to compare the safety and nutritional quality of biotechnologically processed foods with their conventional counterparts and assess any potential differences resulting from genetic modification or biotechnological processing. Comparative studies evaluate parameters such as nutrient composition, allergenicity, toxicological profiles, and agronomic traits to identify any unintended effects or unexpected outcomes associated with biotechnological interventions [117].

g. Post-market surveillance: Post-market surveillance is an integral part of safety assessment protocols for biotechnologically processed foods to monitor their safety and performance after commercialization. Regulatory agencies conduct routine inspections, sampling, and testing of biotechnologically processed foods to detect any adverse effects, allergen contamination, or safety concerns arising from long-term consumption or unintended effects of genetic modification [118].

h. Transparency and communication: Transparency and communication are essential aspects of safety assessment protocols for biotechnologically processed foods to foster public trust, address consumer concerns, and ensure informed decision-making. Regulatory agencies, industry stakeholders, and research institutions engage in transparent communication, risk communication, and stakeholder dialogue to provide accurate information, address misconceptions, and build public confidence in the safety and regulatory oversight of biotechnologically processed foods [119].

i. Continuous improvement: Safety assessment protocols for biotechnologically processed foods undergo continuous improvement and refinement to incorporate advances in scientific knowledge, technological innovations, and regulatory standards. Regulatory agencies collaborate with international organizations, research institutions, and industry stakeholders to update safety assessment methodologies, harmonize regulatory requirements, and ensure the safety and integrity of biotechnologically processed foods in the global marketplace [120].

3.10.3 Challenges and future perspectives in regulation

The regulation of hypoallergenic foods and biotechnologically processed foods faces several challenges and uncertainties, necessitating ongoing adaptation and innovation to ensure effective oversight and consumer protection [121]:

a. Rapid technological advancements: One of the primary challenges in regulatory frameworks for hypoallergenic foods and biotechnologically processed foods is keeping pace with rapid technological advancements and innovations in food science and biotechnology. Emerging technologies such as genome editing, synthetic biology, and nanotechnology present new opportunities for food innovation but also pose novel regulatory challenges related to risk assessment, safety evaluation, and labeling requirements [119].

b. Complexity of allergen management: Allergen management in food production and processing is inherently complex, involving multiple stakeholders, diverse food matrices, and evolving scientific knowledge. Regulatory agencies face challenges in harmonizing allergen labeling regulations, establishing threshold limits for allergenic proteins, and addressing cross-contamination issues in shared manufacturing facilities [122].

c. Consumer awareness and education: Consumer awareness and education play a crucial role in ensuring the success of regulatory frameworks for hypoallergenic foods and biotechnologically processed foods. Challenges include addressing misconceptions, providing accurate information, and enhancing consumer understanding of biotechnological processes, safety assessments, and labeling regulations [123].

d. Global harmonization efforts: Achieving global harmonization of regulatory standards and guidelines for hypoallergenic foods and biotechnologically processed foods remains a significant challenge. Discrepancies in regulatory requirements, divergent risk assessment methodologies, and varying public attitudes toward biotechnology hinder efforts to harmonize regulations and facilitate international trade [124].

e. Risk communication and trust building: Effective risk communication and trust-building strategies are essential for fostering public confidence in the safety and regulatory oversight of hypoallergenic foods and biotechnologically processed foods. Challenges include addressing public perceptions, addressing consumer concerns, and building trust in regulatory agencies, industry stakeholders, and scientific expertise [125].

f. Ethical and social considerations: Regulatory frameworks for hypoallergenic foods and biotechnologically processed foods must navigate ethical and social considerations related to food safety, environmental sustainability, and societal values. Challenges include balancing the benefits and risks of biotechnology, addressing concerns about genetic engineering, and ensuring equitable access to safe and nutritious food for all populations [126].

g. Regulatory capacity and resources: Regulatory agencies face challenges in building regulatory capacity, acquiring technical expertise, and allocating resources to effectively oversee the safety and labeling of hypoallergenic foods and biotechnologically processed foods. Limited funding, personnel constraints, and competing priorities may impede regulatory efforts to keep pace with technological advancements and address emerging issues [127].

h. Adaptation to emerging risks: Regulatory frameworks must adapt to emerging risks and uncertainties associated with hypoallergenic foods and biotechnologically processed foods, including the potential for unintended effects, allergen cross-reactivity, and environmental impacts. Challenges include developing predictive models, surveillance systems, and risk management strategies to anticipate and mitigate emerging risks in food production and consumption [124].

i. Collaborative governance and stakeholder engagement: Addressing the challenges and complexities of regulating hypoallergenic foods and biotechnologically processed foods requires collaborative governance and stakeholder engagement. Regulatory agencies, industry stakeholders, consumer advocacy groups, and research institutions must collaborate to identify common goals, address shared concerns, and develop effective regulatory strategies that promote food safety, innovation, and public trust [128].

3.11 Concluding remarks

The advancement of hypoallergenic food production represents a significant stride in addressing the dietary and health needs of individuals with food allergies. This chapter has explored the multifaceted approaches utilized in this domain, from analytical methods for allergen detection and molecular characterization of allergens to innovative biotechnological techniques such as nanotechnology, enzyme modification, and genetic engineering. These strategies, alongside rigorous safety assessments and evolving regulatory frameworks, underscore the commitment to creating safer and more inclusive food systems. Despite notable progress, challenges persist, including technological limitations, regulatory complexities, and public acceptance. However, the potential of these advancements to revolutionize food safety and enhance the quality of life for individuals with allergies is immense. Future efforts must focus on interdisciplinary collaboration, technological innovation, and harmonized global regulations to overcome existing barriers. By fostering a synergy between science, industry, and policy, the development of hypoallergenic foods can achieve new milestones, paving the way for a more inclusive and secure food landscape. This holistic approach not only addresses the immediate concerns of allergen management but also sets a foundation for long-term advancements in food biotechnology and safety.

References

[1] Warren, C. M., Jiang, J., & Gupta, R. S. Epidemiology and burden of food allergy. Current Allergy and Asthma Reports 2020, 20, 1–9.

[2] Crealey, M. R. Food Allergy among Irish Children: An Exploration of the Risk and Impact of Accidental Allergic Reactions (Doctoral Dissertation), School of Medicine, Trinity College Dublin, 2022.

[3] Cardwell, F. S., Elliott, S. J., Chin, R., Pierre, Y. S., Ben-Shoshan, M., Chan, E. S., . . . & Clarke, A. E. Economic burden of food allergy in Canada: Estimating costs and identifying determinants. Annals of Allergy, Asthma and Immunology 2022, 129(2), 220–230.

[4] Rai, M., Ingle, A. P., Yadav, A., Golińska, P., Trzcińska-Wencel, J., Rathod, S., & Bonde, S. Nanotechnology as a promising approach for detection, diagnosis, and treatment of food allergens. Current Nanoscience 2023, 19(1), 90–102.

[5] Ball, D. J., Ball-King, L. N., & McMichael, A. J. 7.1 Allergenic pollen emissions from vegetation – Threats and prevention 195 Åslög Dahl, Matilda van den Bosch, and Thomas Ogren 7.4 Risk and the perception of risk in interactions with nature 215. In: Oxford Textbook of Nature and Public Health: The Role of Nature in Improving the Health of a Population, 2018, 193.

[6] Senarathna, S., Mel, R., & Malalgoda, M. Utilization of cereal-based protein ingredients in food applications. Journal of Cereal Science 2024, 103867.

[7] Gargano, D., Appanna, R., Santonicola, A., De Bartolomeis, F., Stellato, C., Cianferoni, A., . . . & Iovino, P. Food allergy and intolerance: A narrative review on nutritional concerns. Nutrients 2021, 13(5), 1638.

[8] Fu, L., Cherayil, B. J., Shi, H., Wang, Y., Zhu, Y., Fu, L., . . . & Zhu, Y. Food processing to eliminate food allergens and development of hypoallergenic foods. Food Allergy: From Molecular Mechanisms to Control Strategies 2019, 123–146.

[9] Ekezie, F. G. C., Cheng, J. H., & Sun, D. W. Effects of nonthermal food processing technologies on food allergens: A review of recent research advances. Trends in Food Science and Technology 2018, 74, 12–25.

[10] Ladics, G. S. Assessment of the potential allergenicity of genetically engineered food crops. Journal of Immunotoxicology 2019, 16(1), 43–53.

[11] Tavano, O. L., Berenguer-Murcia, A., Secundo, F., & Fernandez-Lafuente, R. Biotechnological applications of proteases in food technology. Comprehensive Reviews in Food Science and Food Safety 2018, 17(2), 412–436.

[12] Jagtiani, E. Advancements in nanotechnology for food science and industry. Food Frontiers 2022, 3(1), 56–82.

[13] Anagnostou, A., Lieberman, J., Greenhawt, M., Mack, D. P., Santos, A. F., Venter, C., . . . & Brough, H. A. The future of food allergy: Challenging existing paradigms of clinical practice. Allergy 2023, 78(7), 1847–1865.

[14] Gordon, A., & Williams, R. The role and importance of packaging and labeling in assuring food safety, quality and regulatory compliance of export products II: Packaging and labeling considerations. In: Food Safety and Quality Systems in Developing Countries, Academic Press, 2020, 285–341.

[15] Aguilera, J. M. The food matrix: Implications in processing, nutrition, and health. Critical Reviews in Food Science and Nutrition 2019, 59(22), 3612–3629.

[16] Borisjuk, N., Kishchenko, O., Eliby, S., Schramm, C., Anderson, P., Jatayev, S., . . . & Shavrukov, Y. Genetic modification for wheat improvement: From transgenesis to genome editing. BioMed Research International 2019, 2019.

[17] Hathwar, S. C., Rai, A. K., Modi, V. K., & Narayan, B. Characteristics and consumer acceptance of healthier meat and meat product formulations – A review. Journal of Food Science and Technology 2012, 49, 653–664.

[18] Ahmed, I., Chen, H., Li, J., Wang, B., Li, Z., & Huang, G. Enzymatic crosslinking and food allergenicity: A comprehensive review. Comprehensive Reviews in Food Science and Food Safety 2021, 20(6), 5856–5879.

[19] Lee, C. C., Suttikhana, I., & Ashaolu, T. J. Techno-functions and safety concerns of plant-based peptides in food matrices. Journal of Agricultural and Food Chemistry 2024.

[20] Yetisgin, A. A., Cetinel, S., Zuvin, M., Kosar, A., & Kutlu, O. Therapeutic nanoparticles and their targeted delivery applications. Molecules 2020, 25(9), 2193.

[21] Shavronskaya, D. O., Noskova, A. O., Skvortsova, N. N., Adadi, P., & Nazarova, E. A. Encapsulation of hydrophobic bioactive substances for food applications: Carriers, techniques, and biosafety. Journal of Food Processing and Preservation 2023, 2023.

[22] Lokya, V., Parmar, S., Pandey, A. K., Sudini, H. K., Huai, D., Ozias-Akins, P., . . . & Pandey, M. K. Prospects for developing allergen-depleted food crops. The Plant Genome 2023, 16(4), e20375.

[23] Jain, S., Rustagi, A., Kumar, D., Yusuf, M. A., Shekhar, S., & Sarin, N. B. Meeting the challenge of developing food crops with improved nutritional quality and food safety: Leveraging proteomics and related omics techniques. Biotechnology Letters 2019, 41, 471–481.

[24] Popov-Raljić, J., Aleksić, M., & Janković, V. Food allergens – Food safety hazard. Scientific Journal Meat Technology 2022, 63(1), 11–25.

[25] Ngwuluka, N. C., Abu-Thabit, N. Y., Uwaezuoke, O. J., Erebor, J. O., Ilomuanya, M. O., Mohamed, R. R., . . . & Ebrahim, N. A. Natural polymers in micro-and nanoencapsulation for therapeutic and diagnostic applications: Part II-polysaccharides and proteins. Nano and Microencapsulation Techniques and Applications 2021, 8, 9.

[26] Cunha, M. L., Vieira, V. R. M., Santana, A. R., & Anastácio, L. R. Food allergen labeling: Compliance with the mandatory legislation in Brazil. Food Science and Technology 2020, 40, 698–704.

[27] Kucuksezer, U. C., Ozdemir, C., Cevhertas, L., Ogulur, I., Akdis, M., & Akdis, C. A. Mechanisms of allergen-specific immunotherapy and allergen tolerance. Allergology International 2020, 69(4), 549–560.

[28] Muthukumar, J., Selvasekaran, P., Lokanadham, M., & Chidambaram, R. Food and food products associated with food allergy and food intolerance–An overview. Food Research International 2020, 138, 109780.

[29] Ardoin, R., & Prinyawiwatkul, W. Consumer perceptions of insect consumption: A review of western research since 2015. International Journal of Food Science and Technology 2021, 56(10), 4942–4958.

[30] Shah, F., Shi, A., Ashley, J., Kronfel, C., Wang, Q., Maleki, S. J., . . . & Zhang, J. Peanut allergy: Characteristics and approaches for mitigation. Comprehensive Reviews in Food Science and Food Safety 2019, 18(5), 1361–1387.

[31] Dhesi, A., Ashton, G., Raptaki, M., & Makwana, N. Cow's milk protein allergy. Paediatrics and Child Health 2020, 30(7), 255–260.

[32] Zhu, Y., Vanga, S. K., Wang, J., & Raghavan, V. Impact of food processing on the structural and allergenic properties of egg white. Trends in Food Science and Technology 2018, 78, 188–196.

[33] Shan, D., Yu, H., Lyu, B., & Fu, H. Soybean β-conglycinin: Structure characteristic, allergenicity, plasma lipid-controlling, prevention of obesity and non-alcoholic fatty liver disease. Current Protein and Peptide Science 2021, 22(12), 831–847.

[34] Sharma, N., Bhatia, S., Chunduri, V., Kaur, S., Sharma, S., Kapoor, P., . . . & Garg, M. Pathogenesis of celiac disease and other gluten related disorders in wheat and strategies for mitigating them. Frontiers in Nutrition 2020, 7, 6.

[35] Ruethers, T., Taki, A. C., Johnston, E. B., Nugraha, R., Le, T. T., Kalic, T., . . . & Lopata, A. L. Seafood allergy: A comprehensive review of fish and shellfish allergens. Molecular Immunology 2018, 100, 28–57.
[36] Sheehan, W. J., Taylor, S. L., Phipatanakul, W., & Brough, H. A. Environmental food exposure: What is the risk of clinical reactivity from cross-contact and what is the risk of sensitization. The Journal of Allergy and Clinical Immunology: In Practice 2018, 6(6), 1825–1832.
[37] Vojdani, A., Gushgari, L. R., & Vojdani, E. Interaction between food antigens and the immune system: Association with autoimmune disorders. Autoimmunity Reviews 2020, 19(3), 102459.
[38] Shamji, M. H., Valenta, R., Jardetzky, T., Verhasselt, V., Durham, S. R., Würtzen, P. A., & Van Neerven, R. J. The role of allergen specific IgE, IgG and IgA in allergic disease. Allergy 2021, 76(12), 3627–3641.
[39] Tontini, C., & Bulfone-Paus, S. Novel approaches in the inhibition of IgE-induced mast cell reactivity in food allergy. Frontiers in Immunology 2021, 12, 613461.
[40] Vitte, J., Vibhushan, S., Bratti, M., Montero-Hernandez, J. E., & Blank, U. Allergy, anaphylaxis, and nonallergic hypersensitivity: IgE, mast cells, and beyond. Medical Principles and Practice 2022, 31(6), 501–515.
[41] Kauppinen, K., & Kariniemi, A. L. Clinical manifestations and histological characteristics. In: Skin Reactions to Drugs, CRC Press, 2020, 25–50.
[42] Renz, H., Allen, K. J., Sicherer, S. H., Sampson, H. A., Lack, G., Beyer, K., & Oettgen, H. C. Food allergy. Nature Reviews Disease Primers 2018, 4(1), 1–20.
[43] Singer, S. D., Laurie, J. D., Bilichak, A., Kumar, S., & Singh, J. Genetic variation and unintended risk in the context of old and new breeding techniques. Critical Reviews in Plant Sciences 2021, 40(1), 68–108.
[44] Sen, P., Lata, C., Kiran, K., & Mondal, T. K. RNA interference (RNAi) in functional genomics of wheat. Genome Engineering for Crop Improvement 2021, 239–264.
[45] Camerlengo, F., Frittelli, A., Sparks, C., Doherty, A., Martignago, D., Larré, C., . . . & Masci, S. CRISPR-Cas9 multiplex editing of the α-amylase/trypsin inhibitor genes to reduce allergen proteins in durum wheat. Frontiers in Sustainable Food Systems 2020, 4, 104.
[46] Van Esse, H. P., Reuber, T. L., & van der Does, D. Genetic modification to improve disease resistance in crops. New Phytologist 2020, 225(1), 70–86.
[47] Neugebauer, R., Ed. Biological Transformation, Springer Vieweg, 2020.
[48] Pekar, J., Ret, D., & Untersmayr, E. Stability of allergens. Molecular Immunology 2018, 100, 14–20.
[49] Lasekan, A. O. Attenuating the Antibody Reactivity of the Shrimp Major Allergen (Tropomyosin) Using Food Processing Methods, The University of Maine, 2017.
[50] Gaspar, A. L. C., & de Góes-Favoni, S. P. Action of microbial transglutaminase (MTGase) in the modification of food proteins: A review. Food Chemistry 2015, 171, 315–322.
[51] Rodrigues, R. C., Ortiz, C., Berenguer-Murcia, Á., Torres, R., & Fernández-Lafuente, R. Modifying enzyme activity and selectivity by immobilization. Chemical Society Reviews 2013, 42(15), 6290–6307.
[52] Tan, M., Nawaz, M. A., & Buckow, R. Functional and food application of plant proteins–a review. Food Reviews International 2023, 39(5), 2428–2456.
[53] Remington, B., Broekman, H. C. H., Blom, W. M., Capt, A., Crevel, R. W., Dimitrov, I., . . . & Constable, A. Approaches to assess IgE mediated allergy risks (sensitization and cross-reactivity) from new or modified dietary proteins. Food and Chemical Toxicology 2018, 112, 97–107.
[54] Rogers, H., Dora, M., Tsolakis, N., & Kumar, M. Plant-Based Food Supply Chains: Recognising Market Opportunities and Industry Challenges of Pea Protein. Available at SSRN 4803579.
[55] Virkud, Y. V., Wang, J., & Shreffler, W. G. Enhancing the safety and efficacy of food allergy immunotherapy: A review of adjunctive therapies. Clinical Reviews in Allergy and Immunology 2018, 55(2), 172–189.

[56] Di Gioacchino, M., Petrarca, C., Gatta, A., Scarano, G., Farinelli, A., Della Valle, L., . . . & Di Giampaolo, L. Nanoparticle-based immunotherapy: State of the art and future perspectives. Expert Review of Clinical Immunology 2020, 16(5), 513–525.

[57] Deng, Z., Kalin, G. T., Shi, D., & Kalinichenko, V. V. Nanoparticle delivery systems with cell-specific targeting for pulmonary diseases. American Journal of Respiratory Cell and Molecular Biology 2021, 64(3), 292–307.

[58] Anwer, A. H., Ahtesham, A., Shoeb, M., Mashkoor, F., Ansari, M. Z., Zhu, S., & Jeong, C. State-of-the-art advances in nanocomposite and bio-nanocomposite polymeric materials: A comprehensive review. Advances in Colloid and Interface Science 2023, 102955.

[59] Arora, K. Advances in nano-based biosensors for food and agriculture. Nanotechnology, Food Security and Water Treatment 2018, 1–52.

[60] Kumari, R., Suman, K., Karmakar, S., Mishra, V., Lakra, S. G., Saurav, G. K., & Mahto, B. K. Regulation and safety measures for nanotechnology-based agri-products. Frontiers in Genome Editing 2023, 5, 1200987.

[61] Handford, C. E., Dean, M., Henchion, M., Spence, M., Elliott, C. T., & Campbell, K. Implications of nanotechnology for the agri-food industry: Opportunities, benefits and risks. Trends in Food Science and Technology 2014, 40(2), 226–241.

[62] Terrell, G. C., & Hernandez-Jover, M. Meat and meat products. In: Food Safety Management, Academic Press, 2023, 141–184.

[63] Senyuva, H. Z., Jones, I. B., Sykes, M., & Baumgartner, S. A critical review of the specifications and performance of antibody and DNA-based methods for detection and quantification of allergens in foods. Food Additives and Contaminants: Part A 2019, 36(4), 507–547.

[64] Villa, C., Costa, J., & Mafra, I. Detection and quantification of milk ingredients as hidden allergens in meat products by a novel specific real-time PCR method. Biomolecules 2019, 9(12), 804.

[65] Linacero, R., Sanchiz, A., Ballesteros, I., & Cuadrado, C. Application of real-time PCR for tree nut allergen detection in processed foods. Critical Reviews in Food Science and Nutrition 2020, 60(7), 1077–1093.

[66] Birse, N., Burns, D. T., Walker, M. J., Quaglia, M., & Elliott, C. T. Food allergen analysis: A review of current gaps and the potential to fill them by matrix-assisted laser desorption/ionization. Comprehensive Reviews in Food Science and Food Safety 2023, 22(5), 3984–4003.

[67] Muralidharan, S., Zhao, Y., Taylor, S. L., & Lee, N. A. Detection of food allergen residues by immunoassays and mass spectrometry. Food Allergy 2017, 229–282.

[68] Zhou, J., Qi, Q., Wang, C., Qian, Y., Liu, G., Wang, Y., & Fu, L. Surface plasmon resonance (SPR) biosensors for food allergen detection in food matrices. Biosensors and Bioelectronics 2019, 142, 111449.

[69] Zheng, C., Zhu, L., & Wang, J. A review on rapid detection of modified quartz crystal microbalance sensors for food: Contamination, flavour and adulteration. TrAC Trends in Analytical Chemistry 2022, 157, 116805.

[70] Mattarozzi, M., & Careri, M. The role of incurred materials in method development and validation to account for food processing effects in food allergen analysis. Analytical and Bioanalytical Chemistry 2019, 411, 4465–4480.

[71] Perazzio, S. F., Palmeira, P., Moraes-Vasconcelos, D., Rangel-Santos, A., De Oliveira, J. B., Andrade, L. E. C., & Carneiro-Sampaio, M. A critical review on the standardization and quality assessment of nonfunctional laboratory tests frequently used to identify inborn errors of immunity. Frontiers in Immunology 2021, 12, 721289.

[72] Benedé, S., Lozano-Ojalvo, D., Cristobal, S., Costa, J., D'Auria, E., Velickovic, T. C., . . . & Molina, E. New applications of advanced instrumental techniques for the characterization of food allergenic proteins. Critical Reviews in Food Science and Nutrition 2022, 62(31), 8686–8702.

[73] Mostashari, P., Marszałek, K., Aliyeva, A., & Mousavi Khaneghah, A. The impact of processing and extraction methods on the allergenicity of targeted protein quantification as well as bioactive peptides derived from egg. Molecules 2023, 28(6), 2658.
[74] Sagu, S. T., Huschek, G., Homann, T., & Rawel, H. M. Effect of sample preparation on the detection and quantification of selected nuts allergenic proteins by LC-MS/MS. Molecules 2021, 26(15), 4698.
[75] Frossard, M. (2020). Novel Analytical Methods for Allergy-Related Studies (No. 7271).
[76] Plum, M. Dissection of the IgE Interactome on a Molecular Level (Doctoral Dissertation), Staats-und Universitätsbibliothek Hamburg Carl von Ossietzky, 2011.
[77] Gul, M., Ayan, E., Destan, E., Johnson, J. A., Shafiei, A., Kepceoğlu, A., . . . & DeMi̇rci̇, H. Rapid and efficient ambient temperature X-ray crystal structure determination at Turkish Light Source. Scientific Reports 2023, 13(1), 8123.
[78] Solanki, D., Mandaliya, V., & Georrge, J. J. Allergen bioinformatics: Repositories and tools to predict allergic proteins. Recent Trends in Science and Technology-2020 2020, 162–172.
[79] Guryanova, S. V., & Ovchinnikova, T. V. Immunomodulatory and allergenic properties of antimicrobial peptides. International Journal of Molecular Sciences 2022, 23(5), 2499.
[80] Xie, Q., & Xue, W. IgE-Mediated food allergy: Current diagnostic modalities and novel biomarkers with robust potential. Critical Reviews in Food Science and Nutrition 2023, 63(29), 10148–10172.
[81] Gaspan, D. S., & Tolentino, M. P. S. In silico identification and characterization of potential red seaweed allergens. Open Journal of Bioinformatics and Biostatistics 2023, 7(1), 001–017.
[82] Maurer-Stroh, S., Krutz, N. L., Kern, P. S., Gunalan, V., Nguyen, M. N., Limviphuvadh, V., . . . & Gerberick, G. F. AllerCatPro – Prediction of protein allergenicity potential from the protein sequence. Bioinformatics 2019, 35(17), 3020–3027.
[83] Sundararaj, R., Mathimaran, A., Prabhu, D., Ramachandran, B., Jeyaraman, J., Muthupandian, S., & Asmelash, T. In silico approaches for the identification of potential allergens among hypothetical proteins from Alternaria alternata and its functional annotation. Scientific Reports 2024, 14(1), 6696.
[84] Khan, Y. D., Alzahrani, E., Alghamdi, W., & Ullah, M. Z. Sequence-based identification of allergen proteins developed by integration of PseAAC and statistical moments via 5-step rule. Current Bioinformatics 2020, 15(9), 1046–1055.
[85] Muhammed, M. T., & Aki-Yalcin, E. Homology modeling in drug discovery: Overview, current applications, and future perspectives. Chemical Biology and Drug Design 2019, 93(1), 12–20.
[86] Negi, S. S., & Braun, W. Cross-React: A new structural bioinformatics method for predicting allergen cross-reactivity. Bioinformatics 2017, 33(7), 1014–1020.
[87] Dhanushkumar, T., Santhosh, M. E., Selvam, P. K., Rambabu, M., Dasegowda, K. R., Vasudevan, K., & Doss, C. G. P. Advancements and hurdles in the development of a vaccine for triple-negative breast cancer: A comprehensive review of multi-omics and immunomics strategies. Life Sciences 2023, 122360.
[88] Cao, Z., Li, Q., Li, Y., & Wu, J. Identification of plasma protein markers of allergic disease risk: A mendelian randomization approach to proteomic analysis. BMC Genomics 2024, 25(1), 503.
[89] Sabbadini, S., Capocasa, F., Battino, M., Mazzoni, L., & Mezzetti, B. Improved nutritional quality in fruit tree species through traditional and biotechnological approaches. Trends in Food Science and Technology 2021, 117, 125–138.
[90] Brackett, N. F., Pomés, A., & Chapman, M. D. New frontiers: Precise editing of allergen genes using CRISPR. Frontiers in Allergy 2022, 2, 821107.
[91] Hefferon, K. L. Nutritionally enhanced food crops; progress and perspectives. International Journal of Molecular Sciences 2015, 16(2), 3895–3914.
[92] Waters, S., Ramos, A., Culler, A. H., Hunst, P., Zeph, L., Gast, R., . . . & Goodwin, L. Recommendations for science-based safety assessment of genetically modified (GM) plants for food and feed uses. Journal of Regulatory Science 2021, 9(1), 16–21.

[93] Goodman, R. E. (2014). Biosafety: Evaluation and regulation of genetically modified (GM) crops in the United States.
[94] De Mesmaeker, M., Tran, D., Verbeecke, V., Ameye, F., Dubaere, P., Strobbe, S., . . . & De Steur, H. Belgian dietitians' knowledge, perceptions and willingness-to-recommend of genetically modified food and organisms. Journal of Human Nutrition and Dietetics 2024, 37(1), 142–154.
[95] Kuddus, M. Enzymes in food biotechnology. Production, Applications and Future Prospects 2019, 883.
[96] Meng, S., Tan, Y., Chang, S., Li, J., Maleki, S., & Puppala, N. Peanut allergen reduction and functional property improvement by means of enzymatic hydrolysis and transglutaminase crosslinking. Food Chemistry 2020, 302, 125186.
[97] Zimmer, J., Vieths, S., & Kaul, S. Standardization and regulation of allergen products in the European Union. Current Allergy and Asthma Reports 2016, 16, 1–11.
[98] Baker, M. T., Lu, P., Parrella, J. A., & Leggette, H. R. Consumer acceptance toward functional foods: A scoping review. International Journal of Environmental Research and Public Health 2022, 19(3), 1217.
[99] Neethirajan, S., Weng, X., Tah, A., Cordero, J. O., & Ragavan, K. V. Nano-biosensor platforms for detecting food allergens–New trends. Sensing and Bio-sensing Research 2018, 18, 13–30.
[100] Sanità, G., Carrese, B., & Lamberti, A. Nanoparticle surface functionalization: How to improve biocompatibility and cellular internalization. Frontiers in Molecular Biosciences 2020, 7, 587012.
[101] Aquino, A., & Conte-Junior, C. A. A systematic review of food allergy: Nanobiosensor and food allergen detection. Biosensors 2020, 10(12), 194.
[102] Park, J., Wu, Y., Li, Q., Choi, J., Ju, H., Cai, Y., . . . & Oh, Y. K. Nanomaterials for antigen-specific immune tolerance therapy. Drug Delivery and Translational Research 2023, 13(7), 1859–1881.
[103] Satitsuksanoa, P., Głobińska, A., Jansen, K., van de Veen, W., & Akdis, M. Modified allergens for immunotherapy. Current Allergy and Asthma Reports 2018, 18, 1–13.
[104] Ragelle, H., Danhier, F., Préat, V., Langer, R., & Anderson, D. G. Nanoparticle-based drug delivery systems: A commercial and regulatory outlook as the field matures. Expert Opinion on Drug Delivery 2017, 14(7), 851–864.
[105] Mukherjee, A., Maity, A., Pramanik, P., Shubha, K., Joshi, D. C., & Wani, S. H. Public perception about use of nanotechnology in agriculture. In: Advances in Phytonanotechnology, Academic Press, 2019, 405–418.
[106] Liu, H., Bai, Y., Huang, C., Wang, Y., Ji, Y., Du, Y., . . . & Bligh, S. W. A. Recent progress of electrospun herbal medicine nanofibers. Biomolecules 2023, 13(1), 184.
[107] Cummings, C., Selfa, T., Lindberg, S., & Bain, C. Identifying public trust building priorities of gene editing in agriculture and food. Agriculture and Human Values 2024, 41(1), 47–60.
[108] Monaci, L., De Angelis, E., Montemurro, N., & Pilolli, R. Comprehensive overview and recent advances in proteomics MS based methods for food allergens analysis. TrAC Trends in Analytical Chemistry 2018, 106, 21–36.
[109] Pourmadadi, M., Ostovar, S., Ruiz-Pulido, G., Hassan, D., Souri, M., Manicum, A. L. E., . . . & Pandey, S. Novel epirubicin-loaded nanoformulations: Advancements in polymeric nanocarriers for efficient targeted cellular and subcellular anticancer drug delivery. Inorganic Chemistry Communications 2023, 110999.
[110] World Health Organization. (2022). Risk Assessment of Food Allergens. Part 1: Review and validation of Codex Alimentarius priority allergen list through risk assessment: Meeting report.
[111] Dadgarnejad, M., Kouser, S., & Moslemi, M. Genetically modified foods: Promises, challenges and safety assessments. Applied Food Biotechnology 2017, 4(4), 193–202.
[112] Picó, Y. Mass spectrometry in food quality and safety: An overview of the current status. Comprehensive Analytical Chemistry 2015, 68, 3–76.

[113] Bøgh, K. L., van Bilsen, J., Głogowski, R., López-Expósito, I., Bouchaud, G., Blanchard, C., . . . & O'mahony, L. Current challenges facing the assessment of the allergenic capacity of food allergens in animal models. Clinical and Translational Allergy 2016, 6, 1–13.
[114] Zhou, F., He, S., Sun, H., Wang, Y., & Zhang, Y. Advances in epitope mapping technologies for food protein allergens: A review. Trends in Food Science and Technology 2021, 107, 226–239.
[115] Samarasiri, M., Chai, K. F., & Chen, W. N. Forward-looking risk assessment framework for novel foods. Food and Humanity 2023.
[116] Mazzucchelli, G., Holzhauser, T., Cirkovic Velickovic, T., Diaz-Perales, A., Molina, E., Roncada, P., . . . & Hoffmann-Sommergruber, K. Current (food) allergenic risk assessment: Is it fit for novel foods? Status quo and identification of gaps. Molecular Nutrition and Food Research 2018, 62(1), 1700278.
[117] Giraldo, P. A., Shinozuka, H., Spangenberg, G. C., Cogan, N. O., & Smith, K. F. Safety assessment of genetically modified feed: Is there any difference from food? Frontiers in Plant Science 2019, 10, 486827.
[118] Fritsche, E., Elsallab, M., Schaden, M., Hey, S. P., & Abou-El-Enein, M. Post-marketing safety and efficacy surveillance of cell and gene therapies in the EU: A critical review. Cell and Gene Therapy Insights 2019, 5(11), 1505–1521.
[119] Meijer, G. W., Lähteenmäki, L., Stadler, R. H., & Weiss, J. Issues surrounding consumer trust and acceptance of existing and emerging food processing technologies. Critical Reviews in Food Science and Nutrition 2021, 61(1), 97–115.
[120] McHughen, A. A critical assessment of regulatory triggers for products of biotechnology: Product vs. process. GM Crops and Food 2016, 7(3–4), 125–158.
[121] Hallerman, E. M., Bredlau, J. P., Camargo, L. S. A., Dagli, M. L. Z., Karembu, M., Ngure, G., . . . & Wray-Cahen, D. Towards progressive regulatory approaches for agricultural applications of animal biotechnology. Transgenic Research 2022, 31(2), 167–199.
[122] Walker, M. J., & Gowland, M. H. Food allergy: Managing food allergens. Analysis of Food Toxins and Toxicants 2017, 711–742.
[123] Rez Esteban, P., De Faria Catarina, T., & Conrado, C. Allergen management as a key issue in food safety. In: Food Safety and Protection, CRC Press, 2017, 195–241.
[124] Kedar, O., Golberg, A., Obolski, U., & Confino-Cohen, R. Allergic to bureaucracy? Regulatory allergenicity assessments of novel food: Motivations, challenges, compromises, and possibilities. Comprehensive Reviews in Food Science and Food Safety 2024, 23(2), e13300.
[125] Carreto, C., & Carreto, R. Design as an enhancer of the circular economy in fashion. Human Factors for Apparel and Textile Engineering 2022, 32, 106.
[126] Varzakas, T., & Antoniadou, M. A holistic approach for ethics and sustainability in the food chain: The gateway to oral and systemic health. Foods 2024, 13(8), 1224.
[127] Soyombo, D. A. Product management challenges and innovations in baby food: The Nigerian and US Market. GSC Advanced Research and Reviews 2024, 19(1), 4–15.
[128] Mullins, E., Bresson, J. L., Dalmay, T., Dewhurst, I. C., Epstein, M. M., . . . & Moreno, F. J. Scientific Opinion on development needs for the allergenicity and protein safety assessment of food and feed products derived from biotechnology. EFSA Journal 2022, 20(1), e07044.

Omotola Folake Olagunju*, Blessing Nwokocha,
and Abiola Folakemi Olaniran

Chapter 4
Safety and commercialization of gene-edited foods

Abstract: Gene editing (GE) as a biotechnological tool offers distinct improvements in agricultural production over genetic modification. These include lower cost, shorter time for development and lower risk, which may be attributed to the exclusion of foreign genetic materials in the technique. Despite the many benefits that GE presents, there are still concerns about the toxicity and allergenicity of gene-edited foods, and the safety impacts on consumer health and the environment. Whilst recognizing these legitimate concerns, the technology is gaining rapid development by the day, and more information is available on GE and its impacts on public health and environmental sustainability. Successful commercialization of GE products will largely depend on transparency and effectiveness of regulatory framework and well-structured societal introduction. Furthermore, considerations will need to be given to market dynamics, ethical and social concerns, and technological advancements. This chapter presents relevant information on the safety and commercialization of GE foods, with the objective of providing a balance to public perceptions of the complexity of GE technology.

Keywords: Gene editing, safety, commercialization, regulation, consumer health

4.1 Introduction

The rapid increase in world population, notable economic development, and constant changes in dietary preferences have resulted in shifts in the demand for food. According to the European Commission Farm to Fork (F2F) strategy, innovations in the areas of agriculture, particularly plant breeding and crop production can enhance the attainment of a more sustainable food system and simultaneously minimize the impact of food production on the environment [18]. Gene editing (GE) is a breeding technique

***Corresponding author: Omotola Folake Olagunju,** Department of Science, School of Health and Life Sciences, Teesside University, Middlesbrough, England, UK, e-mail: O.Olagunju@tees.ac.uk
Blessing Nwokocha, Department of Science, School of Health and Life Sciences, Teesside University, Middlesbrough, England, UK
Abiola Folakemi Olaniran, Landmark University SDG 12 (Responsible Consumption and Production research Group); Department of Food Science and Nutrition, College of Pure and Applied Sciences, P.M.B., 1001, Landmark University, Omu-Aran, Kwara, Nigeria

https://doi.org/10.1515/9783111441238-005

with the potential to revolutionize agricultural biotechnology [48]. New food products and improved consumer products are some of the benefits offered by advanced genome editing tools and techniques in the food and agricultural sectors [12]. Some of the GE techniques include the clustered regularly interspaced short palindromic repeats/CRISPR-associated 9 (CRISPR/Cas 9), transcription activator-like effector nuclease (TALEN) and zinc finger nucleases (ZFN) [10, 20].

GE involves altering the genome of a particular species or of those with known sexual compatibility (cisgenic editing) and is quite different to genetic modification [12]. As the earliest genomic editing strategy, ZFN uses custom DNA endonucleases largely based on the *Fok*I restriction enzyme fused to a zinc finger DNA-binding domain engineered to target a specific DNA sequence. It is limited by its high cost, high off-target effects created by site-specific nucleases and its possible toxicity to cells. TALENs, in addition to performing many of the roles of ZFNs offer high specificity and a lower probability of off-target effects. However, TALENs are large-sized and more complex to construct. Perhaps a clearer and less expensive GE technology is CRISPR/Cas, the specificity is based on DNA fitness and excludes several protein engineering steps. Genetically engineered crops present several advantages. These include conferring resistance to insect pests or plant viruses as well as providing hybrid seed production. These crops have enhanced nutrient content and provide tolerance to specific herbicides which is an important trait for good weed control [13]. In March 2019, a gene-edited type of soybean with potential to produce more shelf-stable and trans fat-free cooking oil, made its way into the marketplace, as a non-regulated article [12]. Gene-edited foods are more cheaply and more rapidly produced than traditional selective breeding of transgenic modification. GE tools lead to the development of new products with varied agricultural benefits and may contribute to food security, while also promoting environmental sustainability [5, 12].

Despite the economic, environmental, and nutritional benefits of GE, only a few GE crop traits have been approved for commercialization, including soybean, canola, mushroom, camelina and maize. As a market-driven random process, the commercialization stage relies on several factors including the product's total sales, revenue, and resulting net returns. To achieve success in GE food production, careful attention is required to maximize the key elements that bring about success in its commercialization. This chapter presents information that is germane to the safety and commercialization of GE foods.

4.2 Gene editing as a biotechnology tool in food production

Food safety in agrifood systems can be a significant concern for human health. Biofortification had been seen as advantageous and humanitarian, but the adaptation of foreign DNA into plants and some animals in genetically modified (GM) foods frequently generated concerns. GE is invariably a more accurate method than prior induced mutagenesis and genetic engineering procedures, and some of the safety concerns regarding the adaptation of foreign genes into organisms are no longer considered. GE technologies have potentials for application in agriculture, food security, and food processing [52].

Extrinsic factors such as social trends and cultural settings, as well as intrinsic factors such as educational background, play significant roles in the acceptability of foods [58]. A consumer's demand for naturalness, which might include traits like freshness and minimal processing, influence their decision-making process in both extrinsic and intrinsic ways. This desire arises from the belief that natural products are safer for individuals and the environment. Based on the assumption that natural products are healthy, and that GMOs do not occur through natural evolutionary processes, then they must be unnatural and therefore it was assumed unhealthy [54]. The progression of GE technology and its application in food industry, as well as the legal rules and disposition of various countries toward gene-edited food, provides insights on the prospect of gene-edited foods [70]. It becomes important to encourage the extensive use of GE foods, increase safety studies, harmonize international legislation, and raise public awareness.

4.3 Genetically modified (GM) versus gene-edited (GE) products

Controversies have continued to trail the launch of GM organism (GMO) products, these arose largely from concerns about the ethical considerations of practices engaged in their development, and of course the safety of these products, not minding their numerous contributions to consumers and target groups [6, 12, 27]. The history of GMOs seems to be intricately linked to how well GE food products will fare. Criticisms have been raised on the relatedness of GE foods to GMOs, concerns about the ethics of its practices, concerns on the environmental and the social implications as well as the scientific intricacy involved [7, 11, 12, 60]. In a bid to reduce the controversies associated with GMO foods, studies are on-going on the potential risks associated with GE foods but the acceptance by consumers may require more assurance than immediate product benefits and less technical risks [12]. GMO acceptance was characterized by risk and safety concerns amid the public. Avoiding scientific discourse, promoting openness in GE food practices as well as encouraging shared values between developers and consumers are target actions in practice to improve public acceptance of GE food products [12].

The technological processes of GM products and GE products are different; likewise, the time and financial expenses involved in their technical processes differ [7, 8,

11]. These financial and time factors are of major importance in trait development as genetic modification is mostly applied to crops that are cultivated on large hectares of land with traits that are highly valued and widely adopted. Newly developed traits have not been commercialized for barley and some other cereals which require smaller areas [8]. GE is regarded as an alternative that is of lower cost, requiring shorter time for development; however, there are uncertainties regarding the mode of regulation of their application in agricultural practices and food processing [8, 50].

[8] examined the financial and risk factors involved in developing traits through GM and GE technologies. Their analysis revealed that the acreage needed to break even for a GE-developed trait was 96% lower than for a comparable GM-developed trait with the same market potential. For example, at a trait that has a value of $25 for one acre, a trait that is GM-developed would require 62 million acres to recover research and development expenses, whereas a GE-developed trait would need only 2.3 million acres. This significant contrast implies that GE technology could be more practical for introducing traits in crops with smaller cultivation areas, where GM methods may be less feasible.

In addition, the speed of development of GE foods is notable. GE technology is marked for its faster time to market, low risk, and lower cost required in the research and development phase [56]. This lower research and development cost for GE than GM encourages greater entry into the field by different brands, not only the well-known investors in biotechnology. The high cost of GM research and development has hindered entry and participation of some other medium and large organizations, leaving the field to be dominated by specific large organizations [8]. With the introduction of GE, this is not the case, this low cost of research and development, accompanied with less time enhances the participation of other organizations and this increases the commercialization of GE foods.

4.4 Safety of gene-edited foods

The consumers' primary concern about genetically engineered foods is food safety, with fears that the nutritional value and safety of the food may be negatively impacted when foreign fragments are introduced. As a result, many research outcomes are yet to be transferred into industry due to legal and regulatory constraints, despite significant time and effort devoted by many nations [1]. The progressive advancement of these technologies demonstrates not just the breadth and depth of scientific inquiry, but also the boundless possibility of future applications of GE. Using precise editing techniques, it is possible to quickly obtain desired kinds without fear of introducing genes from different species [11]. In some ways, precise GE is analogous to spontaneous mutations observed in nature. As a result, the development of these rigorous methodologies has provided scientists with unprecedented epiphanies while also paving the way for future strategies in food innovation. Potential impact on food safety can in turn affect the environment, nature, and the health of consumers. The assessment of gene-edited products require suitable methods, which are important in ensuring the safety of these products without assigning issues of regulations on small scale producers and middle scale industries [21].

The safety of genome editing depends, in part, on whether the changes are directed to predetermined sites or to targets. The former would reduce or eliminate unintended changes, also referred to as off-target mutations, while the latter would eliminate unintended effects of the intended changes [3]. The precision and efficiency of genome editing is expected to lower the frequency of some sources of unwanted downstream events, and therefore to yield fewer potential hazards at the product level. However, for foods such as wheat, barley, or maize, with large and complex genomes, off-target editing is more likely to occur [2]. Like varieties derived from chemical or radiation mutagenesis, unexpected risks and negative externalities (i.e., possible harm to consumer health and the environment) may still be encountered. At present, there is limited knowledge on the safety of genome editing in plants. Also, the present detection and identification techniques (e.g. bioinformatics) and new analytical tools (e.g. next generation sequencing) face potential drawbacks. This is challenging because there is still a lack of reliable and harmonized procedures for the detection of unintended editing [2].

The potential food safety dangers of agricultural goods, even those produced traditionally, include both toxicity and allergenicity of food components. Allergenicity SDN-3 has been utilized to purposely express novel proteins, resulting in a transgenic creature. Similar to other transgenic crops, SDN-3 uses may result in new or crossover allergenicity due to the introduction of novel proteins. However, for SDN-1/2, the allergy-related potential risk could be an increase in intrinsic allergens, which is also possible with mutants derived from techniques including radiation, chemical mutagenesis, or somaclonal variation, particularly if the host crop is a known source of food allergen. Toxin, allergen, and vitamin levels can all fluctuate naturally, as mutations. Traditional allergenicity assessment for intentionally introduced proteins (transgenics and SDN-3) has relied on several methods, including bioinformatic analysis, looking for sequence homology to known allergens and by using a threshold homology. Such approaches are more suitably adapted to detecting potential cross-reactivity with pre-existing food allergies. Thermal and protease digestibility are two complementary approaches that are utilized in conjunction with bioinformatics to determine the probability of de novo allergenicity in introduced proteins. Another investigation employing IgE binding of sera from patients with known allergies to a protein is needed only if the novel protein comes from an allergy-causing organism or if the bioinformatic analysis yields a positive match.

Toxicity regulation of endogenous toxicants is feasible using SDN-1/2 GE because of its emergence in new types obtained through spontaneous or induced mutation in celery and potato. However, in other procedures based on mutagenesis, the breeder often overlooks the mutation that results in the new feature as well as the presence of any off-target mutations. In contrast, GE provides more understanding, which may help breeders eliminate potentially dangerous mutations more reliably than prior procedures. Off-target edits and unintended DNA insertions with SDN-1 and SDN-2 modifications can occur and this should be considered for determining the safety implications of food products. Though the anticipated alteration might be safe and could even be present in the food supply, unintended insertions, or off-target edits, the food's composition might change in term of toxins content, nutrients or allergens. Considering the availability of more affordable and more efficient DNA sequencing, whole genome sequencing (WGS), in addition to bioinformatic approaches, has been suggested to identify and determine changes in gene-edited foods [32]. It was reported that during and after the regulatory review of the polled gene-edited cattle, WGS detected the insertion of transgenic antibiotic markers [67]. [26] suggested that the most important factor for reducing CRISPR-Cas off-targeting in plants is careful selection of target sequences, which can be helped by using various software tools.

Reviews on gene-edited foods carried out by regulatory organizations have shown considerable data related to its safety. For example, the United States Food and Drug Administration (FDA) received compound equivalence data for a high-oleic acid soybean (edited with transcription activator-like effector nucleases, TALENs), a reduced alpha-gal sugar pig (edited with older transposon methods), and thermotolerant cattle (edited with CRISPR). It was proved unnecessary to conduct feeding trials on the entire gene-edited food using animals or humans, and while some variations were noted using compositional assessment, they were determined to be insignificant for nutrition or safety when compared to conventional food alternatives. Perhaps a different case was the neomycin antibiotic resistance transgene that was inserted into the GalSafe pig (a transgenic animal), which was found to pose a potential microbiological food safety risk. In this case, the pigs needed to be reared without neomycin so that evolution of bacterial resistance to that class of antibiotics would be reduced and human health would not be put at risk [34, 35, 59]. During the FDA examination of thermotolerant cattle, WGS revealed evidence of unexpected genomic sequence alterations caused by GE. Though, comparative bioinformatic study revealed that these did not alter protein expression or have an impact on food safety when compared to their traditional equivalents. There have been arguments on whether whole-food feeding studies are necessary to assess first generation GM crop's safety, and revised guidelines for carrying out such GM food studies have been developed [49].

The risks of gene introgression from GM crops into wild relatives are a source of concern for both GM and gene-edited crops, as well as traditionally bred and mutagenesis variants. Gene flow can minimize differences between populations as well as diversity within a population, hence having a broad impact on biodiversity. Gene introgression into wild cousins may provide greater direct dangers, as determined by the new characteristic. For example, several varieties of spontaneous/somaclonal/ chemical mutagenesis, gene-edited and GM herbicide-tolerant crops were cleared from regulation in the United States of America and elsewhere [64]. They included grasses able to cross pollinate with neighboring wild relatives, as well as commodity and food crops. Herbicide-resistant weeds have emerged from the use of GM and conventionally bred herbicide-tolerant crops due to overuse of the companion herbicide, and herbicide-tolerant crops such as creeping bentgrass and rice have cross pollinated with wild relatives, transferring herbicide resistance [47]. Invasiveness of gene-edited animals, particularly those with features that increase their fitness (ability to survive, reproduce, feed, and persist in ecosystems), may pose a risk to ecosystems if they escape or are deployed in mismanaged conditions. However, there is extensive experience in selecting new animal breeds with spontaneous genetic changes that have minimal invasive potential. For example, the double muscular, smooth, and hornless features in cattle, which are now being gained using SDN-1/2 GE, were previously chosen from spontaneous mutations, and there is extensive evidence that these changes did not enhance invasiveness. Prior experience with GM animals may be relevant for SDN3-edited animals [37]. For example, issues have been raised in the case of GM salmon modified to grow faster than non-modified salmon, but the developer later showed that a faster growth rate presented a fitness disadvantage [25]. There is also the likelihood for hybridization of farmed GM salmon with wild salmon if they escape from their controlled surroundings and occupy nearby waters that contain native salmon. GM salmon have mainly been farmed at inland facilities, although some contained growth is occurring near shores where there are wild salmon populations [28]. CRISPR-based GE technologies have been investigated for bio-confinement of farmed salmon to prevent reproduction if they escape from containment facilities (germ-cell free farmed salmon). Recently, transgenic pet fish with green fluorescent protein (GFP) escaped into open bodies of water. The effects of that escape have yet to be determined, although the intruder may have a direct impact on local species by competing for food, but this is unlikely to differ significantly from the escape of wild-type, non-fluorescent pet fish of the same species. Overall, it is impossible to predict what happens to the ecosystem with the introduction of gene-edited animals, nor the level of introgression of their genes. However, this might not be an issue peculiar to gene-edited animals, as many corresponding traits can be derived by conventional breeding, given the time [55].

Genome editing is an important new technology that has the potential to significantly increase agricultural output. Though GE technology poses a problem for present detection practices and food-labeling systems, and some genome editing techniques will possibly generate off-target mutations other than the desired mutation, these obstacles will be resolved as the technology makes progressive advances, offering the potential for ensuring more safety for gene-edited food products.

4.5 Commercialization of gene-edited foods

4.5.1 Regulatory framework

The commercialization of gene-edited (GE) crops is shaped by international regulatory systems, particularly whether GE products are treated like genetically modified organisms (GMOs), which face lengthy and costly approval processes [31, 40]. Most countries follow the Cartagena Protocol on Biosafety to guide biotechnology regulation. A central issue is whether regulations focus on the process (how the product was made) or the product (its final characteristics) [14, 17, 22, 23, 44, 57]. Process-based systems typically impose stricter oversight on GE technologies.

4.5.2 Regulatory approaches across nations

In certain countries, GE foods undergo less rigorous regulation than GMOs, particularly when no foreign DNA is introduced [8, 41]. For example, under the Genetic Technology **(Precision Breeding) Act 2023**, England regulates precision-bred organisms separately from GMOs, with oversight by the Department for Environment, Food & Rural Affairs (Defra). These products are exempt as long as the product of the genetic changes can occur by natural means or via conventional breeding. Canada uses a product-based approach, regulating all biotech crops—including GE—based on the traits of the final product, not the method of development [16, 40, 56]. Similarly, in the United States, GE plants without foreign DNA or with changes indistinguishable from natural mutations are not classified as GMOs [29, 40, 61]. Globally, many countries do not apply GMO-level scrutiny to GE products, allowing faster and more cost-effective development. However, concerns remain around safety and the need for some oversight due to risks such as off-target mutations and unintended environmental impacts [48].

4.5.3 The role of risk assessment in improving commercialization of gene-edited foods

Generally, GE foods are referred to as nature identical, therefore they cannot be easily identified and distinguished from the natural variations and normal plants. Therefore, there are suggestions that risk assessment is essential for GE foods since their modifications could be unnoticeable [28, 53]. In other words, using metabolite-based safety assessments for gene-edited crops will enhance existing risk assessment methods for biotech crops and result in outcomes that can address public concerns about their acceptance [53].

A major drive for this risk assessment is due to the reservations consumers have based on their ethical concerns about biotechnology. Some consumers have a cautious disposition when considering new products derived from biotechnology [31, 33]. This is as a result of not being familiar with the properties of these biotechnology products. This occurs across nations and is an important aspect as it affects the willingness of the public to purchase these novel foods derived from biotechnology. There are new and existing local, international and trade frictions regarding the use of these novel techniques for innovations in agrifood business [30]. Therefore, there is a need for investigations to be carried out to observe whether there could be risks on the health of consumers or the environment through the consumption of these novel foods. This could serve as a tool in improving the perception and acceptance of the public regarding these products [30].

4.6 Factors influencing the commercialization of gene-edited foods

To achieve success in the commercialization of GE foods, careful observation of the concerns of the public toward GE products is required with strategic solutions to manage them. This is mainly related to their safety and environmental impact since these were the key concerns that resulted in the resistance toward GM foods in the past [41, 66]. Countries can gain speed in the commercialization of GE food products through a structured societal introduction of the GE techniques. The clarification of regulatory approaches can serve as a major tool in achieving this as seen in Argentina and Japan; thus, showing that when regulations and guidelines are carefully done, with transparency, they assist in increasing public acceptance of the techniques [36, 40, 41]. The major elements to be considered for successful commercialization of GE foods include their regulatory framework, market acceptance, ethical and social concerns, and technological advancements [30, 43, 61]. Consideration will also need to be given to policy development, and changes in business model and strategies as necessary elements to successfully engage all stakeholders in the interest of GE products' trade.

4.6.1 Market acceptance

A major concern in market acceptance of GE foods is the question: whether consumers will treat GE foods as GM foods were treated and are still being treated particularly when considering different countries [3, 38, 48]. There are questions surrounding how the consumers will be informed; what level of information is to be given to them and what will be the influence of the information on their preferences [38]. To rightly encourage market acceptance of GE foods, careful consideration is being given to GM foods market acceptance to observe how things can be done better. Resistance to GM foods have been observed both at consumer and regulatory levels. Literatures exist on the willingness as well as the resistance to purchase GMOs [42]. The concern is, will GE foods be treated like GM foods? Therefore, measures are being investigated to ensure GE foods are accepted by the public to facilitate their commercialization. In a study conducted by [30], most participants from the workshop hold the opinion that GE foods should be regulated separately from GM foods, giving reasons that since they are two separate techniques, they should be treated as such. These respondents opined that treating them as same can confuse the public and likely reduce the chances of the people learning about their differences; thus people who are averse to GM foods might treat GE foods in the same manner, thereby reducing the market acceptance and hindering commercialization if not well separated categorically [30].

4.6.2 Willingness to pay

Some studies have provided information on the willingness of consumers to pay for new products developed using gene-edited technology. While other studies have emphasized the notable significant discount for GE foods when compared to conventional ones and /or organic ones which were similar to the reluctance shown for GMOs in the past [9, 38, 46, 68]. Although the willingness to pay for GE foods is low, when compared to conventional/organic ones, it is still slightly higher than the GM ones. However, GE foods still face lower consumer acceptance than traditional, non-GM, and organic alternatives. This serves as a light sense of hope, that if there is increase in awareness of the GE techniques, its nearness to conventional foods and its benefits, there could be an increase in its market acceptance. Based on the observations of [19] when adequate information about GE technologies was given to university students, their willingness to purchase GE foods increased from 24% to 41% [57].

4.6.3 Labeling

Labeling is one of the major ways of influencing the market acceptance of a product [69]. In a study on randomized group approach to identifying label effects [42], investigated the labeling of bioengineered food products; products labeled “gene edited” or “genetically modified” were not preferred when compared to products with no label. From the findings made, it implies that, for consumers to derive the same indirect utility from GE products as with the conventional and organic ones, the selling price of the GE products should be reduced [42].

[38] showed that consumers value and prefer labeling of food products from genetic engineering. The study carried out by [30] observed that consumers preferred that while labeling these foods, the full term “gene edited” should be used to keep consumers informed that ingredients derived from GE are contained in the products. The participants emphasized that this is a means of being transparent which is necessary to allow consumers make their own choices whether to buy and to build consumer trust in gene-edited foods. Some participants in the study stated that GE foods should be labeled differently

from GM foods. This is because some participants preferred GE foods to GM foods and were of the opinion that if they are labeled differently, it will increase their chances of buying GE foods since they can spot the difference. These participants held the opinion that labeling GM foods and leaving out GE foods might be misleading, and this can negatively influence consumer trust which might lead to a distrust in terms of biotech foods in general, thus having a negative impact of the market acceptance of GE foods [30].

4.6.4 Consumer perception and awareness

It is noteworthy that there is a low awareness of the benefits of GM and GE foods [11]. Globally, there is a poor understanding, misconception, and lack of familiarity of plant gene technologies among consumers. This covers all forms of genetic engineering found in agricultural production such as plant gene technologies in agriculture including GMO, new breeding techniques (NBTss) and GE [18]. Consumers find it more acceptable for GE or gene modification to be applied to plants than to animals. So, although GE foods are preferred to GM foods generally, consumers still have some reservations regarding their production and consumption due to some perception that they could be of risk to their health and the welfare of animals [30]. Information contributes toward increasing the acceptance of novel foods derived from GE by the public.

The acceptance of novel foods from GE can be affected by various factors such as knowledge of the consumer, expected benefits, suspected risks, conceptions about technology and trust [11]. Considering age and knowledge as determinant factors for acceptance of GE products [30], conducted a survey and observed that respondents who were male, aged 16–34, and those with higher education levels were much more likely to report being informed about genome editing and to support the sale of gene-edited foods. After a workshop on the awareness of GE foods, some participants viewed it as a way of speeding up an otherwise natural process [30]. Other participants preferred GE to the use of chemicals to induce mutation. Although the process is lab-based, some persons viewed it as a more natural process than they expected because the process only involved changes in the organism's own DNA [30].

In addition to the impact of awareness of these GE technologies among consumers, it is essential to note that other factors contribute to the disposition of consumers to GE products and some other science technologies. Although increase in knowledge of these novel technologies lead to improved acceptance, however, emotions, ideologies, beliefs as well as norms and values of individuals contribute majorly to their attitudes toward the purchase of GE foods [18]. Consumers have identified the factors that contribute to the increased acceptability of GE foods. Some key areas that can facilitate the acceptance of GE food products include: improvement in animal welfare, such as resistance against diseases that incur pain on the animals, added value to human health in terms of allergen-safe foods, increased food security such as increasing affordability and availability, and benefiting the environment via reduced impacts of climate change [30].

4.6.5 Ethical and social concerns

Technological advancements have yielded great benefits in food production and ensuring food security while lowering cost. For example, some of these benefits can be seen in the application of porcine genetics. This has resulted in faster growth rate, improved feed conversion, larger litter sizes, increased food production and lowering cost to the farmer and consumer. Amid these benefits, it is essential to achieve a balance in the desired outcome of tech advancement to maintain acceptability by the society as well as commercial credibility [39]. For GM foods, various questions have emerged, reflecting public anxiety and curiosity, particularly concerning the safety of both existing and future generations of GM crops, how these changes will influence lifestyles in agricultural lands, slaughterhouses, and so on [48]. A major concern among consumers that can affect the commercialization of these GE foods is the idea that the development of disease-resistant traits can result in the increased concentration of animals in confined spaces that are of poor sanitary standards, which is feared to lead to the proliferation of new pandemics. This in turn could lead to more calls for further interventions through GE, thereby leading to an endless cycle [48].

[39] have demonstrated that intensive selection for traits intended to enhance the production of milk, feed efficiency, and meat quality usually leads to a corresponding decline in the animal's fitness functions. There is a tendency to adopt a poorly understood biological system where disease resilience becomes an internal trade-off. Therefore, when addressing social and ethical concerns, it is essential to balance technological advancement with transparency regarding the traits being developed. Genome editing has been noted as an invaluable method for achieving this balance and enhancing the commercial potential of outputs. For example, it has been used to address porcine reproductive and respiratory syndrome (PRRS) in pigs [39].

4.6.6 Technological advancements

New biotechnology tools including CRISPR and TALEN have emerged over time for the development of agricultural products with specific traits and novel food products with added value [15] [44]. The aim of these new techniques is to maximize productivity in agriculture, profitability, and sustainability; the main aim is to solve the global hunger challenge and meet increasing demands for feed and fuel [38]. An example of technological advancement can be seen in the advanced treatment of PRRS (porcine reproductive and respiratory syndrome), a disease found in pigs that causes a compromise in the respiratory immune system of the pig. This disease can cause increased severity of other diseases in the animal [62]. Although it is possible to develop selection screens to produce pigs that are resistant to PRRSV by adopting traditional techniques for animal breeding; they have very low efficiency in their outcome compared to GE techniques. Rather than relying on the generation and selection of random newly arising mutations in the genome to enhance disease resistance in these animals, advanced techniques offer a more transformative approach [38]. Instead of generating and selecting random de novo mutations in the genome to improve the ability of these animals to resist this disease, advanced techniques are more revolutionary. This is because they target the candidate genes that enables the introduction of allelic variations [39]. This advantage of GE tools changes the trajectory of animal breeding, and their commercialization is largely dependent on the benefits they confer.

4.6.7 Development of policies for increased use and sale of GE foods

One way to encourage the acceptance of GE foods is by issuing policies that encourage their production and consumption [33]. When steps are taken to ensure the clarity of the regulatory requirements for GE products, their safety and influence on the environment, consumers can be encouraged to develop a positive response to the GE products. This method was adopted by the Japanese government where success has been recorded so far [41]. For the Japanese government, the law requires the assessment of the impact of genetic engineered living organisms on biodiversity. Under this law the GE organisms are classified based on the technique applied. According to the Cartagena Act, genetically engineered organisms are exempt from regulation if foreign DNA or RNA is not present in them, or if any introduced genetic material can no more be found in the final organism; while genetically engineered organisms that contain foreign DNA or RNA are subject to regulation [45]. However, this goes beyond the regulations, efforts are made to ensure that every developer of a GE product even though not subject to regulation, is required to provide information in advance about the product to the government agencies. The Ministry of Environment (MOE) then provides a summary of this information on a specific website for public access [41]. The same principle applies to the impact on food safety wherein regulations need to be based on whether the specificity or mutations in the GE foods can be compared with respect to their safety to those derived from natural variations or the changes that occur in traditional breeding techniques.

4.6.8 Improving public acceptance of GE foods through balanced disclosure of regulatory approach

Public acceptance of GE foods can be improved by disclosing the details of the regulation of GE novel food products in comprehensive ways to the public. This includes the environmental impact assessment of these GE products and any associated food safety concerns [65]. Another method is by ensuring that the regulatory approaches selected provide a balance between the scientific discussions on GE foods and the public demand [63]. Taking a proactive step of social implementation of these technologies can enhance the commercialization of these products. These steps can help to address the concerns of the fraction of the public that believe that GE foods should undergo similar strict regulations as GM foods [41]. The Japanese government does not only encourage research and development, but it carries the public along and keeps them informed about their current practices. For example, the GABA-enriched tomato which was developed by the University of Tusuba was formed originally from the first social implementation program. In subsequent stages, websites were created for the dissemination of info and to enhance the public awareness and consequently, the acceptance of GE foods.

4.6.9 Improved commercialization through strategic developments and adaptations in business models

A good business model would be useful in the commercialization of GE foods as seen in different technological developments [4]. Traditionally, the business-to-business (B to B) model was mostly used for GM food products. In other words, international companies undertake the development of main crops, which are grown on commercial scales after which they are distributed to wholesale companies and retail stores. However, in Japan, the development of novel products from GE was carried out by indigenous universities. The business model that was adapted for the sale of these novel GE products is known as "D-to-C" which is the direct-to-consumer model. In this model, the start-ups from these universities delivered their product to consumers who needed it directly [41]. A key aspect to improving commercialization is the conscious target of sales strategies and observing consumer attitudes. Instead of adhering to traditional methods that might not be favorable for some novel technologies, the use of internet and crowdfunding to sell to people/companies or businesses who are interested has proven to be successful in many product sales [4]. This also provides a platform for the public to learn more about their products online. Some companies such as Sanatech Seed Ltd have adopted this strategy for the sale of their novel GE products online at premium prices [41].

4.6.10 Highlighting the benefits of GE to encourage public acceptance

The knowledge of the benefits of GE in comparison with GM techniques and in some ways to conventional breeding techniques can assist in improving its acceptance. Such benefits include high precision, flexibility, low cost, and relative ease of application and potency, particularly CRISPR/Cas [41]. Consumers have a higher tendency to support the production and consumption of novel foods from GE technology if there are obvious benefits in the safety and nutrition of these products to the consumers [4,

24, 33]. It is estimated that the market share for GE foods could go beyond 15% in few years if the public is enlightened on the advantages of GE foods with emphasis on the consumer health benefits and positive environmental impact [9].

4.7 Conclusion

Given the transformative potential of GE in agricultural biotechnology, it is crucial to provide structures that could enhance its commercialization. Public perception of biotech innovations, particularly genetically engineered (GE) foods, has often been unfavorable, which has hindered their market acceptance. Additionally, varying regulatory approaches across countries have slowed the global commercialization of this technology. To address these challenges, it is important to educate the public about GE and its benefits. This should include raising awareness about the advantages of GE technology for consumers and its positive environmental impact. Furthermore, nations should consider separate regulatory frameworks for GE, distinct from those for GMOs, to reduce costs and expedite the approval process. Although GE technology is far more precise than traditional genetic engineering, there are still potential side effects and hazards. Unintended mutations may occur throughout the GE process, affecting crop quality, nutritional value, and safety. As a result, detailed and rigorous safety investigations are required before advertising GE foods. In addition, establishing stringent quality control and risk assessment mechanisms to guarantee the safety of genetically edited foods is necessary.

References

[1] Abbott, A. Europe's genetically edited plants stuck in legal limbo. Nature 2015, 528(7582), 319–320.

[2] Agapito, S. Z. PEG-Delivered CRISPR-Cas9 Ribonucleoproteins System for Gene-Editing Screening of Maize Protoplasts. 2020.

[3] Agapito-Tenfen, S. Z., Okoli, A. S., Bernstein, M. J., Wikmark, O. G., & Myhr, A. I. Revisiting risk governance of GM plants: The need to consider new and emerging gene-editing techniques. Frontiers in Plant Science 2018, 9, 1874.

[4] Baden-Fuller, C., & Haefliger, S. Business models and technological innovation. Long Range Planning 2013, 46(6), 419–426.

[5] Bartkowski, B., et al. Snipping around for food: Economic, ethical and policy implications of CRISPR/Cas genome editing. Geoforum 2018, 96, 172–180.

[6] Blancke, S., et al. Fatal attraction: The intuitive appeal of GMO opposition. Trends in Plant Science 2015, 20(7), 414–418.

[7] Brief, I. Global status of commercialized biotech/GM crops: 2016. International Service for the Acquisition of Agri-biotech Applications Br 2019, 54, 3–13.

[8] Bullock, D. W., Wilson, W. W., & Neadeau, J. Gene editing versus genetic modification in the research and development of new crop traits: An economic comparison. American Journal of Agricultural Economics 2021, 103(5), 1700–1719.

[9] Caputo, V., Lusk, J., & Kilders, V. Consumer Acceptance of Gene Edited Foods: A nationwide survey on US consumer beliefs, knowledge, understanding, and willingness to pay for gene-edited foods under different treatments. FMI Foundation Report 2020.
[10] Clark, L. F., & Hobbs, J. E. What is gene editing? In: International Regulation of Gene Editing Technologies in Crops: Current Status and Future Trends, Cham: Springer Nature Switzerland, 2024, 15–29.
[11] Cui, K., & Shoemaker, S. P. Public perception of genetically-modified (GM) food: A nationwide Chinese consumer study. npj Science of Food 2018, 2(1), 10.
[12] Cummings, C., & Peters, D. J. Who trusts in gene-edited foods? Analysis of a representative survey study predicting willingness to eat-and purposeful avoidance of gene edited foods in the United States. Frontiers in Food Science and Technology 2022. doi: 10.3389/frfst.2022.858277.
[13] Delaney, B., Goodman, R. E., & Ladics, G. S. Food and feed safety of genetically engineered food crops. Toxicological Sciences 2018, 162(2), 361–371.
[14] Eckerstorfer, M. F., et al. Plants developed by new genetic modification techniques – Comparison of existing regulatory frameworks in the EU and non-EU countries. Frontiers in Bioengineering and Biotechnology 2019a, 7, 26.
[15] Eckerstorfer, M., Zanon Agapito-Tenfen, S., & Kleter, G. A. (2023). Genome edited organisms for agriculture—challenges and perspectives for development and regulation. Frontiers in Genome Editing, 5, 1287973.
[16] Ellens, K. W., et al. Canadian regulatory aspects of gene editing technologies. In: Transgenic Research, Springer, 2019.
[17] Entine, J., et al. Regulatory approaches for genome edited agricultural plants in select countries and jurisdictions around the world. Transgenic Research 2021, 30(4), 551–584.
[18] Ewa, W., et al. Public perception of plant gene technologies worldwide in the light of food security. GM Crops & Food 2022, 13(1), 218.
[19] Farid, M., et al. Exploring factors affecting the acceptance of genetically edited food among youth in Japan. International Journal of Environmental Research and Public Health 2020, 17(8), 2935.
[20] Fraser, P. D., et al. Metabolomics should be deployed in the identification and characterization of gene-edited crops. The Plant Journal 2020, 102(5), 897–902. Available at: https://doi.org/10.1111/tpj.14679.
[21] Gao, C., Kikulwe, E. M., Kuzma, J., Lema, M., Lidder, P., Robinson, J., Wessler, J., & Zhao, K. Gene editing and agrifood systems. In: FAO, vol. 2022, Rome, 2022. https://doi.org/10.4060/cc3579en,1-86.
[22] Gatica-Arias, A. The regulatory current status of plant breeding technologies in some Latin American and the Caribbean countries. Plant Cell, Tissue and Organ Culture (PCTOC) 2020, 141(2), 229–242.
[23] Gelinsky, E., & Hilbeck, A. European Court of Justice ruling regarding new genetic engineering methods scientifically justified: A commentary on the biased reporting about the recent ruling. Environmental Sciences Europe 2018, 30(1), 52.
[24] Giones, F., & Brem, A. Crowdfunding as a tool for innovation marketing: Technology entrepreneurship commercialization strategies. In: Handbook of Research on Techno-Entrepreneurship, Third Edition, Edward Elgar Publishing, 2019, 156–174.
[25] Gutási, A., Hammer, S. E., El-Matbouli, M., & Saleh, M. Recent applications of gene editing in fish species and aquatic medicine. Animals 2023, 13(7), 1250.
[26] Hahn, F., & Nekrasov, V. CRISPR/Cas precision: Do we need to worry about off-targeting in plants? Plant Cell Reports 2019, 38(4), 437–441.
[27] Hanif, M., Abbas, W., Iqbal, M. A., Khawaja, M. S., Waheed, M., Sarwar, S., . . . Basharat, A. Status of GMOs in Pakistan; need, acceptability, development and regulation status: A comprehensive review. Pak-Euro Journal of Medical and Life Sciences 2023, 6(3), 313–326.
[28] Hindar, K., Bodin, J. E., Duale, N., Jevnaker, A. M. G., Garseth, Å. H., Malmstrøm, M., Rydz, K., Sipinen, V. E., Thorstad, E. B., Berg, P. R., Dalen, K. T., & Velle, G. Environmental risk assessment of genetically modified sterile VIRGIN® Atlantic salmon for use in research trials in aquaculture sea-cages. 2023.

[29] Hoffman, N. E. Revisions to USDA biotechnology regulations: The SECURE rule. Proceedings of the National Academy of Sciences 2021, 118(22), e2004841118.

[30] Ipsos, M. Consumer perceptions of genome edited food. Food Standards Agency UK 2021.

[31] Jenkins, D., et al. Impacts of the regulatory environment for gene editing on delivering beneficial products. Vitro Cellular & Developmental Biology-Plant 2021, 57(4), 609–626.

[32] Karkute, S. G., Singh, A. K., & Singh, P. M. Potential of transgenic and genome editing technologies for improvement of underutilized vegetable crops. 2022.

[33] Kilders, V., & Ali, A. Understanding the influence of end-users on the acceptance of gene edited foods and sensitivity to information. Food Quality and Preference 2024, 120, 105238. https://10.1016/j.foodqual.2024.105238.

[34] Kondo, K., & Taguchi, C. Japanese regulatory framework and approach for genome-edited foods based on latest scientific findings. Food Safety 2022, 10(4), 113–128.

[35] Kuzma, J. Regulating gene-edited crops. Issues in Science and Technology 2018, 35(1), 80–85.

[36] Lema, M. A. Regulatory aspects of gene editing in Argentina. Transgenic Research 2019, 28(Suppl 2), 147–150.

[37] Leony Petersen, G. E., Buntjer, J., Hely, F. S., Byrne, T. J., Whitelaw, B., & Doeschl-Wilson, A. Gene editing in farm animals: A step change for eliminating epidemics on our doorstep? bioRxiv 2021, 2021–2024.

[38] Marette, S., Disdier, A., & Beghin, J. C. A comparison of EU and US consumers' willingness to pay for gene-edited food: Evidence from apples. Appetite 2021, 159, 105064.

[39] Mark Cigan, A., & Knap, P. W. Technical considerations towards commercialization of porcine respiratory and reproductive syndrome (PRRS) virus resistant pigs. CABI Agriculture and Bioscience 2022, 3(1), 34.

[40] Marone, D., Mastrangelo, A. M., & Borrelli, G. M. From transgenesis to genome editing in crop improvement: Applications, marketing, and legal issues. International Journal of Molecular Sciences 2023, 24(8), 7122.

[41] Matsuo, M., & Tachikawa, M. Implications and lessons from the introduction of genome-edited food products in Japan. Frontiers in Genome Editing 2022, 4, 899154.

[42] McFadden, B. R., et al. A randomized group approach to identifying label effects. Journal of Choice Modelling 2023, 48, 100435. Available at: https://doi.org/10.1016/j.jocm.2023.100435.

[43] Menz, J., et al. Genome edited crops touch the market: A view on the global development and regulatory environment. Frontiers in Plant Science 2020, 11, 586027.

[44] Metje-Sprink, J., Sprink, T., & Hartung, F. Genome-edited plants in the field. Current Opinion in Biotechnology 2020, 61, 1–6.

[45] Montreal. Secretariat of the Convention on Biological Diversity. 'Secretariat of the Convention on Biological Diversity. Cartagena Protocol on Biosafety to the Convention on Biological Diversity: Text and annexes.', Available at: https://bch.cbd.int/protocol/text/ (accessed on 30 August 2024).

[46] Muringai, V., Fan, X., & Goddard, E. Canadian consumer acceptance of gene-edited versus genetically modified potatoes: A choice experiment approach. Canadian Journal of Agricultural Economics/Revue Canadienne D'Agroeconomie 2020, 68(1), 47–63.

[47] Platani, M., Sokefun, O., Bassil, E., & Apidianakis, Y. Genetic engineering and genome editing in plants, animals and humans: Facts and myths. Gene 2023, 856, 147141.

[48] Policante, A., & Borg, E. CRISPR futures: Rethinking the politics of genome editing. Human Geography 2024, 17(1), 76–82.

[49] Popova, J., Bets, V., & Kozhevnikova, E. Perspectives in genome-editing techniques for livestock. Animals 2023, 13(16), 2580.

[50] Purnhagen, K. P., & Wesseler, J. H. Maximum vs minimum harmonization: What to expect from the institutional and legal battles in the EU on gene editing technologies. Pest Management Science 2019, 75(9), 2310–2315.

[51] Randhawa, S., & Sengar, S. The evolution and history of gene editing technologies. Progress in Molecular Biology and Translational Science 2021, 178, 1–62.
[52] Rausser, G., Gordon, B., & Davis, J. Recent developments in the California food and agricultural technology landscape. ARE Update 2018, 21(4), 5–8.
[53] Razzaq, A., et al. Advances in metabolomics-driven diagnostic breeding and crop improvement. Metabolites 2022, 511. Available at: https://doi.org/10.3390/metabo12060511.
[54] Roman, S., Sánchez-Siles, L. M., & Siegrist, M. The importance of food naturalness for consumers: Results of a systematic review. Trends in Food Science & Technology 2017, 67, 44–57.
[55] Silva, A. R., Ed. Assisted Reproduction in Wild Mammals of South America, CRC Press, 2024.
[56] Smyth, S. J. Canadian regulatory perspectives on genome engineered crops. GM Crops & Food 2017, 8(1), 35–43.
[57] Sprink, T., et al. Regulatory hurdles for genome editing: Process-vs. product-based approaches in different regulatory contexts. Plant Cell Reports 2016, 35, 1493–1506.
[58] Suzuki, T., Asami, M., Patel, S. G., Luk, L. Y., Tsai, Y. H., & Perry, A. C. Switchable genome editing via genetic code expansion. Scientific Reports 2018, 8(1), 1–12.
[59] Turnbull, C., Lillemo, M., & Hvoslef-Eide, T. A. Global regulation of genetically modified crops amid the gene edited crop boom–a review. Frontiers in Plant Science 2021, 12, 630396.
[60] Urnov, F. D., Ronald, P. C., & Carroll, D. A call for science-based review of the European court's decision on gene-edited crops. Nature Biotechnology 2018, 36(9), 800–802.
[61] USDA Animal and Plant Health Inspection Service (USDA APHIS). Regulated Article Letters of Inquiry, Available at: https://www.aphis.usda.gov/aphis/ourfocus/biotechnology/am-i-regulated/Regulated_Article_Letters_of_Inquiry.
[62] Valdes-Donoso, P., & Jarvis, L. S. Combining epidemiology and economics to assess control of a viral endemic animal disease: Porcine Reproductive and Respiratory Syndrome (PRRS). PLoS One 2022, 17(9), e0274382.
[63] Van der Meer, P., et al. The status under EU law of organisms developed through novel genomic techniques. European Journal of Risk Regulation 2023, 14(1), 93–112.
[64] Verma, V., Negi, S., Kumar, P., & Srivastava, D. K. Global status of genetically modified crops. In: Agricultural Biotechnology: Latest Research and Trends, Singapore: Springer Nature Singapore, 2022, 305–322.
[65] Voytas, D. F., & Gao, C. Precision genome engineering and agriculture: Opportunities and regulatory challenges. PLoS Biology 2014, 12(6), e1001877.
[66] Yamaguchi, T., & Suda, F. Changing social order and the quest for justification: GMO controversies in Japan. Science, Technology, & Human Values 2010, 35(3), 382–407.
[67] Young, A. E., Mansour, T. A., McNabb, B. R., Owen, J. R., Trott, J. F., Brown, C. T., & Van Eenennaam, A. L. Genomic and phenotypic analyses of six offspring of a genome-edited hornless bull. Nature Biotechnology 2020, 38(2), 225–232.
[68] Yunes, M. C., et al. Is gene editing an acceptable alternative to castration in pigs? PloS One 2019, 14(6), e0218176.
[69] Zhang, J., Zhai, L., Osewe, M., & Liu, A. Analysis of factors influencing food nutritional labels use in Nanjing, China. Foods 2020, 9(12), 1796.
[70] Zhang, S., & Zhu, H. Development and prospect of gene-edited fruits and vegetables. Food Quality and Safety 2024, 8, fyad045.

Yusuf B.O. and Saheed Sabiu*

Chapter 5
Molecular, biochemical, and metabolic approaches to functional foods

Abstract: Functional foods offer health benefits in addition to basic nutrition, with a view to lessen the risk of chronic ailments. Since its birth in 1984, the term "functional food" has faced challenges particularly on what qualifies a food to be functional and the bogus claim of the health benefits. Despite the hurdles encountered, functional food market continues to grow due to increased awareness of the need to consume food as medicine without necessarily replacing medicine but with a view to improve wellness. This chapter explores the molecular, biochemical, and metabolic mechanisms underlying the therapeutic actions of functional foods. At the molecular level, bioactive compounds from plant and animal sources, including polyphenols, flavonoids, and omega-3 fatty acids, relate with cellular pathways to alter gene expression, enzyme activity, and cell signaling. These interactions can result in important biochemical effects, including anti-inflammatory, antioxidant, and immunomodulatory. The metabolism of functional foods reveals their ability to modify gut microbiota components, improving nutrient absorption as well as metabolic vigor. The comprehension of these important mechanisms aids in the making of targeted dietary tactics for prevention of chronic ailments and health promotion, as well as paving the way for future improvements in disease management and prevention, targeting a healthier world population.

Keywords: Bioactive compounds, catechin, functional foods, inflammation, omega-3 fatty acid, oxidative stress, polyphenol

Acknowledgments: The support of the National Research Foundation of South Africa and the Directorate of Research and Postgraduate Support, Durban University of Technology, for their generous support of this research project under reference number PSTD2204274997 is well appreciated.

***Corresponding author: Saheed Sabiu**., Department of Biotechnology and Food Science, Faculty of Applied Sciences, Durban University of Technology, Durban, South Africa, e-mail: sabius@dut.ac.za
Yusuf B.O., Department of Biotechnology and Food Science, Faculty of Applied Sciences, Durban University of Technology, Durban, South Africa

https://doi.org/10.1515/9783111441238-006

5.1 Introduction

The concept of "functional foods," which, contrary to mainstream foods, contain bioactive compounds (BCs) from plant and animal sources, that impact physiological activities positively, came into being when it was first used in Japan in 1984 with the idea of producing foods for benefits other than meeting basic nutritional requirements. Several efforts have since been made to give universally accepted definition for this term; however, all the definitions are not without one limitation or the other. Nevertheless, all the definitions tend to describe functional food as a food that promotes health in addition to the nutritional values. The most recent definition referred to functional foods as any natural or processed foods that contain biologically active compounds that are safe for consumption at specific amounts and have been clinically proven to improve wellness either via management, prevention, or treatment of chronic diseases, which can be measured using specific biomarkers [1]. This definition gained wide acceptance but excludes some categories of foods referred to as light foods [2]. Although lack of universally acceptable definition has debarred the attainment of universal standard legislation in manufacturing and marketing of functional foods for the benefit of all, Europe, the United States of America, and Japan have regulations that ensured verification of the claimed health-related advantages of the available functional foods. The birth of this concept following the emergent comprehension of the relationship between diet and health has made functional foods a vital focus in nutrition science and technology, as the global interest in its development continues to rise. It has been said to contribute to the managements of improved health particularly as it concerns chronic diseases like heart diseases, cancer, and diabetes [3, 4]. The development and commercialization of these products, however, remain complex, expensive, and fraught with challenges [5]. The existing literature on functional foods has primarily focused on their practical applications, consumer acceptance, and product development [6]. However, a deeper understanding of the underlying molecular, biochemical, and metabolic mechanisms that drive the therapeutic potential of functional foods through technological and interdisciplinary approaches remains an important research gap to which the future of functional foods advancement relies.

This chapter therefore aims to provide insight into mechanisms underlying the therapeutic potential of functional foods and provide insight into the future perspective leveraging on interdisciplinary collaboration that can advance nutrition-health relationships to deliver reliable and reproducible products that support the management of chronic diseases.

5.2 Molecular mechanisms of therapeutic action of functional foods

Several functional foods contain phytochemicals (Figure 5.1), which are BC naturally found in plants, playing an important function beyond their basic purpose of production, particularly in human health. They can be broadly categorized into groups including polyphenols, alkaloids, terpenes, and more, each uniquely promoting health benefits. These molecules, collectively known as secondary metabolites, are produced in abundance by the herbs, fruits seeds, and vegetables and are consumed raw or modified by man to fit numerous remedies by way of antioxidant, and as modulators of signal transduction and gene expression pathways as well as activator and inhibitor of enzymes [7].

The health benefits of phytochemicals are diverse and significant; and one of the most crucial ways in which they contribute to health is through their antioxidant activity. Antioxidants are compounds that protect cells and tissues from the detrimental impacts of reactive oxygen species (ROS) and free radicals. Their molecular mechanism of action relies largely on their ability to donate electrons or hydrogen atoms resulting in slowing down or inhibiting the oxidative processes [8]. Common sources of antioxidants that are known to contribute immensely to health include fruits, vegetables, berries, apples, oats, wheat, almonds, walnuts, flaxseeds, chia seeds, sage, oregano, thyme, green tea, red wine, certain types of coffee, and legumes [9].

Signal transduction pathways (mitogen-activated protein kinase pathway, nuclear factor kappa B (NF-κB) pathway, phosphoinositide 3-kinase pathway, and janus kinase (JAK) pathway) play vital roles in facilitating the impacts of functional foods on cellular mechanisms. These pathways consist of complex sequences of molecular events triggered by a signal, typically a BC, binding to a receptor. This binding initiates a chain reaction of intracellular activities, ultimately resulting in a targeted cellular response.

The NF-κB pathway is essential for regulating immune responses, cell proliferation, inflammation, and survival. Dysregulation of this pathway is closely linked with various pathological disorders, such as cancer, autoimmune diseases, and chronic inflammation. Typically, NF-κB is inactive in the cytoplasm, bound to IκB inhibitors. However, external stimuli like cytokines or stress trigger the phosphorylation of IκB, resulting to its breakage and the subsequent release of NF-κB that moves to the nucleus, binds to DNA, and initiates the transcription of target genes [10, 11]. Curcumin, a functional food, has been found to inhibit NF-κB activation. This compound prevents the breakdown of IκB, effectively blocking the translocation of NF-κB to the nucleus. This inhibition reduces inflammation and prevents the growth of cancer cells, indicating the significant anti-inflammatory and anti-cancer properties of these polyphenols [11]. Secoisolariciresinol diglucoside, found in flaxseed, and O-3FA, commonly

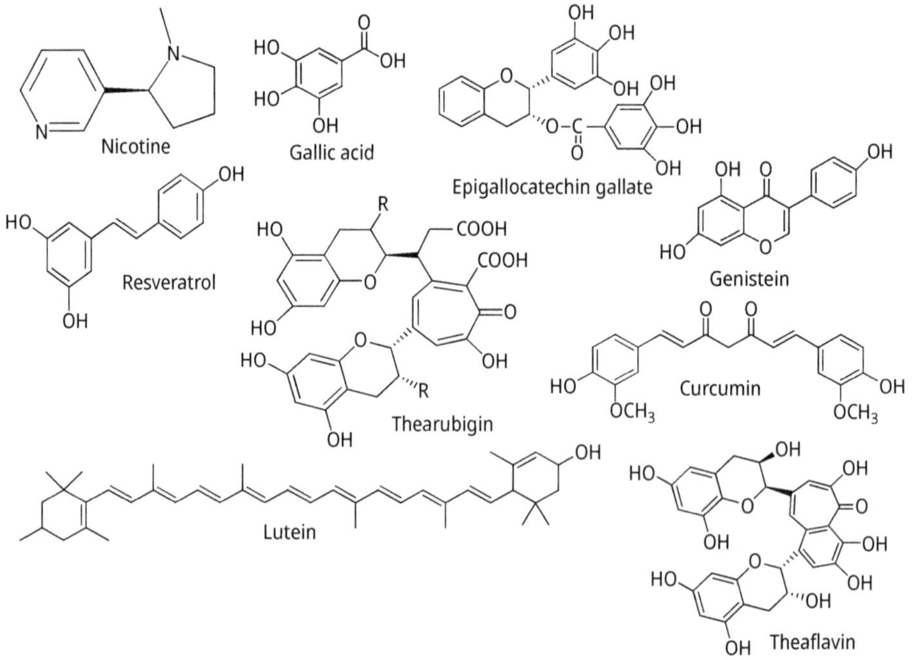

Figure 5.1: Structures of some phytochemicals isolated from functional foods.

present in fish oil and flaxseed, are other examples of functional foods that respectively shown anticancer property via modulation of signal transducer and activator of transcription, JAK, and PI3K/Akt pathways [12].

Gene expression modulation is the manipulation of specific genes, affecting the production of proteins and molecules essential for cellular function. BCs can influence this process through nuclear receptors, transcription factors, and epigenetic modifications [13]. Nuclear receptors, such as peroxisome proliferator-activated receptors (PPARs), play a significant role in mediating the effects of dietary nutrients on gene expression. Transcription factors, which bind to DNA strands, regulate the transcription of genetic information from DNA to RNA. Epigenetic modifications, such as the methylation of DNA, histone modification, and the action of microRNAs (miRNAs), are heritable modifications in gene expression that can be repressed or regulated by BC [13, 14]. All these processes can be achieved through the modulation of the affected pathway by the BC [15].

Enzymes play key roles in regulating biochemical pathways and cellular processes. Functional foods, rich in BC, impact overall health by activating or inhibiting enzymes. Of particular interest is the inhibition of α-glucosidase by quercetin that results in postprandial blood glucose reduction [16]. Inhibition of cyclooxygenase-2

(COX-2) by O-3FA decreases the synthesis of pro-inflammatory prostaglandins and help manage inflammatory conditions [17]. Ursolic acid enhances glucose uptake and the sensitivity of insulin by inhibiting protein tyrosine phosphatases that negatively regulate insulin signaling pathways [18]. EGCG inhibits lipase, reducing fat absorption and potentially aiding in weight management [19].

5.3 Biochemical mechanisms of therapeutic action of functional foods

5.3.1 Impact on metabolic processes

The biochemical mechanism of action of functional foods can best be explained using metabolic processes and their regulations. BCs in functional foods have been proven to exert therapeutic impacts that cumulates into improved wellness of man on metabolic processes including lipids, carbohydrates, and protein metabolism. Among the numerous BC with therapeutic properties for lipid metabolism includes polyphenols, lecithins, probiotics, O-3FA, and plant-based meals, all contributing to a healthy lipid profile and promoting overall well-being via different mechanisms.

Polyphenols, for instance, provide protection against lipid peroxidation and exert an influence on lipid metabolism by modulating enzymes responsible for lipid production and breakdown [20]. Lecithins, a combination of lipoproteins, found in vegetables and soybeans among others have been reported to lower cholesterol, maintain metabolic health, and prevent cardiometabolic diseases by stabilizing lipid droplets and lipoproteins, facilitating their transportation and metabolism, particularly in patients with hyperlipidemia [21]. Probiotics, such as *Lactobacillus plantarum*, can alter microbiota in the gut and boost the synthesis of short-chain fatty acids (SCFAs). These acids are important for fat processing. O-3FA reported to reduce the amount of triglyceride by inhibiting the production of very-low-density lipoprotein (VLDL) in the liver and promoting fatty acid oxidation. Phytosterols are known for their ability to reduce cholesterol. They work by competing with cholesterol for absorption in the intestines, thereby reducing the amount of cholesterol absorbed into the bloodstream [22].

Functional foods impact carbohydrate metabolism by stimulating insulin secretion, influencing the hepatic glucose metabolism, and enhancing the activity of insulin receptors and glucose transporters. Certain polyphenols, such as those derived from *Artemisia dracunculus* L. (tarragon), are effective at enhancing insulin secretion and modulating enzymes involved in gluconeogenesis and glycogenolysis. This can be particularly beneficial for individuals with type 2 diabetes or insulin resistance (IR) as it helps maintain glucose homeostasis and reduces hepatic glucose production [23, 24].

5.3.2 Regulation of biochemical pathways

Functional foods can significantly impact metabolic pathways, such as glycolysis, the tricarboxylic acid cycle, gluconeogenesis, and oxidative phosphorylation, through various mechanisms that include the modulation of enzyme activities, gene expression, and enhancement of mitochondrial function. These metabolic pathways are critical for maintaining cellular energy balance and overall metabolic health [25]. Glycolysis and gluconeogenesis are two opposing pathways that manage glucose metabolism. Functional foods can influence these pathways through diverse mechanisms. BC like polyphenols and flavonoids, commonly found in functional foods, can alter the activity of enzymes involved in glycolysis and gluconeogenesis [26, 27]. For example, cinnamon extract has been reported to prevent IR by modulating enzymes such as glucokinase, which is crucial for glycolysis, and glucose-6-phosphatase, vital for gluconeogenesis [28]. Additionally, gene expression of enzymes associated with glycolysis and gluconeogenesis can be influenced by dietary components. Polyphenolic compounds derived from *Canna indica* L. have been demonstrated to stimulate glucose transport in muscle cells by modulating these pathways at the transcriptional level [29].

The Krebs cycle is essential for energy production through the oxidation of acetyl-CoA. Functional foods can enhance the efficiency of this cycle by upregulating the enzymes involved and providing substrates that fuel the cycle [30]. BC in functional foods, such as those found in maca, can enhance the activity of Krebs cycle enzymes through the upregulation of gene expression and improved availability of substrates like acetyl-CoA and oxaloacetate [31]. Also, functional foods rich in O-3FA and certain polyphenols promote mitochondrial biogenesis by activating key regulators like PGC-1α. This leads to an increased number of mitochondria, thereby enhancing the overall capacity for oxidative metabolism and the Krebs cycle. This enhancement in mitochondrial biogenesis not only supports the Krebs cycle but also improves overall cellular energy production [32].

Oxidative phosphorylation is the final stage of cellular respiration, where ATP is produced through the electron transport chain (ETC). Functional foods contain potent antioxidants and can as such regulate this process by protecting the components of the ETC from oxidative damage and ensure optimal and effective ATP production [33].

5.4 Metabolic mechanisms of therapeutic action of functional foods

5.4.1 Influence on metabolic disorders

5.4.1.1 Diabetes mellitus

Diabetes mellitus (DM) is a complex metabolic derangement characterized by chronically elevated levels of sugar resulting from abnormalities in either insulin production or function. If left untreated, it can disrupt the metabolism of fats, proteins, and carbohydrates and lead to severe complications [34]. Functional foods, owing to their BC, have emerged as potential therapeutic agents in managing DM. These foods offer antioxidant, anti-inflammatory, and insulin-sensitizing properties, which are critical in mitigating the metabolic dysregulation associated with diabetes [35].

Oxidative stress significantly influences the pathogenesis of DM by exacerbating IR and damaging pancreatic β-cells, impairing their ability to secrete insulin [36, 37]. Functional foods rich in antioxidants can nullify free radicals, thereby lowering oxidative stress. This reduction improves insulin sensitivity and conserves β-cell function [36]. For instance, polyphenols, found abundantly in whole grains, vegetables, and fruits exhibit strong antioxidant properties. These compounds hunt ROS, protecting cells from oxidative damage. By mitigating oxidative stress, these antioxidants help in upholding the integrity and functionality of pancreatic β-cells, thus promoting better insulin secretion and action [38].

Chronic inflammation is intently connected to the evolvement of type 2 DM and IR. Reduced insulin sensitivity occurs due to the interference of inflammatory cytokines with insulin signaling pathways, namely tumor necrosis factor-alpha (TNF-α) and interleukin-6 (IL-6) [39, 40]. Significant anti-inflammatory effects are demonstrated by BC components found in functional foods, such as polyphenols, 11, and O-3FA. These substances reduce inflammation by inhibiting the synthesis of pro-inflammatory cytokines [41]. Flavonoids, present in berries and green tea, can curb the activation of NF-κB, which is a critical regulator of inflammatory responses. These BC substances enhance insulin sensitivity and contribute to the effective regulation of glucose metabolism by reducing inflammation [42].

Functional foods can enhance insulin sensitivity and promote insulin secretion through various mechanisms. Certain BC mimic insulin action or stimulate pancreatic β-cells to secrete insulin. Bitter melon (*Momordica charantia*), for instance, contains BC such as charantin and polypeptide-p, which have insulin-like activities. These compounds enhance glucose uptake and utilization by tissues, thereby lowering blood glucose levels. Additionally, they inhibit hepatic gluconeogenesis, the process by which glucose is produced in the liver, further contributing to blood glucose regulation [43].

The development of diabetes and metabolic health is strongly influenced by the gut flora. Functional foods containing prebiotics and probiotics have the capacity to modify the gut microbiota and enhance the growth of beneficial bacteria [44]. Prebiotics, such as fructooligosaccharides and inulin, serve as substrates for favorable gut bacteria, thereby promoting their growth and functionality. Probiotics, such as *Lactobacillus* and *Bifidobacterium* strains [45, 46], are directly administered to the stomach to introduce beneficial bacteria.

This modulation improves gut barrier function, reduces systemic inflammation, and positively affects glucose metabolism and insulin sensitivity. A healthy gut microbiota composition is associated with improved metabolic outcomes, including better glycemic control and reduced IR [47, 48].

Dyslipidemia, branded by raised amounts of triglycerides and low-density lipoprotein (LDL) cholesterol, is common in DM [49]. Functional foods can help regulate lipid metabolism by reducing triglycerides and LDL cholesterol levels while improving high-density lipoprotein (HDL) cholesterol levels [50]. O-3FA, found in fish oil and flaxseeds, are remarkably effective in lowering triglyceride levels [51]. Soluble fibers, such as β-glucan from oats, can reduce LDL cholesterol by binding to bile acids and promoting their excretion [52].

5.4.1.2 Obesity

Functional foods can influence various metabolic pathways and mechanisms involved in obesity, offering a safer alternative to pharmacological treatments. One of the primary ways functional foods can aid in obesity management is by promoting satiety and reducing appetite. Foods high in fiber, protein, and certain BC can delay gastric emptying and enhance the feeling of fullness, which in turn reduces overall energy intake [53, 54]. Components such as dietary fiber and proteins modulate gut hormones like ghrelin, peptide YY, and glucagon-like peptide-1 (GLP-1), which play key roles in regulating hunger and satiety [55, 56].

In addition to regulating appetite, certain dietary components in functional foods can increase thermogenesis, the process by which the body generates heat from metabolizing food, thus enhancing energy expenditure [57]. For example, capsaicin in chili peppers and catechins in green tea have been recognized to boost metabolic rate and promote fat oxidation. These thermogenic effects contribute to a negative energy balance, aiding in weight management [57]. Functional foods also influence lipid metabolism and adipocyte function, which are critical aspects of obesity management. O-3FA, found in fish oil, and conjugated linoleic acid have been reported to reduce fat accumulation by enhancing lipid oxidation and reducing Lipogenesis. Additionally, polyphenols, such as those found in berries, are effective against fat cell development and boost the systemic elimination of the unwanted fat cells, thereby reducing fat mass [58].

Fat storage and energy homeostasis are heavily influenced by the gut microbiota. Probiotics and prebiotics, when included in functional meals, can effectively modulate the gut microbiota, and promote the growth of beneficial bacteria that contribute to weight management [59]. These alterations in the microbiota have the potential to enhance the production of SCFAs, thereby improving energy metabolism and reducing inflammation, thus aiding in the treatment of obesity [60].

5.4.1.3 Influence of functional foods on cardiovascular disease

Modulating lipid metabolism is a key mechanism by which functional foods impact cardiovascular health. Dyslipidemia, a common risk factor for cardiovascular diseases, is branded by raised triglyceride and LDL levels, accompanied by low HDL levels [61]. Evidence shows that functional foods, including nuts, fatty fish, and whole grains, can improve lipid profiles. Lower levels of LDL and total cholesterol observed from the consumption of whole grains have been attributed to the role of soluble fiber in preventing cholesterol absorption in the intestines, thereby exerting a significant effect on reducing lipid levels [62]. Nuts that are rich in fiber, unsaturated fats, and plant sterols, such as walnuts and almonds, have been shown to effectively lower total cholesterol and LDL-C levels [63]. Fatty fish, including salmon and mackerel, are excellent sources of O-3FA, which can help reduce triglycerides and potentially raise HDL-C levels. O-3FA also possess anti-inflammatory properties and can inhibit the synthesis of VLDL [64].

Inflammation and oxidative stress play key roles in the etiology of cardiovascular disorders. Antioxidant-rich functional meals can slow down these processes and lower the risk of cardiovascular disease. Carotenoids, polyphenols, vitamins C and E, and other antioxidants can all be found in abundance in fruits and vegetables. Regular fruit and vegetable consumption is linked to lower oxidative stress indicators and decreased inflammation, both of which promote cardiovascular health [65, 66]. Green tea polyphenols – s in particular – have potent antioxidant qualities. Green tea polyphenols—Epigallocatechin-3-Gallate in particular—has been reported not as a potent antioxidant alone but as a promoter of cardiovascular health. [67]. Flavonoids in dark chocolate have been shown to enhance endothelial function, reduce oxidative stress, and improve lipid profiles. However, these benefits are contingent on moderate consumption due to the calorie content of chocolate [68].

Chronic inflammation is a known promoter of atherosclerosis and other cardiovascular diseases [69]. Functional foods can exert anti-inflammatory effects, thereby reducing cardiovascular risk. Olive oil, rich in monounsaturated fats and polyphenols, has potent anti-inflammatory properties. Its regular consumption is a major component of the Mediterranean food, which is linked with a lesser prevalence of cardiovascular maladies [70]. Legumes are high in fiber, protein, and numerous BC that exhibit anti-inflammatory effects. Regular consumption of legumes has been linked to improved cardiovascular health by reducing indicators of inflammation such as C-reactive protein (CRP) [71].

Hypertension is a foremost risk factor for cardiovascular ailments. Certain functional foods can help regulate blood pressure through various mechanisms. Fermented dairy products like yogurt contain BC peptides that have been shown to exert antihypertensive effects by inhibiting angiotensin-converting enzyme (ACE) activity. Regular intake of these foods is linked with lower blood pressure and lowered cardiovascular threat. Garlic contains allicin, which has vasodilatory properties and can help lower blood pressure. Its consumption has been linked to improvements in both systolic and diastolic blood pressure [72].

5.4.2 Role in metabolic syndrome

5.4.2.1 Insulin resistance

The underlying cause of metabolic syndrome, a cluster of disorders that significantly promote the menace of type 2 diabetes, heart disease, and stroke, is IR [73]. When insulin-sensitive tissues, such as skeletal muscle, liver, and adipose tissue, fail to retort adequately to insulin motivation, a complex disease ensues [74]. The result is a physiological paradox: elevated blood glucose levels coexisting with compensatory hyperinsulinemia. This maladaptive response sets in motion a cascade of metabolic disturbances that characterize the syndrome [75].

IR manifests at the cellular level as an intricate network of dysfunctional signaling pathways. Impairments occur in the insulin signaling cascade, which is typically inducted by insulin binding to its receptor and subsequent phosphorylation events involving phosphatidylinositol 3-kinase (PI3K) and insulin receptor substrates (IRSs) [75, 76]. As a result of this disruption, there is reduced glucose absorption in muscle and adipose tissues since glucose transporter type 4 (GLUT4) fails to translocate to the cell membrane [77]. Simultaneously, disruption of lipid metabolism is significant as unwarranted lipid aggregation in non-adipose tissues activates stress-related kinases, such as protein kinase C (PKC). These kinases further compromise the efficacy of insulin signaling by phosphorylating IRS proteins on serine residues [78].

IR is exacerbated by chronic low-rating inflammation, often rooted in obesity. Inflammatory cytokines such as TNF-α and IL-6 activate kinases that disrupt insulin signaling through similar mechanisms. This inflammatory state creates a self-perpetuating cycle, where inflammation begets more inflammation, continuously aggravating IR [79]. Moreover, recent research has implicated gut microbiota dysbiosis in this process, suggesting that imbalances in intestinal bacteria can lead to increased gut permeability and systemic endotoxemia, further fueling the inflammatory milieu associated with IR.

Mitochondrial dysfunction adds additional layer of intricacy to IR pathophysiology. Exacerbated oxidative phosphorylation and heightened production of ROS not only provoke OS but also activate stress-related kinases that inhibit insulin signaling [80, 81].

Functional meals play a significant role in increasing insulin sensitivity and moderating the effects of IR through various processes. BC polyphenols found in functional foods such as citrus fruits, green tea, and chocolate offer lipid-modulating, antioxidant, and anti-inflammatory benefits [82]. Specifically, green tea polyphenols, including epigallocatechin gallate (EGCG), have been proven to enhance insulin signaling pathways and regulate glucose absorption, thereby increasing insulin sensitivity. These polyphenols also possess the capacity to suppress inflammatory pathways and stress-related kinases, ensuring the proper functioning of IRS proteins and facilitating efficient insulin signaling [83, 84].

Cocoa, rich in flavanols, also contributes to improved insulin sensitivity through its antioxidant properties, which reduce oxidative stress and inflammation. By mitigating oxidative damage, cocoa flavanols help maintain the integrity of insulin signaling pathways and enhance glucose uptake in peripheral tissues [85]. Additionally, dietary fibers present in functional foods like whole grains, legumes, and certain fruits can improve insulin sensitivity by modulating gut microbiota composition. These fibers serve as prebiotics, promoting the growth of beneficial gut bacteria that produce SCFAs. SCFAs have anti-inflammatory properties and can enhance insulin sensitivity by modulating gut barrier function and reducing systemic endotoxemia. The consumption of dietary fibers also helps in weight management, which is crucial for improving insulin sensitivity and decreasing the threat of metabolic syndrome [86].

5.4.2.2 Dyslipidemia

Dyslipidemia, characterized by abnormal levels of lipids in the blood, is a crucial factor in the development and progression of metabolic syndrome. Dyslipidemia involves disruptions in lipid metabolism that result in elevated levels of triglycerides, LDL, total cholesterol, along with reduced levels of HDL [87]. The liver is central to lipid metabolism, regulating the synthesis and clearance of various lipoproteins. In dyslipidemia, there is often an increase in VLDL and LDL particles. VLDL, synthesized in the liver, is a precursor to LDL. In states of IR, elevated insulin levels promote the hepatic synthesis of VLDL. Once secreted into the bloodstream, VLDL is converted into LDL, which, when present in excess, leads to cholesterol accumulation in arterial walls and promotes atherosclerosis. Conversely, HDL participates in reverse cholesterol transport, a critical process that transports cholesterol from peripheral tissues back to the liver for excretion. Reduced HDL levels, commonly observed in metabolic syndrome, impair this process and contribute to cholesterol accumulation in blood vessels [88].

Dyslipidemia is primarily caused by various molecular pathways, including those involving enzymes and receptors that control lipid metabolism. Transcription factors known as sterol regulatory element-binding proteins (SREBPs), direct the expression of genes involved in lipid production and absorption. They represent one of the pathways involved in these genes. Overactivation of SREBPs in dyslipidemia can lead to increased hepatic production of triglycerides and cholesterol [89]. Another significant mechanism is the regulation of LDL receptor levels on hepatocytes by proprotein convertase subtilisin/kexin type 9 (PCSK9). Elevated PCSK9 levels result in reduced LDL receptor levels, impeding the removal of LDL from the bloodstream and worsening hypercholesterolemia [90]. Furthermore, PPARs, which are nuclear receptors, play a crucial role in lipid metabolism. Activating PPARγ in adipose tissue enhances insulin sensitivity and lipid storage, consequently reducing lipid levels. On the other hand, activating PPARα in the liver promotes fatty acid oxidation, leading to a decrease in triglyceride levels [91].

Functional foods have demonstrated significant potential in managing dyslipidemia and its associated risks within metabolic syndrome. O-3FA are known to reduce hepatic triglyceride synthesis and VLDL secretion while promoting the clearance of triglyceride-rich lipoproteins [92]. Dietary fibers increase bile acid excretion and reduce cholesterol absorption, leading to lower serum cholesterol levels [93]. Polyphenols not only enhance the excretion of cholesterol and bile acids but also exhibit antioxidant properties that protect against lipid peroxidation [94]. Furthermore, the consumption of soy protein has been linked with improved lipid profiles through mechanisms such as increased bile acid excretion and modulation of hormone levels that influence lipid metabolism [95].

5.4.2.3 Hypertension

High blood pressure, otherwise called hypertension, is a vital component of metabolic syndrome. Elevated blood pressure levels [96] are indicative of hypertension in the context of metabolic syndrome, and these values can be associated with various interconnected factors. One significant contributing factor is hyperinsulinemia caused by IR, typical of the metabolic syndrome. Increased blood volume and blood pressure may result from the kidneys retaining salt due to elevated insulin levels [97]. Angiotensin 2 and aldosterone influence the vasoconstriction and salt retention, and the interaction between insulin and the renin-angiotensin-aldosterone system enhances their effects [98].

Another crucial factor contributing to hypertension in metabolic syndrome is oxidative stress. Oxidative stress can damage endothelial cells, leading to impaired nitric oxide (NO) production and dysfunction, endothelial, key factors in the development of hypertension. ROS activate inflammatory pathways and increase the expression of pro-inflammatory cytokines such as TNF-α and IL-6, exacerbating vascular inflammation and hypertension. The reduced bioavailability of NO due to oxidative damage leads to increased vascular resistance, further elevating blood pressure [99]. Chronic low-grade inflammation is also prevalent in metabolic syndrome and contributes to hypertension by promoting endothelial dysfunction and arterial stiffness. Inflammatory mediators, including CRP and various interleukins, reduce the bioavailability of NO, leading to increased vascular resistance and blood pressure. This chronic inflammation creates a vicious cycle where endothelial dysfunction and inflammation perpetuate each other, maintaining elevated blood pressure levels [100].

Functional foods play a crucial role in managing hypertension through various metabolic mechanisms. Polyphenols, such as flavonoids and resveratrol, have potent antioxidant properties that mitigate oxidative stress and improve endothelial function. These compounds enhance the production of NO and reduce the activity of NADPH oxidase, an enzyme that generates ROS. Improved NO availability leads to vasodilation and lower blood pressure. By counteracting oxidative stress and promoting endothelial health, polyphenols address a fundamental aspect of hypertension in metabolic syndrome [101].

Dietary fibers also contribute to managing hypertension. Soluble dietary fibers can improve lipid profiles and glycemic control, reducing the risk of hypertension. Fibers influence gut microbiota composition, leading to the production of SCFAs that have anti-inflammatory effects and improve insulin sensitivity [86]. O-3FA have antihypertensive effects. These fatty acids inhibit the production of pro-inflammatory eicosanoids and enhance the synthesis of anti-inflammatory prostaglandins, contributing to lower blood pressure and reduced inflammation [102].

5.5 Mechanisms of action of selected functional foods

5.5.1 Green tea

Green tea's popularity is majorly due to its varied health benefits and is known to contain catechin and its derivatives. The green tea's BC acts through various mechanisms. The ability of EGCG for instance, to impede fat accumulation and effect decrease in adipose tissue mass, and body weight relies on its tendencies to regulate gene expression. This compound also targets the activation of AMPK pathway to improve lipid profile and minimize the danger of becoming obese. The AMPK pathway is a crucial energy sensor in cells, and its activation by EGCG promotes energy-consuming catabolic pathways including β-oxidation and inhibits anabolic pathways such as lipid synthesis. This dual action of AMPK activation improves lipid profiles and reduces adiposity [103, 104].

EC, CGC, and ECG are powerful antioxidants that function through scavenging of radical, lipid peroxidation inhibition, and metal ion chelation through hydrogen atoms or electrons donation, thereby nullifying free radicals and averting the distortion of biomolecules that can result in chronic ailments [105, 106]. The Nrf2 and NF-κB metabolic pathways are also vital for the antioxidant properties of catechins. The surge in the expression of antioxidant enzymes and the decrease in the production of pro-inflammatory cytokines and oxidative stress have been linked to the activation of the Nrf2 pathway and the inhibition of the NF-κB pathway, respectively, by catechins. These events result in detoxification of ROS and thus protect against oxidative damage and inflammation [105, 107].

5.5.2 Blueberries

Blueberries, rich in antioxidants, are known to contain two major anthocyanins (malvidin-3-glucoside and malvidin-3 galactoside) and flavonoids which have been attributed to their antioxidant properties and anti-inflammatory [108]. Anthocyanins mediate their influences via the NF-κB pathway [109, 110]. IκBα is an inhibitory protein that retains NF-κB in the cytoplasm. Inflammatory stimuli such as TNF-α, degrade IκBα allowing NF-κB to translocate to the nucleus and activate pro-inflammatory genes [111]. Mv-3-glc and Mv-3-gal prevent the degradation of IκBα, thereby preventing p65, a component of NF-κB, from entering the nucleus. By blocking this pathway, these anthocyanins effectively reduce the synthesis of inflammatory mediators [112]. These molecules function by inhibiting the production of the three important markers of inflammatory response including monocyte chemotactic protein-1 (MCP-1), intercellular adhesion molecule-1 (ICAM-1), and vascular cell adhesion molecule-1 (VCAM-1) and thereby improving inflammation [109].

5.5.3 Fish oil

Fish oil has a stockpile of O-3FA (docosahexaenoic acid and eicosapentaenoic acid are inclusive) that is known to improve various health conditions, including cardiovascular and inflammation. O-3FA acts via the stimulation of PPAR-α that inhibits the hepatic synthesis of VLDL which in turn accelerates the oxidation of fatty acid and thus lowers blood lipid levels [113, 114]. Also, the key to preserve potential of fish oil on abrupt cardiac death is due to the stabilization of the heart's electrical activity through the modulation of calcium and sodium ion channels, which are crucial for cardiac conductivity and excitability [115, 116].

The anti-inflammatory effects of O-3FA have been established through various metabolic pathways including the inhibition of arachidonic acid-derived eicosanoids. The competition of EPA with arachidonate for lipoxygenase (LOX) and cyclooxygenase

(COX) enzymes leads to production of less inflammatory eicosanoids (276). Prostaglandin E3 (PGE3) and leukotriene B5 (LTB5) are less effective EPA-derived eicosanoids in promoting inflammation compared to their arachidonic acid-derived equivalents, PGE2 and LTB4 [117]. The shift in eicosanoid production reduces inflammatory reactions and lessens symptoms associated with chronic inflammation. In addition, the ability of EPA and DHA to function as precursors to specialized pro-resolving lipid mediators (SPMs), including maresins and protectins, helps in resolving inflammation by decreasing neutrophil infiltration, promoting the clearance of apoptotic cells, and enhancing the resolution phase of inflammation [118, 119]. Also, O-3FA reduce the expression of pro-inflammatory cytokines, such as IL-6 and TNF-α via regulation of the NF-kB B pathway, which is a critical regulator of inflammatory responses [120].

5.5.4 Challenges and future directions

Lack of universally acceptable definition, a harmonized global standard regulatory policy as well as trust issues from consumers are the major challenges facing functional food field. Overcoming these tribulations requires globally agreed definitions and criteria that clearly outline what constitutes a functional food and the standardization of research methodologies to ensure the reproducibility that guarantees the reliability of functional food claims. Also, there is urgent need to develop rigorous procedures for assessment of effectiveness, bioavailability, and safety of functional foods. Such protocols should not be limited only to in vitro or in vivo but also clinical methodologies where necessary. Consistency in BC of functional foods must be strictly ensured via detailed and effective quality control. A strong relationship such as collaborative research among academic researchers, industry experts, and regulatory agencies is the only pivot of this anticipated development in the field of functional foods. The regulatory bodies must ensure a strict guideline on labeling of functional foods to include only the verified health claims to safeguard the consumers interest and trust.

To better the future of functional foods, personalized nutrition focusing on dietary recommendations on genetic, metabolic, and lifestyle factors heralds a promising direction. To achieve this, huge investment in genomic and metabolomics research becomes necessary to unravel the science behind varied individuals' responses to functional foods. Future research that targets biomarkers that are specific for certain BC will be invaluable. Development of algorithms and digital tools including mobile apps that can make monitoring of dietary intake vis-à-vis the health impact will contribute immensely to human well-being and disease prevention.

5.6 Conclusion

Advancement in nutritional research that focuses improved public health requires a good understanding of the molecular, biochemical, and metabolic pathways associated with the purported benefits of the functional foods. The intricates of how functional foods exert their therapeutic effects require a vast knowledge of interaction of BC with the metabolic pathways. A sizable number of BCs that were consumed directly or modified and their health benefits confirmed include O-3FA, carotenoids, and polyphenols that engaged with enzymes and receptors within cells to initiate signaling cascades that influence gene expression, regulate inflammatory responses, and fortify antioxidant defenses. Curcumin exerts its effect by obstructing the NF-κB pathway and thus reducing inflammation and impeding the growth of cancer cells; while O-3FA modify the PI3K/Akt pathway, prompting cancer cells to undergo apoptosis and curb the risk of heart-related illness. Understanding these mechanisms elucidates the health benefits of functional foods and provides a scientific basis for their inclusion in health promotion and disease prevention strategies.

These findings have significant implications for advancement of health and avoidance of diseases. A practical, natural, and readily available approach to addressing the growing prevalence of chronic illnesses is through the incorporation of functional meals. Inclusion of functional foods into the schedule of regular diet potentiates a significant reduction in reliance on pharmacological treatments. Berries are rich in antioxidant principles and can protect pancreatic β-cells from oxidative damage, thereby assisting in the regulation of diabetes. Leavy greens and fatty fish are potent in managing inflammatory bowel disease and arthritis. Future research directions are crucial for unlocking the full potential of functional foods.

Funding

The authors particularly acknowledge the financial support provided by the National Research Foundation of South Africa and the Directorate of Research and Postgraduate Support, Durban University of Technology to B.O. Yusuf (PSTD2204274997) and National Research Foundation Research Development Grant for Rated Researcher (SRUG2204193723) to S. Sabiu.

References

[1] Gur, J., Mawuntu, M., & Martirosyan, D. FFC's advancement of functional food definition. Functional Food in Health and Disease 2018, 8(7), 385–397.

[2] Alongi, M., & Anese, M. Re-thinking functional food development through a holistic approach. Journal of Functional Foods 2021, 81, 104466.

[3] Alissa, E. M., & Ferns, G. A. Functional foods and nutraceuticals in the primary prevention of cardiovascular diseases. Journal of Nutrition and Metabolism 2012, 2012(1), 569486. Martirosyan DM, Singh J. A new definition of functional food by FFC: What makes a new definition unique? Vol. 5, Functional Foods in Health and Disease. 2015. p. 209–23.

[4] European Food Safety Authority (EFSA). Scientific opinion on the substantiation of a health claim related to barley beta-glucans and lowering of blood cholesterol and reduced risk of (coronary) heart disease pursuant to Article 14 of Regulation (EC) No 1924/2006. EFSA Journal 2011, 9(12), 2470.

[5] Cabello-Olmo, M., Krishnan, P. G., Araña, M., Oneca, M., Díaz, J. V., Barajas, M., & Rovai, M. Development, analysis, and sensory evaluation of improved bread fortified with a plant-based fermented food product. Foods 2023, Jul 25, 12(15), 2817.

[6] Singh, H. Nanotechnology applications in functional foods; opportunities and challenges. Preventive Nutrition and Food Science 2016, Mar, 21(1), 1.

[7] Banwo, K., Olojede, A. O., Adesulu-Dahunsi, A. T., Verma, D. K., Thakur, M., Tripathy, S., et al. Functional importance of bioactive compounds of foods with Potential Health Benefits: A review on recent trends. Food Bioscience 2021, 43, 101320.

[8] Akbari, B., Baghaei-Yazdi, N., Bahmaie, M., & Mahdavi Abhari, F. The role of plant-derived natural antioxidants in reduction of oxidative stress. BioFactors 2022, 48, 611–633.

[9] Serventi, L., & Vinola Dsouza, L. Bioactives in legumes. In: Upcycling Legume Water: From Wastewater to Food Ingredients, Springer Cham, 2020, 139–153.

[10] Yu, H., Lin, L., Zhang, Z., Zhang, H., & Hu, H. Targeting NF-κB pathway for the therapy of diseases: Mechanism and clinical study. Signal Transduction and Targeted Therapy 2020, 5, 209.

[11] Khan, H., Ullah, H., Castilho, P. C. M. F., Gomila, A. S., D'Onofrio, G., Filosa, R., et al. Targeting NF-κB signaling pathway in cancer by dietary polyphenols. Critical Reviews in Food Science and Nutrition 2020, 60(16), 2790–2800.

[12] Çetin, Z., Saygili, E. İ., Benlier, N., Ozkur, M., & Sayin, S. Omega-3 polyunsaturated fatty acids and cancer. In: Jafari, S. M., Nabavi, S. M., & Silva, A. S., editors. Nutraceuticals and Cancer Signaling Food Bioactive Ingredients, Cham: Springer, 2021, 591–631. 34. Hussain S, Sajjad A, Butt SZ, Muazzam MA. An overview of antioxidant and pharmacological potential of common fruits. Sci Inq Rev. 2021;5(1):1–18.

[13] Mierziak, J., Kostyn, K., Boba, A., Czemplik, M., Kulma, A., & Wojtasik, W. Influence of the bioactive diet components on the gene expression regulation. Nutrients 2021, 13, 3673.

[14] Eichten, S. R., Schmitz, R. J., & Springer, N. M. Epigenetics: Beyond chromatin modifications and complex genetic regulation. Plant Physiology 2014, 165(3), 933–947.

[15] Jalili, M., Pati, S., Rath, B., Bjørklund, G., & Singh, R. B. Effect of diet and nutrients on molecular mechanism of gene expression mediated by nuclear receptor and epigenetic modulation. Open Nutraceuticals Journal 2013, 6(1), 27–34.

[16] Dhanya, R. Quercetin for managing type 2 diabetes and its complications, an insight into multitarget therapy. Biomedicine and Pharmacotherapy 2022, 146, 112560.

[17] Das, P., Dutta, A., Panchali, T., Khatun, A., Kar, R., Das, T. K., et al. Advances in therapeutic applications of fish oil: A review. Measurement: Food 2024, 13, 100142.

[18] González-Garibay, A. S., López-Vázquez, A., García-Bañuelos, J., Sánchez-Enríquez, S., Sandoval-Rodríguez, A. S., Del Toro Arreola, S., et al. Effect of ursolic acid on insulin resistance and

hyperinsulinemia in rats with diet-induced obesity: Role of adipokines expression. Journal of Medicinal Food 2020, 23(3).

[19] Bustos, A. S., Håkansson, A., Linares-Pastén, J. A., Peñarrieta, J. M., & Nilsson, L. Interaction of quercetin and epigallocatechin gallate (EGCG) aggregates with pancreatic lipase under simplified intestinal conditions. PLoS One 2020, 15(4), e0224853.

[20] Huang, J., Zhang, Y., Zhou, Y., Zhang, Z., Xie, Z., Zhang, J., et al. Green tea polyphenols alleviate obesity in broiler chickens through the regulation of lipid-metabolism-related genes and transcription factor expression. Journal of Agricultural and Food Chemistry 2013, 61(36), 8565–8572.

[21] Robert, C., Couëdelo, L., Vaysse, C., & Michalski, M. C. Vegetable lecithins: A review of their compositional diversity, impact on lipid metabolism and potential in cardiometabolic disease prevention. Biochimie 2020, 169, 121–132.

[22] Zhang, R., Han, Y., McClements, D. J., Xu, D., & Chen, S. Production, characterization, delivery, and cholesterol-lowering mechanism of phytosterols: A review. Journal of Agricultural and Food Chemistry 2022, 70, 2483–2494.

[23] Ribnicky, D. M., Poulev, A., Watford, M., Cefalu, W. T., & Raskin, I. Antihyperglycemic activity of Tarralin™, an ethanolic extract of *Artemisia dracunculus* L. Phytomedicine 2006, 13(8), 550–557.

[24] Govorko, D., Logendra, S., Wang, Y., Esposito, D., Komarnytsky, S., Ribnicky, D., et al. Polyphenolic compounds from Artemisia dracunculus L. inhibit PEPCK gene expression and gluconeogenesis in an H4IIE hepatoma cell line. American Journal of Physiology Endocrinology and Metabolism 2007, 293(6), E1503–10.

[25] Zhang, H., Ma, J., Tang, K., & Huang, B. Beyond energy storage: Roles of glycogen metabolism in health and disease. FEBS Journal 2021, 288, 3772–3783.

[26] Zhao, C., Yang, C., Wai, S. T. C., Zhang, Y., Portillo, P., Paoli, M., et al. Regulation of glucose metabolism by bioactive phytochemicals for the management of type 2 diabetes mellitus. Critical Reviews in Food Science and Nutrition 2019, 59, 830–547.

[27] Golovinskaia, O., & Wang, C. K. The hypoglycemic potential of phenolics from functional foods and their mechanisms. Food Science and Human Wellness 2023, 12, 986–1007.

[28] Senevirathne, B. S., Jayasinghe, M. A., Pavalakumar, D., & Siriwardhana, C. G. *Ceylon cinnamon*: A versatile ingredient for futuristic diabetes management. Journal of Future Foods 2022, 2, 125–142.

[29] Sarje, S. K., Ingole, K., Angad, S., Priya, B., & Ghiware, N. B. A pharmacognostic and pharmacological review on *Canna indica* Linn. International Journal of Research in Pharmaceutical Chemistry 2019, 9(3), 61–77.

[30] Vásquez-Reyes, S., Velázquez-Villegas, L. A., Vargas-Castillo, A., Noriega, L. G., Torres, N., & Tovar, A. R. Dietary bioactive compounds as modulators of mitochondrial function. Journal of Nutritional Biochemistry 2021, 96, 108768.

[31] Wan, W., Li, H., Xiang, J., Yi, F., Xu, L., Jiang, B., et al. Aqueous extract of black maca prevents metabolism disorder via regulating the glycolysis/gluconeogenesis-TCA cycle and PPARα signaling activation in golden hamsters fed a high-fat, high-fructose diet. Frontiers in Pharmacology 2018, 9, 333.

[32] Vaughan, R. A., Mermier, C. M., Bisoffi, M., Trujillo, K. A., & Conn, C. A. Dietary stimulators of the PGC-1 superfamily and mitochondrial biosynthesis in skeletal muscle. A mini-review. Journal of Physiology and Biochemistry 2014, 70, 271–284.

[33] Ramis, M., Esteban, S., Miralles, A., Tan, D. X., & Reiter, R. Protective effects of melatonin and mitochondria-targeted antioxidants against oxidative stress: A review. Current Medicinal Chemistry 2015, 22(22), 2690–2711.

[34] Yusuf, B. O., Yakubu, M. T., & Akanji, M. A. Chromatographic fractions from Chrysophyllum albidum stem bark boost antioxidant enzyme activity and ameliorate some markers of diabetes complications. Journal of Traditional and Complementary Medicine 2021, Jul 1, 11(4), 336–342.

[35] Sharma, M., Vidhya, C. S., Ojha, K., Yashwanth, B. S., Singh, B., Gupta, S., et al. The role of functional foods and nutraceuticals in disease prevention and health promotion. European Journal of Nutrition & Food Safety 2024, 16(2), 61–83.
[36] Eguchi, N., Vaziri, N. D., Dafoe, D. C., & Ichii, H. The role of oxidative stress in pancreatic β cell dysfunction in diabetes. International Journal of Molecular Sciences 2021, 22, 1509.
[37] Bhatti, J. S., Sehrawat, A., Mishra, J., Sidhu, I. S., Navik, U., Khullar, N., et al. Oxidative stress in the pathophysiology of type 2 diabetes and related complications: Current therapeutics strategies and future perspectives. Free Radical Biology and Medicine 2022, May, 184, 114–134. Available from: https://linkinghub.elsevier.com/retrieve/pii/S0891584922001228.
[38] Rudrapal, M., Khairnar, S. J., Khan, J., Bin, D. A., Ansari, M. A., Alomary, M. N., et al. Dietary polyphenols and their role in oxidative stress-induced human diseases: Insights into protective effects, antioxidant potentials and mechanism(s) of action. Frontiers in Pharmacology 2022, 13, 806470.
[39] Rehman, K., Akash, M. S. H., Liaqat, A., Kamal, S., Qadir, M. I., & Rasul, A. Role of interleukin-6 in development of insulin resistance and type 2 diabetes mellitus. Critical Reviews in Eukaryotic Gene Expression 2017, 27(3), 229–236.
[40] Chen, L., Chen, R., Wang, H., & Liang, F. Mechanisms linking inflammation to insulin resistance. International Journal of Endocrinology 2015, 2015, 508409.
[41] Jayarathne, S., Koboziev, I., Park, O. H., Oldewage-Theron, W., Shen, C. L., & Moustaid-Moussa, N. Anti-Inflammatory and anti-obesity properties of food bioactive components: Effects on adipose tissue. Preventive Nutrition and Food Science 2017, 22, 251.
[42] Egbuna, C., Awuchi, C. G., Kushwaha, G., Rudrapal, M., Patrick-Iwuanyanwu, K. C., Singh, O., et al. Bioactive compounds effective against type 2 diabetes mellitus: A systematic review. Current Topics in Medicinal Chemistry 2021, 21(12), 1067–1095.
[43] Mahwish, S. F., Tauseef Sultan, M., Riaz, A., Ahmed, S., Bigiu, N., et al. Bitter melon (*Momordica charantia* l.) fruit bioactives charantin and vicine potential for diabetes prophylaxis and treatment. Plants 2021, 10(4), 730.
[44] Peng, M., Tabashsum, Z., Anderson, M., Truong, A., Houser, A. K., Padilla, J., et al. Effectiveness of probiotics, prebiotics, and prebiotic-like components in common functional foods. Comprehensive Review in Food Science Food Safety 2020, 19(4), 1908–1933.
[45] Wieërs, G., Belkhir, L., Enaud, R., Leclercq, S., Philippart de Foy, J. M., Dequenne, I., et al. How Bioactive compounds affect the microbiota. Frontiers in Cellular and Infection Microbiology 2020, 9, 454.
[46] Lordan, C., Thapa, D., Ross, R. P., & Cotter, P. D. Potential for enriching next-generation health-promoting gut bacteria through prebiotics and other dietary components. Gut Microbes 2020, 11, 1–20.
[47] Salles, B. I. M., Cioffi, D., & Ferreira, S. R. G. Probiotics supplementation and insulin resistance: A systematic review. Diabetology and Metabolic Syndrome 2020, 12, 1–24.
[48] Li, H. Y., Zhou, D. D., Gan, R. Y., Huang, S. Y., Zhao, C. N., Shang, A., et al. Effects and mechanisms of probiotics, prebiotics, synbiotics, and postbiotics on metabolic diseases targeting gut microbiota: A narrative review. Nutrients 2021, 13, 3211.
[49] Feingold, K. R., & Grunfeld, C. Diabetes and dyslipidemia. In: Johnstone, M., & Veves, A., editors. Diabetes And Cardiovascular Disease, Humana: Springer, 2023, 425–472.
[50] Sandner, G., König, A., Wallner, M., & Weghuber, J. Functional foods – Dietary or herbal products on obesity: Application of selected bioactive compounds to target lipid metabolism. Current Opinion in Food Science 2020, 34, 9–20.
[51] Liu, Y. X., Yu, J. H., Sun, J. H., Ma, W. Q., Wang, J. J., & Sun, G. J. Effects of omega-3 fatty acids supplementation on serum lipid profile and blood pressure in patients with metabolic syndrome: A systematic review and meta-analysis of Randomized Controlled Trials. Foods 2023, 12, 725.

[52] Nie, Y., & Luo, F. Dietary fiber: An opportunity for a global control of hyperlipidemia. Oxidative Medicine and Cellular Longevity 2021, 2021, 5542342.

[53] Munekata, P. E. S., Pérez-Álvarez, J. Á., Pateiro, M., Viuda-Matos, M., Fernández-López, J., & Lorenzo, J. M. Satiety from healthier and functional foods. Trends in Food Science and Technology 2021, 113, 397–410.

[54] Boix-Castejón, M., Roche, E., Olivares-Vicente, M., Álvarez-Martínez, F. J., Herranz-López, M., & Micol, V. Plant compounds for obesity treatment through neuroendocrine regulation of hunger: A systematic review. Phytomedicine 2023, 113.

[55] Kabisch, S., Weickert, M. O., & Pfeiffer, A. F. H. The role of cereal soluble fiber in the beneficial modulation of glycometabolic gastrointestinal hormones. Critical Reviews in Food Science and Nutrition 2024, 64, 4331–4347.

[56] Pizarroso, N. A., Fuciños, P., Gonçalves, C., Pastrana, L., & Amado, I. R. A review on the role of food-derived bioactive molecules and the microbiota–gut–brain axis in satiety regulation. Nutrients 2021, 13, 632.

[57] Reguero, M., de Cedrón, M. G., Wagner, S., Reglero, G., Quintela, J. C., & de Molina, A. R. Precision nutrition to activate thermogenesis as a complementary approach to target obesity and associated-meta-bolic-disorders. Cancers 2021, 13, 866.

[58] Chang, Y. C., Yang, M. Y., Chen, S. C., & Wang, C. J. Mulberry leaf polyphenol extract improves obesity by inducing adipocyte apoptosis and inhibiting preadipocyte differentiation and hepatic lipogenesis. Journal of Functional Foods 2016, 21, 249–262.

[59] Green, M., Arora, K., & Prakash, S. Microbial medicine: Prebiotic and probiotic functional foods to target obesity and metabolic syndrome. International Journal of Molecular Sciences 2020, 21, 2890.

[60] Morrison, D. J., & Preston, T. Formation of short chain fatty acids by the gut microbiota and their impact on human metabolism. Gut Microbes 2016, 7, 189–200.

[61] Pereira, T. Dyslipidemia and cardiovascular risk: Lipid ratios as risk factors for cardiovascular disease. Dyslipidemia – From Prevention to Treatment 2012, 279–302.

[62] Tosh, S. M., & Bordenave, N. Emerging science on benefits of whole grain oat and barley and their soluble dietary fibers for heart health, glycemic response, and gut microbiota. Nutrition Reviews 2021, 78, 13–20.

[63] Berryman, C. E., Preston, A. G., Karmally, W., Deckelbaum, R. J., & Kris-Etherton, P. M. Effects of almond consumption on the reduction of LDL-cholesterol: A discussion of potential mechanisms and future research directions. Nutrition Reviews 2011, 69, 171–185.

[64] Bhat, S., Sarkar, S., Zaffar, D., Dandona, P., & Kalyani, R. R. Omega-3 fatty acids in cardiovascular disease and diabetes: A review of recent evidence. Current Cardiology Reports 2023, 25, 51–65.

[65] Granato, D. Functional foods to counterbalance low-grade inflammation and oxidative stress in cardiovascular diseases: A multilayered strategy combining food and health sciences. Current Opinion in Food Science 2022, 47, 100894.

[66] dos Santos, J. L., de Quadros, A. S., Weschenfelder, C., Garofallo, S. B., & Marcadenti, A. Oxidative stress biomarkers, nut-related antioxidants, and cardiovascular disease. Nutrients 2020, 12, 682.

[67] Li, Y., Karim, M. R., Wang, B., & Peng, J. Effects of green tea (−)-epigallocatechin-3-gallate (EGCG) on cardiac function – A review of the therapeutic mechanism and potentials. Mini-Reviews in Medicinal Chemistry 2022, 22(18), 2371–2382.

[68] de Paula Silva, T., Silva, A. A., Toffolo, M. C. F., & de Aguiar, A. S. The action of phytochemicals present in cocoa in the prevention of vascular dysfunction and atherosclerosis. Journal of Clinical and Translational Research 2022, 8(6), 509.

[69] Henein, M. Y., Vancheri, S., Longo, G., & Vancheri, F. The role of inflammation in cardiovascular disease. International Journal of Molecular Sciences 2022, 23, 12906.

[70] Bucciantini, M., Leri, M., Nardiello, P., Casamenti, F., & Stefani, M. Olive polyphenols: Antioxidant and anti-inflammatory properties. Antioxidants 2021, 10(7), 1044.

[71] Grdeń, P., & Jakubczyk, A. Health benefits of legume seeds. Journal of the Science of Food and Agriculture 2023, 103, 5213–5220.

[72] Ried, K., & Fakler, P. Potential of garlic (*Allium sativum*) in lowering high blood pressure: Mechanisms of action and clinical relevance. Integrated Blood Pressure Control 2014, 7, 71–82.

[73] Masenga, S. K., Kabwe, L. S., Chakulya, M., & Kirabo, A. Mechanisms of oxidative stress in metabolic syndrome. International Journal of Molecular Sciences 2023, 24, 7898.

[74] Chandrasekaran, P., & Weiskirchen, R. Cellular and molecular mechanisms of insulin resistance. Current Tissue Microenvironment Reports 2024, 1–12.

[75] Petersen, M. C., & Shulman, G. I. Mechanisms of insulin action and insulin resistance. Physiological Reviews 2018, 98(4), 2133–2223.

[76] Khalilov, R., & Abdullayeva, S. Mechanisms of insulin action and insulin resistance. Advances in Biology & Earth Sciences 2023, 8(2), 165–179.

[77] van Gerwen, J., Shun-Shion, A. S., & Fazakerley, D. J. Insulin signalling and GLUT4 trafficking in insulin resistance. Biochemical Society Transactions 2023, 51, 1057–1069.

[78] Zhao, X., An, X., Yang, C., Sun, W., Ji, H., & Lian, F. The crucial role and mechanism of insulin resistance in metabolic disease. Frontiers in Endocrinology 2023, 14, 1149239.

[79] Rehman, K., & Akash, M. S. H. Mechanisms of inflammatory responses and development of insulin resistance: How are they interlinked? Journal of Biomedical Science 2016, 23(1), 1–18.

[80] Rocha, M., Rovira-Llopis, S., Banuls, C., Bellod, L., Falcon, R., Castello, R., et al. Mitochondrial dysfunction and oxidative stress in insulin resistance. Current Pharmaceutical Design 2013, 19(32), 5730–5741.

[81] Rains, J. L., & Jain, S. K. Oxidative stress, insulin signaling, and diabetes. Free Radical Biology and Medicine 2011, 50, 567–575.

[82] Granato, D., Barba, F. J., Bursać Kovačević, D., Lorenzo, J. M., Cruz, A. G., & Putnik, P. Functional foods: Product development, technological trends, efficacy testing, and safety. Annual Review of Food Science and Technology 2020, 11, 93–118.

[83] Mokra, D., Joskova, M., & Mokry, J. Therapeutic effects of green tea polyphenol (–)-epigallocatechin -3-gallate (EGCG) in relation to molecular pathways controlling inflammation, oxidative stress, and apoptosis. International Journal of Molecular Sciences 2023, 24, 340.

[84] James, A., Wang, K., & Wang, Y. Therapeutic activity of green tea epigallocatechin-3-gallate on metabolic diseases and non-alcoholic fatty liver diseases: The current updates. Nutrients 2023, 15, 3022.

[85] González-Garrido, J. A., García-Sánchez, J. R., López-Victorio, C. J., Escobar-Ramírez, A., & Olivares-Corichi, I. M. Cocoa: A functional food that decreases insulin resistance and oxidative damage in young adults with class II obesity. Nutrition Research & Practice 2023, 17(2), 228–240.

[86] Canfora, E. E., Jocken, J. W., & Blaak, E. E. Short-chain fatty acids in control of body weight and insulin sensitivity. Nature Reviews Endocrinology 2015, 11, 577–591.

[87] Stadler, J. T., & Marsche, G. Obesity-related changes in high-density lipoprotein metabolism and function. International Journal of Molecular Sciences 2020, 21, 8985.

[88] Ofori, E. K. Lipids and lipoprotein metabolism, dyslipidemias, and management. In: Current Trends in the Diagnosis and Management of Metabolic Disorders, 1st edition, CRC Press, 2023, 150–170.

[89] Malhotra, P., Gill, R. K., Saksena, S., & Alrefai, W. A. Disturbances in cholesterol homeostasis and non-alcoholic fatty liver diseases. Frontiers in Medicine 2020, 7, 467.

[90] Melendez, Q. M., Krishnaji, S. T., Wooten, C. J., & Lopez, D. Hypercholesterolemia: The role of PCSK9. Archives of Biochemistry and Biophysics 2017, 625–626, 39–53.

[91] Pawlak, M., Lefebvre, P., & Staels, B. Molecular mechanism of PPARα action and its impact on lipid metabolism, inflammation and fibrosis in non-alcoholic fatty liver disease. Journal of Hepatology 2015, 62, 720–733.

[92] Qi, K., Fan, C., Jiang, J., Zhu, H., Jiao, H., Meng, Q., et al. Omega-3 fatty acid containing diets decrease plasma triglyceride concentrations in mice by reducing endogenous triglyceride synthesis and enhancing the blood clearance of triglyceride-rich particles. Clinical Nutrition 2008, 27(3), 424–430.

[93] Korcz, E., Kerényi, Z., & Varga, L. Dietary fibers, prebiotics, and exopolysaccharides produced by lactic acid bacteria: Potential health benefits with special regard to cholesterol-lowering effects. Food and Function 2018, 9, 3057–3068.

[94] Kardum, N., & Glibetic, M. Polyphenols and their iteractions with other dietary compounds: Implications for human health. In: Toldrá, F., editor. Advances in Food and Nutrition Research, Academic Press, 2018, 103–144.

[95] Torres, N., Torre-Villalvazo, I., & Tovar, A. R. Regulation of lipid metabolism by soy protein and its implication in diseases mediated by lipid disorders. Journal of Nutritional Biochemistry 2006, 17, 365–373.

[96] Katsimardou, A., Imprialos, K., Stavropoulos, K., Sachinidis, A., Doumas, M., & Athyros, V. Hypertension in metabolic syndrome: Novel insights. Current Hypertension Reviews 2019, 16(1), 12–18.

[97] Brosolo, G., Da Porto, A., Bulfone, L., Vacca, A., Bertin, N., Scandolin, L., et al. Insulin resistance and high blood pressure: Mechanistic insight on the role of the kidney. Biomedicines 2022, 10, 2374.

[98] Underwood, P. C., & Adler, G. K. The renin angiotensin aldosterone system and insulin resistance in humans. Current Hypertension Reports 2013, 15, 59–70.

[99] Sinha, N., & Dabla, P. Oxidative stress and antioxidants in hypertension–A current review. Current Hypertension Reviews 2015, 11(2), 134–142.

[100] Bisoendial, R. J., Boekholdt, S. M., Vergeer, M., Stroes, E. S. G., & Kastelein, J. J. P. C-reactive protein is a mediator of cardiovascular disease. European Heart Journal 2010, 31, 2087–2091.

[101] Kumar, G., Dey, S. K., & Kundu, S. Functional implications of vascular endothelium in regulation of endothelial nitric oxide synthesis to control blood pressure and cardiac functions. Life Science 2020, 259, 118377.

[102] Tortosa-Caparrós, E., Navas-Carrillo, D., Marín, F., & Orenes-Piñero, E. Anti-inflammatory effects of omega 3 and omega 6 polyunsaturated fatty acids in cardiovascular disease and metabolic syndrome. Critical Reviews in Food Science and Nutrition 2017, 57(16), 3421–3429.

[103] Hardie, D. G. Sensing of energy and nutrients by AMP-activated protein kinase. American Journal of Clinical Nutrition 2011, 891S–896S.

[104] Marín-Aguilar, F., Pavillard, L. E., Giampieri, F., Bullón, P., & Cordero, M. D. Adenosine monophosphate (AMP)-activated protein kinase: A new target for nutraceutical compounds. International Journal of Molecular Sciences 2017, 18, 288.

[105] Baranwal, A., Aggarwal, P., Rai, A., & Kumar, N. Pharmacological actions and underlying mechanisms of catechin: A review. Mini-Reviews in Medicinal Chemistry 2021, 22(5).

[106] Sadžak, A., Mravljak, J., Maltar-Strmečki, N., Arsov, Z., Baranović, G., Erceg, I., et al. The structural integrity of the model lipid membrane during induced lipid peroxidation: The role of flavonols in the inhibition of lipid peroxidation. Antioxidants 2020, 9(5), 430.

[107] Talebi, M., Talebi, M., Farkhondeh, T., Mishra, G., İlgün, S., & Samarghandian, S. New insights into the role of the Nrf2 signaling pathway in green tea catechin applications. Phytotherapy Research 2021, 35, 3078–3112.

[108] Ma, Z., Du, B., Li, J., Yang, Y., & Zhu, F. An insight into anti-inflammatory activities and inflammation related diseases of anthocyanins: A review of both *in vivo* and *in vitro* investigations. International Journal of Molecular Sciences 2021, 22, 11076.

[109] Huang, W. Y., Liu, Y. M., Wang, J., Wang, X. N., & Li, C. Y. Anti-inflammatory effect of the blueberry anthocyanins malvidin-3-glucoside and malvidin-3-galactoside in endothelial cells. Molecules 2014, 19(8), 12827–12841.

[110] Vendrame, S., & Klimis-Zacas, D. Anti-inflammatory effect of anthocyanins via modulation of nuclear factor-κB and mitogen-activated protein kinase signaling cascades. Nutrition Reviews 2015, 73(6), 348–358.
[111] Rius-Pérez, S., Pérez, S., Martí-Andrés, P., Monsalve, M., & Sastre, J. Nuclear factor kappa B signaling complexes in acute inflammation. Antioxidants and Redox Signaling 2020, 33, 145–165.
[112] Guan, R., Van Le, Q., Yang, H., Zhang, D., Gu, H., Yang, Y., et al. A review of dietary phytochemicals and their relation to oxidative stress and human diseases. Chemosphere 2021, 271, 129499. 264. Matos AL, Bruno DF, Ambrósio AF, Santos PF. The benefits of flavonoids in diabetic retinopathy. Vol. 12, Nutrients. 2020. p. 3169.
[113] Calvo, M. J., Martínez, M. S., Torres, W., Chávez-Castillo, M., Luzardo, E., Villasmil, N., et al. Omega-3 polyunsaturated fatty acids and cardiovascular health: A molecular view into structure and function. Vessel Plus 2017, 1, 116–128.
[114] Qiu, Y. Y., Zhang, J., Zeng, F. Y., & Zhu, Y. Z. Roles of the peroxisome proliferator-activated receptors (PPARs) in the pathogenesis of nonalcoholic fatty liver disease (NAFLD). Pharmacological Research 2023, 192, 106786.
[115] Xiao, L., Chen, X. J., Feng, J. K., Li, W. N., Yuan, S., & Hu, Y. Natural products as the calcium channel blockers for the treatment of arrhythmia: Advance and prospect. Fitoterapia 2023, 169, 105600.
[116] Muiti, B. Impact of omega-3 fatty acid supplementation on cardiovascular health. American Journal of Food Sciences and Nutrition 2024, 6(1), 43–53.
[117] Wall, R., Ross, R. P., Fitzgerald, G. F., & Stanton, C. Fatty acids from fish: The anti-inflammatory potential of long-chain omega-3 fatty acids. Nutrition Reviews 2010, 68, 280–289.
[118] Mason, R. P., Sherratt, S. C. R., & Eckel, R. H. Omega-3-fatty acids: Do they prevent cardiovascular disease? Best Practice and Research: Clinical Endocrinology and Metabolism 2023, 37, 101681.
[119] Sherratt, S. C. R., Libby, P., Budoff, M. J., Bhatt, D. L., & Mason, R. P. Role of omega-3 fatty acids in cardiovascular disease: The debate continues. Current Atherosclerosis Reports 2023, 25, 1–17.
[120] Komprda, T. Eicosapentaenoic and docosahexaenoic acids as inflammation-modulating and lipid homeostasis influencing nutraceuticals: A review. Journal of Functional Foods 2012, 4, 25–38.

Oladunni Mary Ayodele and Saheed Sabiu*

Chapter 6
Functional foods and nutraceuticals: health benefits and innovations

Abstract: Besides fulfilling their nutritional needs, foods can influence bodily functions and impact health outcomes. Hence, the reason for the two terms "functional foods and nutraceuticals." Functional foods are everyday foods with added health benefits, while nutraceuticals are concentrated or isolated compounds used as supplements for improved health and disease prevention, so it is safe to infer that functional foods are primarily food, whereas nutraceuticals are food-derived substances utilized for medicinal purposes. There are evidence-based health benefits of various bioactive components in functional foods and nutraceuticals. Here, we provide information on the health benefits and innovations such as delivery systems including nanoencapsulation, microencapsulation, and liquid encapsulation technologies which helps to enhance the bioavailability and stability of bioactive compounds in functional foods and supplements. Understanding structure–function relationships of food bioactives and development of novel formulation and delivery techniques are recommended as they are crucial to fully harness the health benefits of functional foods and nutraceuticals.

Keywords: Bioavailability, bioactive compounds, functional foods, health benefits, nutraceuticals

Abbreviations

EPA	Eicopentaenoic acid
DNA	Deoxyribonucleic acid
AI	Artificial intelligence
C-III	Demethioxylated curcumin
CBD	Cannabidiol
GSHPx	Glutathione peroxidase
PH	Potential of hydrogen

*__Corresponding author__: **Saheed Sabiu**, Department of Biotechnology and Food Science, Faculty of Applied Sciences, Durban University of Technology, P.O. Box 1334, Durban, South Africa, e-mail: sabius@dut.ac.za
Oladunni Mary Ayodele, Department of Biotechnology and Food Science, Faculty of Applied Sciences, Durban University of Technology, P.O. Box 1334, South Africa

https://doi.org/10.1515/9783111441238-007

6.1 Introduction

The significance of food and comprehension of its importance dates to the origins of humanity itself. Food is defined as any substance consumed by an organism for nutritional support. It provides essential nutrients such as carbohydrates, fats, proteins, vitamins, and minerals [1]. Food while meeting our nutritional needs can also influence various bodily functions and impact health outcomes. Dr. Stephen DeFelice coined the term "nutraceutical" in 1989 by combining "nutrition" and "pharmaceutical"; it refers to bioactive compounds found in fortified foods, dietary supplements, or herbal products. The concept of "functional foods" emerged in the mid-1980s, with its roots in Japan where foods with added health benefits were first recognized by the government as "Foods for Specified Health Use." The two terms have been often used interchangeably hence confusion often set in but the fundamental distinction between functional foods and nutraceuticals lies in their nature: functional foods are essentially regular foods meant to be consumed as a part of one's regular diet, for example, fruits, vegetables, nuts, seeds, and grains fortified with vitamins, minerals, probiotics, or fiber, while nutraceuticals are concentrated or isolated forms used as medicinal supplements to enhance health, combat aging, prevent chronic illnesses, and extend life expectancy [2]. Essentially, functional foods are everyday foods with added health benefits, while nutraceuticals are concentrated or isolated compounds used as supplements for improved health and disease prevention; inferably, functional foods are primarily food, whereas nutraceuticals are food-derived substances utilized for medicinal purposes. While a lot has been reported about functional foods and nutraceuticals, enhancing bioavailability of food bioactives in humans is not fully elucidated. There is therefore an urgent need to understand the science behind bioavailability and interactions of food bioactive with one another to efficiently develop new innovative techniques to enhance their bioavailability. Hence, this chapter provides an understanding on their classes and further discusses how innovations have helped the bioavailability of certain nutraceuticals and functional foods for improved health benefits.

6.2 Understanding the classes of functional foods and nutraceuticals

Since both functional foods and nutraceuticals aim to promote health and wellness through their specific compositions and properties, inferentially, they stem from "medicinal food." Functional foods can be natural or modified foods that contain bioactive compounds and can be classified based on their origin which can be plant-based (like fruits or fiber-rich grains), animal-based, or even produced by fermentation with specific microbial cultures (probiotics, prebiotics) (Figure 6.1). They can also be categorized by their targeted body function, such as foods that aim to improve cardiovascular health, enhance the immune system, or support gastrointestinal health [3]. Nutraceuticals, on the other hand, are products that may range from isolated nutrients, dietary supplements, or herbal products to specific diets or processed foods such as cereals, soups, and beverages that contain added bioactive compounds. The defining characteristic of nutraceuticals is that they are not conventional foods but rather products that have been enhanced to provide specific medical or health benefits, including the prevention and treatment of disease. Nutraceuticals can be classified according to their natural source, with some being derived from plants (phytosterols, phenolics), animals (omega-3 fatty acids, conjugated linoleic acid), or microorganisms (amino acids and eicopentaenoic acid (EPA) by bacteria) [3] and those that are produced artificially, such as enhanced and genetically modified products [4]. They can also be classified by their potential health benefits – such as anti-inflammatory, antioxidant, or cardioprotective – or their physical form (pills, capsules, powders). Functional foods and Nutraceuticals can be natural (traditional) or modified (artificial) and they both can originate from plant, animal, or microbial sources, but nutraceuticals can further be classified based on their physical form such as pills, powder, and capsules (Figure 6.1).

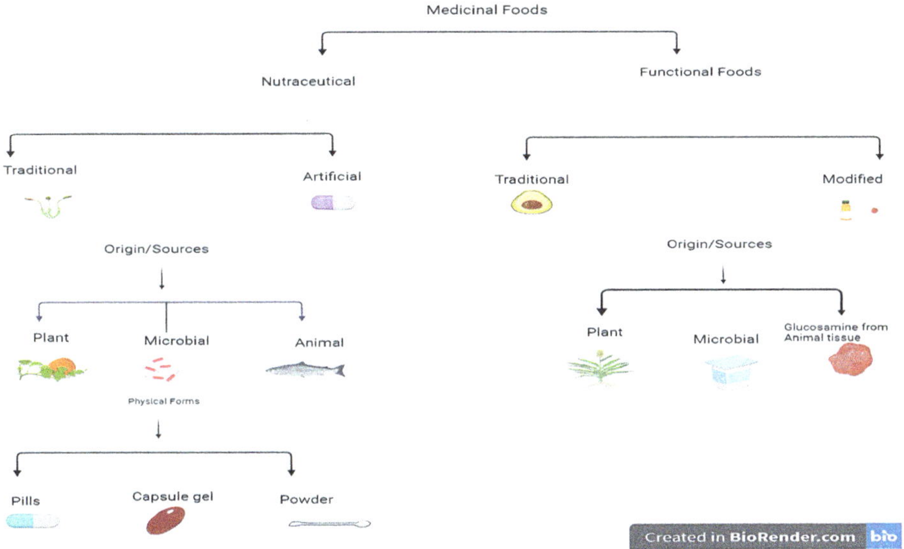

Figure 6.1: Classification of functional foods and nutraceuticals.

6.3 Health benefits and innovations in functional foods and nutraceuticals

There are evidence-based health benefits of various bioactive components in functional foods and nutraceuticals and a few of these are presented in Table 6.1. Certain functional foods, such as fruits (kiwi, apples, berries), not only support regular bowel movements and add bulk to stool but also serve as good antioxidants due to their cellulose and hemicellulose content. Vegetables like brassica, carrots, eggplant, onions, and seaweed, along with β-glucans found in oats, help reduce cholesterol levels. Additionally, various bioactive compounds found in foods have diverse health benefits. Bioactives such as pectin, insulin, konjac glucomannan, ginseng polysaccharides, acanthopanax, ulvans, lentinan, anthocyanins, beta-carotene, catechins, flavonoids, lutein, lycopene, quercetin, resveratrol, rutin, zeaxanthin, zingiberene, alliin, aloins, and curcumin have been reported to lower blood sugar, fight cancer, and offer antimicrobial, antioxidant, and immunomodulatory effects. Animal sources of nutraceutical and functional foods, aside from being rich in protein, provide bioactive compounds like chitin, heparin, hyaluronic acid, omega-3 fatty acids, polyketides, polyunsaturated fatty acids, selenium, and sulfur. These compounds, found in diverse animal sources such as insects, shellfish, fish oil, sea animals, and poultry, are known for their antioxidant, antimicrobial, anti-inflammatory properties, and ability to counteract free radicals and fight degenerative diseases (Table 6.1).

Table 6.1: Selected functional foods and nutraceuticals from plants, animals, and microbes.

Functional foods/ nutraceuticals	Bioactive compounds	Health benefits	References
Kiwi, apple, prunes, wheat bran, almond, green beans, and Brussels sprouts	Cellulose and hemicelluloses	They promote regular bowel movement and add bulk to stool; antioxidant activity, antithrombotic activity, immunomodulating activity, lower cholesterol, and eliminate free radical	[5, 6]
Brassica vegetables, berries, eggplants and okra, and carrot juice	Pectins	They have activities that modify the immune response, effect of reducing cholesterol levels, and slowing down the process of stomach emptying and the transit time in the small intestine	[7]

Table 6.1 (continued)

Functional foods/ nutraceuticals	Bioactive compounds	Health benefits	References
Brown rice, oats, and barley	β-Glucans	They show reduction in cholesterol levels, regulation of lipid levels, control of blood glucose, lowering of elevated blood pressure, and immunostimulation	[8, 9]
Potatoes, rice, green bananas, and legumes that have been cooked and allowed to cool	Resistant starch	They lower blood sugar levels, reduce cholesterol level, guarding against colon cancer, play roles as prebiotics, preventing fat accumulation, and promoting better mineral absorption	[10]
Gums from locust bean, gum arabic, and guar gum	Gums like galactan, xylan, xyloglucan, glucuronic mannan, and galacturonic rhamnosan type	They postpone the rise in blood sugar and fats after a meal, alter the composition of lipoproteins, enhance the feeling of fullness, slow down the process of stomach emptying, effect of lowering cholesterol levels, and the effect of reducing triglyceride levels	[11]
Root of chicory, onion bulb, garlic, and wheat	Inulin	It lowers lipid level and enhancement of mineral absorption such as calcium and magnesium	[11]
Konjac plant	Konjac glucomannan	It has hypocholesterolemic effect and weight reduction	[12]
Panax ginseng root	Ginseng polysaccharides; ginsenosides and panaxosides	It has antirotavirus activity	[13, 14]
Acanthopanax senticosus leaves	*Acanthopanax senticosus* polysaccharides	It has antioxidant activity and immunobiological activity	[13]
Seaweeds, brown seaweeds, and red marine algae	Carbohydrates, polysaccharides (alginates, fucoidans, and laminarans), and phlorotannins	They have antioxidant, antiviral, antidiabetic, anticancer, and radioprotective activities; healing of injuries and transport of proteins	[15, 16]

Table 6.1 (continued)

Functional foods/ nutraceuticals	Bioactive compounds	Health benefits	References
Marine algae	Sulfated polysaccharides (phlorotannins and diterpenes)	They have antiviral properties, combat tumors, including cancer, along with the promotion of prebiotic health	[13]
Green seaweeds (*Ulva*), (*Monostroma latissimum*), (*Codium*), and *Caulerpa*	Ulvans, sulfated rhamnan, sulfated arabinogalactans, sulfated galacotans, and sulfated mannans	They have antioxidant, antiviral, antitumor, and immunomodulating activities	[17–20]
Lichens	β-Glucans lichenan	They have immunomodulatory effects	[21]
Fungi (*Schizophyllum commune*)	Fungi (*Schizophyllum commune*)	They have immunostimulatory effects	[22]
Fungi (*Athelia rolfsii*)	Scleroglucan	It has immunostimulatory effects	[23]
Mushroom (*Lentinula edodes*)	Lentinan	It has antitumor and immunomodulatory effects	[24]
Red cabbage, berries, radishes, plums, and eggplant; black rice; and bananas	Anthocyanins	They have reduction of cholesterol, prevention of cardiovascular diseases; antiobesity or weight loss; cancer prevention; enhance cognitive function	[25–29]
Yellow- and orange-colored vegetables and fruits (sweet potato, carrots, etc.)	Beta-carotene	It converts vitamin A in the body; it prevents the oxidation of lipids triggered by free radicals (acting as an antioxidant); it aids in the prevention of sunburns and damage caused by sunlight, as well as conditions like porphyria and psoriasis	[30–35]
Tea (green), apples, apricots, cherries, peaches, berries, blackberries, and grapes	Catechins	They act as an antioxidant and reduce both blood pressure and levels of low-density lipoprotein	[36–38]

Table 6.1 (continued)

Functional foods/ nutraceuticals	**Bioactive compounds**	**Health benefits**	**References**
Citrus fruits, berries, purple grapes, teas, cocoa, legumes, soy, and whole grains	Flavonoids	They act as an antioxidant, lower inflammation, and may protect against heart disease, stroke, and cancer	[39–42]
Vegetables (green leafy): kale, collards, spinach, romaine and broccoli; yellow corn	Lutein	It protects against eye diseases and may decrease the likelihood of age-related macular degeneration	[43, 44]
Tomatoes, watermelon and pink grapefruit, pink guava, papaya, seabuckthorn, and rosehip	Lycopene	It decreases the risk for prostate cancer and cardiovascular diseases; cholesterol-lowering action, protection against metabolic syndrome; strong antioxidant effect	[45–52]
Fruits and vegetables (yellow): apples/pears, citrus fruits, onions, berries, and tea	Quercetin	It lowers blood pressure, serves as antioxidants, decreases inflammation, and may protect from heart disease, stroke, and cancer	[53, 54]
Red/grapes juice and wine, cocoa seeds/powder; dark chocolate; and peanuts	Resveratrol	It enhances heart health; reduces blood pressure and body mass index	[55]
Buckwheat; asparagus; elderflower tea; apples (with skin); and figs	Rutin (vitamin P or rutoside)	They are anti-inflammatory, neuroprotection against brain ischemia, enhance nitric oxide production to improve endothelial function	[44, 56–61]
Kale, collards, spinach, romaine, and broccoli; corns (yellow)	Zeaxanthin	It shields against eye diseases by soaking up harmful blue light that penetrates the eye; it could potentially lower the chances of developing macular degeneration associated with aging	[43, 44]

Table 6.1 (continued)

Functional foods/ nutraceuticals	Bioactive compounds	Health benefits	References
Ginger	Zingiberene and gingerols	They lower high blood sugar levels, long-term inflammation of the bronchi, substances that increase physiological activity, and reduce discomfort in the throat	[62]
Garlic bulb	Alliin and allicin	They combat bacteria, reduce inflammation, fight against fungi, alleviate gout, prevent blood clot formation, lower blood pressure, and reduce high levels of lipids in the blood	[63]
Aloe vera gel	Aloin and aloesin	They have properties that expand small blood vessels, reduce inflammation, soften and soothe the skin, and promote the healing of wounds	[64]
Turmeric rhizome	Curcumin	It protects against cancer; lowers inflammation and antiseptic properties; and prevents arthritis	[65]
Onion bulb	Allicin and alliin	It helps to lower blood sugar levels, properties that combat bacterial infections, and prevention of atherosclerosis (hardening and narrowing of the arteries)	[66]
Soybeans	Vegetable proteins: casein and whey	The help in body building	[67]
Most fruits especially citrus and berries, tomato, coffee, white bread, and hop oil	Prenol lipids	They shield the membranes of cells; help in maintaining the stability of cellular proteins and bolstering the body's immune defenses; play a crucial role in the metabolism of cells; serve as antioxidants; and act as the initial stage in the production of vitamin A	[68–70]

Table 6.1 (continued)

Functional foods/ nutraceuticals	Bioactive compounds	Health benefits	References
Edible mushroom *Cortinarius caerulescens*	Polyketides	They act as antimicrobial, antiparasitic, and anticancer agents; they help intracellular communication	[68, 71]
Oils from olives, avocados, sunflower canola, macadamia, grape seed, groundnut sesame, almond, and hemp, and cashews	Monounsaturated fatty acids	They prevent neurodegenerative diseases and promote good cholesterol	[72, 73]
Flax seeds, sunflower oil, and walnuts	Polyunsaturated fatty acids	They help in lowering the risk of hypertension, and improve heart function and maintenance	[73, 74]
Brazil nuts, mushrooms, sunflower seeds, beans, oats, and brown rice	Selenium	They counter free radicals and promote a robust immune system, and these substances also aid in thyroid hormone metabolism	[75, 76]
Cruciferous vegetables (broccoli, cauliflower, cabbage, kale, Brussels sprouts, turnips, bok choy, and kohlrabi), (nuts and legumes), allium vegetables (garlic, onions, leeks, and chives), and eggs	Sulfur	They participate in energy metabolism through their role in the electron transport chain, act as contributor to the slowing of aging process, a detoxifier of the liver and cells, an enhancer of mental focus and clarity, an improver of sleep quality, skin health, and athletic performance, and a component of antioxidants	[75, 76]
Crustaceans, insect cuticles, and shells of mollusks	Chitin and chitosan	They have bacteriostatic and fungi static influences, antiviral, drug encapsulation, fat absorber, and wound dressing materials	[77]
Golgi of animal cells	Heparin/heparan sulfate	They have anticoagulating, signaling and development, antimicrobial, anti-inflammatory, and anticancer activities	[78]

Table 6.1 (continued)

Functional foods/ nutraceuticals	Bioactive compounds	Health benefits	References
Animal tissues	Hyaluronic acid	They have chondroprotective effects, and are immunomodulatory	[79]
Animal tissues from porcine intestine and shark cartilage	Chondroitin sulfate/ dermatan sulfate	They have extracellular matrix support and anti-inflammatory effects	[80, 81]
Fish oil	Omega-3 fatty acids	They help reduce inflammation, stabilize the electrical activity in heart muscle cells; they have been associated with both anticancer and heart-protective effects; they thin the blood and can increase bleeding, especially in patients who are on blood-thinning medications like warfarin	[82–86]
Sea urchin and mollusks	Polyketide (Pks2)	They are employed for their antimicrobial, antiparasitic, and anticancer properties, and serve as conveyors of information in communication between cells	[87]
Sea foods	Polyunsaturated fatty acids	They play anti-inflammation role, prevention of heart disease, anticancer, anti-arthritis, anti-depression, and anti-Alzheimer	[74]
Oysters, tuna; meats, eggs, ham, milk, and yogurt	Selenium	They act to counteract free radicals and bolster a healthy immune system and play a role in the metabolism of thyroid hormones	[88]
Fish, poultry, meats, and eggs	Sulfur	As a source of protein it helps energy metabolism and aging process, liver and cell detoxification; improves mental focus and clarity	[89]

6.4 Innovative techniques for enhancing bioavailability of nutraceuticals and functional foods

Nutraceuticals and functional foods have gained significant attention due to their potential health benefits driving innovation in the food and beverage industry. With the rise in popularity of plant-based diets, there's a surge in plant-based functional foods and nutraceuticals. These products offer alternatives to traditional animal-derived ingredients while providing health benefits such as improved heart health, better digestion, and weight management. Beyond traditional food products, for instance, there is a growing market for functional beverages targeting specific health needs [90]. These include fortified waters, herbal teas, functional juices, and energy drinks enhanced with vitamins, minerals, antioxidants, cannabidiol (CBD)-infused products (reducing anxiety, alleviating pain, and improving sleep) and other bioactive compounds. Functional snacks fortified with protein, fiber, vitamins, and minerals are gaining popularity as alternatives to traditional, less nutritious snacks. These compounds are then used in the development of new nutraceuticals and functional food

However, various factors largely determine their effectiveness, bioavailability, and safety. Understanding these factors is crucial to the development and consumption of effective and safe nutraceuticals and functional foods consequentially providing insights to researchers, healthcare professionals, and consumers alike. Bioavailability refers to the percentage of a nutrient that the body can absorb and use [91]. Factors affecting bioavailability include the chemical form of nutrients, presence of other compounds in the food matrix, and individual physiological factors such as age, health status, and genetic makeup [92, 93]. Other essential elements that can influence bioavailability encompass food processing methods like heat processing, mixture techniques (homogenization), various cooking techniques, and storage methods. For instance, polyphenols in certain fruits and vegetables diminish when stored in refrigeration [94], or when subjected to mechanical treatments such as grating and cutting and enzyme activation. Notably, environmental factors such as sunlight, wavelength of light in artificial environments, farming processes and post-harvest activities also factors in bioavailability.

Moreover, to ensure bioactive components in food are effective, it is important to assess how well nutrients are released. Nutrients are classified into macro (carbohydrates, proteins, and fats) which are nutrients needed in copious amounts and micronutrients such as vitamins, minerals, and antioxidants (nutrients required in smaller amounts) (Figure 6.2). Typically, the body effectively absorb and utilize over 90% of the macronutrients consumed (Figure 6.3). Conversely, micronutrients exhibit varying degrees of absorption and utilization upon consumption. This underscores the importance of considering nutrient bioavailability, effects of enzymes in the intestine, binding and uptake by the intestinal mucosa transfer across the gut wall, movement into the blood or lymphatic circulation, distribution and storage, metabolic use and excretion which is the final stage of nutrient utilization. Largely because extreme pH, the mucus layer, epithelium, and gastrointestinal enzymes restrict the delivery of food components via the oral route, that a particular food is rich in nutrients does not mean it can be depended solely for the required nourishment. For example, spinach, a green leafy vegetable, has low bioavailability though it contains the highest calcium content amongst all green leafy vegetables. Sadly, only 5% of its calcium content is absorbable due to the presence of oxalates, an antinutrient that blocks or interferes with the absorption of other nutrients [95]. Likewise, research on the gut microbiome's role in overall health has led to the development of foods and supplements designed to support a healthy gut flora. Probiotics, prebiotics, and symbiotics are increasingly incorporated into functional foods and beverages.

Furthermore, recent research has increasingly focused on how food bioactives interact and interact with other molecules in the body, for instance, the therapeutic use of curcumin is restricted because of its poor bioavailability and elevated metabolic function [96, 97], according to a study, demethoxylated curcumin (C–III), shows a high solubility, polarity and better bioavailability when compared with other curcumin and so exhibits superior activity when it was orally administered in a liposome-encapsulated form. Also, bioavailability of provitamin A carotenoids is significantly more variable than that of preformed vitamin A (retinol) [98, 99]. The bioavailability of phytochemicals, such as lycopene (found in tomatoes) and β–carotene (found in carrots), is increased by fats, promoting health and well-being. Consuming tomatoes with olive oil or salad dressing facilitates the absorption of lycopene [100–102]. The absorption of vitamins is negatively impacted by the intake of caffeine and vices like smoking, and alcoholic drinks [103]. Certain drugs are reported to decrease the levels of vitamin B, D, and C complex present in body cells. Certain diseases, such as 'celiac disease' with 'lactose intolerance', affect food digestion and limits uptake of vitamin. Retinol's bioavailability can be as high 90%, but people suffering from diarrhea and intestinal malady experience a lower absorption rate. Dietary fat and intestinal infections are also implicative in low absorption rate [100–102]. Calcium absorption in individuals depends highly on availability of substances like uronic acid, phytate, and oxalate in the food and the individual's vitamin D status [104, 105]. Heat process also causes a low availability of vitamin B_6 especially in milk [106].

Approximately 90% of dietary iron comes from plant sources (non-heme iron). However, because of its lower bioavailability, only about half of it is absorbed, and this absorption can vary based on an individual's iron levels as well as other foods consumed. Generally, the higher the amount of non-heme iron in the diet, the less efficiently it is absorbed [100, 107]. The most efficient absorption rates are observed with ferrous forms such as fumarate and sulfate due to their water solubility [98].

Conversely, the bioavailability of selenium from our diet depends not just on its absorption but also on its conversion into usable forms within our tissues, such as GSHPx or 5′-deiodinase. Selenium from sources like wheat, Brazil nuts, and beef kid-

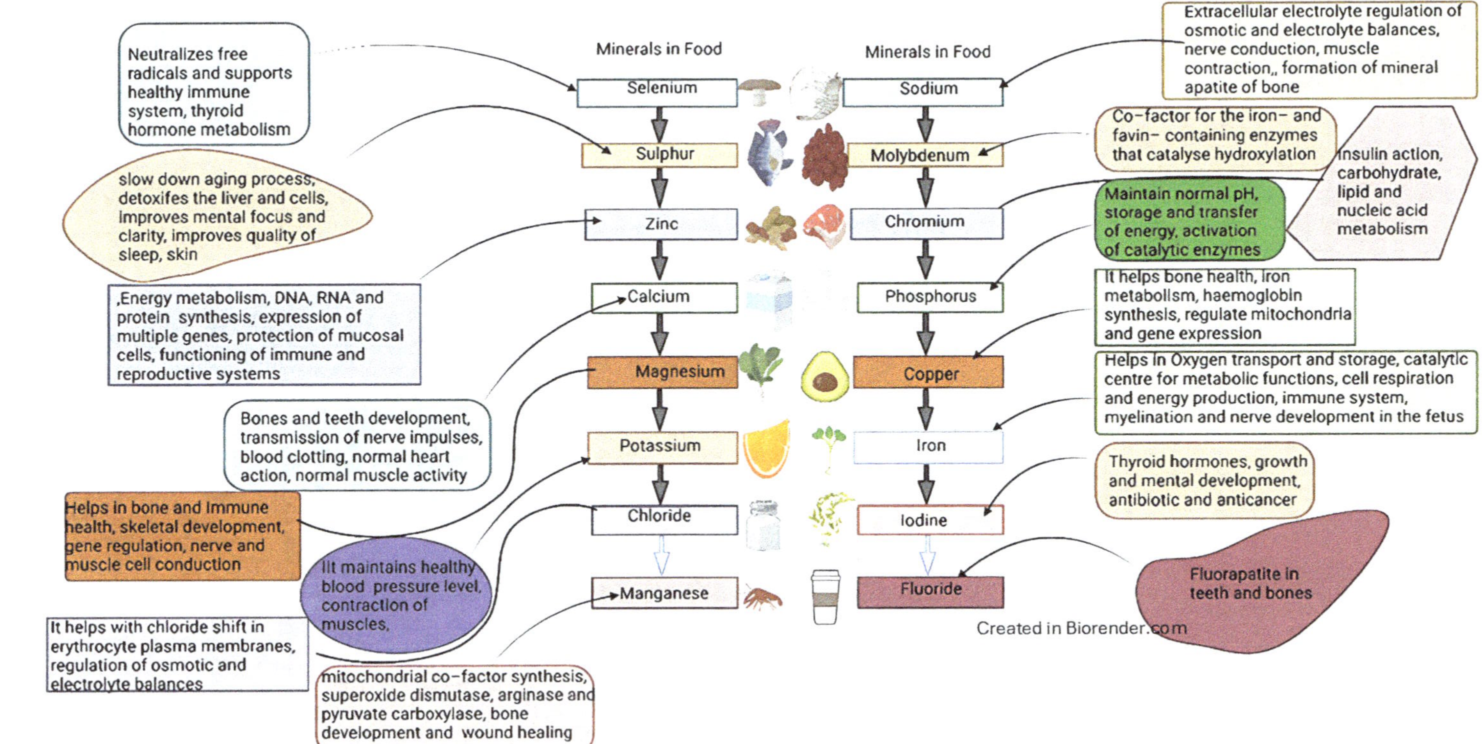

Figure 6.2: A schematic representation of minerals presents in foods and their functions. (here).

ney is highly bioavailable (80–90%), while selenium from tuna fish (as selenite) has lower bioavailability (20–60%) compared to other seafood like shrimp, crab, and Baltic herring [108]. Selenium in mushrooms is less accessible to rats [104]. Chromium's bioavailability can be enhanced by forming complexes with amino acids and other substances, but it can be reduced by zinc and iron. Excessive zinc intake can also negatively affect copper bioavailability [109, 110]. Mineral salts are not fully absorbed and can be lost during food preparation and processing [102].

In addition, protein bioavailability is affected by factors like heat treatment, oxidation, pH, and naturally occurring substances. Heating can form Maillard compounds, reducing lysine availability, while cysteine oxidation decreases tryptophan and threonine bioavailability. Substances like tannins, phytates, and trypsin inhibitors can form during food processing. Lectins, which can be toxic, may form in lentils during genetic modification [98, 111, 112]. Thermal treatment and homogenization can increase carotenoid bio-accessibility, but dietary fiber can reduce it [113, 114]. The food industry uses encapsulation to improve carotenoid bioavailability, though it varies widely due to the carotenoid's structure in the food matrix [115, 116].

In a separate study, it was observed that combining vegetable like broccoli with glucoraphanin powder could enhance the bioavailability of sulforaphane (a glucosinolate). This suggests that regular consumption of this nutrient dense vegetable enriched with glucoraphanin powder can reduce cancer risk more effectively than consuming glucoraphanin powder or sprouts alone [117]. According to an in vitro study, resveratrol shows low bioavailability, leading to insufficient plasma concentrations and limited distribution to target sites, thereby, reducing its pharmacological effects. However, combining it with other phytochemicals can enhance its bioactivity [118–119].

Conclusively, minerals have low bioavailability due to incomplete absorption. Plant-based foods, especially the ones rich in oxalate and phytate, hinder mineral uptake (Figure 6.3). Excess zinc limits iron and copper absorption, while vitamins like ascorbic acid and calciferol enhance mineral uptake. Fiber-rich diets also interfere with mineral bioavailability [118].

All this and more influences the research for better innovative measures for enhancing bioavailability and stability of food bioactive. Consequently, delivery systems for functional foods and nutraceuticals such as nanoencapsulation, microencapsulation and liquid encapsulation technologies [120, 121, 122] is on the rise to enhance the bioavailability and stability of bioactive compounds in functional foods and supplements. These technologies improve absorption rates and allow for controlled release, optimizing the products' efficacy. Technologies such as DNA testing, AI-driven algorithms, synthetic biology, spray-drying technology, 3D printing, nutrigenetics, omnibead technology, and ultrasol technology are set to revolutionize the future of nutraceuticals. Advances in these areas enable more personalized nutrition recommendations, as companies can now develop products tailored to individuals' genetic makeup, lifestyle factors, and health goals [123].

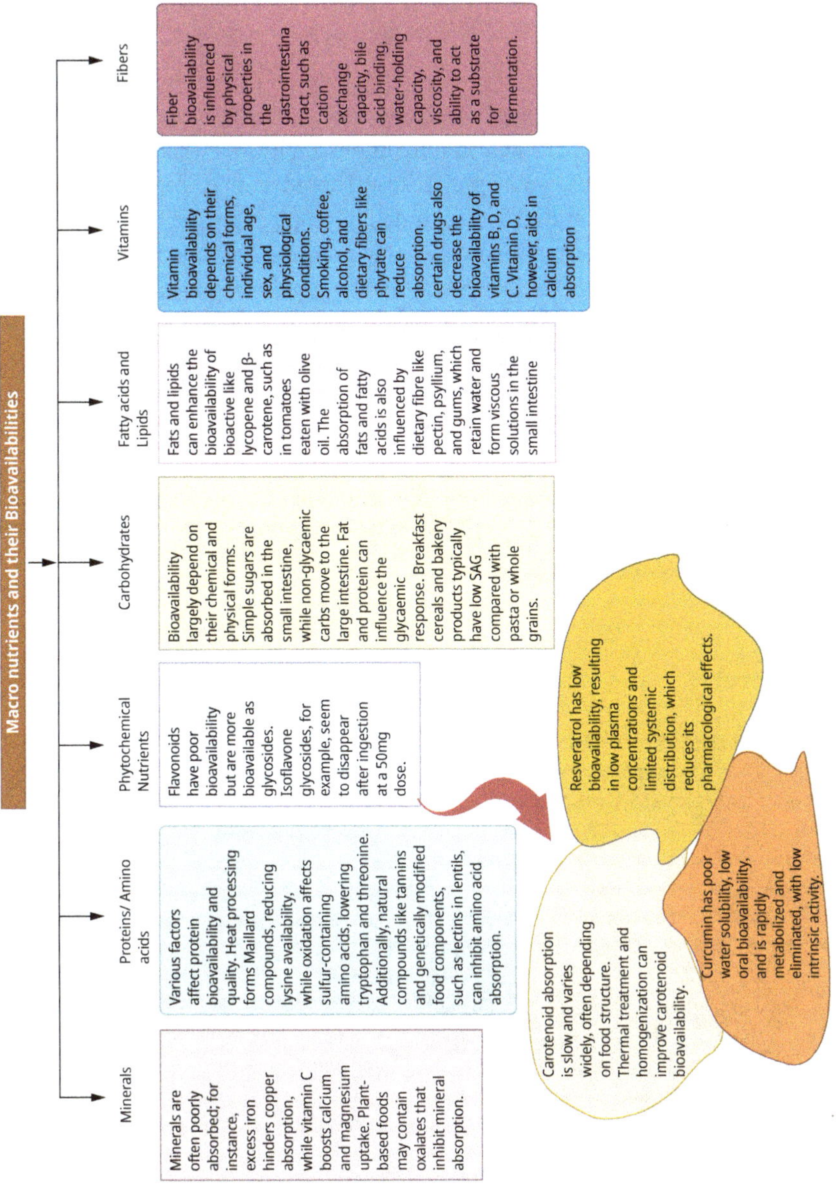

Macro nutrients and their Bioavailabilities
Minerals
Proteins/ Amino acids
Phytochemical Nutrients
Carbohydrates
Fatty acids and Lipids
Vitamins
Fibers
Minerals are often poorly absorbed; for instance, excess iron hinders copper absorption, while vitamin C boosts calcium and magnesium uptake. Plant-based foods may contain oxalates that inhibit mineral absorption.
Various factors affect protein bioavailability and quality. Heat processing forms Maillard compounds, reducing lysine availability, while oxidation affects sulfur-containing amino acids, lowering tryptophan and threonine. Additionally, natural compounds like tannins and genetically modified food components, such as lectins in lentils, can inhibit amino acid absorption.
Flavonoids have poor bioavailability but are more bioavailable as glycosides. Isoflavone glycosides, for example, seem to disappear after ingestion at a 50mg dose.
Bioavailability largely depend on their chemical and physical forms. Simple sugars are absorbed in the small intestine, while non-glycaemic carbs move to the large intestine. Fat and protein can influence the glycaemic response. Breakfast cereals and bakery products typically have low SAG compared with pasta or whole grains.
Fats and lipids can enhance the bioavailability of bioactive like lycopene and β-carotene, such as in tomatoes eaten with olive oil. The absorption of fats and fatty acids is also influenced by dietary fibre like pectin, psyllium, and gums, which retain water and form viscous solutions in the small intestine
Vitamin bioavailability depends on their chemical forms, individual age, sex, and physiological conditions. Smoking, coffee, alcohol, and dietary fibers like phytate can reduce absorption. certain drugs also decrease the bioavailability of vitamins B, D, and C. Vitamin D, however, aids in calcium absorption
Fiber bioavailability is influenced by physical properties in the gastrointestina tract, such as cation exchange capacity, bile acid binding, water-holding capacity, viscosity, and ability to act as a substrate for fermentation.
Resveratrol has low bioavailability, resulting in low plasma concentrations and limited systemic distribution, which reduces its pharmacological effects.
Carotenoid absorption is slow and varies widely, often depending on food structure. Thermal treatment and homogenization can improve carotenoid bioavailability.
Curcumin has poor water solubility, low oral bioavailability, and is rapidly metabolized and eliminated, with low intrinsic activity.

Figure 6.3: A schematic representation of macronutrients and their bioavailability.

6.5 Conclusion

Functional foods and nutraceuticals play significant roles in human health, offering a range of benefits from improved digestion and glucose tolerance to potential disease prevention. Further research is needed to fully understand the mechanisms of these benefits and to explore potential new applications. There is a need for more comprehensive studies to understand the structure–function relationships of food bioactives. To date, there has been no internationally accepted definition of functional foods and nutraceuticals, and no scientific consensus to distinguish between functional foods, nutraceuticals, and dietary supplements. Moreover, despite efforts to regulate, country-specific regulations and health claim substantiation are some of the challenges the functional food and nutraceutical market continues to face. It is therefore recommended that future research and development in functional foods and nutraceuticals should look into understanding structure–function relationships of food bioactives as they are crucial for the success of functional foods. In addition, ways to develop novel formulation and delivery techniques for bioactives to various target tissues, organs, and systems remain an important area for future research.

References

[1] Available from: https://en.wikipedia.org/wiki/Food.

[2] Nasri, H., Baradaran, A., Shirzad, H., & Rafieian-Kopaei, M. New concepts in nutraceuticals as alternative for pharmaceuticals. International Journal of Preventive Medicine 2014, 5(12), 1487.

[3] Ghosh, D., Das, S., Bagchi, D., & Smarta, R. B., editors. Innovation in Healthy and Functional Foods, New York, UK: Taylor and Francis group, 2013.

[4] Alamgir, A., & Alamgir, A. Classification of drugs, nutraceuticals, functional food, and cosmeceuticals; proteins, peptides, and enzymes as drugs. Therapeutic Use of Medicinal Plants and Their Extracts: Volume 1: Pharmacognosy 2017, 125–175.

[5] Wang, H., Liu, Y., Qi, Z., Wang, S., Liu, S., Li, X., et al. An overview on natural polysaccharides with antioxidant properties. Current Medicinal Chemistry 2013, 20(23), 2899–2913.

[6] Wang, H., Hong, T., Li, N., Zang, B., & Wu, X. Soluble dietary fiber improves energy homeostasis in obese mice by remodeling the gut microbiota. Biochemical and Biophysical Research Communications 2018, 498(1), 146–151.

[7] Kamhi, E., Joo, E. J., Dordick, J. S., & Linhardt, R. J. Glycosaminoglycans in infectious disease. Biological Reviews 2013, 88(4), 928–943.

[8] Wang, Q., & Ellis, P. R. Oat β-glucan: Physico-chemical characteristics in relation to its blood-glucose and cholesterol-lowering properties. British Journal of Nutrition 2014, 112(S2), S4–S13.

[9] Xu, J., Inglett, G. E., Chen, D., & Liu, S. X. Viscoelastic properties of oat β-glucan-rich aqueous dispersions. Food Chemistry 2013, 138(1), 186–191.

[10] Toivonen, R. K., Emani, R., Munukka, E., Rintala, A., Laiho, A., Pietilä, S., et al. Fermentable fibres condition colon microbiota and promote diabetogenesis in NOD mice. Diabetologia 2014, 57, 2183–2192.

[11] Uebelhack, R., Busch, R., Alt, F., Beah, Z.-M., & Chong, P.-W. Effects of cactus fiber on the excretion of dietary fat in healthy subjects: A double blind, randomized, placebo-controlled, crossover clinical investigation. Current Therapeutic Research 2014, 76, 39–44.
[12] Tester, R. F., & Al-Ghazzewi, F. H. Mannans and health, with a special focus on glucomannans. Food Research International 2013, 50(1), 384–391.
[13] Gupta, S., & Abu-Ghannam, N. Bioactive potential and possible health effects of edible brown seaweeds. Trends in Food Science & Technology 2011, 22(6), 315–326.
[14] Rajasulochana, P., Krishnamoorthy, P., & Dhamotharan, R. Biochemical investigation on red algae family of Kappahycus Sp. Journal of Chemical and Pharmaceutical Research 2012, 4(10), 4637–4641.
[15] Ueno, M., & Oda, T. Biological activities of alginate. Advances in Food and Nutrition Research 2014, 72, 95–112.
[16] Pangestuti, R., & Kim, S.-K. Biological activities of carrageenan. Advances in Food and Nutrition Research 2014, 72, 113–124.
[17] Zong, A., Cao, H., & Wang, F. Anticancer polysaccharides from natural resources: A review of recent research. Carbohydrate Polymers 2012, 90(4), 1395–1410.
[18] de Jesus Raposo, M. F., De Morais, R. M. S. C., & de Morais, A. M. M. B. Bioactivity and applications of sulphated polysaccharides from marine microalgae. Marine Drugs 2013, 11(1), 233–252.
[19] Shang, Q., Jiang, H., Cai, C., Hao, J., Li, G., & Yu, G. Gut microbiota fermentation of marine polysaccharides and its effects on intestinal ecology: An overview. Carbohydrate Polymers 2018, 179, 173–185.
[20] Udani, J. K., Singh, B. B., Barrett, M. L., & Singh, V. J., editors. Immunomodulatory effects of a proprietary arabinogalactan extract: A randomized, double-blind, placebo-controlled, parallel group study. In: 50th Annual American College of Nutrition Meeting, Resort Lake Buena Vista, Florida, 2009, October 1–4.
[21] Boler, B. M. V., & Fahey, G. C. Jr Prebiotics of plant and microbial origin. In: Direct-Fed Microbials and Prebiotics for Animals: Science and Mechanisms of Action, Springer, 2011, 13–26.
[22] Wang, Q., Wang, F., Xu, Z., & Ding, Z. Bioactive mushroom polysaccharides: A review on monosaccharide composition, biosynthesis and regulation. Molecules 2017, 22(6), 955.
[23] Synytsya, A., & Novák, M. Structural diversity of fungal glucans. Carbohydrate Polymers 2013, 92(1), 792–809.
[24] Zhang, Y., Li, S., Wang, X., Zhang, L., & Cheung, P. C. Advances in lentinan: Isolation, structure, chain conformation and bioactivities. Food Hydrocolloids 2011, 25(2), 196–206.
[25] Mink, P. J., Scrafford, C. G., Barraj, L. M., Harnack, L., Hong, C.-P., Nettleton, J. A., & Jacobs, D. R. Jr Flavonoid intake and cardiovascular disease mortality: A prospective study in postmenopausal women. The American Journal of Clinical Nutrition 2007, 85(3), 895–909.
[26] Qin, Y., Xia, M., Ma, J., Hao, Y., Liu, J., Mou, H., et al. Anthocyanin supplementation improves serum LDL-and HDL-cholesterol concentrations associated with the inhibition of cholesteryl ester transfer protein in dyslipidemic subjects. The American Journal of Clinical Nutrition 2009, 90(3), 485–492.
[27] Bertoia, M. L., Rimm, E. B., Mukamal, K. J., Hu, F. B., Willett, W. C., & Cassidy, A. Dietary flavonoid intake and weight maintenance: Three prospective cohorts of 124 086 US men and women followed for up to 24 years. BMJ 2016, 352.
[28] Azzini, E., Giacometti, J., & Russo, G. L. Antiobesity effects of anthocyanins in preclinical and clinical studies. Oxidative Medicine and Cellular Longevity 2017, 2017.
[29] Lin, B. W., Gong, C. C., Song, H. F., & Cui, Y. Y. Effects of anthocyanins on the prevention and treatment of cancer. British Journal of Pharmacology 2017, 174(11), 1226–1243.
[30] Grune, T., Lietz, G., Palou, A., Ross, A. C., Stahl, W., Tang, G., et al. β-Carotene is an important vitamin A source for humans. The Journal of Nutrition 2010, 140(12), 2268S–85S.
[31] Gul, K., Tak, A., Singh, A., Singh, P., Yousuf, B., & Wani, A. A. Chemistry, encapsulation, and health benefits of β-carotene-A review. Cogent Food & Agriculture 2015, 1(1), 1018696.

[32] Köpcke, W., & Krutmann, J. Protection from Sunburn with β-Carotene – A Meta-analysis. Photochemistry and Photobiology 2008, 84(2), 284–288.
[33] Stahl, W., & Sies, H. β-Carotene and other carotenoids in protection from sunlight. The American Journal of Clinical Nutrition 2012, 96(5), 1179S–84S.
[34] Sassa, S. Modern diagnosis and management of the porphyrias. British Journal of Haematology 2006, 135(3), 281–292.
[35] Rollman, O., & Vahlquist, A. Psoriasis and vitamin A: Plasma transport and skin content of retinol, dehydroretinol and carotenoids in adult patients versus healthy controls. Archives of Dermatological Research 1985, 278, 17–24.
[36] Cabrera, C., Artacho, R., & Giménez, R. Beneficial effects of green tea – A review. Journal of the American College of Nutrition 2006, 25(2), 79–99.
[37] Kim, A. C. A., Barone, M. K., Avino, D., Wang, F., Coleman, C. I., & Phung, O. J. Green tea catechins decrease total and low-density lipoprotein cholesterol: A systematic review and meta-analysis. Journal of the American Dietetic Association 2011, 111(11), 1720–1729.
[38] Khalesi, S., Sun, J., Buys, N., Jamshidi, A., Nikbakht-Nasrabadi, E., & Khosravi-Boroujeni, H. Green tea catechins and blood pressure: A systematic review and meta-analysis of randomised controlled trials. European Journal of Nutrition 2014, 53, 1299–1311.
[39] Brunetti, C., Di Ferdinando, M., Fini, A., Pollastri, S., & Tattini, M. Flavonoids as antioxidants and developmental regulators: Relative significance in plants and humans. International Journal of Molecular Sciences 2013, 14(2), 3540–3555.
[40] Banjarnahor, S. D., & Artanti, N. Antioxidant properties of flavonoids. Medical Journal of Indonesia 2014, 23(4), 239–244.
[41] Serafani, M., Peluso, I., & Raguzzini, A. Antioxidants and the immune system flavonoids as anti-inflammatory agents. Girona, Spanyol. Girona (SP): Nutrition Society Hlm 2009, Okt 21–24, 2010, 273–278.
[42] Pérez-Cano, F. J., & Castell, M. Flavonoids, Inflammation and Immune System, MDPI, 2016, 659.
[43] Johnson, E. J. The role of carotenoids in human health. Nutrition in Clinical Care 2002, 5(2), 56–65.
[44] Ugusman, A., Zakaria, Z., Chua, K. H., Megat Mohd Nordin, N. A., & Abdullah Mahdy, Z. Role of rutin on nitric oxide synthesis in human umbilical vein endothelial cells. The Scientific World Journal 2014, 2014.
[45] Ilic, D., & Misso, M. Lycopene for the prevention and treatment of benign prostatic hyperplasia and prostate cancer: A systematic review. Maturitas 2012, 72(4), 269–276.
[46] Wang, Y., Cui, R., Xiao, Y., Fang, J., & Xu, Q. Effect of carotene and lycopene on the risk of prostate cancer: A systematic review and dose-response meta-analysis of observational studies. PloS One 2015, 10(9), e0137427.
[47] Chen, P., Zhang, W., Wang, X., Zhao, K., Negi, D. S., Zhuo, L., et al. Lycopene and risk of prostate cancer: A systematic review and meta-analysis. Medicine 2015, 94(33).
[48] Rowles, J., Ranard, K., Smith, J., An, R., & Erdman, J. Increased dietary and circulating lycopene are associated with reduced prostate cancer risk: A systematic review and meta-analysis. Prostate Cancer and Prostatic Diseases 2017, 20(4), 361–377.
[49] Cheng, H. M., Koutsidis, G., Lodge, J. K., Ashor, A. W., Siervo, M., & Lara, J. Lycopene and tomato and risk of cardiovascular diseases: A systematic review and meta-analysis of epidemiological evidence. Critical Reviews in Food Science and Nutrition 2019, 59(1), 141–158.
[50] Ried, K., & Fakler, P. Protective effect of lycopene on serum cholesterol and blood pressure: Meta-analyses of intervention trials. Maturitas 2011, 68(4), 299–310.
[51] Senkus, K. E., Tan, L., & Crowe-White, K. M. Lycopene and metabolic syndrome: A systematic review of the literature. Advances in Nutrition 2019, 10(1), 19–29.

[52] Chen, J., Song, Y., & Zhang, L. Effect of lycopene supplementation on oxidative stress: An exploratory systematic review and meta-analysis of randomized controlled trials. Journal of Medicinal Food 2013, 16(5), 361–374.

[53] Serban, M. C., Sahebkar, A., Zanchetti, A., Mikhailidis, D. P., Howard, G., Antal, D., et al. Effects of quercetin on blood pressure: A systematic review and meta-analysis of randomized controlled trials. Journal of the American Heart Association 2016, 5(7), e002713.

[54] Shah, P. M., Priya, V. V., & Gayathri, R. Quercetin-a flavonoid: A systematic review. Journal of Pharmaceutical Sciences and Research 2016, 8(8), 878.

[55] Fogacci, F., Tocci, G., Presta, V., Fratter, A., Borghi, C., & Cicero, A. F. Effect of resveratrol on blood pressure: A systematic review and meta-analysis of randomized, controlled, clinical trials. Critical Reviews in Food Science and Nutrition 2019, 59(10), 1605–1618.

[56] Guardia, T., Rotelli, A. E., Juarez, A. O., & Pelzer, L. E. Anti-inflammatory properties of plant flavonoids. Effects of rutin, quercetin and hesperidin on adjuvant arthritis in rat. Il Farmaco 2001, 56(9), 683–687.

[57] Jadeja, R. N., & Devkar, R. V. Polyphenols in chronic diseases and their mechanisms of action. Polyphen Human Health Disease 2014, 1, 615–623.

[58] Ganeshpurkara, A., & Salujaa, A. Neuroprotective effect of Rutin. Saudi Pharmaceutical Journal 2017, 25(2), 149–164.

[59] Pu, F., Mishima, K., Irie, K., Motohashi, K., Tanaka, Y., Orito, K., et al. Neuroprotective effects of quercetin and rutin on spatial memory impairment in an 8-arm radial maze task and neuronal death induced by repeated cerebral ischemia in rats. Journal of Pharmacological Sciences 2007, 104(4), 329–334.

[60] Khan, M. M., Ahmad, A., Ishrat, T., Khuwaja, G., Srivastawa, P., Khan, M. B., et al. Rutin protects the neural damage induced by transient focal ischemia in rats. Brain Research 2009, 1292, 123–135.

[61] Javed, H., Khan, M., Ahmad, A., Vaibhav, K., Ahmad, M., Khan, A., et al. Rutin prevents cognitive impairments by ameliorating oxidative stress and neuroinflammation in rat model of sporadic dementia of Alzheimer type. Neuroscience 2012, 210, 340–352.

[62] Chauhan, B., Kumar, G., Kalam, N., & Ansari, S. H. Current concepts and prospects of herbal nutraceutical: A review. Journal of Advanced Pharmaceutical Technology & Research 2013, 4(1), 4.

[63] Ibrahim, S. R., El-Halawany, A. M., El-Dine, R. S., Mohamed, G. A., & Abdallah, H. M. Middle Eastern diets as a potential source of immunomodulators. In: Plants and Phytomolecules for Immunomodulation: Recent Trends and Advances, Springer, 2022, 163–190.

[64] Panda, H. Aloe Vera Handbook Cultivation, Research Finding, Products, Formulations, Extraction & Processing, Niir Project Consultancy Services, 2003.

[65] Razavi, B. M., Ghasemzadeh Rahbardar, M., & Hosseinzadeh, H. A review of therapeutic potentials of turmeric (Curcuma longa) and its active constituent, curcumin, on inflammatory disorders, pain, and their related patents. Phytotherapy Research 2021, 35(12), 6489–6513.

[66] Augusti, K. T. Therapeutic and Medicinal Values of Onions and Garlic. Onions and Allied Crops, CRC press, 2020, 93–108.

[67] Sadia Hassan, C. E., Tijjani, H., Jonathan, C., Ifemeje, M. C. O., Kingsley, C., Patrick-Iwuanyanwu, P. C. O., & Ephraim-Emmanuel, A. B. C. Dietary Supplements: Types, Health Benefits, Industry and Regulation. Functional Foods and Nutraceuticals, Switzerland 23 AG: Springer Nature, 2020, 23.

[68] Messias, M. C. F., Mecatti, G. C., Priolli, D. G., & De oliveira carvalho, P. Plasmalogen lipids: Functional mechanism and their involvement in gastrointestinal cancer. Lipids in Health and Disease 2018, 17, 1–12.

[69] Fahy, E., Subramaniam, S., Brown, H. A., Glass, C. K., Merrill, A. H., Murphy, R. C., et al. A comprehensive classification system for lipids1. Journal of Lipid Research 2005, 46(5), 839–861.

[70] Fahy, E., Cotter, D., Sud, M., & Subramaniam, S. Lipid classification, structures and tools. Biochimica et Biophysica Acta (BBA)-molecular and Cell Biology of Lipids 2011, 1811(11), 637–647.

[71] Song, C., Wu, M., Zhang, Y., Li, J., Yang, J., Wei, D., et al. Bioactive monomer and polymer polyketides from edible mushroom Cortinarius caerulescens as glutamate dehydrogenase inhibitors and antioxidants. Journal of Agricultural and Food Chemistry 2022, 70(3), 804–814.

[72] Martinez-Lacoba, R., Pardo-Garcia, I., Amo-Saus, E., & Escribano-Sotos, F. Mediterranean diet and health outcomes: A systematic meta-review. European Journal of Public Health 2018, 28(5), 955–961.

[73] FAO. Fats and fatty acids in human nutrition, Report of Joint FAO/WHO expert consultation. In: FAO, Geneva, 2010. Report No.: ISBN 978-92-5-106733-8.

[74] Villegas, A. S., & Sanchez-Taínta, A. The Prevention of Cardiovascular Disease through the Mediterranean Diet, Academic Press, 2017.

[75] Marcus, J. B. Culinary Nutrition: The Science and Practice of Healthy Cooking, Academic Press, 2013.

[76] Combet, E., & Buckton, C. Micronutrient deficiencies, vitamin pills and nutritional supplements. Medicine 2015, 43(2), 66–72.

[77] Paul, T., Halder, S. K., Das, A., Ghosh, K., Mandal, A., Payra, P., et al. Production of chitin and bioactive materials from Black tiger shrimp (Penaeus monodon) shell waste by the treatment of bacterial protease cocktail. 3 Biotech 2015, 5, 483–493.

[78] Nikitovic, D., Mytilinaiou, M., Berdiaki, A., Karamanos, N., & Tzanakakis, G. Heparan sulfate proteoglycans and heparin regulate melanoma cell functions. Biochimica et Biophysica Acta (Bba)-general Subjects 2014, 1840(8), 2471–2481.

[79] Lam, J., Truong, N. F., & Segura, T. Design of cell–matrix interactions in hyaluronic acid hydrogel scaffolds. Acta Biomaterialia 2014, 10(4), 1571–1580.

[80] Zhang, B., Xiao, W., Qiu, H., Zhang, F., Moniz, H. A., Jaworski, A., et al. Heparan sulfate deficiency disrupts developmental angiogenesis and causes congenital diaphragmatic hernia. The Journal of Clinical Investigation 2014, 124(1), 209–221.

[81] Zhao, W., Liu, W., Li, J., Lin, X., & Wang, Y. Preparation of animal polysaccharides nanofibers by electrospinning and their potential biomedical applications. Journal of Biomedical Materials Research Part A 2015, 103(2), 807–818.

[82] Sierra, S., Lara-Villoslada, F., Olivares, M., Jiménez, J., Boza, J., & Xaus, J. La expresión de IL-10 interviene en la regulación de la respuesta inflamatoria por los ácidos grasos omega 3. Nutrición Hospitalaria 2004, 19(6), 376–382.

[83] Leaf, A., Kang, J. X., & Xiao, Y.-F. Fish oil fatty acids as cardiovascular drugs. Current Vascular Pharmacology 2008, 6(1), 1–12.

[84] Gogos, C. A., Skoutelis, A., & Kalfarentzos, F. The effects of lipids on the immune response of patients with cancer. The Journal of Nutrition, Health & Aging 2000, 4(3), 172–175.

[85] Harris, W. S., & Isley, W. L. Clinical trial evidence for the cardioprotective effects of omega-3 fatty acids. Current Atherosclerosis Reports 2001, 3(2), 174–179.

[86] Gross, B. W., Gillio, M., Rinehart, C. D., Lynch, C. A., & Rogers, F. B. Omega-3 fatty acid supplementation and warfarin: A lethal combination in traumatic brain injury. Journal of Trauma Nursing 2017, 24(1), 15–18.

[87] Torres, J. P., Lin, Z., Winter, J. M., Krug, P. J., & Schmidt, E. W. Animal biosynthesis of complex polyketides in a photosynthetic partnership. Nature Communications 2020, 11(1), 2882.

[88] Olsen, N. Healthline [Internet]. Butler N, editor 2023. Available from: https://www.healthline.com/health/selenium-foods.

[89] Levy, J. Dr Axe [Internet] 2022. Available from: https://draxe.com/nutrition/foods-high-in-sulfur/.

[90] Marissa, F. The grosery aisle. 12 functional food & beverage trends. Functional-Foods.jpg June 28, 2023. Available on: https://www.wholefoodsmagazine.com/articles/16208-12-functional-food-and-beverage-trends (accessed on 28th July, 2024).

[91] Wood, R. Bioavailability: Definition, general aspects and fortificants. In: Encyclopedia of Human Nutrition, 2nd ed, Oxford: Elsevier Ltd., 2005.

[92] Sharma, K., Tayade, A., Singh, J., & Walia, S. Bioavailability of nutrients and safety measurements. Functional Foods and Nutraceuticals: Bioactive Components, Formulations and Innovations 2020, 543–593.
[93] Pressman, P., Clemens, R. A., & Hayes, A. W. Bioavailability of micronutrients obtained from supplements and food: A survey and case study of the polyphenols. Toxicology Research and Application 2017, 1, 2397847317696366.
[94] Gliszczynska-Swiglo, A., & Tyrakowska, B. Quality of commercial apple juices evaluated on the basis of the polyphenol content and the TEAC antioxidant activity. Journal of Food Science 2003, 68(5), 1844–1849.
[95] Mirajkar, K. K. Nutrova [Internet] 2021.
[96] Agrawal, S., & Goel, R. K. Curcumin and its protective and therapeutic uses. National Journal of Physiology, Pharmacy and Pharmacology 2016, 6(1), 1.
[97] Cavaleri, F., & Jia, W. The true nature of curcumin's polypharmacology. Journal of Preventive Medicine 2017, 2(5).
[98] HC, S., Pretorius, B., & Hall, N. Bioavailability of nutrients. In: Caballero, B., Finglas, P., & Toldrá, F., editors. The Encyclopedia of Food and Health, vol. 1, 2016, 401–406.
[99] Ravisankar, P., Reddy, A. A., Nagalakshmi, B., Koushik, O. S., Kumar, B. V., & Anvith, P. S. The comprehensive review on fat soluble vitamins. IOSR Journal of Pharmacy 2015, 5(11), 12–28.
[100] Gibney, M. J., Lanham-New, S. A., Cassidy, A., & Vorster, H. H. Introduction to Human Nutrition, John Wiley & Sons, 2013.
[101] Mahan, L. K. Krause's Food & the Nutrition Care Process-E-Book, Elsevier Health Sciences, 2016.
[102] Human Nutrition [Press Release], Honolulu: University of Hawaii at Manoa Food Science and Human Nutrition Program, 2018.
[103] Aslam, F., Muhammad, S. M., Aslam, S., & Irfan, J. A. Vitamins: Key role players in boosting up immune response-a mini review. Vitamins & Minerals 2017, 6(01).
[104] Khokhar, S., Garduño-Diaz, S., Marletta, L., Shahar, D., Ireland, J., Jansen-van Der Vliet, M., & De Henauw, S. Mineral composition of commonly consumed ethnic foods in Europe. Food & Nutrition Research 2012, 56(1), 17665.
[105] Allen, L. H. Calcium bioavailability and absorption: A review. The American Journal of Clinical Nutrition 1982, 35(4), 783–808.
[106] Da silva, V. R., Russell, K. A., & Gregory, J. F. III Vitamin B6. Present Knowledge in Nutrition 2012, 307–320.
[107] Monsen, E. R. Iron nutrition and absorption: Dietary factors which impact iron bioavailability. Journal of the American Dietetic Association 1988, 88(7), 786–790.
[108] Contempre, B., Le Moine, O., Dumont, J. E., Denef, J.-F., & Many, M.-C. Selenium deficiency and thyroid fibrosis. A key role for macrophages and transforming growth factor β (TGF-β). Molecular and Cellular Endocrinology 1996, 124(1–2), 7–15.
[109] Wapnir, R. A. Copper absorption and bioavailability. The American Journal of Clinical Nutrition 1998, 67(5), 1054S–60S.
[110] Platel, K., & Srinivasan, K. Bioavailability of micronutrients from plant foods: An update. Critical Reviews in Food Science and Nutrition 2016, 56(10), 1608–1619.
[111] Gilani, G. S., Xiao, C. W., & Cockell, K. A. Impact of antinutritional factors in food proteins on the digestibility of protein and the bioavailability of amino acids and on protein quality. British Journal of Nutrition 2012, 108(S2), S315–S32.
[112] Ribarova, F. Amino Acids: Carriers of nutritional and biological value foods. In: Food Processing for Increased Quality and Consumption, Elsevier, 2018, 287–311.
[113] Cilla, A., Bosch, L., Barberá, R., & Alegría, A. Effect of processing on the bioaccessibility of bioactive compounds–A review focusing on carotenoids, minerals, ascorbic acid, tocopherols and polyphenols. Journal of Food Composition and Analysis 2018, 68, 3–15.

[114] Fernández-García, E., Carvajal-Lérida, I., Jarén-Galán, M., Garrido-Fernández, J., Pérez-Gálvez, A., & Hornero-Méndez, D. Carotenoids bioavailability from foods: From plant pigments to efficient biological activities. Food Research International 2012, 46(2), 438–450.
[115] Afreen, A., Ahmed, Z., & Anjum, N. Novel nutraceutical compounds. In: Therapeutic, Probiotic, and Unconventional Foods, Elsevier, 2018, 201–226.
[116] Boonlao, N., Ruktanonchai, U. R., & Anal, A. K. Enhancing bioaccessibility and bioavailability of carotenoids using emulsion-based delivery systems. Colloids and Surfaces B: Biointerfaces 2022, 209, 112211.
[117] Cramer, J. M., Teran-Garcia, M., & Jeffery, E. H. Enhancing sulforaphane absorption and excretion in healthy men through the combined consumption of fresh broccoli sprouts and a glucoraphanin-rich powder. British Journal of Nutrition 2012, 107(9), 1333–1338.
[118] Wang, S., Su, R., Nie, S., Sun, M., Zhang, J., Wu, D., & Moustaid-Moussa, N. Application of nanotechnology in improving bioavailability and bioactivity of diet-derived phytochemicals. The Journal of Nutritional Biochemistry 2014, 25(4), 363–376.
[119] De Vries, K., Strydom, M., & Steenkamp, V. Bioavailability of resveratrol: Possibilities for enhancement. Journal of Herbal Medicine 2018, 11, 71–77.
[120] Vesely, O., Baldovska, S., & Kolesarova, A. Enhancing bioavailability of nutraceutically used resveratrol and other stilbenoids. Nutrients 2021, 13(9), 3095.
[121] Mark-Herbert, C. Innovation of a new product category – functional foods. Technovation 2004, 24(9), 713–719.
[122] Weiss, J., Takhistov, P., & McClements, D. J. Functional materials in food nanotechnology. Journal of Food Science 2006, 71(9), R107–R16.
[123] BaK, V. The Future of Tech in Nutraceuticals Industry [Internet]. India 2023. Podcast. Available from: https://www.worldpharmatoday.com/white-papers/the-future-of-tech-in-nutraceuticals-industry/#.

Melania Dandadzi, Tinotenda Nhovoro, Morleen Muteveri, and Faith Matiza Ruzengwe*

Chapter 7 Food microbiome and related biological systems: metabolism for food and health enhancement

Abstract: The chapter explores the relationship between the food microbiome, related biological systems, and overall human health. The focus is on how the complex ecosystem of microorganisms influences metabolism and, ultimately, human health. The composition of the gut microbiome is significantly influenced by the food microbiome, which plays a crucial role in digestion, nutrient absorption, immune function, and overall health.

Key components like prebiotics, probiotics, and synbiotics are key to modulating the gut microbiome and promoting beneficial bacterial growth. Personalized nutrition based on the individual gut microbiome profile can improve human health. Food microbiome composition and dynamics are influenced by processing and preservation techniques such as fermentation. There is a need for advancements in microbiome engineering, precision fermentation, and bacteriophage engineering to manipulate further and enhance the food microbiome for improved health outcomes.

Keywords: Food microbiome, metabolism, biological systems, prebiotics, probiotics, food processing

7.1 Introduction

Microbiomes are various microorganisms that live in a functional relationship with their specific environments. These include bacteria, viruses, fungi, and archaea [1]. This diverse, complex community can be present on or within food products. Mi-

*Corresponding author: Faith Matiza Ruzengwe, Department of Livestock, Wildlife and Fisheries, Gary Magadzire School of Agriculture and Engineering, Great Zimbabwe University, P.O. Box 1235, Masvingo, Zimbabwe, e-mail: faithruzengwe@gmail.com
Melania Dandadzi, Department of Food Science and Technology, Chinhoyi University of Technology, P. Bag 7724, Chinhoyi, Zimbabwe
Tinotenda Nhovoro, Department of Livestock, Wildlife and Fisheries, Gary Magadzire School of Agriculture and Engineering, Great Zimbabwe University, P.O. Box 1235, Masvingo, Zimbabwe
Morleen Muteveri, Department of Physics, Geography and Environmental Science, School of Natural Sciences, Great Zimbabwe University, P.O. Box 1235, Masvingo, Zimbabwe

https://doi.org/10.1515/9783111441238-008

crobes are key in food safety, food preservation, bioavailability of micronutrients, flavor and taste development. The gut microbiome refers to the entire microbial community of more than 100 trillion microbial cells (or 10^{13}–10^{14}) populating the mammalian gastrointestinal (GI) tract [2, 3]. Most of these microbes reside in the colon (3.8×10^{13} microorganisms). The human GI tract can be colonized by 1150 species of microbes, and at least 160 species can be hosted by each individual [2]. There are five significant phyla for the human gut microbiota, namely, Firmicutes, Bacteroidetes, Actinobacteria, Proteobacteria, and Verrucomicrobe, with the two dominating of these, Firmicutes and Bacteroidetes, representing 90% of the gut microbiota [4]. The diet affects the composition and function of the gut microbiota and directly affects the host homeostasis and biological processes [5]. As such, there has been increased interest beyond the gut, including the food microbiome and the diverse microbial populations associated with our food. Thus, links between the microbes in the food and within the human body have been revealed and shown to impact an individual's health [6].

A diet abundant in fiber, fruits, vegetables, and whole grains fosters the proliferation of beneficial microbiota. In contrast, a high intake of processed foods disrupts the equilibrium of the microbiome, resulting in dysbiosis. Nutrients for the microbiome can be obtained from fibers (oatmeal and beans), phenols (blueberries, red peppers, and purple cabbage), fermented foods (sauerkraut, kimchu, yoghurt, and kombucha) and healthy fats. The microbes break down food components, extracting nutrients and influencing metabolism [6]. These microbes work together to break down food components like carbohydrates, proteins, and fibers that the human body fails to digest independently. Through fermentation, gut microbes produce short-chain fatty acids (SCFAs) like acetate, propionate, and butyrate. These SCFAs play a vital role in metabolism by:

- **Fueling gut cells:** provide energy for intestinal epithelial cells, strengthening the mucosal barrier, thus promoting gut health
- **Regulating blood sugar:** enhance insulin sensitivity to help regulate blood sugar levels.
- **Reducing inflammation:** It has anti-inflammatory properties, potentially reducing the risk of metabolic disorders [3].

Gut microbes are crucial in extracting nutrients from food by breaking down complex carbohydrates the human body cannot digest into simpler forms readily absorbed by the gut. Certain gut bacteria can synthesize essential vitamins, such as vitamins K and B, thereby contributing to the host's nutritional status. By modulating the gastrointestinal environment, gut microbiota can significantly influence the absorption of essential minerals (e.g., calcium and iron). When present in appropriate proportions, various strains of gut bacteria play a vital role in promoting human health [2, 3]. The short-chain fatty acids decrease inflammation, enhance cognitive function, regulate metabolism and mood, and support the immune system [6].

7.2 The food microbiome: a complex ecosystem

The food microbiome is a dynamic and complex ecosystem made up of many trillions of microorganisms. These include bacteria, fungi, archaea, and viruses, present in the food [7]. This diverse community of indigenous and native microorganisms that reside on a food product is acquired from the environment during harvest, processing, and storage. The archaea are less abundant, although they are significant members, including methanogens such as *Methanobrevibacter smithii.* Fungi (yeasts and molds) are also prevalent members of the food microbiome, and viruses that infect bacteria (bacteriophages) can influence bacterial populations and gene transfer. The community can be composed of beneficial and pathogenic microbiota influencing the microbiome differently. The beneficial bacteria, like lactic acid (LAB), could potentially benefit the food's microbial safety while, the pathogenic members, such as *Escherichia coli* or *Listeria monocytogenes*, can harm human health [8, 9]. The food microbiome exhibits considerable dynamism, with its composition being affected by factors, such as temperature, salinity, pH, water activity, nutrient composition, and interactions with other organisms [10]. A change in any one of these parameters can result in a shift in the overall population proportions. Furthermore, these dynamics can be influenced by adding new microorganisms. In natural habitats, such as the food environment, microbes are not found as isolated cells; instead, they exist within complex assemblages. In these contexts, species interactions and communication significantly influence population dynamics and functional activities [11]. Food products have several microenvironments, making the existence of various microorganisms possible, and the spatial and temporal distribution of these microorganisms in food follows a heterogeneous pattern.

A food microbiome comprises an ecological community of microorganisms interacting at different levels. The commensal, symbiotic, and pathogenic microorganisms engage in various interactions that significantly influence microbial fitness, population dynamics, and functional capabilities within the microbiome [12, 13]. The interactions among microbiota may occur between same or different species, genera, families, and other domains of life, and these microbial interactions can be positive, negative, and neutral. Positive interactions encompass mutualism, synergism, and commensalism, whereas negative microbial interactions include amensalism, parasitism, predation, antagonism, and competition. Neutral interactions are characterized by the absence of observable effects on the interacting species' functionality, capabilities, or fitness [13]. The key components of a food microbiome include prebiotics and synbiotics.

7.3 Food as a tool for microbiome modulation

7.3.1 Prebiotics

The Global Prebiotic Association (GPA) provides a clear definition of prebiotics. According to the GPA, prebiotics are ingredients that undergo selective fermentation in the gastrointestinal microbiota, leading to distinct alterations in composition and/or activity. These changes benefit the host's overall health [14]. It is essential to acknowledge that the human body does not metabolize prebiotics; instead, they serve as substrates for beneficial gut bacteria. By altering these microbes' types and/or activities, prebiotics promote the dominance of bacteria that promote good health. Prebiotics ultimately improve digestion, immunity, and metabolism [15].

Prebiotics must meet the following criteria:

- Resistance to gastric acidity, susceptibility to hydrolysis by mammalian enzymes, and mechanisms of gastrointestinal absorption: Prebiotics should not be broken down or absorbed by the human digestive system, ensuring that they remain intact until they reach the targeted site (large intestine) [16].
- Fermentation by intestinal microbiota: Prebiotics must serve as a food source for beneficial gut bacteria, which will promote their growth and/or activity [17].
- Selective stimulation of the growth and/or activity of intestinal microbiota linked to health and well-being: Prebiotics should specifically enhance the proliferation and/or activity of beneficial microbial populations, such as bifidobacteria and lactobacilli, while simultaneously inhibiting the growth of pathogenic bacteria [18].

7.3.1.1 Major types and sources of prebiotics

7.3.1.1.1 Fructans

Fructans are a class of prebiotics characterized by chains of fructose molecules linked together. These compounds occur naturally in various plant sources, including chicory root, Jerusalem artichoke, garlic, onions, leeks, asparagus, bananas, and wheat [19]. The commonly used fructans as prebiotics are described.

Inulin is a form of soluble fiber naturally found in various plants, including chicory root, Jerusalem artichoke, garlic, onions, and leeks. It is recognized for its capacity to promote the proliferation of beneficial bifidobacteria [20].

Fructooligosaccharides (FOS): FOS consists of shorter chains of fructose molecules than inulin. It is naturally found in onions, garlic, bananas, asparagus, and wheat. FOS is also commercially produced and added to foods as a prebiotic sweetener [21]. **Oligofructose**: Oligofructose is a type of FOS often commercially produced from inulin [22].

7.3.1.1.2 Galactooligosaccharides (GOS)

GOS are complex sugars synthesized from lactose. They are commonly found in dairy products and infant formula and promote the growth of beneficial *bifidobacteria* and *lactobacilli* in the gut [23].

7.3.1.1.3 Resistant starch

Resistant starch is not digested in the small intestine. Instead, it passes into the large intestine, where it acts as a substrate for beneficial bacteria [24]. Resistant starch is present in unripe bananas, cooked and cooled potatoes, legumes, and certain whole grains.

7.3.1.1.4 Other prebiotic fibers

Additional prebiotic fibers include pectin, a soluble fiber found in apples and citrus fruits, and beta-glucan, which is present in oats and barley. Human milk oligosaccharides (HMOs) are complex sugars exclusive to human milk and are considered crucial prebiotics for infants, playing a significant role in shaping their developing gut microbiome [25].

7.3.1.2 Impact of prebiotics on beneficial gut bacteria

Prebiotics are important in modulating the gut microbiome by selectively enhancing growth and activity of beneficial bacterial populations [26]. Unlike digestible carbohydrates broken down in the small intestine, they remain undigested until they are in large intestines. This allows specific bacterial groups, such as *bifidobacteria* and *lactobacilli*, to thrive, as they possess the necessary enzymes to ferment these complex carbohydrates. The fermentation process nourishes growth and proliferation of these bacteria and produces various metabolites that significantly impact gut health and overall well-being.

Prebiotics maintain their structure as they transit to the large intestine, where they act as a selective nutrient source for beneficial bacteria, including *bifidobacteria* and *lactobacilli*.

These beneficial bacteria ferment the prebiotics, increasing their numbers and dominance in the gut microbiome. The fermentation of prebiotics by beneficial bacteria produces Short Chain Fatty Acids (SCFAs), including butyrate, propionate, and acetate. SCFAs play a crucial role in gut health by performing the following functions:

- Nourishing the cells lining the colon: Among the SCFAs, butyrate is especially important as it is the primary energy source for these cells, promoting the integrity of the gut barrier.
- Reducing inflammation: SCFAs possess anti-inflammatory properties, which can help protect against inflammatory bowel diseases and other related conditions.

- Regulating gene expression: SCFAs can influence gene expression in gut cells, affecting various physiological processes, including immunity and metabolism.

Additionally, prebiotics play a significant role in indirectly enhancing the gut barrier by fostering the proliferation of beneficial bacteria synthesizing SCFAs. A robust gut barrier is a protective mechanism, preventing the translocation of dangerous substances into the bloodstream. This phenomenon subsequently mitigates inflammation and bolsters overall immune function. By augmenting the population of beneficial bacteria, prebiotics establish an environment that is less favorable for the proliferation of pathogenic bacteria. This reduces their ability to colonize the gut and cause infections. Prebiotics again indirectly influence the immune system by interacting with immune cells and modulating the production of immune signaling molecules. This modulation promotes a balanced immune response, mitigating the risk of allergies and autoimmune disorders.

7.3.2 Probiotics

Probiotics are live microorganisms that confer health benefits to the host organism when ingested in the correct quantities. These advantageous microbes, predominantly comprising bacteria and yeasts, are naturally present in fermented foods such as yoghurt, sauerkraut, and kimchi. Additionally, they are available as dietary supplements [27]. Probiotics are classified according to their genus, species, and strain. The two most frequently used genera in probiotic products are *Lactobacillus* and *Bifidobacterium*, although other bacteria and yeasts may also be utilized [28]. Below is a more comprehensive overview of some commonly found types of probiotics.

Lactobacillus species are prevalent in different fermented foods, such as yoghurt, kimchi, sauerkraut, and kefir [29]. They maintain gastrointestinal health and confer numerous benefits. *Lactobacillus acidophilus* supports digestive health, reduces irritable bowel syndrome (IBS) symptoms, and boosts immunity [30]. *Lactobacillus rhamnosus* GG: Prevents and treats diarrhea, particularly in children [31] and may prevent allergies and eczema. *Lactobacillus plantarum*: Improves gut barrier function, reduces inflammation, and enhances immune responses [32]. *Lactobacillus reuteri*: Reduces inflammation, assists with colic in infants, and may aid in diarrhea management.

Bifidobacterium species form another significant group of probiotic bacteria. They are present in high numbers in the gut microbiota of breastfed infants, for the gut's early development and the immune system's functioning. Several commonly known *Bifidobacterium* strains include *Bifidobacterium lactis*, which alleviates constipation, improves lactose intolerance and modulates immune responses. *Bifidobacterium longum*: Improves gut barrier function, reduces inflammation, and may benefit mental health. *Bifidobacterium bifidum*: Supports digestion, strengthens the immune system, and may help prevent infections.

Other bacteria include *Streptococcus thermophilus,* which is found in yoghurt and cheese, aids in lactose digestion, and may boost the immune system [33]. *Bacillus coagulans:* Reduces inflammation, improves gut health, and may help with IBS and rheumatoid arthritis [34]. Yeasts such as *Saccharomyces boulardii,* a unique yeast that prevents and treats antibiotic-associated diarrhea and reduces the risk of *Clostridioides difficile* infection [35].

7.3.3 Synbiotics

Synbiotics represent a synergistic combination of probiotics and prebiotics collaboratively functioning to enhance gastrointestinal health. This combination offers a superior approach to promoting gut health compared to using probiotics or prebiotics individually [36]. In a synbiotic product, prebiotics selectively nourishes the probiotic bacteria. This specific nourishment aids the survival of the probiotics in the harsh, acidic environment of the stomach and small intestine, enabling them to reach the large intestine in higher quantities. Once in the large intestine, the prebiotics further support growth and activity of the probiotic bacteria, resulting in a healthier and more diverse gut microbiota [18].

The prebiotic component of a synbiotic serves as a food source for the probiotic bacteria, promoting growth and activity in the gut. This results in a higher concentration of beneficial bacteria and a more favorable gut environment. Prebiotics aid in survival of probiotic bacteria against harsh conditions of the stomach and small intestine, enabling them to reach the large intestine and colonize effectively. The synergistic impact of probiotics and prebiotics may lead to enhanced health benefits compared to using either alone. These benefits may include improved digestion and gut health, enhanced immune function, reduced inflammation, and prevention and treatment of certain diseases (e.g., diarrhea, irritable bowel syndrome) [37].

Personalized nutrition, informed by an individual's gut microbiome composition, represents an emerging field that seeks to tailor dietary recommendations to the distinct microbial profile of each person. This concept acknowledges that everyone's gut microbiome is distinct, similar to a fingerprint, and reacts differently to different foods and nutrients. By analyzing an individual's gut microbiome composition, personalized nutrition seeks to enhance health outcomes by offering dietary advice that aligns with their gut microbiota's specific requirements and characteristics [38]. Personalized nutrition can be achieved by conducting a microbiome analysis followed by personalized recommendations. The process typically begins with analyzing a stool sample using DNA sequencing technology. This allows for identifying and quantifying the diverse bacterial populations in the gut. The analysis provides a detailed breakdown of the individual's gut microbiome composition, including the following aspects:

- **Diversity**: This measures the various bacterial species in the gut. A higher diversity is generally considered beneficial for health.
- **Relative Abundance:** This refers to the proportions of different bacterial species. Some species are beneficial, while others may be harmful. The analysis can reveal imbalances or dysbiosis in the gut microbiome.
- **Specific Bacterial Strains**: In some cases, the analysis can identify specific strains of bacteria associated with certain health conditions. This information can be used to develop more targeted dietary recommendations.

Researchers are also investigating alternative non-invasive techniques for analyzing the gut microbiome, including breath tests and analyzing microbial metabolites in stool samples. Nevertheless, DNA sequencing analysis of stool samples remains the current benchmark.

After conducting a microbiome analysis, personalized dietary recommendations are developed. These recommendations may specify which foods to include or avoid and suggest prebiotic or probiotic supplements. The goal is to promote the growth and activity of beneficial bacteria while inhibiting harmful ones. This way, gut health can be optimized, improving overall well-being [38].

7.4 Food microbiome and health enhancement

Food components in the human diet supply the essential nutrients required by the body and various substrates that support the mutualistic microbial flora in the gastrointestinal tract, referred to as the gut microbiome [3]. Undigested food components are metabolized to a diverse array of metabolites. Undigested food components are metabolized into a diverse variety of metabolites. The diet significantly influences the gut microbiome's structure, composition, and function, which interacts with the intestinal epithelium and the mucosal immune system, thereby maintaining intestinal homeostasis in a healthy state. Dysbiosis, or shifts in the gut microbiome, has been implicated in various diseases, including inflammatory bowel disease (IBD), cancers, type I and type II diabetes, obesity, allergies and neurological disorders [39–41]. There has been a growing interest in targeted nutritional interventions for the gut microbiome in IBD. Studies on dietary effects on the composition and patterns in the gut microbiome have increased in the past decade [42, 43]. Understanding the impact of significant food components and the interactions between the gut microbiome and the host immune system has advanced our comprehension of how diet influences the interaction between these two entities [44–47]. Modulation of the food microbiome may help prevent weight gain, diabetes, heart diseases, and risk of cancer.

7.5 Food processing and preservation techniques to enhance the food microbiome

Food processing and preservation methods can significantly affect food's microbiome either positively or negatively. Using specific techniques can increase the quantities and effectiveness of helpful microorganisms, resulting in better food quality, safety, and potential health advantages for consumers.

7.5.1 Techniques to enhance the food microbiome

7.5.1.1 Fermentation

Fermentation, an ancient technique, involves the controlled growth of beneficial microorganisms such as bacteria, yeasts, or molds. These microorganisms convert carbohydrates into acids, alcohols, or other compounds. The process of fermentation not only preserves food by establishing an acidic environment that suppresses the proliferation of spoilage and pathogenic bacteria, but it also enriches the food with a variety of beneficial byproducts, including probiotics [48]. Additionally, fermentation produces enzymes that break down complex molecules in food, making it easier to digest and absorb nutrients. Moreover, fermentation contributes to fermented food's sensory characteristics and overall quality by creating organic acids, volatile flavor compounds, and other bioactive substances. Examples of fermented foods include yoghurt, sauerkraut, kimchi, and kombucha.

7.5.1.2 Controlled spoilage

Controlled spoilage entails deliberately manipulating environmental conditions to foster the multiplication of specific beneficial microorganisms while simultaneously suppressing the growth of spoilage-causing or pathogenic microbes. This can be done by adjusting temperature, humidity, and oxygen levels or adding starter cultures. Examples of this approach include the production of certain cheeses, dry-cured meats, and fermented sausages.

7.5.1.3 High-pressure processing

High-pressure processing utilizes high pressure to deactivate harmful bacteria while maintaining food quality and nutritional benefits. This non-thermal method can also selectively promote the growth of beneficial microorganisms, such as lactic acid bacteria, which contribute to a healthier food microbiome.

7.5.1.4 Pulsed electric field

Pulsed electric field (PEF) is another non-thermal preservation technique. It works by using short, high-voltage pulses to disrupt the cell membranes of microorganisms, effectively rendering them inactive. By employing PEF, it is possible to selectively preserve beneficial microbes while simultaneously inactivating spoilage or pathogenic bacteria. This process ultimately enhances food safety and quality.

7.5.1.5 Ultraviolet irradiation

Ultraviolet light can be utilized to deactivate surface microorganisms present in food products, thereby reducing the likelihood of spoilage and foodborne illnesses. However, it is crucial to exercise meticulous control to prevent any detrimental effects on the nutritional quality of the food and to safeguard beneficial microbes selectively.

7.5.1.6 Modified atmosphere packaging

Modified atmosphere packaging (MAP) is a method used to maintain the quality and freshness of a food product by altering the composition of gases within the packaging. MAP effectively inhibits the proliferation of pathogenic microorganisms responsible for spoilage and disease by modulating the concentrations of oxygen, carbon dioxide, and nitrogen. Furthermore, this technique can establish an environment that promotes the proliferation of beneficial bacteria.

7.5.1.7 Hurdle technology

Uses multiple gentle preservation techniques to achieve a synergistic effect ensures food safety and quality while minimizing any adverse impact on the microbiome. By utilizing various methods such as mild heat treatment, pH adjustment, and reducing water activity, it becomes possible to control manage the growth of microorganisms effectively and selectively promote the development of beneficial ones.

By utilizing food processing and preservation techniques, there is the ability to develop a new line of food products that are safe and nutritious and contribute to a healthier gut microbiome. Enhancing the food microbiome can be achieved through a range of techniques, including both traditional and modern approaches. The goal is to support beneficial microorganisms' growth while preventing harmful ones' proliferation. These techniques can be implemented at various stages, such as during food processing, preservation, or even at the agricultural level [48].

7.6 Applications of the food microbiome in food technology and nutrition

7.6.1 Development of novel food and beverages with enhanced nutritional and health benefits

The dramatic increase in chronic metabolic diseases resulting from an imbalance in gut microbiota has caused a growing interest in developing novel food products that target gut health or as therapy for treating some metabolic disorders [50]. Knowledge of how the trillions of microbes in the gut interact with the food consumed has led researchers to create exciting possibilities for personalized nutrition and improving overall well-being. Dietary components supply essential nutrients as well as substrates for the mutualistic microbial flora residing in the gastrointestinal tract. In a state of balance, these microbes contribute nutrients and energy to the host through the fermentation of indigestible dietary components within the large intestine, thereby maintaining equilibrium with the host's metabolic processes and immune system [50]. However, in the presence of dysbiosis, microorganisms can serve as sources of inflammation and infection, contribute to gastrointestinal diseases, and potentially play a role in the development of diabetes mellitus and obesity. Fermented foods, fibers, phenols, and healthy fats have been known to promote good gut bacteria such as *Bifidobacterium* and *Clostridium* spp.

7.6.2 Fermented foods

Fermentation is the oldest food and beverage preservation methods and enhancement of nutritional value. It facilitates promoting and regulating microorganisms and their metabolic activities [51]. Acetic acid, lactic acid, bacteriocins, carbon dioxide, ethanol, hydrogen peroxide, and antimicrobial peptides are metabolites produced during fermentation. These compounds, whether acting individually or synergistically, inhibit the growth of spoilage and pathogenic organisms, thereby extending the shelf life of the products. In cereals, the fermentation process reduces the level of carbohydrates, non-digestible polysaccharides, and oligosaccharides [52]. Fermenting cereals with lactic acid bacteria (LAB) reduces the levels of tannins and phytic acid while enhancing iron absorption. Additionally, the enzymatic activities of amylase, phytase, hemicellulase, and protease are significantly increased, leading to improvements in shelf life, digestibility, and the nutritional value of the fermented food.

Fermented dairy products, including yoghurt, kefir, and amasi, are recognized for their health benefits, attributed not only to their nutritional composition but also to their associated microbiome, which performs several probiotic functions that enhance consumer health. The microbiome plays a crucial role in regulating intestinal health and may contribute to the treatment or prevention of inflammatory bowel diseases [53]. Additionally, milk fermented by *Lactobacillus* species has been shown to have favorable effects on the management of cardiovascular diseases related to hypertension [54]. The proteolytic activity of lactic acid bacteria during the fermentation of milk increases the availability of amino acids and peptides, resulting in a higher solubility of proteins in the final product [55].

Fermentation of meat alters raw meat into a distinct product characterized by new organoleptic properties. This process enhances protein coagulation, thereby improving flavor and the tenderness of the tissue while simultaneously reducing water activity. Consequently, the resulting product is highly nutritious due to its enzymatic activity and provides safety from pathogens through its protective mechanisms. One of the primary benefits of consuming fermented foods is the introduction of probiotics into the gut, which are beneficial in restoring the balance of the gut microbiome thereby redressing dysbiosis [56]. Fermented foods provide a means to improve the digestibility of complex carbohydrates and proteins through the enzymatic breakdown of starch into oligosaccharides and polypeptides into amino acids [57, 58]. They facilitate the concentration of key nutrients such as in cheese making by the removal of water resulting in increased bioavailability of calcium which is important for skeletal health [59]. Fermentation also transforms food to be tolerated by consumers who were once intolerant of the original product. This occurs because, during fermentation, lactic acid bacteria (LAB) metabolize lactose, resulting in a significant reduction of lactose levels in the fermented food product [48]. Fermented products result in increased bioavailability of polyphenols which have antioxidant properties, and they promote pathogen inhibition [60]. Studies suggest that consuming fermented foods may help to improve immune function, reduce the risk of certain types of cancer, and even protect against cardiovascular disease [61].

7.7 Future perspectives

7.7.1 Microbiome engineering

This process entails intentionally manipulating the food microbiome by introducing particular strains of beneficial microbes or utilizing prebiotics to enhance their growth. This approach can result in enhanced food safety, quality, and nutritional value.

7.7.2 Precision fermentation

Precision fermentation is a technique that utilizes genetically engineered microorganisms to produce targeted compounds with higher accuracy and efficiency compared to traditional fermentation methods. It allows for the creation of innovative ingredients like probiotics, prebiotics, and postbiotics [49]. These ingredients can be customized to meet specific health requirements and dietary preferences.

7.7.3 Bacteriophage technology

Bacteriophages are a class of viruses that possess the capacity to specifically infect and kill bacterial cells. They can be utilized to effectively control harmful bacteria in food products, providing a potential alternative to antibiotics.

7.7.4 Personalized nutrition

Further development of personalized nutrition approaches based on individual gut microbiome profiles has the potential to improve human health. This involves the use of advanced technologies to analyze individual microbiomes and provide tailored dietary recommendations for optimal health.

7.7.5 Artificial intelligence (AI) and machine learning integration

AI and machine learning algorithms can be used to analyze the vast amount of data on the food microbiome, human health, and dietary patterns. Additionally, predictive models can be developed to identify personalized dietary interventions for improving gut health.

7.7.6 Ethical and regulatory considerations

The formulation of ethical guidelines and regulatory frameworks is essential for the safe and responsible implementation of microbiome-based technologies in food production and human health.

7.8 Conclusion

The chapter examines food microbiomes, their relationship with the gut microbiome, and their overarching role in human health. The composition of the gut microbiome is influenced by food microbiomes, and consequently, a balanced microbiome is essential for digestion, nutrient absorption, immune function, and well-being. Some of the key components of the food microbiome are the prebiotics and probiotics as well as the synbiotics. These are essential in promoting the growth of beneficial bacteria in the gut. Food processing and preservation techniques have an impact on the food microbiome. Some of these techniques such as fermentation and precision fermentation amongst others have the potential to create a new generation of food products able to contribute to a balanced gut microbiome. Researchers can focus on personalized nutrition, basing it on the individual's gut microbiome this is anticipated to subsequently impact the health of the individual. Another area for further exploration is the development of innovative food products that offer specific health benefits derived from the food microbiome.

References

[1] Berg, G., Rybakova, D., Fischer, D., Cernava, T., Vergès, M. C., Charles, T., Chen, X., Cocolin, L., Eversole, K., Corral, G. H., & Kazou, M. Microbiome definition re-visited: Old concepts and new challenges. Microbiome 2020, Dec, 8, 1–22.

[2] Sender, R., Fuchs, S., & Milo, R. Revised estimates for the number of human and bacteria cells in the body. PLoS Biology 2016, Aug 19, 14(8), e1002533.

[3] Zhang, P. Influence of foods and nutrition on the gut microbiome and implications for intestinal health. International Journal of Molecular Sciences 2022, Aug 24, 23(17), 9588.

[4] Arumugam, M., Raes, J., Pelletier, E., Le Paslier, D., Yamada, T., Mende, D. R., Fernandes, G. R., Tap, J., Bruls, T., Batto, J. M., & Bertalan, M. Enterotypes of the human gut microbiome. Nature 2011, May 12, 473(7346), 174–180.

[5] Elechi, J. O., Sirianni, R., Conforti, F. L., Cione, E., & Pellegrino, M. Food system transformation and gut microbiota transition: Evidence on advancing obesity, cardiovascular diseases, and cancers – A narrative review. Foods 2023, Jun 6, 12(12), 2286.

[6] Franck, M., de Toro-Martín, J., & Vohl, M. C. Eco-evolutionary dynamics of the human gut microbiota symbiosis in a changing nutritional environment. Evolutionary Biology 2022, Sep, 49(3), 255–264.

[7] Brul, S., Kallemeijn, W., & Smits, G. Functional genomics for food microbiology: Molecular mechanisms of weak organic acid preservative adaptation in yeast. CABI Reviews 2008, Apr 9, 2008, 14.

[8] Gibbons, J. G., & Rinker, D. C. The genomics of microbial domestication in the fermented food environment. Current Opinion in Genetics & Development 2015, Dec 1, 35, 1–8.

[9] Di Lorenzo, F., De Castro, C., Silipo, A., & Molinaro, A. Lipopolysaccharide structures of Gram-negative populations in the gut microbiota and effects on host interactions. FEMS Microbiology Reviews 2019, May, 43(3), 257–272.

[10] Bokulich, N. A., Chung, J., Battaglia, T., Henderson, N., Jay, M., Li, H., D. Lieber, A., Wu, F., Perez-Perez, G. I., Chen, Y., & Schweizer, W. Antibiotics, birth mode, and diet shape microbiome maturation during early life. Science Translational Medicine 2016, Jun 15, 8(343), 343ra82.
[11] Bassler, B. L. Small talk: Cell-to-cell communication in bacteria. Cell 2002, May 17, 109(4), 421–424.
[12] Banerjee, S., Schlaeppi, K., & van der Heijden, M. G. Keystone taxa as drivers of microbiome structure and functioning. Nature Reviews Microbiology 2018, Sep, 16(9), 567–576.
[13] Kerns, K. A. Microbially Induced Inflammation Results in Host and Microbial Changes in Contralateral Healthy Tissues within the Oral Cavity and is Correlated with Clinical Responder Phenotypes, University of Washington, 2021.
[14] Gibson, G. R., Hutkins, R., Sanders, M. E., Prescott, S. L., Reimer, R. A., Salminen, S. J., Scott, K., Stanton, C., Swanson, K. S., Cani, P. D., & Verbeke, K. Expert consensus document: The International Scientific Association for Probiotics and Prebiotics (ISAPP) consensus statement on the definition and scope of prebiotics. Nature Reviews Gastroenterology & Hepatology 2017, Aug, 14(8), 491–502.
[15] Davani-Davari, D., Negahdaripour, M., Karimzadeh, I., Seifan, M., Mohkam, M., Masoumi, S. J., Berenjian, A., & Ghasemi, Y. Prebiotics: Definition, types, sources, mechanisms, and clinical applications. Foods 2019, Mar 9, 8(3), 92.
[16] Bedu-Ferrari, C., Biscarrat, P., Langella, P., & Cherbuy, C. Prebiotics and the human gut microbiota: From breakdown mechanisms to the impact on metabolic health. Nutrients 2022, May 17, 14(10), 2096.
[17] Lordan, C., Thapa, D., Ross, R. P., & Cotter, P. D. Potential for enriching next-generation health-promoting gut bacteria through prebiotics and other dietary components. Gut Microbes 2020, Jan 2, 11(1), 1–20.
[18] Markowiak, P., & Śliżewska, K. Effects of probiotics, prebiotics, and synbiotics on human health. Nutrients 2017, Sep 15, 9(9), 1021.
[19] Kolida, S., & Gibson, G. R. Prebiotic capacity of inulin-type fructans. The Journal of Nutrition 2007, Nov 1, 137(11), 2503S–6S.
[20] Mudannayake, D. C., Jayasena, D. D., Wimalasiri, K. M., Ranadheera, C. S., & Ajlouni, S. Inulin fructans–food applications and alternative plant sources: A review. International Journal of Food Science & Technology 2022, Sep, 57(9), 5764–5780.
[21] Kherade, M., Solanke, S., Tawar, M., & Wankhede, S. Fructooligosaccharides: A comprehensive review. Journal of Ayurvedic and Herbal Medicine 2021, 7, 193–200.
[22] Wilson, B., & Whelan, K. Prebiotic inulin-type fructans and galacto-oligosaccharides: Definition, specificity, function, and application in gastrointestinal disorders. Journal of Gastroenterology and Hepatology 2017, Mar, 32, 64–68.
[23] Ambrogi, V., Bottacini, F., Cao, L., Kuipers, B., Schoterman, M., & van Sinderen, D. Galacto-oligosaccharides as infant prebiotics: Production, application, bioactive activities and future perspectives. Critical Reviews in Food Science and Nutrition 2023, Feb 28, 63(6), 753–766.
[24] Raigond, P., Ezekiel, R., & Raigond, B. Resistant starch in food: A review. Journal of the Science of Food and Agriculture 2015, Aug 15, 95(10), 1968–1978.
[25] Kumari, A., KG, R., Sudhakaran, V. A., Warrier, A. S., & Singh, N. K. Unveiling the health benefits of prebiotics: A comprehensive review. Indian Journal of Microbiology 2024, Mar 18, 1–3.
[26] Cunningham, M., Azcarate-Peril, M. A., Barnard, A., Benoit, V., Grimaldi, R., Guyonnet, D., Holscher, H. D., Hunter, K., Manurung, S., Obis, D., & Petrova, M. I. Shaping the future of probiotics and prebiotics. Trends in Microbiology 2021, Aug 1, 29(8), 667–685.
[27] Maftei, N. M., Raileanu, C. R., Balta, A. A., Ambrose, L., Boev, M., Marin, D. B., & Lisa, E. L. The potential impact of probiotics on human health: An update on their health-promoting properties. Microorganisms 2024, Jan 23, 12(2), 234.
[28] Lukjancenko, O., Ussery, D. W., & Wassenaar, T. M. Comparative genomics of Bifidobacterium, Lactobacillus and related probiotic genera. Microbial Ecology 2012, Apr, 63, 651–673.

[29] Rezac, S., Kok, C. R., Heermann, M., & Hutkins, R. Fermented foods as a dietary source of live organisms. Frontiers in Microbiology 2018, Aug 24, 9, 1785.

[30] Sadrin, S., Sennoune, S., Gout, B., Marque, S., Moreau, J., Zinoune, K., Grillasca, J. P., Pons, O., & Maixent, J. M. A 2-strain mixture of Lactobacillus acidophilus in the treatment of irritable bowel syndrome: A placebo-controlled randomized clinical trial. Digestive and Liver Disease 2020, May 1, 52(5), 534–540.

[31] Szajewska, H., Kołodziej, M., Gieruszczak-Białek, D., Skorka, A., Ruszczyński, M., & Shamir, R. Systematic review with meta-analysis: Lactobacillus rhamnosus GG for treating acute gastroenteritis in children–a 2019 update. Alimentary Pharmacology & Therapeutics 2019, Jun, 49(11), 1376–1384.

[32] Wang, J., Ji, H., Wang, S., Liu, H., Zhang, W., Zhang, D., & Wang, Y. Probiotic Lactobacillus plantarum promotes intestinal barrier function by strengthening the epithelium and modulating gut microbiota. Frontiers in Microbiology 2018, Aug 24, 9, 1953.

[33] Vemuri, P. K., Velampati, R. H., & Tipparaju, S. L. Probiotics: A novel approach in improving the values of human life. International Journal of Pharmacy and Pharmaceutical Science 2014, 6(1), 41–43.

[34] Majeed, M., Nagabhushanam, K., Arumugam, S., Majeed, S., & Ali, F. Bacillus coagulans MTCC 5856 for the management of major depression with irritable bowel syndrome: A randomised, double-blind, placebo controlled, multi-centre, pilot clinical study. Food & Nutrition Research 2018, 62.

[35] Szajewska, H., & Kołodziej, M. Systematic review with meta-analysis: Saccharomyces boulardii in the prevention of antibiotic-associated diarrhoea. Alimentary Pharmacology & Therapeutics 2015, Oct, 42(7), 793–801.

[36] Swanson, K. S., Gibson, G. R., Hutkins, R., Reimer, R. A., Reid, G., Verbeke, K., Scott, K. P., Holscher, H. D., Azad, M. B., Delzenne, N. M., & Sanders, M. E. The International Scientific Association for Probiotics and Prebiotics (ISAPP) consensus statement on the definition and scope of synbiotics. Nature Reviews Gastroenterology & Hepatology 2020, Nov, 17(11), 687–701.

[37] Kolida, S., & Gibson, G. R. Synbiotics in health and disease. Annual Review of Food Science and Technology 2011, Apr 10, 2(1), 373–393.

[38] Pedroza Matute, S., & Iyavoo, S. Exploring the gut microbiota: Lifestyle choices, disease associations, and personal genomics. Frontiers in Nutrition 2023, Oct 5, 10, 1225120.

[39] Halfvarson, J., Brislawn, C. J., Lamendella, R., Vázquez-Baeza, Y., Walters, W. A., Bramer, L. M., D'amato, M., Bonfiglio, F., McDonald, D., Gonzalez, A., & McClure, E. E. Dynamics of the human gut microbiome in inflammatory bowel disease. Nature Microbiology 2017, Feb 13, 2(5), 1–7.

[40] Niederreiter, L., Adolph, T. E., & Tilg, H. Food, microbiome and colorectal cancer. Digestive and Liver Disease 2018, Jul 1, 50(7), 647–652.

[41] Oliphant, K., & Allen-Vercoe, E. Macronutrient metabolism by the human gut microbiome: Major fermentation by-products and their impact on host health. Microbiome 2019, Dec, 7, 1–5.

[42] Martín, R., Miquel, S., Ulmer, J., Langella, P., & Bermudez-Humaran, L. G. Gut ecosystem: How microbes help us. Beneficial Microbes 2014, Sep 1, 5(3), 219–233.

[43] Valdes, A. M., Walter, J., Segal, E., & Spector, T. D. Role of the gut microbiota in nutrition and health. BMJ 2018, Jun 13, 361.

[44] Hou, X., Zheng, Z., Wei, J., & Zhao, L. Effects of gut microbiota on immune responses and immunotherapy in colorectal cancer. Frontiers in Immunology 2022, Nov 8, 13, 1030745.

[45] Schwabe, R. F., & Jobin, C. The microbiome and cancer. Nature Reviews Cancer 2013, Nov, 13(11), 800–812.

[46] Halmos, T., & Suba, I. Physiological patterns of intestinal microbiota. The role of dysbacteriosis in obesity, insulin resistance, diabetes and metabolic syndrome. Orvosi Hetilap 2016, Jan 1, 157(1), 13–22.

[47] Chen, Z., Radjabzadeh, D., Chen, L., Kurilshikov, A., Kavousi, M., Ahmadizar, F., Ikram, M. A., Uitterlinden, A. G., Zhernakova, A., Fu, J., & Kraaij, R. Association of insulin resistance and type 2 diabetes with gut microbial diversity: A microbiome-wide analysis from population studies. JAMA Network Open 2021, Jul 1, 4(7), e2118811.

[48] Leeuwendaal, N. K., Stanton, C., O'toole, P. W., & Beresford, T. P. Fermented foods, health and the gut microbiome. Nutrients 2022, Apr 6, 14(7), 1527.

[49] Augustin, M. A., Hartley, C. J., Maloney, G., & Tyndall, S. Innovation in precision fermentation for food ingredients. Critical Reviews in Food Science and Nutrition 2023, Jan 6, 1–21.

[50] Wastyk, H. C., Fragiadakis, G. K., Perelman, D., Dahan, D., Merrill, B. D., Feiqiao, B. Y., Topf, M., Gonzalez, C. G., Van Treuren, W., Han, S., & Robinson, J. L. Gut-microbiota-targeted diets modulate human immune status. Cell 2021, Aug 5, 184(16), 4137–4153.

[51] McNeil, B., Harvey, L. M., Rowan, N. J., & Giavasis, I. Fermentation monitoring and control of microbial cultures for food ingredient manufacture. In: Microbial Production of Food Ingredients, Enzymes and Nutraceuticals, Woodhead Publishing, 2013, Jan 1, 125–143.

[52] van Zonneveld, M., Kindt, R., Solberg, S. Ø., N'Danikou, S., & Dawson, I. K. Diversity and conservation of traditional African vegetables: Priorities for action. Diversity and Distributions 2021, Feb, 27(2), 216–232.

[53] Shen, J., Zuo, Z. X., & Mao, A. P. Effect of probiotics on inducing remission and maintaining therapy in ulcerative colitis, Crohn's disease, and pouchitis: Meta-analysis of randomized controlled trials. Inflammatory Bowel Diseases 2014, Jan 1, 20(1), 21–35.

[54] Rodrigues, R. R., Greer, R. L., Dong, X., DSouza, K. N., Gurung, M., Wu, J. Y., Morgun, A., & Shulzhenko, N. Antibiotic-induced alterations in gut microbiota are associated with changes in glucose metabolism in healthy mice. Frontiers in Microbiology 2017, Nov 22, 8, 2306.

[55] Wu, Z., Wu, J., Cao, P., Jin, Y., Pan, D., Zeng, X., & Guo, Y. Characterization of probiotic bacteria involved in fermented milk processing enriched with folic acid. Journal of Dairy Science 2017, Jun 1, 100(6), 4223–4229.

[56] Hill, J. M., Bhattacharjee, S., Pogue, A. I., & Lukiw, W. J. The gastrointestinal tract microbiome and potential link to Alzheimer's disease. Frontiers in Neurology 2014, Apr 4, 5, 43.

[57] Shah, A. M., Tarfeen, N., Mohamed, H., & Song, Y. Fermented foods: Their health-promoting components and potential effects on gut microbiota. Fermentation 2023, Jan 26, 9(2), 118.

[58] Çabuk, B., Nosworthy, M. G., Stone, A. K., Korber, D. R., Tanaka, T., House, J. D., & Nickerson, M. T. Effect of fermentation on the protein digestibility and levels of non-nutritive compounds of pea protein concentrate. Food Technology and Biotechnology 2018, Jun, 56(2), 257.

[59] Rosenberg, H. F., Masterson, J. C., & Furuta, G. T. Eosinophils, probiotics, and the microbiome. Journal of Leucocyte Biology 2016, Nov, 100(5), 881–888.

[60] Zhou, F., Li, Y. L., Zhang, X., Wang, K. B., Huang, J. A., Liu, Z. H., & Zhu, M. Z. Polyphenols from Fu brick tea reduce obesity via modulation of gut microbiota and gut microbiota-related intestinal oxidative stress and barrier function. Journal of Agricultural and Food Chemistry 2021, Nov 9, 69(48), 14530–14543.

[61] Marco, M. L., Heeney, D., Binda, S., Cifelli, C. J., Cotter, P. D., Foligné, B., Gänzle, M., Kort, R., Pasin, G., Pihlanto, A., & Smid, E. J. Health benefits of fermented foods: Microbiota and beyond. Current Opinion in Biotechnology 2017, Apr 1, 44, 94–102.

Abiola Folakemi Olaniran*, Oluwatobi Victoria Obayomi, and Omotola F. Olagunju

Chapter 8 The gut microbiota, nutrigenomics, and digestive health

Abstract: The gut microbiota plays a crucial role in maintaining the physiological function of the gut, which is involved not only in digestion and absorption of nutrients, but also in major physiological processes such as the immune system and protection against pathogens. Nutrigenomics examines the interaction between dietary nutrients and genes and provides new insights into the relationship among diet, host genetics and gut microbiota, and how they interact to influence digestive health. This chapter looked into nutrigenomics, the gut microbiota, and how the two factors interact to influence digestive health. The role of the gut microbiota in human health and the composition and function of the gut microbiota are described first. The basic principles of nutrigenomics and how nutrigenomics influences gut microbiota composition and function are then discussed. The influence of the gut microbiota on nutrigenomics and how dietary intervention can influence gut microbiota composition and gene expression are also discussed. This chapter ends with a discussion about current research and the perspective of future studies on gut microbiota, nutrigenomics, and digestive health, highlighting possible directions for future research.

Keywords: diet, nutrigenomics, gut microbiota, digestive health

***Corresponding author: Abiola Folakemi Olaniran**, Landmark University SDG 12 (Responsible Consumption and Production Research Group), Omu-Aran, Kwara, Nigeria; Department of Food Science and Nutrition, College of Pure and Applied Sciences, P.M.B. 1001, Landmark University, Omu-Aran, Kwara, Nigeria; Department of Microbiology, College of Pure and Applied Sciences, P.M.B. 1001, Landmark University, Omu-Aran, Kwara, Nigeria, e-mails: olaniran.abiola@lmu.edu.ng, abiolaolaniran@gmail.com

Oluwatobi Victoria Obayomi, Department of Food Science and Nutrition, College of Pure and Applied Sciences, P.M.B. 1001, Landmark University, Omu-Aran, Kwara, Nigeria; Department of Microbiology, College of Pure and Applied Sciences, P.M.B. 1001, Landmark University, Omu-Aran, Kwara, Nigeria

Omotola F. Olagunju, Department of Science, School of Health and Life Sciences, Teesside University, Middlesbrough, TS1 3BX North Yorkshire, England

https://doi.org/10.1515/9783111441238-009

8.1 Introduction

Over the past decade, significant progress has been made with respect to the intestinal microbiota, nutrigenomics, and digestive system that leave a lot of possibilities for deconstruction or construction of health and disease. The directed gut microbiome refers to approximately one hundred trillions of organisms that inhabit the gastrointestinal tracts and therefore comprises the gut microbiome [1]. Mostly composed of bacteria, with a minor fraction of archaea, fungi and viruses, it is involved in nutrient metabolism, immune modulation, and key pathogen defense responses [2]. Disruptions (dysbiosis) in the gut microbiota are associated with many gastrointestinal illnesses, highlighting its importance in preserving general health [3]. Nutrigenomics, a growing discipline at the forefront of nutrition and genetics, reveals how food components communicate with the genome to affect gene expression and physiological responses [4]. Nutrigenomics is the interplay between nutrients and the genome to regulate gene expression and the resulting physiological responses is termed nutrigenomics and it is an emerging field where diet meets genetics [5]. Understanding the principles of nutrigenomics is paramount in unraveling the intricate mechanisms underlying the interplay between diet, genetics, and gut microbiota in modulating digestive health. Dietary components, such as fiber, polyphenols, and fatty acids, serve as substrates for microbial metabolism, producing metabolites that influence host gene expression and epigenetic modifications [6]. Conversely, host genetics can dictate the composition of the gut microbiota, shaping its response to dietary interventions. Understanding these bidirectional interactions is crucial for devising personalized dietary strategies aimed at optimizing digestive health and preventing gastrointestinal disorders [7]. This chapter aims to provide a comprehensive analysis of the complex relationship between nutrigenomics, gut microbiota, and their impact on digestive health. By thoroughly examining the definition, makeup, and roles of the gut microbiota and delving into the concepts of nutrigenomics, this review provides a basis for comprehending the mutual relationships between these two domains and how they impact the host's physiological health.

8.2 The overview of gut microbiota composition

Microbiome of the gastrointestinal tract or gut is a community of microorganisms like bacteria, archaea, fungi and viruses resident within an animal's alimentary canal [8]. This microbiota is particularly rich in the human intestine and is involved in different functions including colonization and protection against pathogens, supporting the integrity of the epithelial layer, controlling the metabolism of nutrients and drugs, and impacting the activity of the immune defense and the enteric nervous system through the gut-brain axis [9]. There is distinction in the microbial associations of the gut microbiota along the alimentary canal, given that the colon holds the biggest number of microbes and around 300–1,000 microbial species in average [10]. Proteobacteria, Actinobacteria, Firmicutes, and Bacteroidetes are the most dominant phyla in the gut microbiota with Proteobacteria being highly abundant in human gut and *Escherichia* and *Helicobacter* are the most dominant genera making up to 60% of the dry fecal mass [11]. Although, the latter is the most prevalent, there are equally high numbers of aerobic bacteria within the human cecum [12]. Scientifically, it is assumed that the human gut hosts the largest variety of bacterial species and types of all regions of the body [13]. The cell lining of the intestines also known as the target of the gut flora produces nutrients and creates a barrier against pathogenic microbes, usually, built up in the first two years after birth [14]. This may be caused by the factors such as location, diet, lifestyle, age, genetics and medications as highlighted in Figure 8.1. Some of the major bacterial divisions identified in the human gastrointestinal tract include Firmicutes, Bacteroidetes, Actinobacteria, and Proteobacteria with both Firmicutes and Bacteroidetes are dominant in relative abundance [15].

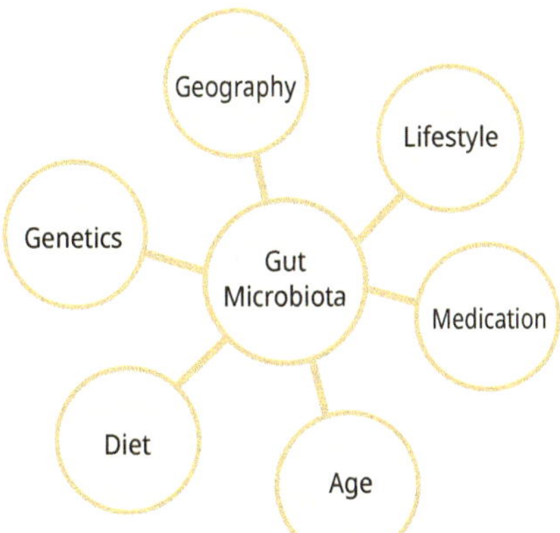

Figure 8.1: Factors that influence gut microbiota composition.

8.3 Functions of gut microbiota

8.3.1 Digestion and nutrient metabolism

Complex polysaccharides which are prevalent in human diets are intricate structures of polysaccharides which can only be broken down with difficulty by human enzymes [16]. The fermentation of polysaccharides, fibers, and other substances not easily consumed by enzymes in healthy human intestines is carried out by gut microbiota [3, 17]. Acetate, propionate, and butyrate which are short-chain fatty acids (SCFAs) are synthesized when plant polysaccharides such as cellulose and xylan are subdued by certain bacterial enzymes, including cellulase and xylanase [18]. Apart from the basic function that supplies colonocytes with the necessary energy, these SCFAs affect the host's metabolism in various ways. The effects of SCFAs produced by microorganisms include certain beneficial effects on the metabolism of the host organism. It is now known that colonocytes actively transport and incorporate acetate, propionate, and butyrate as energy substrates into their cell for metabolism [19]. Butyrate specifically has been identified for its anti-inflammatory effects and indeed is an important agent needed to support the gut barrier [20]. SCFAs are also allow out the release of gut hormone associated with appetite and glucose metabolism [21]. Gut microbes also play a major role in production of several vitamins that are vital in human body. Obligate anaerobes present in the human gastrointestinal tract and those facultative anaerobes that live in symbiosis with essential vitamin-producing bacteria include the following: vitamin K, vitamin B_7 (biotin), vitamin B_9 (folic acid), and the last is vitamin B_{12}/cobalamin [22]. For instance, vitamin K aids in blood coagulation, and biotin and folate are linked with metabolism and biosynthesis of amino acids and formation of DNA, respectively [23].

Polyphenols are bioactive compounds belonging to plant tissues present in food of plant origin such as fruits, vegetables, and leguminous crops. Polyphenols are compounds that undergo a chemical reaction in the gut to produce smaller and more soluble forms which may be utilized by the host [24]. Microbial metabolism of polyphenols has been linked to health benefits, such as the antioxidant activity and anti-inflammation activity [25]. Individual bacteria possess proteolytic factors that are capable of breaking proteins into smaller molecules such as peptides and amino acids. These peptides and amino acids synthesized by microbes are in turn can be transported to the host tissue and used for a variety of purposes ranging from protein synthesis to synthesis of neurotransmitters and maintaining the immune system of the host [26]. Gut microbes influence host lipid metabolism through several mechanisms. Conjugated linoleic acid (CLA) and secondary bile acids are examples of the bioactive lipid metabolites that they can produce [27]. They can also alter the expression of genes involved in lipid absorption and synthesis. They can also metabolize dietary fats and bile acids [28]. They also have the bifunctional capacity for the metabolism of both diets and bile acids. It has been identified that the alterations in gut microbiota are associated with the variations in lipid metabolism and increase susceptibility to develop obesity and dyslipidemia [21].

8.3.2 Immune regulation

The gut microbiota plays a decisive role in the development of the immune system. The good bacteria help maintain the integrity of a barrier within the intestines, which prevents the passage of unwanted pathogens and toxins into the bloodstream [29]. Additionally, they regulate immune cell function to integrate the immune response and to protect against inflammation diseases such as allergies and inflammatory bowel disease (IBD) [30]. Particularly in the early stages of life, the gut microbiota is essential to the growth and maturation of the host immune system. Studies on germ-free animals have shown that the lack of gut microbes causes significant immune system development defects, such as underdeveloped lymphoid tissues, decreased antibody production, and changed immune cell populations [31]. Several studies have also highlighted the importance of colonization with commensal microbes, which are necessary for the shaping of immunity, so that the immune system of a host responds appropriately to pathogens while not reacting adversely to non-harmful antigens [3, 32]. It can be explained that the gut microbiota is essential in sustaining the structural and functional structure of the host's intestinal barrier, which is beneficent in preventing the propagation of pathogens and noxious compounds [33]. Commensal bacteria contribute to the maintenance of tight junctions between intestinal epithelial cells, preventing the translocation of bacteria and toxins from the gut lumen into the bloodstream [34, 35].The phenomenon where the integrity of the gut lining is compromised, known as leaky gut, results in heightened immune response, and consequently contributes to the unfoldment of IBD and other inflammatory diseases [36]. Symbiotic gut microbes communicate with the various players of host immune response and therefore can directly associate with inhabitants of innate immune system including macrophages, dendritic cells, natural killer T and B cells [37]. Commensal bacteria influence the activation, differentiation, and function of these immune cells through direct interactions with pattern recognition receptors (PRRs) such as toll-like receptors (TLRs) and NOD-like receptors (NLRs), as well as through the production of metabolites such as SCFAs and microbial-associated molecular patterns (MAMPs) [38]. Tregs are a functional type of T lymphocytes that are responsible for immune regulation and suppression of self-recognizing responses. Gut microbes have been shown to promote the generation and expansion of Tregs within the intestinal mucosa, contributing to the maintenance of immune homeostasis and tolerance to commensal bacteria and dietary antigens [39]. This process is mediated through various mechanisms, including the production of immunomodulatory metabolites such as SCFAs and the induction of tolerogenic dendritic cells [40]. Additionally symbiotic bacteria such as Bifidobacteria, Firmicutes, Proteobacteria, and Actinobacteria in the gut send signals to the host's immune system and form determinants of how the body responding to particular infections [41]. It is now appreciated that gut microbiota plays a role in shaping immune responses that are systemic instead of restricted to the gastrointestinal tract. Abnormal composition of the resident gut microbiota also plays a significant role in many immune disorders such as autoimmune diseases, allergies, and metabolic syndromes [42].

8.3.3 Protection against pathogens

In addition to promoting immune tolerance, the gut microbiota also plays a crucial role in protecting against invading pathogens. Symbiotic bacteria have the ability to fight other bacteria as they share the same locality, environment, and immediately available resources that are found in the GIT in an effort to expel any invading pathogenic agents [43]. Additionally, some of the commensal bacteria release molecules like antimicrobial peptides (AMPs) or bacteriocins, which affect the growth of the pathogens directly [44]. These commensal microscopic organisms compete with pathogenic bacteria to obtain nutrients like carbohydrates, amino acids as well as vitamins and elements found in the gastrointestinal tract [45]. For instance, they have well-characterized metabolic capabilities of making the best out of what is available to them in a way that may help to starve and eventually outcompete the more undesirable and pathogenic counterparts. Moreover, some bacteria release substances like enzymes and metabolic products which have a direct impact on the pathogens or hamper the effective working of toxins [46]. Bacteriocins, defensins, and organic acids are some antimicrobial components which are released by commensal bacteria and inhibits the growth of pathogenic bacteria [47]. For instance, some of the beneficial bacteria include lactic acid bacteria which are *lactobacilli* and *Bifidobacteria* that produce lactic acid that creates an unfavorable condition in the gut environment as it reduces the pH making it difficult for pathogens to survive [48]. Additionally, yeasts produce hydrogen peroxide, which is a strong oxidizing agent, and facultative anaerobes like *lactobacilli* produce AMPs that target specific bacterial species [49]. The gut microbiota plays a role in the modulation of the intestinal barrier which is composed of the intestinal mucosa and epithelial cells; the barrier prevents invading pathogens from accessing the body. The beneficial microbes can reduce loss of barrier integrity in the epithelial cells lining the gut wall through tight junction proteins, mucus, and AMP production [50]. Thus, gut microbes help to sustain the integrity of the gut barrier, which is barrier against pathogen and other dangerous molecules in the gut lumen that may spread in the bloodstream and cause systemic infection and inflammation [51].

8.3.4 Metabolic regulation

Emerging evidence suggests that the gut microbiota plays a role in regulating host metabolism and energy balance. Dysbiosis, or disruption of the gut microbial community, has been associated with metabolic disorders such as obesity, insulin resistance, and type 2 diabetes [21, 52]. Specific commensal bacteria encode factors that regulate host energy harvest from the diet and lipid metabolism and govern the inflammatory state of fat stores [53]. Through its various processes, gut microbiota has been known to impact host lipid metabolism in the following ways. They can digest fat and products of bile origin, act on genes that are involved in absorbing and synthesizing lipids and fatty substances, and generate useful lipid byproducts including CLA and secondary bile substances [54]. There is growing evidence that dysbiosis of the gut microbiota is involved in the disruption of lipid metabolism and development of obesity, dyslipidemia, and non-alcoholic fatty liver disease (NAFLD) [55]. Gut microbiota interacts with the host and influences both glucose homeostasis and insulin levels in the host organism [56]. The gut microbiota plays a role in regulating host glucose metabolism and insulin sensitivity. Certain gut bacteria produce SCFAs such as propionate, which can serve as substrates for gluconeogenesis in the liver and contribute to the regulation of blood glucose levels. Additionally, gut microbes can influence host glucose metabolism through the production of metabolites such as trimethylamine-*N*-oxide (TMAO), which has been implicated in insulin resistance and cardiovascular disease [57]. There are multiple ways in which gut microbes are capable of modulating host appetite and energy expenditure. Two aspects related to digestive health include intestinal microflora and interactions between nutrients and genes on one side, and host physiology and disease risk on the other side. Refinement of nutrition-based therapy tailored to the individual microbiome profile is attributed to the complex interactions between host genetics, diet, and the microbiota [58]. New approaches that concentrate on the topicality of gut microbiota and nutrigenomics appear to enhance global well-being and help in the treatment of gastrointestinal diseases as the subject is further studied. To understand the complexity of how diet, genes, and microbiota work together in the future directions in this field, new types of diets will be examined as well as advancements in the sequencing of the gut microbiome, and multiomics approaches. Therefore, addressed to microbiota profiles, self-organized, synergistic interactions may be designed between gut microbiota and nutrigenomics to pave the way for personalized nutrition interventions. This will in the long run enhance the results of individuals suffering from poor digestion and enhanced well-being [59]. Additionally, gut microbes help to control the host energy expenditure and diet energy extraction efficiency; alternatively, they involve metabolic rate alteration in distant organs and tissues, including muscle and fat tissue [53].

8.4 Nutrigenomics

Defining it as the study of how foods and other nutrients affect gene expression and function, including genetic susceptibility to diseases, nutrigenomics is the scientific specialty that explores the connection between nutrition and genetics. This explained for a better understanding how the difference in one's genetic code influences his or her ability to metabolize food and react to diets and on how various factors, aspects to do with diets influence cellular metabolism, gene signals, and, consequently, health.

8.4.1 Metabolomics

Metabolomics is the largest omics area of focus that is involved in the comprehensive analysis of low molecular weight biomolecules called metabolites in tissue, cell, biofluid samples like blood, urine, or saliva, etc. These metabolites provide the final end products of cellular processes and, therefore, represent the final effect of gene-protein-environment combinations [60]. Metabolomics is the study of metabolome or the comprehensive identification and measurement of small molecules or metabolites involved in a biological system, a body, or an organism to understand the metabolic pathways or networks and physiological condition of an organism [61]. Metabolomics that deals with the exhaustive analysis of metabolites present in a given biological sample is one of the emerging technologies that hold a promise for the improvement of the existing personalized medicine strategies. Metabolomics represents a particular analytical problem since the target compositions encompass species with very different physical characteristics such as, for example, highly water-soluble organic acids on one hand, and nonpolar lipids on the other hand [62]. Due to this, popular global metabolomic platforms have developed subclassified metabolomic profiles on the basis of polarity or stereotypically related structures that use specialized sample preparation and analytical techniques. Due to the dynamic application of different metabolomic techniques based on the development of analytical technologies, a lot of variations may exist in workflow from laboratory to laboratory, which hinders data exchange and may decrease the confidence about the identification of metabolites. Official projects and tests such as Metabolomics Standards Initiative (MSI) and ring tests also focus on increasing the reliability of the processes and results [63]. However, metabolomics has the potential to give the understanding of the human metabolome and to identify new biomarkers in pathophysiological states since it provides information about the physiological changes induced by external stimuli and disease processes.

As primary products of cellular processes, metabolites can be quickly identified and validated for disease biomarkers and metabolic relations with the environment [64]. Techniques such as liquid chromatography–mass spectrometry (LC-MS) in metabolic profiling helps to quantify the metabolites and use the biomarkers in enriched early enough in the disease in large scale longitudinal cohorts. Studies on metabolomics also shed light on the connections between diet and disease, as seen in the case of the higher levels of branched-chain amino acids and obesity-related insulin resistance [65, 66]. As mentioned earlier, simultaneous measurement of multiple isotopologues in the same sample using techniques such as mass spectrometry enables discussion of the specific changes in metabolism in cancer tissues using stable isotope-labeled tracers represents how tracer studies extend knowledge of metabolic processes in vivo with respect to their perturbation in disease and potential remediation [67]. This is pivotal for the assessments of response to treatments, for diagnosis and staging of cancer. Its usefulness also goes beyond cancer only, to pathologies like inflammation for instance [68]. Stable isotope-containing tracers are nonradioactive tools for metabolic research in living organisms and register high safety levels. They can be used to calibrate reactions by measuring the metabolism of a number of biological molecules that are produced [69]. Techniques like ^{13}C hyperpolarization magnetic resonance enable the visualization of the rate of metabolism in a living organism and in real time [70].

8.4.2 Epigenetics

Epigenetics refers to a non-sequence information regulation of genes; epigenetic modifications that modulate activity of the DNA include reversible changes capable of switching genes on and off and, or altering the profiles of proteins passing through them [71]. This field is integral to understanding the diversity of cellular functions despite a shared genetic blueprint among cells, manifested by various cell types and tissues with distinct gene expression patterns [72]. Epigenetic mechanisms such as DNA methylation, histone modifications, and RNA-associated silencing interact within cells to regulate gene expression [72]. DNA methylation process involves adding a methyl group to DNA, typically at CpG sites, altering the structure and interactions of genes within the cell nucleus. DNA methylation can regulate gene expression and imprinting, distinguishing parental gene copies and influencing inheritance patterns [73]. Histones are proteins where DNA coils to form a chromosome; chromosomes have unique structure called chromatin. Hydrophilic histones bound to chromosomes include the amino terminal tails, and they may be covalently modified with acetyl or methyl groups; this determines the location of the chromatin and thus the transcriptional activity of the DNA [74]. These changes can tell the condition of the chromatin – whether it is in a compacted or condensed state or in the relaxed and accessible state; capable of controlling gene expression [74]. Originally, RNA-associated silencing used to be referred as gene expression regulation. These may include RNA such as antisense transcripts, non-coding RNA, or RNA interferences that can induce heterochromatin formation or particular histone modifications and DNA methylation [75]. Epigenetics plays pivotal roles in normal development and health, but aberrations can contribute to various diseases. Disruptions in epigenetic systems may lead to abnormal gene activation or silencing, implicated in cancer, chromosomal instability syndromes, and mental retardation [76]. Cancer was the first disease linked to epigenetics, with research revealing aberrant DNA methylation in tumor tissues [77]. Hypo- or hypermethylation of CpG islands near gene promoters can activate or silence genes, including tumor suppressors or oncogenes [78]. Such epigenetic alterations are common in human cancers, impacting gene expression and contributing to tumor initiation and progression [79]. Fragile X syndrome, a common cause of inherited mental disability, is an example of disease-related epigenetics [80]. Expansion of CGG repeats in the FMR1 gene promoter causes DNA methylation, which silences the gene and prevents production of the fragile X mental retardation protein

[80]. Other mental retardation disorders, including Rubenstein-Taybi and Rett syndromes, include epigenetic changes affecting gene expression and neurological development [81].

8.4.3 Proteogenomics

Proteogenomics is a cross-disciplinary research sector, involving proteomics and genomics as two interconnected parts to investigate the interaction of an organism's genome and its proteome. It integrates genomics which is a process of sequencing DNA. Transcriptomics is the study of mRNA transcript profile, and proteomics is the large-scale study of proteins profiling and identification. Proteogenomics in its turn is an applied science that aims at enhancing the knowledge of gene expression systems, proteins, and their functions within certain biological processes utilizing the data obtained from comprehensive genomics and proteomics analyses [82]. The integration of protein information and genomics facilitates identification of new proteins and their functions that may not be clearly identified based on genome data. Combining genomics and proteomics, which is the study of proteins, researchers can now pinpoint the encoding regions of proteins in the genome that were earlier undefined or incorrectly classified. This may result in a new protein form, splicing variant or a translating product of gene that play part in certain cellular processes and diseases [83]. Proteogenomics provides insights into dynamic changes in the proteome under different biological conditions, such as development, disease, or drug treatment [84]. By integrating time-course proteomic data with genomic information, researchers can track changes in protein expression, isoform usage, and PTM patterns over time. This helps unravel the molecular mechanisms underlying biological processes and identify key regulators and biomarkers associated with specific physiological states or disease states [85]. Proteomic assists in identifying how molecules are involved in biological processes and releasing the key regulatory molecules and satisfactory biomarkers that occurred with certain physiological or disease conditions [86]. Proteogenomic approaches can help with personalized medicine and biomarker discovery by identifying protein associated with disease subtypes, drug responses, and treatment outcomes [87]. Integrating genomic, transcriptomic, and proteomic data from patient samples allows researchers to identify molecular signatures that predict disease prognosis, stratify patients based on their likelihood of response to therapy, and guide personalized treatment strategies [88].

8.4.4 Transcriptomics

Transcriptomics is the study of an organism's complete transcriptome, which has been described as the summed total of all RNA transcripts, including mRNA, tRNA, rRNA, and other non-coding RNA species, and their respective levels of expression

[89]. More generally, in the context of nutrition and human health, transcriptomics involves the critical role of understanding that nutrient and other components of diet and nutritional status affect the cellular and tissue expression of genes and molecular changes [90]. For example, certain dietary components including vitamins, minerals, fatty acids, and phytocompounds can alter cell activity, the immune system, cell activities [91]. Transcriptomics can be used in the context of dietary intervention to evaluate the effects of certain diets or specific components of diets on genes and their related functions [92, 93]. This is helpful in determining systems biology that relates to the health impact of positive or negative of various dietary patterns [94]. Transcriptomics as a technological approach is often used concurrently with other omics technologies like genomics, proteomics, and metabolomics to elucidate genes' role and their regulatory networks in regard to nutrition and health [95].

8.5 Interactions between gut microbiota, nutrigenomics, and digestive health

Investigations on the relationship of the microbiome with epigenetic regulation have become popular in recent times, due to the understanding of how these microbes can be either pathogenic to the GI tract or exhibit health benefits. For example, changes in luminal nutrient concentrations, for instance, SCFAs produced by the gut microbiota, can induce post-translational modifications of DNA and histones, and therefore have an impact on gene expression without actually altering the genetic code [96]. Such changes can be reversible; nevertheless, some of these changes are often permanent after the cell divisions and might negatively influence the gene regulation procedure [96]. Histone acetylation, methylation, and crotonylation are also some of the pathways that occur within histones through which lysines are converted to acetyl, methyl, or crotonyl groups, respectively, while DNA methyltransferases (DNMTs) are enzymes that further methyleate DNA CpG islands [97]. These procedures have been found to be altered by gut microbes, which in turn will influence the host's physiology. For instance, microbe-generated SCFAs raise histone crotonylation and adjust chromatin accessibility to other modifiers [98]. Numerous researchers report that cells, containing crotonylated histones in the colonic epithelium, have different cycling characteristics important for the S or G2-M phases [99–101]. On the other hand, the level of crotonylation in the colon of the mice treated with antibiotics was reduced significantly [102]. Furthermore, several metal ions, such as cobalt, iodine, zinc, and selenium, which have been involved in the acetylation and methylation of histones, are synthesized by the gut microbes [103]. Thus, histone methylation can lead to either enhancement or repression of gene expression based on the identity of the lysine residue, whereas histone acetylation usually tends to lead to gene expression [103]. It was also observed DNA methylation and the internal environment of the microbiota also plays a role in DNA methylation, and the effects depend on the site of methylation [103]. It is well established that promoter methylation is associated with gene inactivity while body methylation may enhance the transcription of a specific gene on the DNA [104]. There was a link established between DNA promoter methylation and the relative quantity of Firmicutes and Bacteroidetes in the gut; Firmicutes was found to promote genes associated with obesity due to its higher presence in the gut microbiota than that of Bacteroidetes [105]. Through the detection of genes that were differentially methylated in the presence of pathogens such as *Helicobacter pylori*, there is likelihood of disease such as gastric mucosal cell aberrations [106]. However, changes in host epigenetic dynamics may also influence the character of gut microbiota. For instance, mice lacking HDAC3 mice, raised in germ-free conditions, harbored altered gut microbiota [107]. This underscores the importance of gut microbiota in modulating the stability of the intestinal barrier and effect of host epigenetics on the viable microbiota composition.

Another research focused on the interaction between hosts and microbes within neonatal necrotizing enterocolitis (NEC), one serious inflammatory bowel disease that affects premature infants [108–110]. In addition, the study that involved culture of intestinal epithelial cells with probiotics and pathogenic bacteria revealed changes in the DNA of the cells across more than 200 genes. These observations have highlighted that the process involving epigenetics of the host and the composition of the gut microbiome are interrelated in a bidirectional manner [111]. Different diet management options exist as they help modulate composition of the gut microbial population. Some of the nutrients in the common diet can modulate and alter the microbial ecosystem in the gastrointestinal tract within a short period, affecting host physiology. Thus, gut microbes mediate the fate of dietary consumed nutrients, including dietary fiber, glucosinolates, dietary lipids, and polyphenols, and their epigenome affecting the host's phenotype and health [103]. SCFAs, primarily produced by gut microbes, serve as crucial energy sources for colonocytes. They influence host physiology, including energy balance and protection against inflammatory bowel diseases and colon cancer [112]. A major short-chain fatty acid, butyrate, operates through histone deacetylase enzyme activity suppression. This inhibition leads to decreased growth of epithelial cell formation and increased rate of the apoptosis that help in preventing formation of colon cancer and inflammation of the bowel [113]. Isothiocyanates (ITCs) are breakdown products of glucosinolates present in cruciferous vegetables, they suppress and shift gut microbiotas, where they influence host wellness and health [114]. Dietary fat intake can impact host epigenetics and gut microbiota composition, affecting metabolic health [21]. High-fat diets have been linked to obesity and colon cancer [115–117]. According to the research of Sowah et al. [118], caloric restriction alters gut microbiota composition and improves metabolic health. Fasting influences intestinal cell proliferation and may impact tumorigenesis and metabolic health [119].

8.6 Conclusion

Digestive health is determined by the gut microbiota and the host's genotypes and their interaction on how they are influenced by nutrients. This is especially important when it comes to the use of targeting the gut microbiota in personalized nutrition based on people's microbial genotypes since the interactions between host genetics, diet, and microbiota are complex. The future developments of modern research capable of targeting the relation between the gut microbiota and the human nutritional genetics will shed light on new therapeutic approaches that can also help to treat the main digestive diseases and enhance the general well-being of an individual. To further elaborate on the complex interplays between diet, genetics, and gut microbiota the future of this convened field will entail examining other forms of dietary interventions, the advancements in the sequencing techniques, as well as multi-omics. The opportunity to develop a request for targeted nutrition based on the individual's microbiome profiles can be tentatively regarded as the synergistic interaction between the possibilities of the gut microbiota and nutrigenomics. This can eventually enhance the overall health aspect related to digestion and other related abnormalities.

References

[1] Srivastava, A., Prabhakar, M. R., Mohanty, A., & Meena, S. S. Influence of gut microbiome on the human physiology. Systems Microbiology and Biomanufacturing 2021, 1–15.

[2] Wiertsema, S. P., van Bergenhenegouwen, J., Garssen, J., & Knippels, L. M. J. The interplay between the gut microbiome and the immune system in the context of infectious diseases throughout life and the role of nutrition in optimizing treatment strategies. Nutrients 2021, 13(3), 886.

[3] Obayomi, O. V., Olaniran, A. F., & Owa, S. O. Unveiling the role of functional foods with emphasis on prebiotics and probiotics in human health: A review. Journal of Functional Foods 2024, 119, 106337.

[4] Missong, H., Joshi, R., Khullar, N., Thareja, S., Navik, U., Bhatti, G. K., et al. Nutrient-Epigenome interactions: Implications for personalized nutrition against aging-associated diseases. Journal of Nutritional Biochemistry 2024, 109592.

[5] Laddu, D., & Hauser, M. Addressing the nutritional phenotype through personalized nutrition for chronic disease prevention and management. Progress in Cardiovascular Diseases 2019, 62(1), 9–14.

[6] Hullar, M. A. J., & Fu, B. C. Diet, the gut microbiome, and epigenetics. Cancer Journal 2014, 20(3), 170–175.

[7] Org, E., Parks, B. W., Joo, J. W. J., Emert, B., Schwartzman, W., Kang, E. Y., et al. Genetic and environmental control of host-gut microbiota interactions. Genome Research 2015, 25(10), 1558–1569.

[8] Maccaferri, S., Biagi, E., & Brigidi, P. Metagenomics: Key to human gut microbiota. Digestive Diseases 2011, 29(6), 525–530.

[9] Kim, S., Covington, A., & Pamer, E. G. The intestinal microbiota: Antibiotics, colonization resistance, and enteric pathogens. Immunological Reviews 2017, 279(1), 90–105.

[10] Andrews, M. C., Duong, C. P. M., Gopalakrishnan, V., Iebba, V., Chen, W.-S., Derosa, L., et al. Gut microbiota signatures are associated with toxicity to combined CTLA-4 and PD-1 blockade. Natural Medicine 2021, 27(8), 1432–1441.

[11] Adak, A., & Khan, M. R. An insight into gut microbiota and its functionalities. Cellular and Molecular Life Sciences 2019, 76, 473–493.
[12] Smallwood, T., Allayee, H., & Bennett, B. J. Choline metabolites: Gene by diet interactions. Current Opinion in Lipidology 2016, Feb, 27(1), 33–39.
[13] Goodrich, J. K., Davenport, E. R., Clark, A. G., & Ley, R. E. The relationship between the human genome and microbiome comes into view. Annual Review of Genetics 2017, 51, 413–433.
[14] Thursby, E., & Juge, N. Introduction to the human gut microbiota. Biochemical Journal 2017, 474(11), 1823–1836.
[15] Gomaa, E. Z. Human gut microbiota/microbiome in health and diseases: A review. Antonie Van Leeuwenhoek 2020, 113(12), 2019–2040.
[16] Kardum, N., & Glibetic, M. Polyphenols and their interactions with other dietary compounds: Implications for human health. In: Advances in Food and Nutrition Research, Elsevier, 2018, 103–144.
[17] Obayomi, O. V., Olaniran, A. F., & Owa, S. Effects of bioprocessing on elemental composition, physicochemical, techno-functional, storage and sensorial properties of gluten-free flour from fonio and date fruit. Food Bioscience 2024, 105143.
[18] Louis, P., Solvang, M., Duncan, S. H., Walker, A. W., & Mukhopadhya, I. Dietary fibre complexity and its influence on functional groups of the human gut microbiota. Proceedings of the Nutrition Society 2021, 80(4), 386–397.
[19] Sivaprakasam, S., Bhutia, Y. D., Yang, S., & Ganapathy, V. Short-chain fatty acid transporters: Role in colonic homeostasis. Comprehensive Physiology 2017, 8(1), 299.
[20] Bach Knudsen, K. E., Lærke, H. N., Hedemann, M. S., Nielsen, T. S., Ingerslev, A. K., Gundelund Nielsen, D. S., Theil, P. K., et al. Impact of diet-modulated butyrate production on intestinal barrier function and inflammation. Nutrients 2018, 10(10), 1499.
[21] Obayomi, O. V., Olaniran, A. F., Olawoyin, D. C., Falade, O. V., Osemwegie, O. O., & Owa, S. O. Role of enteric dysbiosis in the development of central obesity: A review. Scientific African 2024, e02204. Available from: https://www.sciencedirect.com/science/article/pii/S2468227624001509.
[22] Rudzki, L., Stone, T. W., Maes, M., Misiak, B., Samochowiec, J., & Szulc, A. Gut microbiota-derived vitamins–underrated powers of a multipotent ally in psychiatric health and disease. Progress in Neuro-Psychopharmacology & Biological Psychiatry 2021, 107, 110240.
[23] Rodionov, D. A., Arzamasov, A. A., Khoroshkin, M. S., Iablokov, S. N., Leyn, S. A., Peterson, S. N., et al. Micronutrient requirements and sharing capabilities of the human gut microbiome. Frontiers in Microbiology 2019, 10, 1316.
[24] Aravind, S. M., Wichienchot, S., Tsao, R., Ramakrishnan, S., & Chakkaravarthi, S. Role of dietary polyphenols on gut microbiota, their metabolites and health benefits. Food Research International 2021, 142, 110189.
[25] Kawabata, K., Yoshioka, Y., & Terao, J. Role of intestinal microbiota in the bioavailability and physiological functions of dietary polyphenols. Molecules 2019, 24(2), 370.
[26] Abdallah, A., Elemba, E., Zhong, Q., & Sun, Z. Gastrointestinal interaction between dietary amino acids and gut microbiota: With special emphasis on host nutrition. Current Protein and Peptide Science 2020, 21(8), 785–798.
[27] Zhang, L. S., & Davies, S. S. Microbial metabolism of dietary components to bioactive metabolites: Opportunities for new therapeutic interventions. Genome Medicine 2016, 8, 1–18.
[28] Schoeler, M., & Caesar, R. Dietary lipids, gut microbiota and lipid metabolism. Reviews in Endocrine and Metabolic Disorders 2019, 20, 461–472.
[29] Ge, Y., Wang, X., Guo, Y., Yan, J., Abuduwaili, A., Aximujiang, K., et al. Gut microbiota influence tumor development and Alter interactions with the human immune system. Journal of Experimental & Clinical Cancer Research 2021, 40, 1–9.

[30] Tamburini, B., La Manna, M. P., La Barbera, L., Mohammadnezhad, L., Badami, G. D., Shekarkar Azgomi, M., et al. Immunity and nutrition: The right balance in inflammatory bowel disease. Cells 2022, 11(3), 455.

[31] Kamareddine, L., Najjar, H., Sohail, M. U., Abdulkader, H., & Al-Asmakh, M. The microbiota and gut-related disorders: Insights from animal models. Cells 2020, 9(11), 2401.

[32] Maynard, C. L. The microbiota in immunity and inflammation. In: Clinical Immunology, Elsevier, 2019, 207–219.

[33] Mo, C., Lou, X., Xue, J., Shi, Z., Zhao, Y., Wang, F., et al. The influence of Akkermansia muciniphila on intestinal barrier function. Gut Pathogens 2024, 16(1), 41.

[34] Ogulur, I., Pat, Y., Yazici, D., Ardicli, S., Ardicli, O., Mitamura, Y., et al. Epithelial barrier dysfunction, type 2 immune response, and the development of chronic inflammatory diseases. Current Opinion in Immunology 2024, 91, 102493.

[35] Obayomi, O. V., Olaniran, A. F., & Owa, S. O. Role of food habit and enteric microbes in the development of colon cancer. 2024 International Conference on Science, Engineering and Business for Driving Sustainable Development Goals 2024, 1–7.

[36] Mohebali, N., Ekat, K., Kreikemeyer, B., & Breitrück, A. Barrier protection and recovery effects of gut commensal bacteria on differentiated intestinal epithelial cells in vitro. Nutrients 2020, 12(8), 2251.

[37] Kmieć, Z., Cyman, M., & Ślebioda, T. J. Cells of the innate and adaptive immunity and their interactions in inflammatory bowel disease. Advances in Medical Sciences 2017, 62(1), 1–16.

[38] Liwinski, T., Zheng, D., & Elinav, E. The microbiome and cytosolic innate immune receptors. Immunological Reviews 2020, 297(1), 207–224.

[39] Wang, J., He, M., Yang, M., & Ai, X. Gut microbiota as a key regulator of intestinal mucosal immunity. Life Science 2024, 122612.

[40] Xie, L., Alam, M. J., Marques, F. Z., & Mackay, C. R. A major mechanism for immunomodulation: Dietary fibres and acid metabolites. Seminars in Immunology 2023, 101737.

[41] Demirturk, M., Cinar, M. S., & Avci, F. Y. The immune interactions of gut glycans and microbiota in health and disease. Molecular Microbiology 2024.

[42] Fitzgibbon, G., & Mills, K. H. G. The microbiota and immune-mediated diseases: Opportunities for therapeutic intervention. European Journal of Immunology 2020, 50(3), 326–337.

[43] Dey, P., & Ray Chaudhuri, S. The opportunistic nature of gut commensal microbiota. Critical Reviews in Microbiology 2023, 49(6), 739–763.

[44] Khan, I., Bai, Y., Zha, L., Ullah, N., Ullah, H., Shah, S. R. H., et al. Mechanism of the gut microbiota colonization resistance and enteric pathogen infection. Frontiers in Cellular Infection and Microbiology 2021, 11, 716299.

[45] Tewari, N., & Dey, P. Navigating commensal dysbiosis: Gastrointestinal host-pathogen interplay in orchestrating opportunistic infections. Microbiological Research 2024, 127832.

[46] Nie, Y., Luo, F., & Lin, Q. Dietary nutrition and gut microflora: A promising target for treating diseases. Trends in Food Science & Technology 2018, 75, 72–80.

[47] Shah, T., Baloch, Z., Shah, Z., Cui, X., & Xia, X. The intestinal microbiota: Impacts of antibiotics therapy, colonization resistance, and diseases. International Journal of Molecular Sciences 2021, 22(12), 6597.

[48] Mgomi, F. C., Yang, Y., Cheng, G., & Yang, Z. Lactic acid bacteria biofilms and their antimicrobial potential against pathogenic microorganisms. Biofilm 2023, 5, 100118.

[49] Prosekov, A. Y., Dyshlyuk, L. S., Milentyeva, I. S., Sykhikh, S. A., Babich, O. O., Ivanova, S. A., et al. Antioxidant and antimicrobial activity of bacteriocin-producing strains of lactic acid bacteria isolated from the human gastrointestinal tract. Progress in Nutrition 2017, 19(1), 67–80.

[50] Li, Z., Wan, M., Wang, M., Duan, J., & Jiang, S. Modulation of gut microbiota on intestinal permeability: A novel strategy for treating gastrointestinal related diseases. International Immunopharmacology 2024, 137, 112416.

[51] Dicks, L. M. T., Dreyer, L., Smith, C., & Van Staden, A. D. A review: The fate of bacteriocins in the human gastro-intestinal tract: Do they cross the gut–blood barrier? Frontiers in Microbiology 2018, 9, 412492.
[52] Yang, G., Wei, J., Liu, P., Zhang, Q., Tian, Y., Hou, G., et al. Role of the gut microbiota in type 2 diabetes and related diseases. Metabolism 2021, 117, 154712.
[53] Bohan, R., Tianyu, X., Tiantian, Z., Ruonan, F., Hongtao, H., Qiong, W., et al. Gut microbiota: A potential manipulator for host adipose tissue and energy metabolism. Journal of Nutritional Biochemistry 2019, 64, 206–217.
[54] Badawy, S., Liu, Y., Guo, M., Liu, Z., Xie, C., Marawan, M. A., et al. Conjugated linoleic acid (CLA) as a functional food: Is it beneficial or not? Food Research International 2023, 113158.
[55] Han, H., Jiang, Y., Wang, M., Melaku, M., Liu, L., Zhao, Y., et al. Intestinal dysbiosis in nonalcoholic fatty liver disease (NAFLD): Focusing on the gut–liver axis. Critical Reviews in Food Science and Nutrition 2023, 63(12), 1689–1706.
[56] Ashraf, A., & Hassan, M. I. Microbial endocrinology: Host metabolism and appetite hormones interaction with gut microbiome. Molecular and Cellular Endocrinology 2024, 112281.
[57] Zhen, J., Zhou, Z., He, M., Han, H.-X., Lv, E.-H., Wen, P.-B., et al. The gut microbial metabolite trimethylamine N-oxide and cardiovascular diseases. Frontiers in Endocrinology (Lausanne) 2023, 14, 1085041.
[58] Ordovás, J. A multifaceted approach to precision nutrition: The genome, epigenome, and microbiome in the prevention and therapy of cardiovascular diseases. In: Precision Nutrition, Elsevier, 2024, 181–200.
[59] Petersen, N., Greiner, T. U., Torz, L., Bookout, A., Gerstenberg, M. K., Castorena, C. M., et al. Targeting the gut in obesity: Signals from the inner surface. Metabolites 2022, 12(1), 39.
[60] Muthubharathi, B. C., Gowripriya, T., & Balamurugan, K. Metabolomics: Small molecules that matter more. Molecular Oral Microbiology 2021, 17(2), 210–229.
[61] Damiani, C., Gaglio, D., Sacco, E., Alberghina, L., & Vanoni, M. Systems metabolomics: From metabolomic snapshots to design principles. Current Opinion in Biotechnology 2020, 63, 190–199.
[62] Valdés, A., Alvarez-Rivera, G., Socas-Rodríguez, B., Herrero, M., Ibanez, E., & Cifuentes, A. Foodomics: Analytical opportunities and challenges. Analytical Chemistry 2021, 94(1), 366–381.
[63] Thompson, J. W., Adams, K. J., Adamski, J., Asad, Y., Borts, D., Bowden, J. A., et al. International ring trial of a high resolution targeted metabolomics and lipidomics platform for serum and plasma analysis. Analytical Chemistry 2019, 91(22), 14407–14416.
[64] Cutshaw, G., Uthaman, S., Hassan, N., Kothadiya, S., Wen, X., & Bardhan, R. The emerging role of Raman spectroscopy as an omics approach for metabolic profiling and biomarker detection toward precision medicine. Chemical Reviews 2023, 123(13), 8297–8346.
[65] Zhao, H., Zhang, F., Sun, D., Wang, X., Zhang, X., Zhang, J., et al. Branched-chain amino acids exacerbate obesity-related hepatic glucose and lipid metabolic disorders via attenuating Akt2 signaling. Diabetes 2020, 69(6), 1164–1177.
[66] Huang, C., Luo, Y., Zeng, B., Chen, Y., Liu, Y., Chen, W., et al. Branched-chain amino acids prevent obesity by inhibiting the cell cycle in an NADPH-FTO-m6A coordinated manner. Journal of Nutritional Biochemistry 2023, 122, 109437.
[67] Sellers, K., Fox, M. P., Bousamra, M., Slone, S. P., Higashi, R. M., Miller, D. M., et al. Pyruvate carboxylase is critical for non–small-cell lung cancer proliferation. Journal of Clinical Investigation 2015, 125(2), 687–698.
[68] Hess, S., Blomberg, B. A., Zhu, H. J., Høilund-Carlsen, P. F., & Alavi, A. The pivotal role of FDG-PET/CT in modern medicine. Academic Radiology 2014, 21(2), 232–249.
[69] Renuse, S., Benson, L. M., Vanderboom, P. M., Ruchi, F. N. U., Yadav, Y. R., Johnson, K. L., et al. 13C15N: Glucagon-based novel isotope dilution mass spectrometry method for measurement of glucagon metabolism in humans. Clinical Proteomics 2022, 19(1), 16.

[70] Brindle, K. M. Imaging metabolism with hyperpolarized 13C-labeled cell substrates. Journal of the American Chemical Society 2015, 137(20), 6418–6427.
[71] Vaijayanthi, T., Pandian, G. N., & Sugiyama, H. Chemical control system of epigenetics. Chemical Record 2018, 18(12), 1833–1853.
[72] Al Aboud, N. M., Tupper, C., & Jialal, I. Genetics, epigenetic mechanism. 2018.
[73] Bommarito, P. A., & Fry, R. C. The role of DNA methylation in gene regulation. In: Toxicoepigenetics, Elsevier, 2019, 127–151.
[74] Zhou, B.-R., & Bai, Y. Chromatin structures condensed by linker histones. Essays in Biochemistry 2019, 63(1), 75–87.
[75] Das, M., Renganathan, A., Dighe, S. N., Bhaduri, U., Shettar, A., Mukherjee, G., et al. DDX5/p68 associated lncRNA LOC284454 is differentially expressed in human cancers and modulates gene expression. RNA Biology 2018, 15(2), 214–230.
[76] Schenkel, L. C., Rodenhiser, D., Siu, V., McCready, E., Ainsworth, P., & Sadikovic, B. Constitutional epi/genetic conditions: Genetic, epigenetic, and environmental factors. Journal of Pediatric Genetics 2017, 6(1), 30–41.
[77] Feinberg, A. P., & Vogelstein, B. A technique for radiolabeling DNA restriction endonuclease fragments to high specific activity. Analytical Biochemistry 1983, 132(1), 6–13.
[78] Kaneda, A., Tsukamoto, T., Takamura-Enya, T., Watanabe, N., Kaminishi, M., Sugimura, T., et al. Frequent hypomethylation in multiple promoter CpG islands is associated with global hypomethylation, but not with frequent promoter hypermethylation. Cancer Science 2004, 95(1), 58–64.
[79] Ehrlich, M. DNA hypermethylation in disease: Mechanisms and clinical relevance. Epigenetics 2019, 14(12), 1141–1163.
[80] Gropman, A. L. Epigenetics and pervasive developmental disorders. In: Epigenetics in Psychiatry, Elsevier, 2021, 519–552.
[81] Jakub, T., Quesnel, K., Keung, C., Bérubé, N. G., & Kramer, J. M. Epigenetics in intellectual disability. In: Epigenetics in Psychiatry, Elsevier, 2021, 489–517.
[82] Hernandez-Valladares, M., Vaudel, M., Selheim, F., Berven, F., & Ø, B. Proteogenomics approaches for studying cancer biology and their potential in the identification of acute myeloid leukemia biomarkers. Expert Review of Proteomics 2017, 14(8), 649–663.
[83] Li, J., Qu, L., Sang, L., Wu, X., Jiang, A., Liu, J., et al. Micropeptides translated from putative long non-coding RNAs: Micropeptides translated from lncRNAs. Acta Biochimica et Biophysica Sinica (Shanghai) 2022, 54(3), 292.
[84] Locard-Paulet, M., Pible, O., De Peredo, A., Alpha-Bazin, B., Almunia, C., Burlet-Schiltz, O., et al. Clinical implications of recent advances in proteogenomics. Expert Review of Proteomics 2016, 13(2), 185–199.
[85] Brandi, J., Noberini, R., Bonaldi, T., & Cecconi, D. Advances in enrichment methods for mass spectrometry-based proteomics analysis of post-translational modifications. Journal of Chromatography A 2022, 1678, 463352.
[86] Tong, Y., Sun, M., Chen, L., Wang, Y., Li, Y., Li, L., et al. Proteogenomic insights into the biology and treatment of pancreatic ductal adenocarcinoma. Journal of Hematology & Oncology 2022, 15(1), 168.
[87] Chauvin, A., & Boisvert, F.-M. Clinical proteomics in colorectal cancer, a promising tool for improving personalised medicine. Proteomes 2018, 6(4), 49.
[88] Kwon, Y. W., Jo, H.-S., Bae, S., Seo, Y., Song, P., Song, M., et al. Application of proteomics in cancer: Recent trends and approaches for biomarkers discovery. Frontier Medicine 2021, 8, 747333.
[89] Scherrer, K. Primary transcripts: From the discovery of RNA processing to current concepts of gene expression-review. Experimental Cell Research 2018, 373(1–2), 1–33.

[90] Afman, L., Milenkovic, D., & Roche, H. M. Nutritional aspects of metabolic inflammation in relation to health – Insights from transcriptomic biomarkers in PBMC of fatty acids and polyphenols. Molecular Nutrition & Food Research 2014, 58(8), 1708–1720.
[91] Tourkochristou, E., Triantos, C., & Mouzaki, A. The influence of nutritional factors on immunological outcomes. Frontiers in Immunology 2021, 12, 665968.
[92] Konstantinidou, V., Covas, M.-I., Sola, R., & Fito, M. Up-to date knowledge on the in vivo transcriptomic effect of the Mediterranean diet in humans. Molecular Nutrition & Food Research 2013, 57(5), 772–783.
[93] Osada, J. The use of transcriptomics to unveil the role of nutrients in mammalian liver. International Scholarly Research Notices 2013, 2013(1), 403792.
[94] Castaner, O., Corella, D., Covas, M.-I., Sorlí, J. V., Subirana, I., Flores-Mateo, G., et al. In vivo transcriptomic profile after a Mediterranean diet in high–cardiovascular risk patients: A randomized controlled trial. American Journal of Clinical Nutrition 2013, 98(3), 845–853.
[95] Kussmann, M., Raymond, F., & Affolter, M. OMICS-driven biomarker discovery in nutrition and health. Journal of Biotechnology 2006, 124(4), 758–787.
[96] Rodrigues, B. A. G., Steigleder, K. M., Menta, P. L. R., De Castro, M. M., Milanski, M., & Leal, R. F. The exposome-diet-epigenome axis in inflammatory bowel diseases – A narrative review. Digestive Medicine Research 2024, 7.
[97] Javaid, N., & Choi, S. Acetylation-and methylation-related epigenetic proteins in the context of their targets. Genes (Basel) 2017, 8(8), 196.
[98] Fellows, R., & Varga-Weisz, P. Chromatin dynamics and histone modifications in intestinal microbiota-host crosstalk. Molecular Metabolism 2020, 38, 100925.
[99] Xu, H., Wu, M., Ma, X., Huang, W., & Xu, Y. Function and mechanism of novel histone posttranslational modifications in health and disease. BioMed Research International 2021, 2021(1), 6635225.
[100] Shan, S., Zhang, Z., Nie, J., Wen, Y., Wu, W., Liu, Y., et al. Marine algae-derived oligosaccharide via protein crotonylation of key targeting for management of type 2 diabetes mellitus in the elderly. Pharmacological Research 2024, 107257.
[101] Pant, K., Peixoto, E., Richard, S., & Gradilone, S. A. Role of histone deacetylases in carcinogenesis: Potential role in cholangiocarcinoma. Cells 2020, 9(3), 780.
[102] Fellows, R., Denizot, J., Stellato, C., Cuomo, A., Jain, P., Stoyanova, E., et al. Microbiota derived short chain fatty acids promote histone crotonylation in the colon through histone deacetylases. Nature Communications 2018, 9(1), 105.
[103] Shock, T., Badang, L., Ferguson, B., & Martinez-Guryn, K. The interplay between diet, gut microbes, and host epigenetics in health and disease. Journal of Nutritional Biochemistry 2021, 95, 108631.
[104] Watson, M. M., van der Giezen, M., & Søreide, K. Gut microbiome influence on human epigenetics, health, and disease. In: Handbook of Epigenetics, Elsevier, 2023, 669–686.
[105] Cuevas-Sierra, A., Ramos-Lopez, O., Riezu-Boj, J. I., Milagro, F. I., & Martinez, J. A. Diet, gut microbiota, and obesity: Links with host genetics and epigenetics and potential applications. Advances in Nutrition 2019, 10(suppl_1), S17–S30.
[106] Liu, X., Irfan, M., Xu, X., Tay, C.-Y., & Marshall, B. J. Helicobacter pylori infection induced genome instability and gastric cancer. Genome Instability & Disease 2020, 1, 129–142.
[107] Whitt, J., Woo, V., Lee, P., Moncivaiz, J., Haberman, Y., Denson, L., et al. Disruption of epithelial HDAC3 in intestine prevents diet-induced obesity in mice. Gastroenterology 2018, 155(2), 501–513.
[108] Niño, D. F., Sodhi, C. P., & Hackam, D. J. Necrotizing enterocolitis: New insights into pathogenesis and mechanisms. Nature Reviews Gastroenterology & Hepatology 2016, 13(10), 590–600.
[109] Hunter, C. J., Upperman, J. S., Ford, H. R., & Camerini, V. Understanding the susceptibility of the premature infant to necrotizing enterocolitis (NEC). Pediatric Research 2008, 63(2), 117–123.

[110] Denning, N.-L., & Prince, J. M. Neonatal intestinal dysbiosis in necrotizing enterocolitis. Molecular Medicine 2018, 24(1), 4.

[111] Yeramilli, V., Cheddadi, R., Benjamin, H., & Martin, C. The impact of stress, microbial dysbiosis, and inflammation on necrotizing enterocolitis. Microorganisms 2023, 11(9), 2206.

[112] Carretta, M. D., Quiroga, J., López, R., Hidalgo, M. A., & Burgos, R. A. Participation of short-chain fatty acids and their receptors in gut inflammation and colon cancer. Frontiers in Physiology 2021, 12, 662739.

[113] Silva, J. P. B., Navegantes-Lima, K. C., Oliveira, A. L. B., Rodrigues, D. V. S., Gaspar, S. L. F., Monteiro, V. V. S., et al. Protective mechanisms of butyrate on inflammatory bowel disease. Current Pharmaceutical Design 2018, 24(35), 4154–4166.

[114] Bouranis, J. A., Beaver, L. M., & Ho, E. Metabolic fate of dietary glucosinolates and their metabolites: A role for the microbiome. Frontiers in Nutrition 2021, 8, 748433.

[115] Padidar, S., Farquharson, A. J., Williams, L. M., Kearney, R., Arthur, J. R., & Drew, J. E. High-fat diet alters gene expression in the liver and colon: Links to increased development of aberrant crypt foci. Digestive Diseases & Sciences 2012, 57, 1866–1874.

[116] O'Neill, A. M., Burrington, C. M., Gillaspie, E. A., Lynch, D. T., Horsman, M. J., & Greene, M. W. High-fat Western diet–induced obesity contributes to increased tumor growth in mouse models of human colon cancer. Nutritional Research 2016, 36(12), 1325–1334.

[117] Lee, C., Lee, S., & Yoo, W. Metabolic interaction between host and the gut microbiota during high-fat diet-induced colorectal cancer. Journal of Microbiology 2024, 1–13.

[118] Sowah, S. A., Milanese, A., Schübel, R., Wirbel, J., Kartal, E., Johnson, T. S., et al. Calorie restriction improves metabolic state independently of gut microbiome composition: A randomized dietary intervention trial. Genome Medicine 2022, 14(1), 30.

[119] Weng, M., Chen, W., Chen, X., Lu, H., Sun, Z., Yu, Q., et al. Fasting inhibits aerobic glycolysis and proliferation in colorectal cancer via the Fdft1-mediated AKT/mTOR/HIF1$α$ pathway suppression. Nature Communications 2020, 11(1), 1869.

Temitope Ruth Olopade*, Titilayo Adenike Ajayeoba, Opeyemi Titilayo Lala, Opeyemi Christianah Ogunbiyi, Oluwayomi Christianah Olaoye, and Adetinuke Aina

Chapter 9 Blockchain in the food value chain: enhancing traceability, security, and sustainability

Abstract: The global food industry faces increasing pressure to enhance transparency, safety, and sustainability across complex supply chains. Traditional traceability systems, often reliant on outdated methods, are prone to inefficiencies and fraud, necessitating innovative solutions. This paper explores blockchain technology's transformative potential in revolutionizing food traceability. With a decentralized, immutable, and transparent ledger, blockchain ensures data integrity, fosters trust, and reduces the risks of foodborne illnesses and counterfeit products. This comprehensive analysis examines blockchain's core principles, contrasting its capabilities with the limitations of conventional traceability systems. We delve into real-world implementations, showcasing successful case studies such as IBM Food Trust and TE-Food, which demonstrate tangible improvements in traceability speed, regulatory compliance, and consumer engagement. Furthermore, this study identifies the benefits and challenges of blockchain adoption across key stages of the food value chain, from agricultural production to distribution and retail. Our findings reveal that blockchain, when integrated with technologies like IoT and AI, significantly enhances food safety, minimizes waste, and strengthens brand reputation. However, scalability, energy consumption, and the digital divide present considerable challenges. The chapter concludes by examining emerging trends and policy recommendations, providing insights into the future integration of blockchain technology in the global food industry and its potential to create more resilient and transparent food systems.

Keywords: Blockchain, food traceability, food supply chain, food safety, transparency

---

***Corresponding author: Temitope Ruth Olopade**, Department of Microbiology, College of Basic Medical and Allied Sciences, Salem University, Lokoja, Kogi, Nigeria, e-mails: topefunmi24@yahoo.com, temitope.olopade@salemuniversity.edu.ng

Titilayo Adenike Ajayeoba, Opeyemi Titilayo Lala, Food Science and Nutrition Program, Department of Microbiology, Faculty of Science, Adeleke University, Ede, Osun

Opeyemi Christianah Ogunbiyi, Department of Basic Sciences, Faculty of Science and Technology, Babcock University, Ilishan-Remo, Ogun

Oluwayomi Christianah Olaoye, Department of Animal Nutrition, College of Animal and Livestock Production, Federal University of Agriculture, Abeokuta

Adetinuke Aina, Department of Works and Physical Planning, University of Lagos

https://doi.org/10.1515/9783111441238-010

9.1 Introduction

The global food sector is evolving rapidly, driven by customer demand for greater transparency, safety, and sustainability. As supply chains increase and become more sophisticated, ensuring food traceability from production to consumption has become a priority. Blockchain technology provides a transformative method by improving data integrity, security, and transparency [1]. As a decentralized ledger, blockchain records transactions securely and immutably, eliminating the necessity for intermediaries. This innovation optimizes operations, mitigates fraud, and enhances trust within the food sector. Unlike conventional databases, blockchain relies on cryptographic protocols and consensus mechanisms, ensuring reliability and automation through smart contracts, thus, enabling peer-to-peer interactions enhances efficiency while minimizing reliance on third-party verification [2].

The impact of blockchain extends beyond food supply chains, influencing sectors such as finance, healthcare, and logistics. In the food industry, it improves traceability by delivering real-time, verifiable information, deterring fraudulent practices, and reducing the risks of foodborne illnesses. When combined with technology such as the Internet of things (IoT) and artificial intelligence (AI), blockchain enhances inventory management, reduces food waste, and bolsters consumer trust [3].

Food traceability involves tracking the movement of food products and ingredients throughout the supply chain, playing a critical role in food safety, quality assurance, and regulatory compliance. Efficient traceability systems facilitate the identification of contamination sources during outbreaks, allowing for swift recalls and mitigating health concerns [4]. Additionally, they offer transparency by checking product authenticity and sustainability claims, tackling concerns over ethical sourcing and environmental effect. Certifications like organic, fair trade, and non-GMO labels can be authenticated through blockchain's immutable ledger, enhancing consumer trust and combating food fraud [5].

Traditional food tracking methods have various limitations. Many rely on paper-based records or centralized databases, which are prone to errors, inefficiencies, and data manipulation. Disengaged stakeholders and compartmentalized information obstruct real-time access, postponing responses to food safety incidents [5]. Manual recordkeeping heightens the likelihood of inconsistencies and human mistake, complicating recall procedures and escalating expenses. Small-scale farmers and enterprises frequently encounter significant financial challenges in adopting advanced tracking systems, limiting their involvement in transparent food supply chains. These challenges highlight the need for a robust, digital, and decentralized alternative [5, 6].

This chapter examines the transformative impact of blockchain on food traceability and supply chain management. The analysis commences with an exploration of blockchain's fundamental features and its significance in food monitoring, so developing a basic comprehension of its mechanics. The debate then switches to the drawbacks of current tracing systems, noting their inefficiencies in ensuring transparency and security. Real-world uses of blockchain in food supply chains will be studied, highlighting successful implementations and their impact. Additionally, this chapter will evaluate the advantages and drawbacks of blockchain adoption, examining both its potential benefits and the challenges it poses for large-scale implementation. Finally, emerging trends and policy recommendations will be explored, providing insights into the future integration of blockchain technology in the global food industry.

9.2 Fundamentals of blockchain technology

Blockchain technology has emerged as a transformative force across industries, redefining data security, transparency, and automation. At its core, blockchain operates as a decentralized, immutable, and secure ledger system, enabling trustless transactions and disrupting traditional centralized models [1]. The integration of smart contracts and diverse consensus mechanisms further strengthens its applicability, making it a cornerstone for innovation in sectors like finance, healthcare, and the food value chain [7].

9.3 Core principles of blockchain technology

Blockchain technology is built on key principles that enhance reliability, trust, and security in decentralized systems. Immutability ensures that once data is recorded, it cannot be modified or erased, reinforcing accountability and preventing fraud [1, 6]. Transparency enables real-time tracking of transactions, fostering trust among participants and minimizing disputes. Security is upheld through cryptographic encryption, making blockchain networks resistant to cyber threats and unauthorized access [5].

Despite these strengths, blockchain encounters scalability challenges, as decentralization often leads to slower transaction speeds. Energy consumption remains a concern, particularly in proof of work (PoW) systems, raising sustainability issues. Additionally, while transparency boosts accountability, it introduces privacy risks in sectors handling sensitive information [8]. These limitations require innovative approaches, such as hybrid blockchain frameworks, privacy-enhancing encryption, and energy-efficient consensus models like proof of stake [9]. These advancements help balance decentralization, security, and efficiency, making blockchain more adaptable across industries. Table 9.1 shows a comparative analysis of blockchain-based and traditional food traceability systems, highlighting their respective challenges and sustainable solutions.

Table 9.1: Comparison of blockchain-based and traditional food traceability systems: challenges and sustainable solution.

Factor	Blockchain systems	Existing systems	Challenges	Sustainable Solutions	References
Transparency	Fully transparent and tamper-proof records accessible to all stakeholders in real time	Data is often siloed within organizations, leading to limited transparency	Lack of visibility across the supply chain	Implementing a decentralized and interoperable tracking system	[5, 10]
Data security	Uses cryptographic encryption, making records immutable and resistant to hacking	Centralized databases prone to cyber-attacks and data manipulation	Risk of fraud, data breaches, and tampering	Adoption of blockchain-integrated security protocols	[11]
Traceability efficiency	End-to-end tracking with real-time updates, reducing delays in identifying food origins	Manual and semi-automated tracking systems with potential errors and delays	Slow response times during recalls and contamination events	Automating traceability with IoT and blockchain	[5, 12]
Data integrity	Immutable ledger ensures that recorded transactions cannot be altered or deleted	Prone to errors, manipulation, and loss of critical data	Difficulty in verifying product authenticity	Standardized blockchain protocols for data verification	[13]
Interoperability	Potential for global interoperability if standardized frameworks are adopted	Lack of uniform data formats across different regions and stakeholders	Fragmented systems create inefficiencies	Development of universally accepted blockchain traceability standards	[14]
Cost of implementation	High initial setup cost but lower long-term operational costs	Lower initial costs but higher costs in long-term due to inefficiencies	Investment barriers for small-scale food producers	Public–private partnerships to subsidize blockchain adoption	[15]
Fraud and counterfeiting prevention	Reduces food fraud by verifying product origin, quality, and authenticity	Paper-based and traditional digital systems are vulnerable to fraud	Increase in counterfeit products affecting public health	Smart contracts for automatic verification of product authenticity	[16]

Table 9.1 (continued)

Factor	Blockchain systems	Existing systems	Challenges	Sustainable Solutions	References
Regulatory compliance	Helps meet stringent regulatory standards through automated compliance reporting	Requires extensive manual paperwork and audits for regulatory checks	Compliance inefficiencies and increased operational burden	Blockchain-powered automated compliance verification	[17]
Supply chain collaboration	Encourages trust and cooperation among stakeholders through a shared ledger	Lack of data sharing due to competitive interests and proprietary systems	Resistance to transparency from some industry players	Incentivizing data sharing through blockchain tokenization	[4, 11, 17]
Speed of transactions	Fast and automated verification using smart contracts	Delays due to manual verification and multiple intermediaries	Bottlenecks in supply chain operations.	Integration of AI and blockchain for instant authentication.	[7, 9, 10]
Environmental impact	Some blockchain models consume high energy, but newer ones are more sustainable.	Traditional systems require excessive paper use and inefficient logistics	High carbon footprint from inefficient tracking systems	Adoption of energy-efficient blockchain solutions	[12, 18]
Consumer trust and engagement	Consumers can verify product authenticity via blockchain-powered QR codes	Consumers rely on labels that may be misleading or manipulated	Lack of trust in food safety and quality	Implementing blockchain-based consumer information platforms	[19]

9.4 The role of blockchain in food traceability

The growing complexity of global food supply networks has raised concerns about transparency, accountability, and food safety. Traditional traceability systems frequently use paper-based or centralized digital records, which are prone to inefficiencies, fraud, and data manipulation [9]. Blockchain technology provides a decentralized, immutable, and transparent approach for improving food traceability by storing transactions in a tamper-proof digital ledger [10]. Blockchain enables that all stakeholders, including farmers, processors, distributors, retailers, and consumers, have access to verifiable, real-time information through cryptographic security and consensus methods [13].

9.4.1 Blockchain for transparency and trust in food supply chains

Transparency is a critical requirement in modern food systems, driven by consumer demand for ethical sourcing, regulatory compliance, and food safety assurance. Blockchain technology addresses these concerns by offering an immutable, decentralized ledger that records every transaction across the supply chain in real time [5]. This ensures end-to-end traceability, allowing all stakeholders – farmers, processors, distributors, retailers, and consumers – to access a single, verifiable version of the truth. Unlike conventional supply chain management systems that rely on centralized databases and intermediaries, blockchain enhances accountability by preventing unilateral data modifications [5, 10]. Every entry is cryptographically secured and verified through consensus mechanisms, ensuring authenticity and reliability [3]. This feature is particularly valuable in sectors such as organic, halal, and fair-trade food, where verifying product claims is essential to maintaining consumer trust. When transparent and tamper-proof tracking of food products from origin to consumption is enabled, blockchain reduces fraud, minimizes foodborne disease risks, and enhances recall efficiency. Smart contracts further strengthen accountability by automating compliance with safety and quality standards, reducing manual intervention and potential errors [4, 6, 13]. As the food industry embraces digital transformation, blockchain emerges as a game-changing solution for fostering trust, improving traceability, and ensuring sustainable and ethical food practices [6]. Its implementation paves the way for a more resilient global food system, where transparency is not just an expectation but a guarantee.

9.4.2 Real-time data sharing and secure record-keeping

A significant constraint of conventional traceability systems is the latency in data transmission among supply chain nodes. Blockchain facilitates real-time data sharing, providing immediate access to essential information such production dates, processing conditions, and transit records. This level of transparency guarantees adherence to regulatory standards, enables swift responses to contamination events, and reduces inefficiencies [20]. Moreover, the cryptographic integrity of blockchain records inhibits illegal modifications, hence reducing the dangers of data breaches. Integrating blockchain with the IoT enables the smooth recording of real-time environmental variables, such as temperature and humidity throughout storage and transport, thereby preserving the integrity of perishable commodities [4, 8, 20].

9.4.3 Reducing fraud, counterfeiting, and foodborne disease outbreaks

Food fraud and counterfeiting inflict an annual economic toll exceeding $40 billion globally while posing serious public health risks through adulterated or mislabeled products [21]. Blockchain technology, with its cryptographically secured and decentralized framework, mitigates these risks by eliminating vulnerabilities associated with centralized record-keeping. Its immutability, reinforced by cryptographic hashing and consensus mechanisms, ensures that provenance data, including supplier certifications and handling conditions, remains tamper-proof. For example, blockchain-integrated smart contracts autonomously verify organic certifications and fair-trade labels, reducing counterfeit infiltration in premium markets like seafood, wine, and organic produce by over 60% in pilot studies [22].

From a food safety perspective, blockchain revolutionizes recall procedures by enabling real-time traceability. Conventional systems, hindered by fragmented databases, often take days to pinpoint contaminated batches, increasing health risks and financial losses. In contrast, blockchain's granular tracking – enhanced by IoT-derived parameters such as temperature logs and RFID identifiers – enables targeted recalls of specific production lots within minutes, cutting recall costs by 45–70% [5, 8, 9, 23]. Additionally, blockchain's audit trail aligns with regulatory frameworks such as the EU's General Food Law, ensuring compliance while streamlining administrative processes.

Through the incorporation of decentralized identifiers and zero-knowledge proofs, blockchain can protect sensitive commercial data, facilitating stakeholder collaboration without eliminating competitive interests. These innovations highlight blockchain's dual role as both a fraud deterrent and a driver of sustainable, data-driven food systems, as demonstrated by its integration into EU seafood traceability programs and USDA-backed organic supply chains.

9.4.4 Case studies for some blockchain-powered food traceability

Blockchain technology is revolutionizing food traceability by enhancing transparency, security, and efficiency across global supply chains. Leading innovations such as IBM Food Trust, VeChain, TE-Food, OriginTrail, and AgriDigital demonstrate how blockchain, combined with IoT and AI, improves tracking speed, regulatory compliance, and consumer trust. These systems offer real-time data sharing, fraud prevention, and automated compliance verification, addressing major food safety challenges (Table 9.2). With the integration smart contracts, decentralized governance, and tokenized incentives, blockchain is reshaping food authentication and sustainability.

Table 9.2: Blockchain-enabled food traceability: a comparative analysis of global case studies.

Index	Global food traceability alliance					References
	IBM Food Trust (Walmart/Nestlé)	**VeChain (Walmart China)**	**TE-Food (Vietnam)**	**OriginTrail (European Retailers)**	**AgriDigital (Australia Grain Supply)**	
Traceability speed	Reduced mango traceback from 6 days to 2.2 s via Hyperledger Fabric	Real-time tracking via barcode scans (expanded from 23 to 100+ product lines)	Tracks 12,000 pigs/day with automated logistics	Optimized supply chain tracking for dairy and meat	Blockchain-secured grain transactions in seconds	[13, 24]
Technology infrastructure	Hyperledger Fabric, IoT sensors, QR codes for consumers	VeChainThor blockchain, IoT for seafood/wine, barcode scanning	QR/RFID tools, customizable livestock tracking	Decentralized Knowledge Graph, GS1 standards integration	Ethereum-based smart contracts, cloud integration	[13, 24, 25]
Supply chain complexity	Piloted with 5 suppliers, 25 products	Expanded to 100 + product lines, 50% traceable meat	Manages 6,000 + businesses across farms, slaughterhouses, and retailers	Used across EU retail networks, enhancing global interoperability	Digitized 1,300 + agribusinesses, improving traceability in grain trade	[5, 24, 26]
Stakeholder collaboration	Partnered with GS1, Nestlé for cross-border tracing	Collaborated with PwC, CCFA to integrate government/supplier data	Trained 10,000 + staff, connects 3,100 farms and 2,600 retailers	Works with global food certification bodies	Engaged farmers, traders, and banks to digitize transactions	[10, 27]
Regulatory compliance				Ensured EU compliance for food safety labeling	Complies with Australian export and trade regulations	[13, 17]

Data accuracy and immutability	Prevented pork certificate fraud with cryptographic hashing	IoT + blockchain ensure tamper-proof logistics data	Security seals and reputation scoring prevent manipulation	GS1-compliant blockchain ensures verifiable food provenance	Secured real-time data updates for transparent transactions	[1, 2, 13, 20, 25]
Consumer trust and engagement	QR codes on baby food jars show ingredient origins	Authenticity verification reduced fraud in premium wines/seafood	Mobile app provides freshness data, rewards scans	Enhanced consumer trust through transparent labeling	Farmers receive instant payments, reducing transaction disputes	[4, 6, 19, 24]

9.5 Blockchain integration in key stages of the food value chain

The incorporation of blockchain technology into the food value chain represents a significant change in the tracking, verification, and distribution of agricultural products. Due to inefficiencies, data asymmetry, and a lack of transparency, traditional supply chain systems frequently result in food fraud, safety violations, and financial losses [5, 14, 23, 27]. Blockchain, characterized by its decentralized and irreversible record, provides a formidable response to these issues by improving traceability, guaranteeing regulatory compliance, and cultivating consumer trust. This transformative technology strengthens food traceability and enhances efficiency at every stage of the supply chain, from agricultural production to consumer engagement [28]. When blockchain is integrated into primary production, processing, distribution, and retail, stakeholders can achieve greater transparency, reduce waste, and optimize resource management. The following sections explore how blockchain is revolutionizing each phase of the food value chain, ensuring authenticity, safety, and sustainability in global food systems.

9.5.1 Agriculture and primary production

The utilization of blockchain in agriculture transforms record-keeping for farm inputs, organic certification, and sustainability monitoring. Blockchain guarantees openness in agricultural activities by digitizing and safeguarding data related to soil conditions, seed quality, pesticide application, and water usage. This is especially vital for organic and fair-trade certification, as deceptive assertions erode consumer confidence. Smart contracts enabled by blockchain technology can automate compliance checks, minimizing human errors and corruption in certification procedures. Nonetheless, the digital gap persists as a substantial obstacle, as smallholder farmers in underdeveloped areas may be deficient in infrastructure and technical proficiency necessary for the efficient adoption of blockchain solutions. Furthermore, concerns regarding data ownership and the monopolization of blockchain platforms by major agribusinesses present ethical and equity-related dilemmas.

9.5.2 Processing and manufacturing

In food processing and manufacturing, blockchain improves adherence to safety requirements and enables real-time oversight of food production. Blockchain guarantees compliance with Hazard Analysis and Critical Control Points and Good Manufacturing Practices by documenting critical control points such as temperature, humidity, and sanitation protocols on an immutable ledger. This feature is especially beneficial in high-risk sectors like meat and dairy processing, where contamination can result in significant public health repercussions [24, 29]. Interoperability problems emerge when combining blockchain with current enterprise resource planning systems. The dependence on human input for data entry presents dangers of intentional fraud or inaccuracy, hence requiring the integration of automated IoT devices for impartial data collection [14, 22, 29].

9.5.3 Distribution and logistics

The significance of blockchain in distribution and logistics is crucial for monitoring temperature and storage conditions of perishable items, thus reducing food loss. Integrating blockchain with IoT-enabled cold chain monitoring systems allows stakeholders to obtain real-time data on shipment conditions, thereby assuring adherence to regulatory standards like the U.S. Food Safety Modernization Act [30]. This mitigates occurrences of food spoiling resulting from inadequate handling or deceptive relabeling of expired items. Nonetheless, scalability persists as a significant concern, as blockchain networks necessitate considerable computational power and energy expenditure [15]. Moreover, the challenge of guaranteeing the veracity of initial data inputs – often termed the "garbage in, garbage out" issue – elicits apprehensions over the authenticity of blockchain-verified records.

9.5.4 Retail and consumer engagement

Blockchain-enabled QR codes provide comprehensive traceability, allowing consumers to authenticate product origins, sustainability claims, and quality guarantees. This level of transparency cultivates confidence and enables consumers to make informed purchase choices, especially in the expanding market for ethically sourced and organic products. Additionally, blockchain can enable dynamic pricing models that respond to real-time demand and inventory levels, hence minimizing food waste at the retail stage. The level of consumer engagement with blockchain-traceable items is still uncertain [31]. Although early adopters value transparency, broad acceptance is impeded by insufficient blockchain literacy and possible concern about the security of decentralized networks [7, 24, 31].

9.6 Blockchain adoption in the food sector

Blockchain technology is transforming the food industry by enhancing traceability, transparency, and security. As global food safety regulations tighten and consumers demand greater authenticity, blockchain presents a viable solution to these challenges. However, widespread adoption is hindered by barriers such as high implementation costs, interoperability issues, and environmental concerns.

9.6.1 Benefits of blockchain adoption in the food sector

Blockchain provides an immutable record of food transactions, enabling rapid identification and recall of contaminated products. This capability is particularly crucial in regions with stringent regulations, such as the European Union (General Food Law EC 178/2002) and the United States (Food Safety Modernization Act). Real-time supply chain monitoring facilitates proactive risk management, reducing the frequency and severity of foodborne disease outbreaks [4, 7, 17].

Enhancing transparency in product labeling and sourcing also prevents counterfeit goods from entering the market. In Asia, fraudulent labeling of honey, olive oil, and dairy products has been a long-standing issue, while in the Middle East, blockchain is increasingly used for halal food certification to uphold religious dietary standards [30].

Consumers can verify product origins through blockchain-powered QR codes, fostering trust in organic, halal, and fair-trade markets [19]. In regions such as Africa and South America, blockchain adoption can enhance agricultural exports by improving credibility in international trade, thereby strengthening market competitiveness.

Governments are integrating blockchain into food safety enforcement mechanisms. For example, China has incorporated blockchain into its pork supply chain, while the African Continental Free Trade Area could leverage blockchain-driven quality standardization to facilitate seamless intra-African trade [32].

9.6.2 Challenges of blockchain adoption in the food sector

Developing economies face significant challenges in adopting blockchain due to limited digital infrastructure. Small and medium-sized enterprises in sub-Saharan Africa and Southeast Asia may struggle with the financial burden associated with implementing blockchain-based traceability systems [33]. Variability in tracking systems among supply chain stakeholders creates inconsistencies that hinder blockchain's effectiveness in cross-border trade. The absence of globally recognized standards for blockchain-based traceability further exacerbates these challenges, limiting seamless data exchange [34]. Many farmers, distributors, and retailers accustomed to conventional record-keeping practices may resist blockchain adoption due to concerns about data privacy, transparency, and regulatory control. Additionally, in informal agricultural economies, digital literacy barriers further complicate the transition to blockchain-based systems. Blockchain networks that rely on energy-intensive mechanisms like proof of work contribute to significant carbon emissions. The food sector should explore more sustainable alternatives, such as proof of stake or hybrid blockchain models, to balance transparency with environmental responsibility [10, 17, 30, 35].

9.6.3 Stakeholder perspectives on blockchain adoption

Small-scale farmers face both technological and financial barriers to blockchain adoption. While the technology offers improved market access and fair pricing, equitable participation requires incentives such as blockchain-based rewards for sustainable farming practices. Blockchain enables real-time food traceability, but widespread consumer adoption depends on education and the availability of user-friendly platforms. Tackling concerns related to data security and manipulation is essential to fostering consumer confidence. Regulatory bodies must develop interoperable blockchain standards while balancing transparency with data privacy. Collaborative policymaking with industry leaders is crucial to establishing guidelines that facilitate large-scale adoption. Although integrating blockchain requires significant investment, its long-term benefits – including fraud prevention, streamlined recalls, and enhanced brand trust – offer a competitive advantage. Companies must assess blockchain as a strategic tool for improving consumer engagement and market credibility.

9.6.4 Scalability and adoption barriers in blockchain-based food traceability

While blockchain presents significant promise for food traceability, widespread adoption remains constrained by scalability challenges, high implementation costs, interoperability issues, and regulatory uncertainties. Public blockchains, such as Ethereum and Bitcoin, often experience slow transaction speeds and high fees, making real-time supply chain tracking not visible [10, 31]. Hybrid blockchain models, which integrate public and private blockchains, offer a more viable solution by ensuring transparency while maintaining high-speed transactions among verified participants. Platforms like Hyperledger Fabric and VeChain have demonstrated success in enhancing both scalability and security. Additionally, sidechains and state channels improve transaction efficiency without overloading primary blockchain networks [7, 20, 34, 36].

The adoption of blockchain in food traceability faces challenges, particularly among small-scale farmers and supply chain actors unfamiliar with digital systems. Financial incentives, such as government subsidies, tokenized rewards, and tax benefits, can encourage participation and promote transparency [37]. However, achieving global adoption requires standardized protocols to ensure interoperability, with organizations like the Codex Alimentarius Commission and ISO playing a key role in regulatory alignment. Additionally, legal uncertainties around data ownership, liability, and compliance must be addressed. Governments should establish smart contract regulations, cross-border compliance agreements, and cybersecurity measures to balance transparency with data privacy, ensuring blockchain's long-term success in food traceability [4, 11, 32, 38].

9.7 Prospects of blockchain in the food value chain

Blockchain technology is undergoing a significant transformation through its integration with AI and the IoT, fostering predictive analytics and automation in the food value chain. Additionally, Decentralized Autonomous Organizations (DAOs) and tokenization mechanisms are shaping the governance and sustainability aspects of food systems. However, the absence of global regulatory frameworks remains a challenge, necessitating policy recommendations for standardization and widespread adoption.

9.7.1 AI and IoT integration with blockchain for predictive analytics and automation

The integration of AI, IoT, and blockchain is transforming food traceability and quality control. AI-powered analytics, leveraging machine learning, process vast datasets from IoT sensors embedded in farms, processing units, and cold storage. Combined with blockchain, these technologies enhance supply chain optimization, detect food safety risks early, and prevent fraud. For example, AI can analyze sensor data to predict spoilage risks in perishable goods, enabling timely interventions. Smart contracts further automate quality assurance, reducing reliance on manual checks. However, challenges such as interoperability between AI-IoT systems and blockchain, along with high computational costs, necessitate further technological advancements and policy support [24, 25, 39].

9.7.2 Decentralized autonomous organizations for food governance

Unlike traditional governance structures, DAOs leverage smart contracts and consensus mechanisms, enabling inclusive participation from farmers, processors, regulators, and consumers without centralized control. In food safety and certification,

DAOs facilitate real-time auditing and compliance tracking. For instance, a blockchain-powered DAO could autonomously verify organic farming practices, ensuring immutable and publicly accessible certifications. However, regulatory uncertainty, legal liability, and smart contract enforceability pose challenges to widespread adoption. Clear legal frameworks and cross-border regulatory cooperation are crucial for integrating DAOs into the food sector [10, 26, 40].

9.7.3 Tokenization and blockchain-based incentives for sustainable farming practices

Tokenization, which converts real-world assets into digital tokens on a blockchain, offers a novel approach to promoting sustainable agriculture. Farmers employing eco-friendly methods, such as regenerative agriculture or reduced pesticide use, can earn blockchain-based tokens redeemable for financial incentives, subsidies, or premium market access. Additionally, blockchain-powered carbon credit systems enable farmers to trade verified carbon offsets in global markets, enhancing sustainability efforts. Initiatives like IBM's Food Trust and VeChain's agricultural traceability projects highlight the effectiveness of tokenized incentives. However, challenges related to token valuation stability, financial system integration, and accessibility for smallholder farmers necessitate strategic policies to ensure fair benefits across agricultural sectors [30, 41].

9.7.4 Policy recommendations for global standardization of blockchain in the food value chain

The absence of standardized regulations hinders the widespread adoption of blockchain in global food supply chains. To address this, governments and international organizations must collaborate on unified blockchain protocols that ensure data interoperability, legal recognition of smart contracts, and compliance with food safety laws. Public–private partnerships should also be encouraged to drive infrastructure development and digital literacy, particularly benefiting small-scale farmers. Furthermore, regulatory frameworks must incorporate ethical considerations, including data privacy, cybersecurity, and energy efficiency, to promote sustainable and responsible blockchain adoption in the food industry [10, 13, 32].

9.8 Conclusion

Blockchain technology presents a groundbreaking opportunity to transform food traceability by enhancing transparency, security, and efficiency across the supply chain. Its core principles – immutability, decentralization, and cryptographic security – directly address challenges such as fraud, inefficiencies, and regulatory compliance. The integration of blockchain with AI and IoT further strengthens predictive analytics, real-time monitoring, and automated compliance, reducing risks and ensuring food safety. Additionally, innovations like DAOs and tokenization foster decentralized governance and incentivize sustainable agricultural practices, making food systems more accountable and ethical.

Despite its advantages, blockchain adoption faces barriers, including scalability, interoperability, and high implementation costs. Overcoming these hurdles requires investment in infrastructure, standardized data-sharing protocols, and alignment with existing regulatory frameworks. Public–private partnerships will play a pivotal role in funding pilot projects, educating stakeholders, and facilitating equitable access, particularly for smallholder farmers and developing economies. Furthermore, AI-powered analytics and IoT-enabled sensors can optimize risk detection and compliance verification, reinforcing blockchain's reliability.

To enable widespread adoption, globally recognized regulatory frameworks must define standards for smart contracts, data governance, and blockchain-based food certification. Ethical considerations – such as cybersecurity, data privacy, and environmental impact – must be integrated into policy discussions to ensure responsible implementation.

Ultimately, interdisciplinary collaboration among food scientists, technologists, policymakers, and industry leaders is essential for maximizing blockchain's potential. A strategic, multisectoral approach will drive innovation, enhance food security, and build a resilient, transparent, and sustainable global food system.

References

[1] Emiliaty, A. Blockchain technology: Revolutionizing data integrity and security in digital environments. Journal of Advanced Technological Innovations 2024, 1(1), 1–8.

[2] Kairaldeen, A. R., Abdullah, N. F., Abu-Samah, A., & Nordin, R. Peer-to-peer user identity verification time optimization in IoT Blockchain network. Sensors 2023, 23(4), 2106.

[3] Rejeb, A., Keogh, J. G., Zailani, S., Treiblmaier, H., & Rejeb, K. Blockchain technology in the food industry: A review of potentials, challenges and future research directions. Logistics 2020, 4(4), 27.

[4] Gallo, A., Accorsi, R., Goh, A., Hsiao, H., & Manzini, R. A traceability-support system to control safety and sustainability indicators in food distribution. Food Control 2021, 124, 107866.

[5] Nygaard, A. Authenticity, blockchain technology and green marketing. In: Green Marketing and Entrepreneurship, Springer, 2024, 155–165.

[6] Panwar, A., Khari, M., Misra, S., & Sugandh, U. Blockchain in agriculture to ensure trust, effectiveness, and traceability from farm fields to groceries. Future Internet 2023, 15(12), 404.

[7] Lin, S.-Y., Zhang, L., Li, J., L-I, J., & Sun, Y. A survey of application research based on blockchain smart contract. Wireless Networks 2022, 28(2), 635–690.

[8] Hasan, M. K., Alkhalifah, A., Islam, S., Babiker, N. B., Habib, A. A., Aman, A. H. M., & Hossain, M. A. Blockchain technology on smart grid, energy trading, and big data: Security issues, challenges, and recommendations. Wireless Communications and Mobile Computing 2022, 2022(1), 9065768.

[9] Gulia, P., Gill, N. S., Yahya, M., Gupta, P., Shukla, P. K., & Shukla, P. K. Exploring the potential of blockchain technology in an iot-enabled environment: A review. IEEE Access 2024, 12, 31197–31227.

[10] Akindotei, O., Igba, E., Awotiwon, B. O., & Otakwu, A. Blockchain integration in critical systems enhancing transparency, efficiency, and real-time data security in agile project management, Decentralized Finance (DeFi), and cold chain management. International Journal of Scientific Research and Modern Technology (IJSRMT) 2024, 3.

[11] Wylde, V., Rawindaran, N., Lawrence, J., Balasubramanian, R., Prakash, E., Jayal, A., et al. Cybersecurity, data privacy and blockchain: A review. SN Computer Science 2022, 3(2), 127.

[12] Rashed, M. S., Fakhry, S., Satour, R., & Pathania, S. Blockchain technology in food supply chain management: Enhancing traceability, safety, and quality. In: Food and Industry 50: Transforming the Food System for a Sustainable Future, Springer, 2025, 427–455.

[13] Akter, S., Hussain, M. I., Bhuiyan, M. K. I., Sumon, S. A., Hossain, M. I., & Akhter, A. Enhancing Data Integrity and Traceability in Industry Cyber Physical Systems (ICPS) through Blockchain Technology: A Comprehensive Approach. Available at SSRN 5041397. 2024.

[14] Henninger, A., & Mashatan, A. Distributed interoperable records: The key to better supply chain management. Computers 2021, 10(7), 89.

[15] Chen, Y., & Volz, U. Scaling up sustainable investment through blockchain-based project bonds. Development Policy Review 2022, 40(3).

[16] Maritano, V., Barge, P., Biglia, A., Comba, L., Aimonino, D. R., Tortia, C., & Gay, P. Anti-counterfeiting and fraud mitigation solutions for high-value food products. Journal of Food Protection 2024, 100251.

[17] Abikoye, B. E., Umeorah, S. C., Adelaja, A. O., Ayodele, O., & Ogunsuji, Y. M. Regulatory compliance and efficiency in financial technologies: Challenges and innovations. World Journal of Advanced Research and Reviews 2024, 23(1), 1830–1844.

[18] Pu, S., & Lam, J. S. L. Greenhouse gas impact of digitalizing shipping documents: Blockchain vs. centralized systems. Transportation Research Part D: Transport and Environment 2021, 97, 102942.

[19] Vazquez Melendez, E. I., Bergey, P., & Smith, B. Blockchain technology for supply chain provenance: Increasing supply chain efficiency and consumer trust. Supply Chain Management: An International Journal 2024, 29(4), 706–730.

[20] Udeh, E. O., Amajuoyi, P., Adeusi, K. B., & Scott, A. O. The role of IoT in boosting supply chain transparency and efficiency. Magna Scientia Advanced Research and Reviews 2024, 12(1), 178–197.

[21] Amit, S., & Kafy, -A.-A. Addressing the dollar crisis by investigating underlying causes, effects, and strategic solutions in emerging economies. Research in Globalization 2024, 8, 100187.

[22] Mitra, A., Vangipuram, S. L., Bapatla, A. K., Bathalapalli, V. K., Mohanty, S. P., Kougianos, E., & Ray, C. Everything you wanted to know about smart agriculture. arXiv preprint arXiv:220104754 2022.

[23] van Engelen, E. S. Emerging Technologies: Blockchain of Intelligent Things to Boost Revenues. 2020.

[24] Sri Vigna Hema, V., & Manickavasagan, A. Blockchain implementation for food safety in supply chain: A review. Comprehensive Reviews in Food Science and Food Safety 2024, 23(5), e70002.

[25] Tasic, I., & Cano, M. D. An orchestrated IoT-based blockchain system to foster innovation in agritech. IET Collaborative Intelligent Manufacturing 2024, 6(2), e12109.

[26] Bhat, S. A., Huang, N.-F., Sofi, I. B., & Sultan, M. Agriculture-food supply chain management based on blockchain and IoT: A narrative on enterprise blockchain interoperability. Agriculture 2021, 12(1), 40.
[27] das Nair, R., & Nontenja, N. Digital technology adoption in agro-processing value chains. CCRED: Centre for Competition Regulation and Economic Development 2023, 69.
[28] Esmaeilian, B., Sarkis, J., Lewis, K., & Behdad, S. Blockchain for the future of sustainable supply chain management in Industry 4.0. Resources, Conservation and Recycling 2020, 163, 105064.
[29] Kolikipogu, R., Shivaputra,, Muniyandy, E., Maroor, J. P., Lakshmi, G. V. R., Konduri, B., & Naveenkumar, R. Improving food safety by IoT-based climate monitoring and control systems for food processing plants. Remote Sensing in Earth Systems Sciences 2025, 1–13.
[30] Bhutta, M. N. M., & Ahmad, M. Secure identification, traceability and real-time tracking of agricultural food supply during transportation using internet of things. IEEE Access 2021, 9, 65660–65675.
[31] Aman, A., Jain, V., Radia, A., Awsarkar, R., Garg, B., & Kapoor, R., editors. Safeguarding authenticity: A blockchain-driven approach to Combat Liquor Counterfeiting with QR code and Holograph Seal Integration. In: 2024 IEEE International Conference on Blockchain and Distributed Systems Security (ICBDS), IEEE, 2024.
[32] Lin, C.-F. Blockchainizing food law: Promises and perils of incorporating distributed ledger technologies to food safety, traceability, and sustainability governance. In: Food Safety and Technology Governance, Routledge, 2022, 74–102.
[33] Asante Boakye, E., Zhao, H., & Ahia, B. N. K. Blockchain technology prospects in transforming Ghana's economy: A phenomenon-based approach. Information Technology for Development 2023, 29(2–3), 348–377.
[34] Kshetri, N. Blockchain and sustainable supply chain management in developing countries. International Journal of Information Management 2021, 60, 102376.
[35] Friedman, N., & Ormiston, J. Blockchain as a sustainability-oriented innovation?: Opportunities for and resistance to Blockchain technology as a driver of sustainability in global food supply chains. Technological Forecasting and Social Change 2022, 175, 121403.
[36] Saxena, S., Bhushan, B., & Ahad, M. A. Blockchain based solutions to secure IoT: Background, integration trends and a way forward. Journal of Network and Computer Applications 2021, 181, 103050.
[37] van Hilten, M., Ongena, G., & Ravesteijn, P. Blockchain for organic food traceability: Case studies on drivers and challenges. Frontiers in Blockchain 2020, 3, 567175.
[38] Nurgazina, J., Pakdeetrakulwong, U., Moser, T., & Reiner, G. Distributed ledger technology applications in food supply chains: A review of challenges and future research directions. Sustainability 2021, 13(8), 4206.
[39] Saidu, Y., Shuhidan, S. M., Aliyu, D. A., Aziz, I. A., & Adamu, S. Convergence of blockchain, IoT, and AI for enhanced traceability systems: A comprehensive review. IEEE Access 2025.
[40] Theodorakopoulos, L., Theodoropoulou, A., & Halkiopoulos, C. Enhancing decentralized decision-making with big data and blockchain technology: A comprehensive review. Applied Sciences 2024, 14(16), 7007.
[41] Boçe, M. T., & Hoxha, J. Blockchain technology as a catalyst for sustainable development: Exploring economic, social, and environmental synergies. Academic Journal of Interdisciplinary Studies 2024, 13(2).

Blessing O. Awoyemi, Frank Abimbola Ogundolie*, Charlene Pillay, and Christiana Eleojo Aruwa

Chapter 10 Microbial engineering tools for improved fermentation processes

Abstract: For millennia, people have used food fermentation to enhance its nutritional, sensory, and physicochemical qualities as well as to preserve food. To fully understand the complexities of fermentation processes, and the need for improvement, a more comprehensive investigation of microbial engineering tools for improved fermentation processes is paramount. Precision in strain engineering is pivotal to driving the adoption of innovative tools across industries. Fermentation processes represent a cornerstone of biotechnology, food production, and pharmaceutical manufacturing. This chapter delves into the complexities of fermentation, and the imperatives for precision through microbial engineering tools. Given the multifaceted nature of the fermentative process, the chapter explores past, current, and emerging technologies that enable targeted modifications of microbial species to optimize fermentation efficiency and product quality. It further discusses the advantages and limitations of fermentation, while emphasizing the necessity for tailored approaches to overcome challenges and maximize benefits. The broad spectrum of microbial engineering (synthetic biology, microbial metabolic engineering, genomics, transcriptomics, metabolomics, and other omics) tools used to design and manipulate microbial strains for desired traits, and enhance substrate utilization, product formation, and environmental resilience are expatiated upon. Advancements revolutionizing fermentation engineering (e.g., gas fermentation, cell-free synthetic biology, optogenetics, and fermenter design innovations) are also highlighted. These advancements catalyze cross-industry impacts, bridging the gap between biotechnology and the Fourth to Sixth Industrial Revolutions. Thus, the

Acknowledgments: The authors acknowledge the South African National Research Foundation (NRF)/Department of Science and Innovation (DSI) for the 2023 to 2024 Innovation Postdoctoral Grant (PSTD2204133389) awarded to C.E. Aruwa. Authors also acknowledge the support of the Directorate of Research and Postgraduate Support, Durban University of Technology (DUT), South Africa.

***Corresponding author: Frank Abimbola Ogundolie**, Faculty of Computing and Applied Sciences, Department of Biotechnology, Baze University, Abuja, Nigeria, e-mail: fa.ogundolie@gmail.com
Blessing O. Awoyemi, Department of Microbiology, School of Life Sciences, Federal University of Technology, Akure, PMB 704, Ondo, Nigeria, e-mail: awoyemiblessing6930@gmail.com
Charlene Pillay, Department of Biotechnology and Food Science, Faculty of Applied Sciences, Durban University of Technology, Durban 4001, South Africa, e-mail: CharleneP@dut.ac.za
Christiana Eleojo Aruwa, Department of Biotechnology and Food Science, Faculty of Applied Sciences, Durban University of Technology, Durban 4001, South Africa, e-mail: ChristianaA@dut.ac.za

https://doi.org/10.1515/9783111441238-011

chapter sets the stage for ongoing exploration and innovation in microbial engineering, promising transformative impacts across diverse industrial sectors.

Keywords: Fermentation, microbial engineering, microbial strains, synthetic biology, omics technologies, optogenetics

10.1 Introduction

Microbial engineering tools have revolutionized fermentation processes, offering unprecedented control and optimization in the production of valuable compounds [1]. This field combines microbiology, genetic engineering, and bioprocess engineering to manipulate microbial organisms for enhanced fermentation outcomes [2]. The utilization of microbial engineering tools is pivotal for addressing challenges in industrial biotechnology, including improving yields, product quality, and process efficiency [3]. Fermentation is a fundamental process harnessed across diverse industries, from pharmaceuticals to food and biofuels [4]. Microorganisms like bacteria, yeast, and fungi are employed to convert raw materials into desired products through controlled biochemical pathways. However, inherent limitations often impede optimal production, such as low yields, slow growth rates, and sensitivity to environmental conditions [5].

Microbial engineering offers tailored solutions to overcome these constraints. One key aspect of microbial engineering is genetic modification of strains to enhance desired traits. This involves techniques like gene editing, where specific genes are inserted, deleted, or modified to optimize metabolic pathways. For example, introducing genes encoding enzymes that enhance substrate utilization or product synthesis could significantly boost fermentation efficiency [6, 7]. An interdisciplinary tool like synthetic biology has been pivotal to designing and constructing novel biological systems to perform desired functions such as engineering custom microbes with complex metabolic networks for derivation of a wide array of compounds, including pharmaceuticals, biochemicals, and biofuels. Synthetic biology combines computational and systems biology, bioinformatics, and artificial intelligence (AI) to study and define microbial genomes and their associated metabolic processes for enhanced engineering outcomes. The development of synthetic cellular factories limits the formation of unwanted metabolic by-products, permitting the production of high-value functional food ingredients [8, 9].

Strain development through directed evolution is another engineering method that involves selective pressure application on microbial populations to drive the evolution of desirable traits over successive generations. By subjecting microbes to specific conditions, such as high substrate concentrations or extreme temperatures, researchers can isolate strains with improved fermentation capabilities [10, 11]. Other advanced bioprocess engineering tools used to optimize fermentation conditions include bioreactor design, process monitoring, and control strategies. By integrating microbial engineering with bioprocess engineering, it is now possible to fine-tune fermentation parameters like pH, temperature, and oxygen levels to maximize productivity and product quality [12]. Also, the advent of high-throughput screening, molecular biology, next-generation sequencing (NGS), multi-omics and bioinformatics tools, and advanced statistical approaches have also accelerated microbial engineering research. They have advanced research into microbial composition of fermented food products, and aided full genome sequencing of industrially significant microorganisms [13]. These tools allow for rapid characterization of microbial strains at the molecular level, facilitating the identification of key genetic targets for optimization [14]. In light of the foregoing, the proceeding sections go on to examine how past, current, and emerging microbial engineering tools contribute to enhancing food fermentation processes. Additionally, insights into the innovative and cross-industry impacts of fermentation engineering to stir the Fourth Industrial Revolution (4IR) to Sixth Industrial Revolution (6IR) are also discussed.

10.2 Fermentation process overview

Fermentation is a pivotal biological process utilized in the production of food and beverages, pharmaceuticals, and biofuels, via the transformation of organic compounds (sugars, starches) by fermentative microbes like bacteria, yeast, or fungi into valuable products such as alcohol, acids, or gases in the absence of oxygen [15]. The fermentation process comprises several key steps. Microorganisms utilize specific organic substrates such as glucose, sucrose, starches, or cellulose as their primary energy sources. Through intricate metabolic pathways, like glycolysis followed by fermentation steps, microorganisms break down these substrates to generate energy and produce desired end products [5]. Various types of fermentation exist based on the microorganisms and substrates involved [16]. Multiple factors influence the efficiency and outcomes of fermentation processes. These include the specific microbial strains used, substrate availability and composition, environmental conditions (temperature, pH, and oxygen level) within the fermentation vessel, and the duration of fermentation [17]. Optimal control of these factors is essential for maximizing product yield, quality, and process efficiency [18]. In the food and beverage sector, fermentation is integral to producing beer, wine, bread, yogurt, cheese, and other fermented foods. In the pharmaceutical industry, fermentation processes are employed to produce antibiotics, vaccines, and therapeutic proteins [19]. The intricacies of fermentation processes encompass several fascinating aspects, including metabolic pathways, microbial interactions, and product inhibition, all of which significantly influence the efficiency and outcomes of fermentation.

Fermentation involves intricate metabolic pathways that microorganisms utilize to convert substrates into products. Different microbes possess specific enzymatic systems that catalyze biochemical reactions, leading to the production of desired products. For instance, in alcoholic fermentation by yeast (*Saccharomyces cerevisiae*), glucose undergoes glycolysis to produce pyruvate, which is then converted into ethanol and carbon dioxide. Understanding and optimizing these metabolic pathways are essential for maximizing product yields and controlling the fermentation process [20]. Researchers often employ techniques like metabolic engineering to enhance specific pathways or introduce novel pathways to improve fermentation efficiency and diversify product outputs [21].

Microbial interactions within fermentation systems play a crucial role in shaping the process dynamics. In complex microbial communities, such as those found in sourdough starters or mixed-culture fermentations, interactions between different species influence fermentation outcomes. Synergistic interactions may enhance metabolic efficiency and product formation, while competitive interactions affect community composition and overall process stability. Understanding these interactions is vital for designing controlled fermentation processes. Researchers leverage tools from microbial ecology and systems biology to study and manipulate microbial communities, aiming to optimize fermentation conditions and promote desired outcomes [22, 23]. Also, product inhibition also occurs when the accumulation of fermentation products negatively impacts microbial growth and metabolic activity. As microorganisms produce specific compounds (e.g., ethanol and organic acids), these products may reach concentrations that inhibit their own enzymatic processes or disrupt cellular functions. As such, end-product inhibition could limit fermentation efficiency and reduce overall productivity. Strategies such as continuous fermentation, product removal tools (in situ product recovery), or metabolic engineering approaches (altering enzyme kinetics) could useful to mitigate this challenge. Understanding the mechanisms and kinetics of product inhibition is crucial for designing fermentation processes that maintain microbial activity and sustain high productivity levels [24–26].

The need for enhancement of fermentation processes arises from various factors and challenges associated with current practices. Enhancing fermentation processes is imperative for maximizing productivity, reducing costs, and meeting evolving market demands for value-added products. By optimizing fermentation conditions and leveraging advancements in microbial engineering, industries could achieve higher product yields and improved process efficiency. Genetic engineering plays a crucial role in developing microbial strains with enhanced capabilities, such as increased substrate utilization and higher product formation rates. These advancements not only boost productivity but also contribute to reducing production costs by optimizing resource utilization and minimizing waste generation. Additionally, shortening the processing time through accelerated microbial growth and metabolic activities allows for faster production cycles, enhancing overall throughput and operational efficiency [27, 28]. Moreover, enhancing fermentation processes enables industries to diversify product portfolios and adapt to changing market trends. Hence, businesses that expand into bio-based chemicals, pharmaceutical intermediates, and functional ingredients could also capitalize on emerging opportunities and remain competitive.

10.3 Pros and cons of the fermentative process

The fermentative process, while offering numerous advantages, also comes with certain limitations and challenges that are highlighted in Table 10.1. Part of the advantages (pros) of fermentation include its versatility, low energy requirement, use of renewable energy sources/raw materials, and biochemically diverse microbial species, while difficulties with scale-up and downstream processes, product inhibition, microbial contamination, and slow reaction rate make up some of its drawbacks (cons) [29–34].

Table 10.1: Some pros and cons of the fermentative process.

Pros of fermentation	Cons of fermentation
Versatility: Fermentation can be applied to a wide range of substrates, including sugars, starches, and organic wastes, making it versatile for producing various valuable products such as ethanol, organic acids, enzymes, and pharmaceuticals.	Slow reaction rates: Fermentation processes can be inherently slow compared to chemical reactions, leading to longer production cycles and, potentially, limiting overall productivity.
Renewable resource utilization: Fermentation often utilizes renewable resources like agricultural crops, cellulosic materials, and waste streams, contributing to sustainability, and reducing reliance on finite fossil fuels.	Product inhibition: Accumulation of fermentation products can inhibit microbial activity, impacting process efficiency and requiring strategies such as product removal or metabolic engineering to mitigate.
Low energy requirements: Compared to traditional chemical synthesis processes, fermentation typically requires lower energy inputs, resulting in reduced operational costs and environmental impact.	Sensitive to environmental conditions: Fermentation is sensitive to environmental factors such as pH, temperature, and oxygen levels. Variations in these conditions can affect microbial growth and product formation, necessitating precise control and monitoring.
Biochemical diversity: Microorganisms used in fermentation can produce a diverse array of biochemicals with specific functionalities, making fermentation a valuable tool for specialty chemical and pharmaceutical industries.	Microbial contamination: Contamination by unwanted microorganisms can occur during fermentation, leading to reduced product quality and yield. Strict hygiene practices and aseptic techniques are essential to prevent contamination.

Table 10.1 (continued)

Pros of fermentation	Cons of fermentation
Mild reaction conditions: Fermentation generally occurs under mild reaction conditions (e.g., ambient temperatures and pressures), preserving the integrity of heat-sensitive compounds and reducing the risk of unwanted byproducts.	Complex downstream processing: Purification and recovery of fermentation products from complex fermentation broths can be challenging and costly, particularly for high-value biochemicals and pharmaceuticals.
	Scale-up challenges: Transitioning from laboratory-scale to industrial-scale fermentation presents engineering and operational challenges related to equipment design, process scalability, and cost-effectiveness.

10.4 Factors impacting the fermentative process: a call for "precision" strain engineering

Factors impacting the fermentative process are numerous and varied, and precision in understanding and controlling these factors is indeed crucial for optimizing fermentation outcomes, especially in strain engineering. The choice of microbial strain used in fermentation is foundational, since different strains exhibit varying metabolic capabilities, product yields, and tolerance to environmental conditions. Strain engineering involves modifying these strains genetically to enhance desirable productivity, substrate utilization efficiency, or stress tolerance traits [35, 36]. Secondly, the type and quality of the substrate (or feedstock) significantly influence fermentation. The composition of the substrate, its nutrient content, and any inhibitors present could impact microbial growth and product formation. Precision in substrate selection involves understanding how different substrates affect the fermentation process and choosing the most suitable ones [37–39]. Thirdly, temperature, pH, oxygen availability, and agitation are critical environmental factors affecting fermentation. Microbial strains have specific optimal ranges for these parameters, and deviations may lead to reduced productivity or total failure of the fermentation process. Precision control of these conditions is essential for consistent and efficient fermentation [40–42].

In considering the nutrient availability factor, besides the primary carbon source, fermentation requires other nutrients like nitrogen, phosphorus, vitamins, and minerals. Deficiencies or imbalances in these nutrients limit microbial growth and productivity. Precision in nutrient supplementation ensures that microbial cells have all the essential building blocks for efficient metabolism and product formation [19, 43]. In the case of design and operation of the bioreactor that is integral to fermentation performance, the inaccurate combination of reactor size, mixing efficiency, aeration, and sterilization methods could adversely impact microbial growth and product formation. Precision engineering of bioreactors involves optimizing these parameters to create an ideal environment for the microbial culture [42, 44, 45]. Another key factor to consider is the real-time monitoring of fermentation parameters such as biomass concentration, substrate utilization rate, product formation rate, and by-product accumulation. This is crucial for making timely adjustments during the fermentation process. Precision control systems may be used to automate these adjustments based on predefined targets, ensuring optimal fermentation performance [46]. Lastly, the use of strain engineering aims to improve microbial strains by modifying their genetic makeup to enhance specific metabolic pathways. This may involve introducing genes for desired traits, such as increased production of target compounds or improved tolerance to fermentation inhibitors [47, 48].

10.5 Past, current, and emerging tools for engineering fermentative microbial strains

The engineering of fermentative microbial species has witnessed significant advancements through the integration of past, current, and emerging tools, thus leading to transformative developments in bioprocess engineering [49]. Historically, traditional strain improvement methods like random mutagenesis and adaptive evolution laid the groundwork for enhancing microbial capabilities in fermentation. With advancements in molecular biology, genetic engineering tools such as clustered regularly interspaced short palindromic repeats (CRISPR-Cas9) have revolutionized strain modification, allowing precise genome editing and pathway engineering. Concurrently, systems biology approaches integrating omics technologies (genomics, transcriptomics, proteomics, and metabolomics) have enabled comprehensive understanding of cellular processes, guiding targeted modifications for improved fermentation performance. Furthermore, synthetic biology has emerged as a powerful tool for enabling the design and construction of novel genetic circuits and pathways to achieve tailored metabolic functions in microbial hosts. Looking forward, emerging tools such as machine learning (ML) and high-throughput screening promise to further accelerate strain development and optimization. This convergence of traditional and cutting-edge techniques underscores the dynamic landscape of fermentative microbial engineering, and is poised to deliver innovative solutions for bioproduction and industrial applications [50, 51].

10.5.1 Natural, single versus consortia-induced fermentation

Natural microbial processes epitomize the intricate dynamics within wild, uncultured microbial communities flourishing in their native habitats, and untouched by human intervention. These ecosystems constitute a staggering array of species, each contributing to a complex web of interactions shaped by environmental factors [52]. What sets natural microbial communities apart is their remarkable functional redundancy. Here, multiple species often perform similar metabolic functions, providing a safety net for ecosystem stability. This redundancy ensures that vital processes, such as nutrient cycling or decomposition, persist even if certain species are lost or environmental conditions undergo fluctuations. Moreover, natural microbes have evolved specialized metabolic adaptations over millennia, fine-tuned to their specific ecological niches. These adaptations enable them to thrive and contribute uniquely to their ecosystems, thus highlighting the immense diversity and resilience inherent in natural microbial processes [53].

In contrast to the intricacies of natural ecosystems, single microbial processes entail harnessing the capabilities of a solitary microbial strain for targeted biotechnological applications. This approach offers a streamlined and controlled method for genetic manipulation and metabolic engineering [54]. In industrial settings such as biopharmaceutical production or biofuel synthesis, utilizing a single strain allows for predictable outcomes and simplified process optimization. However, this simplicity comes at a cost, as single strains often lack the adaptability and robustness exhibited by natural microbial communities. Their metabolic repertoire is typically narrower, limiting their versatility in responding to changing environmental conditions or performing complex tasks requiring multifaceted metabolic pathways. Nonetheless, single microbial processes serve as fundamental building blocks in biotechnology, providing a foundation upon which more sophisticated engineering strategies may be developed [55, 56].

Consortia-induced processes represent a nuanced approach that bridges the complexity of natural microbial communities with the controllability of single-strain systems. By assembling cooperative interactions between multiple microbial species or strains engineered to complement each other, a consortium achieves heightened performance and resilience in various applications [57]. These consortia exhibit distributed metabolic functions, enabling them to tackle intricate tasks efficiently, such as bioremediation of diverse pollutants or the synthesis of intricate biomolecules. Designing optimal microbial consortia demands a deep understanding of ecological principles and microbial interactions. As technology advances, leveraging insights from natural ecosystems in consortia design holds immense promise for realizing innovative solutions in synthetic ecology and sustainable biotechnology [58, 59].

10.5.2 Synthetic biology tools

Synthetic biology stands at the intersection of biology, engineering, and computer science, offering innovative approaches to design and manipulate biological systems for specific purposes. A suite of sophisticated tools has emerged to support research and applications in this field. The application of DNA synthesis and assembly technologies enables researchers to custom-design DNA sequences, including genes and genetic circuits, using methods like Gibson Assembly and Golden Gate Assembly [60]. These techniques allow for the building of complex DNA constructs that underpin synthetic biology endeavors [60, 61]. Genome editing tools, particularly CRISPR-Cas systems like CRISPR-Cas9, offer precise methods for modifying DNA sequences within living organisms. This technology has revolutionized genetic engineering by enabling targeted modifications with unprecedented accuracy. Complementary to genome editing, genetic parts libraries provide standardized biological components (promoters, terminators, and coding sequences) that serve as building blocks for constructing synthetic genetic circuits. Computational modeling tools, including bioinformatics software and systems biology models, play essential roles in predicting and analyzing the behavior of synthetic biological systems, thus guiding the design process and optimizing outcomes [62–64].

Protein engineering tools like directed evolution and rational design empower researchers to tailor protein properties for specific functions or improved stability. In the realm of metabolic engineering, synthetic biology tools facilitate the design and optimization of metabolic pathways in microorganisms, enabling the efficient production of valuable compounds such as biofuels and pharmaceuticals [65–67]. Additionally, cell-free systems offer versatile platforms for protein synthesis and metabolic engineering, outside of living cells, accelerating the prototyping and optimization of genetic circuits and biochemical pathways. Together, these synthetic biology tools catalyze groundbreaking advancements across diverse fields, from medicine to biomanufacturing and beyond, by harnessing the power of engineered biological systems [68, 69].

10.5.3 Microbial metabolic engineering

Microbial metabolic engineering is a branch of biotechnology that focuses on modifying the metabolism of microorganisms to produce desired compounds or enhance specific biochemical pathways [70]. This field leverages the natural metabolic capabilities of microbes such as bacteria, yeast, and fungi, by genetically modifying them to optimize their productivity and versatility. The ultimate goal is to design microbial cell factories capable of efficiently synthesizing valuable chemicals, biofuels, pharmaceuticals, and other products [70, 71]. At the core of microbial metabolic engineering lies the manipulation of genetic pathways within microbial cells. This involves identifying and modifying key enzymes and regulatory elements involved in target metabolic pathways. Techniques like genome editing, particularly CRISPR-Cas systems, enable precise modifications of microbial genomes to enhance desired traits. By introducing or enhancing specific genes, deleting or repressing others, or rewiring metabolic networks, researchers can redirect cellular resources toward the synthesis of desired compounds [72–74]. Metabolic engineering strategies often rely on systems biology approaches to understand and optimize microbial metabolism comprehensively. Through iterative cycles of design, build, test, and learn, metabolic engineers fine-tune microbial strains to achieve high product yields, substrate utilization efficiency, and tolerance to environmental stresses [75].

10.5.4 Design of microbial fermenters or cell factories

The design of microbial fermenters or cell factories involves a strategic approach to engineer microorganisms for efficient production of desired compounds through fermentation. This process begins with the selection of a suitable host organism based on its metabolic capabilities, genetic tractability, and compatibility with the target application. Common microbial hosts like bacteria, yeast, and fungi are chosen for their ability to utilize specific substrates and tolerate fermentation conditions [76, 77]. Once the host organism is selected, metabolic pathway engineering is employed to optimize the production of target compounds. This involves modifying or introducing metabolic pathways within the microbial cells to enhance precursor supply, increase flux toward desired products, and minimize the formation of by-products. Techniques such as gene knockout, gene overexpression, and pathway balancing are used to rewire cellular metabolism toward the desired metabolic goals [78, 79]. The genetic tools and engineering strategies employed in microbial fermenter design are crucial for introducing desired genetic modifications into the host organism. Rational design approaches based on metabolic modeling and bioinformatics analysis guide the selection and optimization of genetic modifications, while directed evolution methods can be used to improve enzyme activities and pathway efficiency, iteratively. Genetic constructs such as synthetic DNA sequences and plasmids are introduced into the host organism to implement engineered pathways and regulatory circuits [80, 81].

Optimizing fermentation conditions is a critical aspect of microbial fermenter design to ensure high productivity and yield of target compounds. Factors such as pH, temperature, oxygen levels, and nutrient availability are carefully controlled and optimized to support microbial growth and maximize product formation [82]. Bioreactor design plays a key role in maintaining optimal fermentation conditions at different scales, from laboratory experiments to industrial production. Process integration involves developing downstream processing steps for product recovery, purification, and formulation, ensuring the final product meets quality standards for commercial use. Overall, the design of microbial fermenters integrates microbiology, molecular biology, metabolic engineering, and chemical engineering principles to create efficient and scalable bioprocesses for producing valuable bioproducts through microbial fermentation [82–84].

10.5.5 Gas fermentation

Gas fermentation is a specialized process in biotechnology where microorganisms convert gaseous substrates, such as carbon monoxide (CO), carbon dioxide (CO_2), or hydrogen (H_2), into valuable products through metabolic pathways. This innovative approach leverages the metabolic capabilities of certain microbial species to utilize gases as carbon and energy sources, enabling the sustainable production of biofuels, chemicals, and other commodities [85]. The key to gas fermentation lies in the selection and engineering of microbial strains capable of efficiently metabolizing gaseous substrates. For example, some acetogenic bacteria and archaea have the unique ability to use CO or CO_2 as carbon sources through the Wood-Ljungdahl pathway, producing compounds like acetate, ethanol, and butanol. Similarly, certain hydrogenotrophic microorganisms can utilize H_2 as an electron donor, in combination with CO_2, to synthesize methane or other organic molecules [86]. Advantages of gas fermentation include the potential to utilize abundant and renewable gaseous feedstocks, such as industrial waste gases or syngas from biomass gasification, reducing reliance on traditional carbon sources like sugars or starches. Gas fermentation processes are also less resource-intensive compared to traditional fermentation using complex organic substrates, making them attractive for sustainable bioproduction. However, challenges remain in optimizing microbial strains, engineering metabolic pathways, and scaling up gas fermentation processes for commercial applications [87, 88].

10.5.6 Omics technologies (genomics, transcriptomics, proteomics, and metabolomics)

Omics technologies encompass a suite of powerful tools used to comprehensively study biological systems at the molecular level, offering insights into the structure, function, and dynamics of organisms [89]. Genomics, the study of an organism's complete set of genes (genome), employs high-throughput sequencing technologies (NGS) to decode and analyze DNA sequences. This allows researchers to explore genetic variations, evolutionary relationships, and the genetic basis of traits or diseases through techniques such as gene mapping, comparative genomics, and genome editing using CRISPR-Cas systems [90, 91]. Transcriptomics focuses on the entirety of RNA transcripts produced by cells or organisms under specific conditions. The use of RNA sequencing (RNA-seq) is a key technique in transcriptomics that quantifies and characterizes the transcriptome, revealing gene expression patterns, regulatory networks, alternative splicing events, and responses to environmental stimuli. Transcriptomic data aids in understanding cellular functions and pathways at the transcriptional level, identifying differentially expressed genes associated with specific biological processes or disease states [92, 93].

Metabolomics examines the complete set of small molecules (metabolites) within biological samples using analytical techniques such as mass spectrometry (MS) and nuclear magnetic resonance (NMR) spectroscopy. In profiling metabolites, metabolomics reveals insights into metabolic pathways, biochemical reactions, and cellular metabolism. This allows researchers to identify biomarkers, characterize metabolic phenotypes, and understand how genetic, environmental, or physiological factors influence metabolic regulation and disease mechanisms [94–96]. Integrating omics data from genomics, transcriptomics, and metabolomics enables a systems-level understanding of biological systems. Systems biology approaches leverage multi-omics datasets to construct comprehensive models that elucidate the interactions and dynamics between genes, transcripts, proteins, metabolites, and environmental factors. This integrative analysis is crucial for advancing fields such as personalized medicine, microbial engineering, and agriculture, and provides a holistic understanding of biological processes, which is essential for developing targeted interventions and innovations tailored to individual traits and conditions [97, 98].

As omics technologies continue to evolve and become more accessible, their applications will drive transformative discoveries in biology, medicine, and biotechnology. The comprehensive insights provided by omics data contribute to precision and personalized approaches in various scientific disciplines, accelerating progress toward understanding complex biological phenomena and developing novel therapeutic strategies and biotechnological solutions [99, 100]. Harnessing the power of omics tools, platforms, and databases promises to revolutionize our understanding of life at the molecular level and pave the way for innovative advancements that benefit human health, agriculture, environmental sustainability, and beyond.

10.5.7 Cell-free synthetic biology

Cell-free synthetic biology represents a cutting-edge approach that harnesses the molecular machinery of cells without the need for intact living cells. In this innovative field, biological processes are reconstituted in vitro using cell extracts containing essential cellular components such as enzymes, ribosomes, and transcriptional machinery. This allows researchers to engineer and manipulate biological systems, outside of living cells, offering unique advantages for biotechnological applications [101]. One key aspect of cell-free synthetic biology is the flexibility and modularity it provides for genetic and metabolic engineering. By bypassing the complexity of cellular membranes and regulatory networks, researchers may directly program synthetic genetic circuits and metabolic pathways using purified biomolecules. This enables rapid prototyping of genetic designs and metabolic pathways, facilitating the development of novel biosensors, biocatalysts, and biochemical production platforms [102, 103]. Cell-free systems are particularly valuable for accelerating the design-build-test cycle in synthetic biology. They offer a controlled environment where genetic constructs could be expressed and tested in a high-throughput manner, without the constraints and limitations of living cells. This rapid prototyping capability is instrumental in optimizing biological designs and engineering new functions, leading to innovative solutions for biomanufacturing, biosensing, and biomedical applications [104].

10.5.8 Optogenetics

Optogenetics, originally developed for neuroscience research, has been increasingly applied as a versatile tool in microbial engineering to enable precise control over microbial behavior and gene expression using light-responsive genetic elements [105]. In microbial systems, optogenetics involves the genetic introduction of light-sensitive proteins (opsins) that can modulate cellular processes, in response to specific wavelengths of light. This innovative approach allows researchers to remotely and non-invasively manipulate microbial functions with high spatiotemporal precision, opening up new possibilities for synthetic biology and biotechnology applications [106, 107]. One key application of optogenetics in microbial engineering is the regulation of gene expression. By incorporating light-responsive genetic switches (optogenetic promoters or riboswitches) into microbial genomes, the control of timing, duration, and intensity of gene expression using light pulses may be possible [108]. This precise control over gene expression patterns is valuable for optimizing metabolic pathways, enhancing product yields, and studying complex cellular processes in real-time processes under different environmental conditions [109].

Another area where optogenetics excels in microbial engineering is the manipulation of cellular signaling and behavior. Opsins may be engineered into microbial cells to modulate intracellular signaling pathways or ion fluxes, upon light stimulation. This capability enables the remote triggering of specific cellular responses such as growth arrest, metabolic activation, or biofilm formation by simply exposing the microbial culture to light of a defined wavelength [110, 111]. Moreover, optogenetics offers unique advantages for spatially resolved engineering of microbial communities. By encoding different microbial strains with distinct optogenetic regulators, spatially patterned microbial populations that respond to localized light cues may be created. This approach has implications for designing synthetic consortia, studying microbial interactions, and engineering spatially structured bioprocesses [112, 113].

10.6 Fermentation engineering: innovations and cross-industry impacts

Fermentation engineering, a cornerstone of bioprocessing, is undergoing significant innovation to drive impacts across multiple industries, in a bid to align with the 4IR and 5IR, and the emerging 6IR. These advancements are revolutionizing biomanufacturing, sustainability, and personalized medicine, reshaping the global economic landscape [1]. Fermentation engineering impacts span automation, data analytics, and bioprocess optimization. In the realm of automation and robotics, fermentation processes are transitioning toward increased autonomy through the integration of robotic systems and AI-driven technologies. Automated bioreactors, equipped with sophisticated monitoring devices, enable real-time control and adjustment of fermentation parameters, enhancing production efficiency and consistency, while reducing human intervention. Concurrently, the integration of Internet of things (IoT) sensors and data analytics platforms facilitates continuous monitoring of fermentation processes. This enables predictive maintenance, quality control, and formulation optimization based on real-time data insights, ensuring optimal performance and resource utilization in fermentation operations [114].

Furthermore, bioprocess optimization is revolutionized by computational modeling and simulation tools, aligning fermentation engineering with the principles of Industry 4.0. These tools allow for the systematic optimization of fermentation conditions, microbial strains, and nutrient inputs to maximize yields and minimize waste [115]. By leveraging data-driven insights and predictive analytics, researchers and engineers are able to iteratively refine fermentation processes, accelerating innovation and enabling more sustainable and efficient biomanufacturing practices. The convergence of automation, data analytics, and bioprocess optimization in fermentation engineering epitomizes the transformative impact of the 4IR, driving advancements that enhance productivity, quality, and scalability across diverse industries reliant on fermentation-based technologies [116]. In the context of 5IR, fermentation engineering is poised to make transformative impacts through several key avenues. First, the concept of biological digital twins (DTs) represents a significant advancement, where virtual models of fermentation processes are created and continuously updated using real-time data. This approach enhances process understanding by providing dynamic simulations that reflect the complexities of microbial behavior and bioprocess dynamics. Biological DTs enable predictive control strategies for fermentation-based production, allowing for proactive adjustments and optimizations based on modeled scenarios and observed data, ultimately improving process efficiency and product quality [117].

In the 6IR context, fermentation engineering is experiencing transformative impacts that are reshaping the landscape of biotechnology and food production. Advances in synthetic biology and bio-design are enabling the precise engineering of microbial strains tailored for targeted fermentation processes. This allows for the production of bio-based materials, ingredients, and pharmaceuticals with customized functionalities and properties. In leveraging synthetic biology techniques, scientists are able to design novel microbial pathways that efficiently convert renewable feedstocks into high-value products, hence driving sustainability and innovation in biomanufacturing [117]. Moreover, the convergence of fermentation with emerging technologies such as nanotechnology, AI, and quantum computing is fuelling unprecedented innovation in biomaterials, biopharmaceuticals, and sustainable food production systems. Nanotechnology-enhanced fermentation processes enable the production of nano-scale materials with unique properties and applications, while AI-driven optimization algorithms enhance fermentation efficiency and product yield. Furthermore, quantum computing holds promise for revolutionizing bioprocess modeling and optimization, enabling complex simulations and predictive analytics that accelerate the development of novel fermentation strategies [118]. The 6IR also fosters global collaborative networks among interdisciplinary teams working at the intersection of fermentation science, biotechnology, and digital technologies. This collaborative approach facilitates knowledge-sharing, data exchange, and innovation diffusion on a global scale, leading to collective efforts to address complex societal challenges related to food security, health, and sustainability. By leveraging the transformative potential of 6IR technologies, fermentation engineering is poised to drive impactful advancements that promote economic growth, environmental stewardship, and human well-being in the evolving landscape of biotechnology and industrial fermentation [118].

10.7 Future prospects

The outlook on microbial engineering for improving fermentation processes is promising and driven by advancements in synthetic biology, biotechnology, and data-driven approaches. These tools offer innovative solutions to enhance the efficiency, scalability, and sustainability of fermentation-based bioproduction across diverse industries. Synthetic biology and metabolic engineering represent transformative fields that hold immense promise for the future of fermentation processes. Synthetic biology generates biological systems with custom functionalities that are often achieved through genetic manipulation of microbial strains. In the context of fermentation, synthetic biology enables the creation of novel microbial strains tailored for optimized metabolic pathways. This approach allows for the production of valuable compounds in a sustainable and cost-effective manner, reducing reliance on traditional chemical synthesis methods and fossil fuels. Advancements in synthetic biology are paving the way for the development of next-generation microbial cell factories, optimized for specific bioproduction purposes. By leveraging techniques such as genome editing, pathway engineering, and gene circuit design, scientists are able to tailor microbial strains to exhibit enhanced metabolic capabilities, substrate utilization efficiencies, and product yields [119].

Synthetic biology also enables the creation of biosensors and genetic switches that facilitate dynamic control over gene expression and metabolic flux, enabling fine-tuning of fermentation processes in response to environmental cues or production demands. As the field of synthetic biology continues to evolve, it will drive innovation in fermentation engineering by expanding the repertoire of engineered microbes and metabolic pathways available for bioproduction, ultimately unlocking new opportunities for sustainable and scalable biomanufacturing processes [120, 121]. Furthermore, the integration of high-throughput screening methods with omics technologies represents a significant advancement with promising future prospects. High-throughput screening allows for the rapid evaluation of large microbial libraries to identify strains with desirable traits such as high productivity or specific metabolic capabilities. Coupling this screening approach with omics technologies, including genomics, transcriptomics, and metabolomics, provides a comprehensive understanding of microbial behavior and responses within fermentation environments. Multiomics data analysis enables the identification of genetic targets and metabolic pathways that could be optimized to enhance fermentation performance. This integration of high-throughput screening and omics technologies accelerates strain development and bioprocess optimization, leading to more efficient and tailored microbial systems for diverse biotechnological applications [13, 35].

The application of high-throughput "omics" tools in fermentation engineering enables the characterization of cellular responses under various fermentation conditions. By systematically analyzing gene expression profiles, metabolite concentrations, and metabolic fluxes, researchers gain insights into the molecular mechanisms underlying microbial behavior during fermentation. This knowledge can guide the rational design of genetic modifications and process parameters to optimize fermentation outcomes. The synergy between high-throughput screening and omics technologies empowers researchers to uncover new genetic targets, metabolic pathways, and regulatory networks that drive microbial performance, paving the way for innovative strategies to enhance the efficiency, sustainability, and scalability of fermentation processes in biotechnology and industrial applications [35, 122].

In the case of optogenetics, this cutting-edge technique harnesses light to precisely control biological processes, particularly in microbial engineering. By incorporating light-responsive genetic elements into microbial systems, researchers can achieve unprecedented control over gene expression and metabolic activities. This technology allows for the manipulation of microbial behavior at a level of detail previously unattainable with conventional methods. With optogenetics, specific cellular functions can be triggered or inhibited, simply by exposing the engineered microbes to different wavelengths of light. This level of control not only opens new possibilities for fundamental research into microbial physiology but also has practical applications in optimizing bioproduction processes. For instance, by synchronizing cellular activities through light stimulation, researchers can enhance the efficiency of fermentation processes, leading to improved yields of valuable bio-based products [115]. Light-controlled processes in microbial engineering offer unique advantages, primarily due to their high spatiotemporal resolution. This means that researchers can precisely target specific microbial populations or even individual cells within a culture using focused light signals. Such fine-tuning of microbial activities is invaluable in optimizing bioproduction parameters, where subtle adjustments can have significant impacts on product quality and quantity [123]. Optogenetics also allows for the dynamic modulation of gene expression patterns in response to varying environmental conditions, offering a versatile tool for adaptive microbial engineering. Moreover, by utilizing light as a non-invasive external stimulus, optogenetics minimizes interference with cellular processes, making it a safe and efficient method for controlling microbial behavior [124].

The integration of AI and ML techniques holds immense promise for transforming fermentation process optimization [125]. These technologies empower researchers and engineers to monitor fermentation processes in real time with unprecedented precision and insight. In leveraging AI-driven algorithms, vast amounts of data generated during fermentation may be analyzed comprehensively to identify patterns and correlations that may not be apparent through conventional analysis methods [126, 127]. This data-driven approach enables the development of predictive models that can forecast optimal fermentation conditions and growth parameters, facilitating proactive adjustments to bioreactor operations. Through continuous learning and adaptation, AI systems can dynamically optimize fermentation processes to achieve maximum productivity, while minimizing resource consumption and waste [128]. ML in fermentation process optimization also enhances efficiency by enabling adaptive control strategies. The AI algorithms further learn from historical process data and responds in real time to changing conditions within the bioreactor. This adaptability allows for the automatic adjustment of fermentation parameters, such as temperature, pH, nutrient levels, and agitation rates, to optimize microbial growth and product formation. Furthermore, AI-driven optimization can lead to significant improvements in scalability and reproducibility across different fermentation batches and production scales [129, 130]. Ultimately, the application of AI and ML in fermentation process optimization promises to revolutionize bioprocessing by streamlining operations, improving product quality, and advancing sustainability through smarter resource management.

10.8 Concluding remarks

A marked increase in the global food production demand needs to be met by 2050. However, with the current regimens utilized in food industries, the demands may not be met and require innovative concepts to increase productivity and nutritional quality [131]. The arsenal of microbial engineering tools available today is poised to revolutionize fermentation processes and bioproduction. Techniques such as optogenetics offer precise control over microbial behavior and gene expression with unprecedented accuracy using light-responsive genetic elements [132]. Algorithms driven by AI analyze complex datasets to identify optimal growth parameters, forecast fermentation conditions, and dynamically adjust bioreactor operations [133, 134]. Ultimately, by harnessing microbial engineering tools, humans could realize the full potential of fermentation technology to drive innovation and sustainability in industries ranging from biotechnology and pharmaceuticals to food and agriculture. The array of microbial engineering tools and the integration of cutting-edge technologies and interdisciplinary strategies promise to reshape our approach to fermentative bioprocessing, enabling the development of next-generation, sustainable, efficient, and tailored bioproducts that benefit society and the environment.

References

[1] Sarkale, P. S., Ghewari, P. G., & Sarkale, P. S. Emerging trends in fermentation technology: Implications for food security and environmental sustainability. Naturalista Campano 2024, 28, 252–260.

[2] Rangel, A. E., Gomez Ramirez, J. M., & Gonzalez Barrios, A. F. From industrial by-products to value-added compounds: The design of efficient microbial cell factories by coupling systems metabolic engineering and bioprocesses. Biofuels, Bioproducts and Biorefining 2020, 14, 1228–1238.

[3] Kumar, G., Shekh, A., Jakhu, S., Sharma, Y., Kapoor, R., & Sharma, T. R. Bioengineering of microalgae: Recent advances, perspectives, and regulatory challenges for industrial application. Frontiers in Bioengineering and Biotechnology 2020, 8, 914.

[4] Niakousari, M., Razmjooei, M., Nejadmansouri, M., Barba, F. J., Marszałek, K., & Koubaa, M. Current developments in industrial fermentation processes. Fermentation Processes: Emerging and Conventional Technologies 2021, 23–96.

[5] Mengesha, Y., Tebeje, A., & Tilahun, B. A review on factors influencing the fermentation process of Teff (Eragrostis teff) and other cereal-based Ethiopian injera. International Journal of Food Science 2022, 24, 2022.

[6] Deckers, M., Deforce, D., Fraiture, M. A., & Roosens, N. H. Genetically modified micro-organisms for industrial food enzyme production: An overview. Foods 2020, 9, 326.

[7] Zhao, R., Liu, Y., Zhang, H., Chai, C., Wang, J., Jiang, W., & Gu, Y. CRISPR-Cas12a-mediated gene deletion and regulation in *Clostridium ljungdahlii* and its application in carbon flux redirection in synthesis gas fermentation. ACS Synthetic Biology 2019, 8, 2270–2279.

[8] Lv, X., Wu, Y., Gong, M., Deng, J., Gu, Y., Liu, Y., Li, J., Du, G., Ledesma-Amaro, R., Liu, L., & Chen, J. Synthetic biology for future food: Research progress and future directions. Future Foods 2021, 3, 100025.

[9] Wu, G., Yan, Q., Jones, J. A., Tang, Y. J., Fong, S. S., & Koffas, M. A. Metabolic burden: Cornerstones in synthetic biology and metabolic engineering applications. Trends in Biotechnology 2016, 34, 652–664.

[10] Sanghavi, G., Gupta, P., Rajput, M., Oza, T., Trivedi, U., & Singh, N. K. Microbial strain engineering. In: Engineering of Microbial Biosynthetic Pathways, Singapore: Springer, 2020, 11–32.

[11] Makino, T., Skretas, G., & Georgiou, G. Strain engineering for improved expression of recombinant proteins in bacteria. Microbial Cell Factories 2011, 10, 1–10.

[12] Paulová, L., Patáková, P., & Brányik, T. Advanced fermentation processes. Engineering Aspects of Food Biotechnology 2013, 29, 89–110.

[13] Sari, B., Isik, M., Eylem, C. C., Bektas, C., Okesola, B. O., Karakaya, E., Emregul, E., Nemutlu, E., & Derkus, B. Omics technologies for high-throughput-screening of cell–biomaterial interactions. Molecular Omics 2022, 18, 591–615.

[14] Sarnaik, A., Liu, A., Nielsen, D., & Varman, A. M. High-throughput screening for efficient microbial biotechnology. Current Opinion in Biotechnology 2020, 64, 141–150.
[15] Sharma, R., Garg, P., Kumar, P., Bhatia, S. K., & Kulshrestha, S. Microbial fermentation and its role in quality improvement of fermented foods. Fermentation 2020, 6, 106.
[16] Siddiqui, S. A., Erol, Z., Rugji, J., Taşçı, F., Kahraman, H. A., Toppi, V., Musa, L., Di Giacinto, G., Bahmid, N. A., Mehdizadeh, M., & Castro-Muñoz, R. An overview of fermentation in the food industry-looking back from a new perspective. Bioresources and Bioproducts 2023, 10, 85.
[17] Ezemba, C., Ekwegbalu, E. A., & Ezemba, A. S. Fermentation, types of fermenters, design & uses of fermenters and optimization of the fermentation process. 2022, 1–121.
[18] Du, Y. H., Wang, M. Y., Yang, L. H., Tong, L. L., Guo, D. S., & Ji, X. J. Optimization and scale-up of fermentation processes driven by models. Bioengineering 2022, 9, 473.
[19] Behera, S. S., Ray, R. C., Das, U., Panda, S. K., & Saranraj, P. Microorganism in fermentation. In: Berenjian, A., editor. Essentials in Fermentation Technology, Switzerland: Springer Nature, 2019, 1–39.
[20] Sanchez, S., & Demain, A. L. Metabolic regulation of fermentation processes. Enzyme & Microbial Technology 2002, 31, 895–906.
[21] Peng, Q., Jiang, S., Chen, J., Ma, C., Huo, D., Shao, Y., & Zhang, J. Unique microbial diversity and metabolic pathway features of fermented vegetables from Hainan, China. Frontiers in Microbiology 2018, 9, 399.
[22] Olanbiwoninu, A. A., & Odunfa, S. A. Microbial interaction in selected fermented vegetable condiments in Nigeria. International Food Research Journal 2018, 25, 439–445.
[23] Sarsan, S., Pandiyan, A., Rodhe, A. V., & Jagavati, S. Synergistic interactions among microbial communities. In: Raghvendra, P. S., Geetanjali, M., Kaushik, B., & Hovik, P., editors. Microbes in Microbial Communities: Ecological and Applied Perspectives, Switzerland: Springer, 2021, 1–37.
[24] Maiorella, B., Blanch, H. W., & Wilke, C. R. By-product inhibition effects on ethanolic fermentation by *Saccharomyces cerevisiae*. Biotechnology & Bioengineering 1983, 25, 103–121.
[25] Herrero, A. A. End-product inhibition in anaerobic fermentations. Trends in Biotechnology 1983, 1, 49–53.
[26] Chen, Y., Yin, Y., & Wang, J. Recent advance in inhibition of dark fermentative hydrogen production. International Journal of Hydrogen Energy 2021, 46, 5053–5073.
[27] Huang, Y., Wang, Y., Shang, N., & Li, P. Microbial fermentation processes of lactic acid: Challenges, solutions, and future prospects. Foods 2023, 12, 2311.
[28] Sharma, R., Garg, P., Kumar, P., Bhatia, S. K., & Kulshrestha, S. Microbial fermentation and its role in quality improvement of fermented foods. Fermentation 2020, 6, 106.
[29] Chai, W. Y., Teo, K. T., Tan, M. K., & Tham, H. J. Fermentation process control and optimization. Chemical Engineering & Technology 2022, 45, 1731–1747.
[30] Ammar, E. M., & Philippidis, G. P. Fermentative production of propionic acid: Prospects and limitations of microorganisms and substrates. Applied Microbiology and Biotechnology 2021, 105, 6199–6213.
[31] Esteban, N. Potential and Limitations of Co-fermentation: A Review, University of Barcelona, Master dissertation, 2022, 1–56.
[32] Brethauer, S., & Wyman, C. E. Continuous hydrolysis and fermentation for cellulosic ethanol production. Bioresource Technology 2010, 101, 4862–4874.
[33] Meyer, H. P., Minas, W., & Schmidhalter, D. Industrial-scale fermentation. Industrial Biotechnology: Products and Processes 2017, 1–53.
[34] Makwana, M., & Hati, S. Fermented beverages and their health benefits. Fermented Beverages 2019, 5, 1–29.
[35] Wehrs, M., de Beaumont-Felt, A., Goranov, A., Harrigan, P., de Kok, S., Lieder, S., Vallandingham, J., & Tyner, K. You get what you screen for: On the value of fermentation characterization in high-

throughput strain improvements in industrial settings. Journal of Industrial Microbiology and Biotechnology 2020, 47, 913–927.

[36] Conacher, C. G., Luyt, N. A., Naidoo-Blassoples, R. K., Rossouw, D., Setati, M. E., & Bauer, F. F. The ecology of wine fermentation: A model for the study of complex microbial ecosystems. Applied Microbiology and Biotechnology 2021, 105, 3027–3043.

[37] Waqas, M., Rehan, M., Khan, M. D., & Nizami, A. S. Conversion of food waste to fermentation products. Encyclopedia of Food Security and Sustainability 2019, 1, 501–509.

[38] Zannini, E., Lynch, K. M., Nyhan, L., Sahin, A. W., O'Riordan, P., Luk, D., & Arendt, E. K. Influence of substrate on the fermentation characteristics and culture-dependent microbial composition of water kefir. Fermentation 2022, 9, 28.

[39] Steudler, S., Werner, A., & Walther, T. It is the mix that matters: Substrate-specific enzyme production from filamentous fungi and bacteria through solid-state fermentation. Solid State Fermentation: Research and Industrial Applications 2019, 51–81.

[40] Al Daccache, M., Louka, N., Maroun, R. G., & Salameh, D. Culture condition changes for enhancing fermentation processes. Fermentation Processes: Emerging and Conventional Technologies 2021, 97–116.

[41] Ferreira, L. J., de Souza Gomes, M., de Oliveira, L. M., & Santos, L. D. Coffee fermentation process: A review. Food Research International 2023, 13, 112793.

[42] Karthikeyan, A., Joseph, A., Subramanian, R., & Nair, B. G. Fermenter design. In: Industrial Microbiology and Biotechnology, Singapore: Springer, 2022, 129–167.

[43] Adebo, J. A., Njobeh, P. B., Gbashi, S., Oyedeji, A. B., Ogundele, O. M., Oyeyinka, S. A., & Adebo, O. A. Fermentation of cereals and legumes: Impact on nutritional constituents and nutrient bioavailability. Fermentation 2022, 8, 63.

[44] Shanmugam, M. K., Mandari, V., Devarai, S. K., & Gummadi, S. N. Types of bioreactors and important design considerations. In: Current Developments in Biotechnology and Bioengineering, Elsevier, 2022, 3–30.

[45] Mahdinia, E., Cekmecelioglu, D., & Demirci, A. Bioreactor scale-up. In: Berenjian, A., editor. Essentials in Fermentation Technology, Switzerland: Springer Nature, 2019, 213–236.

[46] Reyes, S. J., Durocher, Y., Pham, P. L., & Henry, O. Modern sensor tools and techniques for monitoring, controlling, and improving cell culture processes. Processes 2022, 18(10), 189.

[47] Jawed, K., Yazdani, S. S., & Koffas, M. A. Advances in the development and application of microbial consortia for metabolic engineering. Metabolic Engineering Communications 2019, 9, e00095.

[48] Dasgupta, A., Chowdhury, N., & De, R. K. Metabolic pathway engineering: Perspectives and applications. Computer Methods and Programs in Biomedicine 2020, 192, 105436.

[49] Ko, Y. S., Kim, J. W., Lee, J. A., Han, T., Kim, G. B., Park, J. E., & Lee, S. Y. Tools and strategies of systems metabolic engineering for the development of microbial cell factories for chemical production. Chemical Society Reviews 2020, 49, 4615–4636.

[50] Begum, P. S., Rajagopal, S., & Razak, M. A. Emerging trends in microbial fermentation technologies. Recent Developments in Applied Microbiology and Biochemistry 2021, 113–119.

[51] Graham, A. E., & Ledesma-Amaro, R. The microbial food revolution. Nature Communications 2023, 14, 2231.

[52] Shade, A., Peter, H., Allison, S. D., Baho, D. L., Berga, M., Bürgmann, H., Huber, D. H., Langenheder, S., Lennon, J. T., Martiny, J. B., & Matulich, K. L. Fundamentals of microbial community resistance and resilience. Frontiers in Microbiology 2012, 3, 417.

[53] Prosser, J. I., Bohannan, B. J., Curtis, T. P., Ellis, R. J., Firestone, M. K., Freckleton, R. P., Green, J. L., Green, L. E., Killham, K., Lennon, J. J., & Osborn, A. M. The role of ecological theory in microbial ecology. Nature Reviews Microbiology 2007, 5, 384–392.

[54] Zhou, S., & Alper, H. S. Strategies for directed and adapted evolution as part of microbial strain engineering. Journal of Chemical Technology and Biotechnology 2019, 94, 366–376.

[55] Luo, H., Yang, R., Zhao, Y., Wang, Z., Liu, Z., Huang, M., & Zeng, Q. Recent advances and strategies in process and strain engineering for the production of butyric acid by microbial fermentation. Bioresource Technology 2018, 253, 343–354.
[56] Becker, J., & Wittmann, C. Advanced biotechnology: Metabolically engineered cells for the bio-based production of chemicals and fuels, materials, and health-care products. Angewandte Chemie International Edition 2015, 54, 3328–3350.
[57] Brenner, K., You, L., & Arnold, F. H. Engineering microbial consortia: A new frontier in synthetic biology. Trends in Biotechnology 2008, 26, 483–489.
[58] Duncker, K. E., Holmes, Z. A., & You, L. Engineered microbial consortia: Strategies and applications. Microbial Cell Factories 2021, 20, 1–3.
[59] Shong, J., Diaz, M. R., & Collins, C. H. Towards synthetic microbial consortia for bioprocessing. Current Opinion in Biotechnology 2012, 23, 798–802.
[60] Khalil, A. S., & Collins, J. J. Synthetic biology: Applications come of age. Nature Reviews Genetics 2010, 11, 367–379.
[61] Adams, B. L. The next generation of synthetic biology chassis: Moving synthetic biology from the laboratory to the field. ACS Synthetic Biology 2016, 5, 1328–1330.
[62] Ansori, A. N., Antonius, Y., Susilo, R. J., Hayaza, S., Kharisma, V. D., Parikesit, A. A., Zainul, R., Jakhmola, V., Saklani, T., Rebezov, M., & Ullah, M. E. Application of CRISPR-Cas9 genome editing technology in various fields: A review. Narra Journal 2023, 3(2).
[63] Zhang, X., Lin, Y., Wu, Q., Wang, Y., & Chen, G. Q. Synthetic biology and genome-editing tools for improving PHA metabolic engineering. Trends in Biotechnology 2020, 38, 689–700.
[64] Kwon, S. W., Paari, K. A., Malaviya, A., & Jang, Y. S. Synthetic biology tools for genome and transcriptome engineering of solventogenic *Clostridium*. Frontiers in Bioengineering and Biotechnology 2020, 8, 282.
[65] Gupta, S., & Pappachan, A. Approaches and tools of protein tailoring for metabolic engineering. In: Current Developments in Biotechnology and Bioengineering, Elsevier, 2022, 131–150.
[66] Marcheschi, R. J., Gronenberg, L. S., & Liao, J. C. Protein engineering for metabolic engineering: Current and next-generation tools. Biotechnology Journal 2013, 8, 545–555.
[67] Savino, S., Desmet, T., & Franceus, J. Insertions and deletions in protein evolution and engineering. Biotechnology Advance 2022, 60, 108010.
[68] Guan, A., He, Z., Wang, X., Jia, Z. J., & Qin, J. Protein engineering for the synthetic cell factory: Recent advance and perspective. Biotechnology Advance 2024, 108366.
[69] Lv, X., Hueso-Gil, A., Bi, X., Wu, Y., Liu, Y., Liu, L., & Ledesma-Amaro, R. New synthetic biology tools for metabolic control. Current Opinion in Biotechnology 2022, 76, 102724.
[70] Montaño López, J., Duran, L., & Avalos, J. L. Physiological limitations and opportunities in microbial metabolic engineering. Nature Reviews Microbiology 2022, 20, 35–48.
[71] Chaudhary, R., Nawaz, A., Fouillaud, M., Dufossé, L., Haq, I. U., & Mukhtar, H. Microbial cell factories: Biodiversity, pathway construction, robustness, and industrial applicability. Microbiological Research 2024, 15, 247–272.
[72] Luo, Q., Ding, N., Liu, Y., Zhang, H., Fang, Y., & Yin, L. Metabolic engineering of microorganisms to produce pyruvate and derived compounds. Molecules 2023, 28, 1418.
[73] Park, S. Y., Yang, D., Ha, S. H., & Lee, S. Y. Metabolic engineering of microorganisms for the production of natural compounds. Advanced Biosystems 2018, 2, 1700190.
[74] Zhu, K., Kong, J., Zhao, B., Rong, L., Liu, S., Lu, Z., Zhang, C., Xiao, D., Pushpanathan, K., Foo, J. L., & Wong, A. Metabolic engineering of microbes for monoterpenoid production. Biotechnology Advance 2021, 53, 107837.
[75] Lian, J., Mishra, S., & Zhao, H. Recent advances in metabolic engineering of Saccharomyces cerevisiae: New tools and their applications. Metabolic Engineering 2018, 50, 85–108.

[76] Gustavsson, M., & Lee, S. Y. Prospects of microbial cell factories developed through systems metabolic engineering. Microbial Biotechnology 2016, 9, 610–617.

[77] Fisher, A. K., Freedman, B. G., Bevan, D. R., & Senger, R. S. A review of metabolic and enzymatic engineering strategies for designing and optimizing performance of microbial cell factories. Computational and Structural Biotechnology Journal 2014, 11, 91–99.

[78] Volke, D. C., & Nikel, P. I. Getting bacteria in shape: Synthetic morphology approaches for the design of efficient microbial cell factories. Advanced Biosystems 2018, 2, 1800111.

[79] Acevedo-Rocha, C. G., Gronenberg, L. S., Mack, M., Commichau, F. M., & Genee, H. J. Microbial cell factories for the sustainable manufacturing of B vitamins. Current Opinion in Biotechnology 2019, 56, 18–29.

[80] Cho, J. S., Kim, G. B., Eun, H., Moon, C. W., & Lee, S. Y. Designing microbial cell factories for the production of chemicals. JACS Au 2022, 2, 1781–1799.

[81] Xia, P. F., Ling, H., Foo, J. L., & Chang, M. W. Synthetic genetic circuits for programmable biological functionalities. Biotechnology Advance 2019, 37, 107393.

[82] Singh, V., Haque, S., Niwas, R., Pasupuleti, M., & Tripathi, C. K. Strategies for fermentation medium optimization: An in-depth review. Frontiers in Microbiology 2017, 7, 227613.

[83] Parekh, S., Vinci, V. A., & Strobel, R. J. Improvement of microbial strains and fermentation processes. Applied Microbiology and Biotechnology 2000, 54, 287–301.

[84] Basiony, M., Ouyang, L., Wang, D., Yu, J., Zhou, L., Zhu, M., Wang, X., Feng, J., Dai, J., Shen, Y., & Zhang, C. Optimization of microbial cell factories for astaxanthin production: Biosynthesis and regulations, engineering strategies and fermentation optimization strategies. Synthetic and Systems Biotechnology 2022, 7, 689–704.

[85] Fackler, N., Heijstra, B. D., Rasor, B. J., Brown, H., Martin, J., Ni, Z., & Köpke, M. Stepping on the gas to a circular economy: Accelerating development of carbon-negative chemical production from gas fermentation. Annual Review of Chemical and Biomolecular Engineering 2021, 12, 439–470.

[86] Bourgade, B., Minton, N. P., & Islam, M. A. Genetic and metabolic engineering challenges of C1-gas fermenting acetogenic chassis organisms. FEMS Microbiology Reviews 2021, 45, fuab008.

[87] Babu, S., Rathore, S. S., Singh, R., Kumar, S., Singh, V. K., Yadav, S. K., & Wani, O. A. Exploring agricultural waste biomass for energy, food and feed production and pollution mitigation: A review. Bioresource Technology 2022, 360, 127566.

[88] Mitra, S., Paliya, S., & Mandpe, A. Microbial engineering in biofuel production – A global outlook, advances, and roadmap. In: Emerging Sustainable Technologies for Biofuel Production, Switzerland, Cham: Springer Nature, 2024, 547–593.

[89] Li, M., & Wen, J. Recent progress in the application of omics technologies in the study of bio-mining microorganisms from extreme environments. Microbial Cell Factories 2021, 20, 178.

[90] Meera, K. B., Khan, M. A., & Khan, S. T. Next-generation sequencing (NGS) platforms: An exciting era of genome sequence analysis. Microbial Genomics in Sustainable Agroecosystems 2019, 2, 89–109.

[91] de Sá, P. H., Guimarães, L. C., das Graças, D. A., de Oliveira, V. A. A., Barh, D., Azevedo, V., & Ramos, R. T. Next-generation sequencing and data analysis: Strategies, tools, pipelines and protocols. In: Omics Technologies and Bio-Engineering, Academic Press, 2018, 191–207.

[92] Chambers, D. C., Carew, A. M., Lukowski, S. W., & Powell, J. E. Transcriptomics and single-cell RNA-sequencing. Respirology 2019, 24, 29–36.

[93] Kulkarni, A., Anderson, A. G., Merullo, D. P., & Konopka, G. Beyond bulk: A review of single cell transcriptomics methodologies and applications. Current Opinion in Biotechnology 2019, 58, 129–136.

[94] Seger, K., Declerck, S., Mangelings, D., Heyden, Y. V., & Eeckhaut, A. V. Analytical techniques for metabolomic studies: A review. Bioanalysis 2019, 11, 2297–2318.

[95] Takis, P. G., Ghini, V., Tenori, L., Turano, P., & Luchinat, C. Uniqueness of the NMR approach to metabolomics. Trends in Analytical Chemistry 2019, 120, 115300.

[96] Letertre, M. P., Dervilly, G., & Giraudeau, P. Combined nuclear magnetic resonance spectroscopy and mass spectrometry approaches for metabolomics. Analytical Chemistry 2020, 93, 500–518.
[97] Jamil, I. N., Remali, J., Azizan, K. A., Nor Muhammad, N. A., Arita, M., Goh, H. H., & Aizat, W. M. Systematic multi-omics integration (MOI) approach in plant systems biology. Frontiers Plant Science 2020, 11, 540561.
[98] Graw, S., Chappell, K., Washam, C. L., Gies, A., Bird, J., Robeson, M. S., & Byrum, S. D. Multi-omics data integration considerations and study design for biological systems and disease. Molecular Omics 2021, 17, 170–185.
[99] Sindelar, R. D. Genomics, other "OMIC" technologies, precision medicine, and additional biotechnology-related techniques. In: Pharmaceutical Biotechnology: Fundamentals and Applications, Switzerland, Cham: Springer International Publishing, 2024, 209–254.
[100] McSweeney, M. A., & Styczynski, M. P. Effective use of linear DNA in cell-free expression systems. Frontiers in Bioengineering and Biotechnology 2021, 9, 715328.
[101] Mohr, B. Advancing a Systems Cell-free Metabolic Engineering Approach to Natural Product Synthesis and Discovery, University of Tennessee, Doctoral thesis, 2019, Dec, 1–144.
[102] Jiang, L., Zhao, J., Lian, J., & Xu, Z. Cell-free protein synthesis enabled rapid prototyping for metabolic engineering and synthetic biology. Synthetic and Systems Biotechnology 2018, 3, 90–96.
[103] Morgado, G., Gerngross, D., Roberts, T. M., & Panke, S. Synthetic biology for cell-free biosynthesis: Fundamentals of designing novel *in vitro* multi-enzyme reaction networks. Synthetic Biology and Metabolic Engineering 2018, 117–146.
[104] Laohakunakorn, N., Grasemann, L., Lavickova, B., Michielin, G., Shahein, A., Swank, Z., & Maerkl, S. J. Bottom-up construction of complex biomolecular systems with cell-free synthetic biology. Frontiers in Bioengineering and Biotechnology 2020, 8, 213.
[105] Hoffman, S. M., Tang, A. Y., & Avalos, J. L. Optogenetics illuminates applications in microbial engineering. Annual Review of Chemical and Biomolecular Engineering 2022, 13, 373–403.
[106] Chia, N., Lee, S. Y., & Tong, Y. Optogenetic tools for microbial synthetic biology. Biotechnology Advance 2022, 59, 107953.
[107] Baumschlager, A., & Khammash, M. Synthetic biological approaches for Optogenetics and tools for transcriptional Light-control in bacteria. Advances in Biology 2021, 5, 2000256.
[108] Pérez, A. L., Piva, L. C., Fulber, J. P., de Moraes, L. M., De Marco, J. L., Vieira, H. L., Coelho, C. M., Reis, V. C., & Torres, F. A. Optogenetic strategies for the control of gene expression in yeasts. Biotechnology Advance 2022, 54, 107839.
[109] Salinas, F., Rojas, V., Delgado, V., Agosin, E., & Larrondo, L. F. Optogenetic switches for light-controlled gene expression in yeast. Applied Microbiology and Biotechnology 2017, 101, 2629–2640.
[110] Lalwani, M. A. Dynamic and Modular Control of Microbial Chemical and Protein Production Using Optogenetics, Princeton University, 2021.
[111] Hair, S. M. Tools for Quantitative Optogenetics in *Bacillus Subtilis*, Doctoral dissertation, Rice University.
[112] Benisch, M., Aoki, S. K., & Khammash, M. Unlocking the potential of optogenetics in microbial applications. Current Opinion in Microbiology 2024, 77, 102404.
[113] Emiliani, V., Entcheva, E., Hedrich, R., Hegemann, P., Konrad, K. R., Lüscher, C., Mahn, M., Pan, Z. H., Sims, R. R., Vierock, J., & Yizhar, O. Optogenetics for light control of biological systems. Nature Reviews Methods Primers 2022, 2, 55.
[114] Vošahlík, J. IoT and measurement of fermentation process of rice wine. Agronomy Research 2023, 21, 1419–1426.
[115] Amore, A., & Philip, S. Artificial intelligence in food biotechnology: Trends and perspectives. Frontiers in Industrial Microbiology 2023, 1, 1255505.
[116] Florea, A., Sipos, A., & Stoisor, M. C. Applying AI Tools for modeling, predicting and managing the white wine fermentation process. Fermentation 2022, 8, 137.

[117] Mudziwapasi, R., Mufandaedza, J., Jomane, F. N., Songwe, F., Ndudzo, A., Nyamusamba, R. P., Takombwa, A. R., Mahla, M. G., Pullen, J., Mlambo, S. S., & Mahuni, C. Unlocking the potential of synthetic biology for improving livelihoods in sub-Saharan Africa. All Life 2022, 15, 1–12.
[118] Simate, G. S., Ndlovu, S., Iyuke, S. E., & Walubita, L. F. Biotechnology and nanotechnology: A means for sustainable development in Africa. Chemistry for Sustainable Development in Africa 2013, 159–191.
[119] García-Granados, R., Lerma-Escalera, J. A., & Morones-Ramírez, J. R. Metabolic engineering and synthetic biology: Synergies, future, and challenges. Frontiers in Bioengineering and Biotechnology 2019, 7, 36.
[120] Ramírez, R. A. A., Swidah, R., & Schindler, D. Microbes of traditional fermentation processes as synthetic biology chassis to tackle future food challenges. Frontiers in Bioengineering and Biotechnology 2022, 10, 982975.
[121] Voigt, C. A. Synthetic biology 2020–2030: Six commercially-available products that are changing our world. Nature Communications 2020, 11, 1–6.
[122] Amer, B., & Baidoo, E. E. Omics-driven biotechnology for industrial applications. Frontiers in Bioengineering and Biotechnology 2021, 9, 613307.
[123] Aggarwal, N., Pham, H. L., Ranjan, B., Saini, M., Liang, Y., Hossain, G. S., Ling, H., Foo, J. L., & Chang, M. W. Microbial engineering strategies to utilize waste feedstock for sustainable bioproduction. Nature Reviews Bioengineering 2024, 2, 155–174.
[124] Wu, H. L. Team for biotechnology organization. International Journal of Innovative Science and Research Technology 2022, 7, 1–6.
[125] Elahi, Y., & Baker, M. A. Light control in microbial systems. International Journal of Molecular Sciences 2024, 25, 4001.
[126] Pouzet, S., Banderas, A., Le Bec, M., Lautier, T., Truan, G., & Hersen, P. The promise of optogenetics for bioproduction: Dynamic control strategies and scale-up instruments. Bioengineering 2020, 7, 151.
[127] Bhardwaj, A., Kishore, S., & Pandey, D. K. Artificial intelligence in biological sciences. Life 2022, 12, 1430.
[128] Yang, C. T., Kristiani, E., Leong, Y. K., & Chang, J. S. Big data and machine learning driven bioprocessing–recent trends and critical analysis. Bioresource Technology 2023, 372, 128625.
[129] Yu, H., Liu, S., Qin, H., Zhou, Z., Zhao, H., Zhang, S., & Mao, J. Artificial intelligence-based approaches for traditional fermented alcoholic beverages' development: Review and prospect. Critical Reviews in Food Science and Nutrition 2024, 64, 2879–2889.
[130] Gargalo, C. L., Udugama, I., Pontius, K., Lopez, P. C., Nielsen, R. F., Hasanzadeh, A., Mansouri, S. S., Bayer, C., Junicke, H., & Gernaey, K. V. Towards smart biomanufacturing: A perspective on recent developments in industrial measurement and monitoring technologies for bio-based production processes. Journal of Industrial Microbiology and Biotechnology 2020, 47, 947–964.
[131] Roell, M. S., & Zurbriggen, M. D. The impact of synthetic biology for future agriculture and nutrition. Current Opinion in Biotechnology 2020, 61, 102–109.
[132] Mondal, P. P., Galodha, A., Verma, V. K., Singh, V., Show, P. L., Awasthi, M. K., Lall, B., Anees, S., Pollmann, K., & Jain, R. Review on machine learning-based bioprocess optimization, monitoring, and control systems. Bioresource Technology 2023, 370, 128523.
[133] Cheng, Y., Bi, X., Xu, Y., Liu, Y., Li, J., Du, G., Lv, X., & Liu, L. Artificial intelligence technologies in bioprocess: Opportunities and challenges. Bioresource Technology 2023, 369, 128451.
[134] Volk, M. J., Lourentzou, I., Mishra, S., Vo, L. T., Zhai, C., & Zhao, H. Biosystems design by machine learning. ACS Synthetic Biology 2020, 9, 1514–1533.

Adetunji Adeoluwa Iyiade* and Adetunji Rose Oluwaseun

Chapter 11 Artificial intelligence and machine learning applications in food system optimization

Abstract: Agri-food systems face significant challenges as a result of issues such as population growth, climate change, and scarce resources. Presently, research studies are investigating methods for improvement in agri-food systems in order to tackle these challenges. The goal is to enhance productivity, effectiveness, and sustainability over time. This chapter review offers a concise summary of the field, emphasizing significant methodologies, possible advantages, and current constraints. Given the intricate nature and difficulties of contemporary food systems, it is imperative to develop innovative approaches that provide effectiveness, environmental friendliness, and the availability of food for a constantly growing global population. Artificial intelligence (AI) and machine learning (ML) are powerful tools that can improve different facets of the agri-food system, ranging from agricultural production to consumption. This chapter explores the capacity of AI and ML to further transform and optimize agri-food systems, with a specific emphasis on essential characteristics and the advantages they offer.

Keywords: Agri-food systems, artificial intelligence, innovations, food security, machine learning, optimization

11.1 Introduction

Agri-food systems are complex systems that encompass the complete journey of food, ranging from its production to its consumption. This encompasses an extensive network of operations pertaining to the growth, processing, distribution, preparation, and consumption of food. According to Tzachor et al. [1], the current state of agri-food systems is not aligned with the Sustainable Development Goals (SDGs) [1]. This can be attributed to substantial obstacles that the entire food system encounters, such as limited resources, climate change, growing population, and escalating food wastage [2]. The optimization of food systems, in accordance with the SDGs, necessitates addressing the challenges pertaining to food security, nutrition, and environmental sustain-

***Corresponding author: Adetunji Adeoluwa Iyiade,** Labworld/Philafrica Foods (Pty) Ltd, 11 Quality Road, Isando, Kempton Park, Gauteng, South Africa, e-mail: iyiade01@gmail.com
Adetunji Rose Oluwaseun, JBS Innovation Lab, Johannesburg Business School, University of Johannesburg, JBS Park, 69 Kingsway Avenue, Auckland Park, South Africa

https://doi.org/10.1515/9783111441238-012

ability. According to Geissdoerfer et al. [3], sustainability can be defined as a continuous effort to uphold a harmonious integration of social, environmental, and economic aspects in order to yield advantages for both present and future generations. Various researches have been conducted, emphasizing the need for innovative strategies toward improving the efficiency of food production, processing, distribution, and consumption.

This chapter focuses on a range of technology advancements that can be effectively employed within the food systems for its optimization and sustainability such as artificial intelligence (AI) and machine learning (ML). Several scholarly publications have reported that the implementation of AI and ML technologies has become important tools in supporting agri-food systems to tackle the issues posed by a rapidly increasing global population, and assuring food security [1, 2, 4]. This is based on their capacity to revolutionize the agri-food business through optimizing production, sustainability, and decision-making processes [4]. This review also highlights the significance of AI-based algorithms in providing timely analytical insights to facilitate proactive decision-making. Furthermore, it provides insights into how these advanced technologies can further enhance productivity in agri-food systems by implementing policies that promote their widespread adoption. Overall, this chapter acknowledges the current application of these technologies across many stages of the food chain, while emphasizing the importance of conducting thorough testing and risk-benefit assessments in order to boost their general adoption.

11.2 Understanding food systems and the challenges of modern food systems

Food systems demonstrate intricate and interdependent structures. As highlighted by Vermeulen et al. [5], food system comprises a series of complex activities (Figure 11.1), including the production and distribution of agricultural inputs such as seeds, animal feed, fertilizers, and pest control. It also encompasses agricultural production, including crops, livestock, fisheries, and edible wildlife. Additionally, it involves the primary and secondary processing of agricultural produce, as well as the packaging, storage, transportation, and distribution of these products [6]. Furthermore, it encompasses marketing and retailing, catering services, domestic food management, and waste management. However, food systems comprise not only the entire sequence of activities as listed, but also the resulting effects and organizational structure of these activities, involving societal, economic, and environmental factors that impact these processes [7, 8].

In the recent past, global food systems have encountered a wide range of challenges, encompassing concerns pertaining to nutrition, environmental deterioration, and climate change [9]. A change in any aspect of the food systems has the potential to lead to a series of interconnected effects throughout the entire system. For example, impact of climate change, leading to the occurrence of drought, could negatively affect agricultural productivity, thus resulting in a rise in food costs [10]. The phenomenon outlined above has the potential to have negative impacts on food security, nutrition, and livelihoods. In view of this, there is a growing consensus among experts that the existing food system is unsustainable in effectively tackling these difficulties, thereby necessitating a fundamental overhaul of agricultural and food systems [9]. Numerous research investigations, detailed literature reviews, and extensive discussions have highlighted the significance of transforming agricultural and food systems in order to effectively tackle the multifaceted concerns pertaining to nutrition, poverty, and environmental degradation.

According to Anderson et al., agroecology has gained significant prominence in scientific, agricultural, and political conversations [11]. As a scientific field, agroecology encompasses a range of activities and social movements that provide avenues for achieving sustainable food systems [12]. This concept is based on the integration of ecological principles and social values employed to establish comprehensive and resilient agri-food systems that effectively tackle the various issues encountered by global food systems [9]. Agroecology approach has the goal of facilitating recycling, minimizing inputs, enhancing soil and animal health, promoting biodiversity, facilitating economic diversification, fostering knowledge co-creation, promoting social values and diets, ensuring fairness, establishing connectivity, governing land and natural resources, and encouraging participation, as highlighted [9]. The significance of agroecology in the transformation of agricultural and food systems toward sustainability has been widely acknowledged [9, 11]. Their review also reiterated the significance of considering soil quality and health as crucial indicators of sustainable agriculture. In addition, research has advocated for a worldwide outlook on sustainable agriculture, considering the efficient application of mineral and organic fertilizers, as well as the incorporation of traditional knowledge and the standardization of evaluation techniques and metrics for measuring effects [12].

11.2.1 Challenges in achieving sustainable agri-food systems

The transition toward sustainable agri-food systems is not without challenges. Lack of integration and coordination among many stakeholders, including agricultural producers, policymakers, researchers, and consumers, poses a significant challenge [14]. This interferes with the achievement of comprehensive and unified strategies toward sustainability of food systems [14]. An additional impediment is the requirement to make trade-offs and maintain a delicate equilibrium in order to attain sustainability across economic, social, and environmental aspects [6, 9]. Wezel et al. [9] indicated that certain agroecological approaches have the potential to enhance biodiversity and ecological services. However, it is important to acknowledge that these approaches

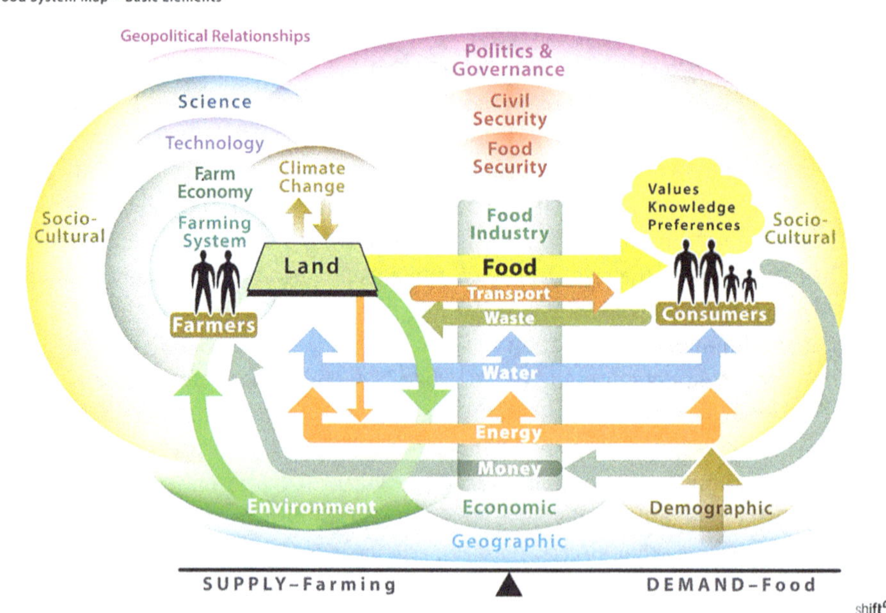

Figure 11.1: Comprehensive representation of the global food systems [13].

may also include trade-offs in relation to productivity or profitability. In addition, the process of promoting sustainable agricultural methods and ensuring their adoption among small-scale farmers in low- and middle-income economies can present difficulties, as a result of limited availability of resources, knowledge, and markets [9]. Furthermore, the impacts of climate change pose a significant challenge to sustainable agri-food systems [15]. The implications encompass heightened occurrence and severity of extreme weather phenomena, alterations in rainfall patterns, and escalating temperatures, all of which have the potential to undermine agricultural productivity and heighten susceptibility to food insecurity.

Another impediment lies in the necessity for further investigation and advancement in sustainable agriculture methodologies [16]. This encompasses the advancement of novel technologies, methodologies, and regulations that have the potential to improve the sustainability of agri-food systems, while simultaneously tackling the distinct obstacles encountered in various geographical areas and circumstances. In addition, the fluctuating preferences and behaviors of consumers present a significant obstacle in the pursuit of sustainable agri-food systems [17]. The pursuit of convenience, affordability, and a diverse range of food choices by consumers, commonly drives unsustainable production methods and exacerbates concerns related to food waste and excessive consumption.

Finally, the agri-food industry's globalized nature and its dependence on international trade pose challenges to attaining sustainability [1]. To address these difficulties, it is necessary to foster collaboration across different fields, garner policy backing, and allocate resources toward research and innovation. This will enable the development of solutions that could contribute to sustainable agri-food systems [14]. In recent years, there has been a notable increase in focus and improvements pertaining to the sustainability of agri-food systems [9]. Nevertheless, the attainment of really sustainable agri-food systems requires addressing a number of challenges [18], thereby increasing the need for significant shifts in agri-food systems in order to attain sustainability. According to Melchior and Newig [18], the proposed modifications encompass the resolution of the trade-offs between biodiversity and ecological services, the promotion of sustainable practices for small-scale farmers, adapting to the impacts of climate change, and the encouragement of sustainable consumer behavior. Further to this, is a notable surge in scientific research and research interest pertaining to sustainable agri-food systems [19]. This development can be attributed to the recognition of knowledge and implementation limitations, as well as a greater emphasis on overcoming obstacles that hinder the attainment of sustainable agri-food systems.

11.2.2 Developments in agri-food sustainability

There have been major developments in addressing challenges to attaining agri-food optimization and sustainability. These developments encompass a greater emphasis on the trade-offs between biodiversity and sustainability, the promotion of sustainable approaches for small-scale farmers, adjusting to the effects of climate change, and the encouragement of sustainable consumer behavior [18]. Some of the key trends and developments include:

1. Increased research and understanding of the trade-offs between biodiversity conservation and ecosystem services in agri-food systems [19].
2. Increased efforts to scale up sustainable practices for smallholder farmers, including the promotion of agroecological approaches and agroforestry systems [9].
3. The emergence of climate change as a key driver of change in agri-food systems, leading to increased research and adaptation efforts [19].
4. The recognition of the importance of sustainable consumer behavior in driving positive change in agri-food systems, leading to increased efforts to promote awareness and education about sustainable food choices [20].

5. The emergence of new technologies and innovations that have the potential to transform agri-food systems, such as precision agriculture, vertical farming, and alternative protein sources such as plant-based and lab-grown meats [9].

However, despite all these developments, there persist notable limitations in understanding challenges and implementing solutions pertaining to the attainment of sustainability in agri-food systems [19]. The gaps consist several aspects, such as a limited comprehension of the trade-offs associated with various sustainability objectives, limited availability of tools and support for small-scale farmers to embrace sustainable practices, difficulties in expanding sustainable initiatives beyond initial pilot projects, insufficient consumer awareness and awareness of sustainable food choices, and the necessity for more comprehensive and integrated strategies to tackle the intricate challenges of agri-food sustainability [9]. Moreover, it is increasingly acknowledged that achieving the goal of agri-food sustainability necessitates more than just technological advancements or modifications in individual behaviors. Instead, it demands comprehensive transformations throughout the entirety of the agri-food system, encompassing policies, institutions, and market frameworks [18].

Furthermore, there has been a notable increase in efforts aimed at extending sustainable practices targeting smallholder farmers. This includes the active promotion of agroecological techniques and agroforestry systems, as highlighted by Wezel et al. [9]. As part of the development drive, climate change serves as a significant catalyst for transformation in agri-food systems, prompting greater research and efforts toward adaptation of innovations [21]. The acknowledgment of the significance of sustainable consumer behavior in facilitating favorable transformations in agri-food systems has potential to promote research and actionable interventions in advancing sustainability of agri-food markets through understanding and training regarding sustainable food choices [20]. Given the issues that have been identified, there has been a significant increase in research and innovation aimed at improving various agricultural practices and technologies, with the objective of optimizing and ensuring the sustainability of agri-food systems [17, 19]. This increased research and innovation in various agricultural practices and technologies, led to the adoption of vertical farming, precision agriculture, organic agriculture, regenerative agriculture, and agroecology, among others [9]. The aforementioned methodologies prioritize the improvement of soil health, biodiversity, and ecological resilience, while concurrently diminishing the reliance for synthetic inputs and promoting resource efficiency.

11.3 Technological advancements of agri-food system optimization and sustainability

Achieving successful sustainable agri-food initiatives demonstrate the potential for transformative change in the sector [22]. One of the approaches to this is through the adoption of intelligent and innovative technologies such as AI and ML [23]. For example, precision agriculture technologies, such as remote sensing and data analytics, can help optimize resource use and minimize environmental impacts [22]. Additionally, the application of blockchain technology in the agri-food sector has the potential to improve traceability and transparency [24, 25], allowing consumers to make informed choices about the products they purchase and their environmental and social impacts. Adoption of AI and ML technologies are revolutionizing the entire agri-food sectors and addressing these challenges. Some of the potentials of AI and ML techniques include optimization of crop management practices, improvement in yield prediction, detection of diseases and weeds, enhancement of livestock production and welfare, management of water resources, and improvement in soil management. While there have been positive transformation of agri-food systems toward sustainability in reducing emissions of greenhouse gases, increasing biodiversity, and improving livelihoods through these technologies, there are still barriers and challenges that need to be addressed for widespread adoption and impact [26]. Notwithstanding, the past decades have shown a growing recognition of the need for sustainability in agri-food systems [18], leading to increased research and innovation in various agricultural practices and technologies, as well as a focus on the impact of technological advancements on sustainability.

Likewise, there has been a significant emphasis on leveraging AI and ML to address issues in food processing [27]. Adoption of these technologies were noted to enable the industry to better monitor and manage quality, nutritional content, and processing methods, ultimately leading to improved products that align with consumer preferences. Across different sectors of the food industry, including dairy, bakery, beverages, and fruit and vegetable processing, AI and ML are being applied to optimize processes and enhance product quality [28, 29]. For example, in dairy processing, these technologies can be used for predictive maintenance of equipment, real-time monitoring of product quality, and even precision farming techniques for better milk production. Also, AI-based technology can assist in recipe development, ingredient sourcing, and even personalized product recommendations based on consumer preferences [30]. Additionally, the integration of robotics in food and beverage production is gaining momentum, offering benefits such as increased efficiency, consistency, and safety [31]. Robots can handle tasks such as sorting, packaging, and even cooking, reducing human error and improving overall productivity. Adoption of technological advancements such as digital technologies and precision agriculture have shown great potential in enhancing agri-food systems' sustainability [32]. These technologies can increase productivity, resource efficiency, and coordination in the food chain, leading to improved sustainability outcomes. However, it is important to consider the potential negative impacts of these technologies on sustainability.

11.3.1 Artificial intelligence application in agri-food system optimization

AI is a broad field of computer science concerned with creating intelligent machines capable of performing tasks that typically require human intelligence [33]. According to Taneja et al. [34], AI is becoming more increasingly adopted in the agri-food industry to enhance productivity, efficiency, and sustainability. The technology has the capacity to revolutionize the agri-food industry in various ways, such as smart/precision agriculture, crop monitoring, predictive analytics, supply chain optimization, food processing, quality control, tailored nutrition, and food safety. Application of AI in the field of agriculture is also referred to as agricultural intelligence (Figure 11.2). Thus, several applications of agricultural intelligence have been created, including smart/precision farming, disease detection, and crop phenotyping [35–37], to mention a few.

AI entails the development of algorithms and computational models that enables machines to efficiently handle and analyze vast quantities of data, detect patterns and correlations, and provide predictions or judgments based on such analysis [34]. This involves applications of tools such as ML, deep learning (DL), image processing, artificial neural networks, wireless sensor network technology, robotics, and the Internet of things (IoT) [23] to optimize agri-food practices and enhance productivity. The integration of these technologies has the potential to reduce the use of chemicals, improve soil fertility, reduce the amount of water usage and ultimately lead to more sustainable and efficient food production methods that can help meet the demands of a growing global population [36]. However, there are several challenges that need to be addressed in order for AI-based technologies to be widely adopted in the agricultural industry [39]. As highlighted, these challenges include uneven distribution of mechanization, the ability of algorithms to process large amounts of data accurately and quickly, and concerns regarding the security and privacy of data collected by AI systems.

Notwithstanding these challenges, recent trend has shown a growing demand for the adoption of these intelligent systems in the agri-food industries due to the increasing global population and the need for efficient production methods [4]. In view of this, AI has an integral part to play in meeting this rising food demand in a sustainable and efficient manner. As reviewed, these intelligent systems have been also deployed in various food production tasks such as food quality determination, quality control tools, classification of food, and prediction purposes [4]. These are based on

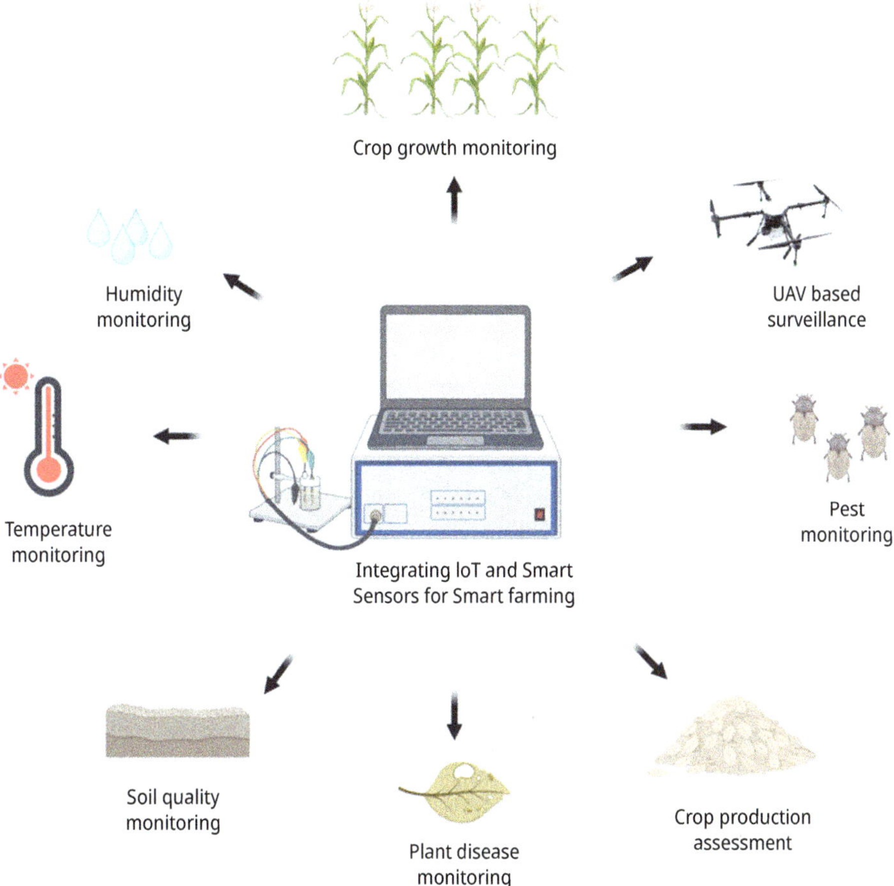

Figure 11.2: An overview of a smart/precision farming operations [38].

the deployment of systems such as ML, DL, image processing, artificial neural networks, and other AI techniques to analyze data and make informed decisions.

11.3.2 Machine learning as a tool in agri-food system optimization

The application of AI techniques to improve the performance of agri-food systems has been increasingly attracting interest. ML is a technique that has been recognized as a valuable tool for aiding decision makers and improving the efficiency of large-scale processes, such as supply chain management [40]. ML algorithms are mostly employed to address complex issues that transcend human skill, such as weather forecasting, spam detection, plant disease identification, and pattern recognition [41]. By employing these ML algorithms, agri-food production businesses can greatly improve their efficiency and optimize several areas of their operations, such as water conservation, soil and plant health, and sustainability [42, 43]. These ML algorithms can also assist in several activities, including predicting demand, optimizing production schedules, enhancing quality control, and minimizing waste, as reviewed [44]. Sharma et al. reviewed several studies that have shown the efficacy of ML algorithms in enhancing the efficiency and productivity of agri-food production operations [41]. These ML algorithms facilitate the analysis of huge volumes of data and identify patterns and trends that humans might miss, resulting in more precise forecasts and well-informed decision-making [45].

Furthermore, ML in agriculture has also demonstrated its potential to address challenges related to climate change and food safety. For example, in precision farming, ML algorithms have been employed to analyze data collected through digital devices, such as sensors, to effectively optimize agricultural practices such as irrigation, fertilization, and pest management [46, 47], as shown in Figure 11.2. Additional areas where ML algorithms have been applied include the prediction of crop yields and the implementation of intelligent harvesting operations [48], enabling farmers to make informed decisions about planting, harvesting times, and resource allocation. Finally, supply chain optimization has been enabled through the application of ML algorithms in analyzing data from various sources in predicting weather conditions, transportation routes, and market demand, and to optimize the logistics and distribution of agri-food products, thereby reducing waste, and ensuring timely delivery [43]. Through the adoption of these innovative technologies in agri-food processing optimization, there is a great capacity to significantly enhance production quality, profitability, and sustainability in the business, while simultaneously tackling urgent global issues by promoting more environmentally friendly and streamlined supply chain management systems [49].

11.3.2.1 Impact of machine learning on agri-food systems' efficiency

ML, as a cutting-edge tool, has the capacity to significantly enhance efficiency in agri-food systems [2]. This is accomplished through the collection and analysis of data from sources such as sensors, satellites, and drones, and then ML algorithms are utilized to analyze the collected data, with the aim of identifying patterns and insights. These are then used to make informed decisions when implementing targeted interventions, ultimately in optimizing various aspects of the processes involved in agri-food systems, such as production scheduling, quality control, supply chain management, and waste management [50] (Figure 11.3). More so, ML has the capability to create predictive models that can precisely predict demand and market trends [51]. By implementing AI, food processing companies can make well-informed decisions regarding production scale, inventory management, and resource allocation. The deployment of ML can lead to substantial cost reductions, enhanced efficiency, and enhanced product quality, thereby benefiting both food processing businesses and consumers. Furthermore, ML algorithms can be utilized to improve food safety by identifying impurities and pathogens in food products, as reviewed by Wang et al. [52]. This aids in mitigating the occurrence of epidemics and guaranteeing the safety of the food distribution network.

The implementation of ML in supply chain management has the capacity to transform and modernize supply chain operations in the food industry. Supply chain management systems utilizing AI can effectively optimize inventory management, save transportation expenses, boost demand forecasting, and improve overall operational efficiency [53]. Food enterprises can utilize AI technologies, like ML, natural language processing, and predictive analytics, to obtain immediate insights into their supply chain operations. This enables them to take proactive measures to reduce risks and address disruptions promptly [54]. Additionally, these AI solutions can be employed to oversee and regulate operations across the entire supply chain, thereby reducing any potential delays and optimizing profit margins. Furthermore, adoption of a ML-based methodology has been successfully deployed to assess the performance of food delivery services with the aim of enhancing operational effectiveness by improving consumer satisfaction [55, 56]. Similarly, these advanced technologies are essential in ensuring food safety by allowing enterprises to analyze and monitor products at each stage of the supply chain for possible hazards or contamination [57, 58]. Therefore, incorporating AI-based technologies throughout the agri-food supply chain can result in significant advantages. These include cost reduction, reduction in errors, enhanced decision-making efficiency and accuracy, and improved overall productivity and sustainability through effective training of personnel [59].

11.3.3 Challenges and opportunities in applying AI and ML to agri-food system optimization

11.3.3.1 Challenges in adoption of AI and ML in agri-food systems

Although ML and AI have significant advantages in optimizing agri-food processing and managing supply chains, there are also obstacles that hinder their successful adoption and deployment. The hurdles and constraints encompass potential biases in the algorithms, the necessity for skilled workers to create and maintain these AI systems, and the prospective implications on employment in the industry [60]. Mavani et al. went further to note that implementing AI and ML technologies in the agriculture and food industry may require significant investment, which may not be feasible for all businesses [4]. A study further identified other problems that must be addressed in order to successfully transition toward sustainability and optimization of agri-food systems, through the adoption of AI and ML technologies [61]. These were

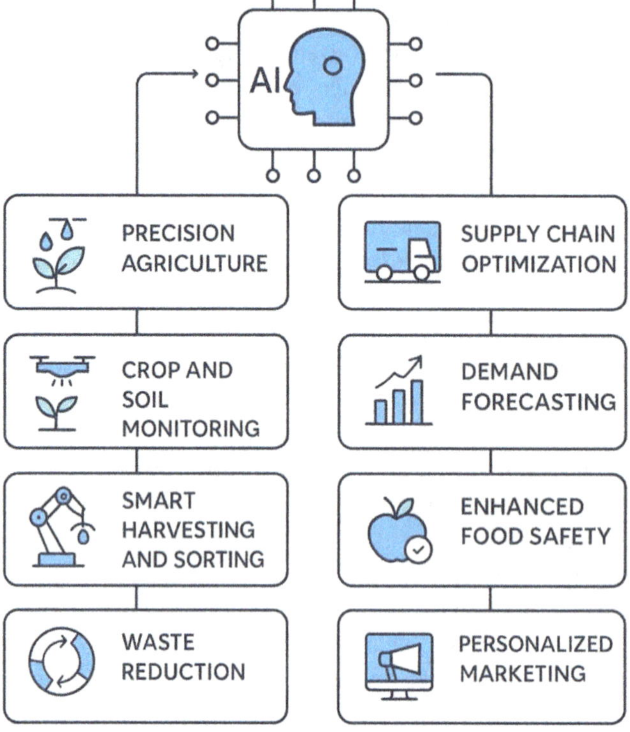

Figure 11.3: Impact of artificial intelligence (AI) and machine learning (ML) in the agri-food and FMCG sector [34].

classified into five primary levels: hardware and infrastructure requirements, data (quality and availability), network, application, security and privacy concerns, and system. This study also observed that among these levels, one level known as the data level, specifically pertains to the integration of AI and ML in agricultural and food systems.

Additionally, the adoption of AI technology in the food industry may be more challenging for small and medium-sized businesses (SMEs) due to financial constraints. Hansen and Bogh noted that the adoption and implementation of AI technology has been mostly by larger organizations [62]. However, they highlighted that small and medium-sized firms (SMEs) are crucial for the economy of many nations. Therefore, it is important to ensure that the SMEs have easy accessibility to these technologies and are able to effectively utilize them. Various factors have been identified to hinder the implementation of AI and ML techniques in SMEs, particularly in sub-Saharan Africa, as reviewed [63]. These factors are highlighted as follows: lack of access to technology, lack of technical expertise, and lack of awareness and understanding of AI. The lack of awareness was attributed to a lack in understanding and knowledge regarding the potential benefits associated with AI and its applicability to SMEs, as emphasized [63]. Two important key factors are extensively elaborated below.

i. **Data quality and availability**: One of the main challenges in applying ML to agri-food processing is the availability of high-quality and relevant data [64]. This is attributed to data collected not always standardized or easily accessible, making it difficult to train accurate ML models. As reviewed, these problems arise from the inherent attributes of big data, including its volume, variety, velocity, and authenticity. An additional factor to take into account is the diverse range of data sources, such as images, videos, sensor data, and expert data [64]. Hence, it is crucial to have a standardized data format, which can accommodate the entire data set, for example, conforming Da, obtained from a data source, DSa, to the same format and representation as the data, Db, obtained from another data source, DSb. However, management of a heterogeneous database becomes feasible with the availability of metadata to understand the source, data type, processing, and modifications performed. The emerging solution for managing diverse databases is through the use of Data Lake [65]. It enables the management of data across various groups, including raw data, partially processed data, and fully processed data. This system ensures continuous data traceability, hence enhancing data quality. As reviewed by Cravero et al., application of this innovative technique is still limited [64].

ii. **ML algorithm complexity and optimization**: Developing ML algorithms that can effectively optimize agri-food processing operations and supply chain management requires expertise in algorithm design and optimization. While big data allows ML algorithms to discover more intricate patterns and generate more fast and precise predictions than ever before, it also poses significant obstacles to ML, such as the scalability of models and the need for distributed computing, as well as data streaming, adaptability, and usability [66]. Coupled with this are the large-scale ML applications requiring generation of a low-rank matrix approximation in order to decrease the high time and space complexities [67]. Cutting-edge techniques such as the standard Nyström method are employed to construct low-rank matrix approximations. Therefore, this further demonstrates the complex nature of ML algorithms and the need for specialized skills in optimizing them. Additionally, there is a need to continually update and refine these algorithms as new data becomes available or as the processing environment changes, thereby increasing the complexity of ML algorithms [66].

11.3.3.2 Opportunities in adoption of AI and ML in agri-food systems

The ability of agri-food businesses to assure the sustainability and to address the associated challenges of the demand-supply cycle necessitated the need for innovative enhancement of human engagement [68]. The application of industrial automation such as AI, ML, and DL algorithms serves as the optimal choice for attaining these goals in the agri-food sectors, as reviewed [69]. These AI-based systems present opportunities for efficient control of agri-food production and delivery, thereby enhancing operational efficiency, sustainability, and profitability. Some key areas of opportunities for adopting AI-based systems in agri-food systems optimizations include the following: precision agriculture, supply chain optimization, quality control and traceability, market intelligence and decision support, and personalized nutrition. Overall, the adoption of AI and ML in agri-food system optimization offers significant opportunities to enhance productivity, sustainability, and resilience throughout the food supply chain, ultimately benefiting farmers, consumers, and the environment alike [70]. Hence, these highlighted opportunities outweigh the challenges associated with the adoption of these innovative technologies in agri-food systems' optimization and its sustainability. Two important areas are extensively elaborated below:

i. **Precision agriculture**: The emergence of agricultural technologies has led to improvements in sustainability and the discovery of more efficient farming methods [71]. This includes all processes of digitalization and automation such as big data, AI, robotics, the IoT, and virtual and augmented reality. AI enables precision farming techniques through the adoption of ML algorithms, which are significantly influencing the global economy by revolutionizing the way data is processed and decisions are made [72]. Their impact on agri-food production systems is substantial, particularly in light of the global food supply issue. Through AI-powered monitoring systems, data on crop growth, soil moisture levels, and pest infestations are obtained and analyzed in real-time, providing farmers with timely recommendations for interventions, and also improving forecasting accuracy of different factors that have great impact on the sector, thereby promoting precision of various agriculture practices such as targeted irrigation, fertilization, and pest management, leading to optimized resource use and increased crop yields [34]. Furthermore, applications of ML algorithms can also predict crop yields and optimize planting schedules, helping farmers make informed decisions to maximize productivity, while minimizing the impact of various farming practices on the environment, ranging from crop management to livestock and water management [72], as highlighted in Figure 11.4.
ii. **Quality control and traceability**: AI systems can optimize agri-food processing operations by identifying patterns and trends in data, improving production efficiency, minimizing waste, and enhancing product quality and safety. For example, adoption of AI and ML technologies can enhance quality control measures by detecting defects, contaminants, and spoilage in food products during processing and packaging [73]. More importantly, there has been a steady rise in food falsification, resulting in significant economic losses and undermined customers' confidence, creating a pressing concern for all parties concerned, including producers, researchers, governments, consumers, and other interested stakeholders [74]. This growing trend necessitates and makes it an essential prerequisite for implementing innovative methods to verify and authenticate the source, quality specifi-

cations, and transportation/storage information associated with food [75], thereby leading to effective means for tracing and verifying the food supply chain in order to identify and manage sources of contamination in the global food supply chain. Furthermore, blockchain technology, combined with AI, has been identified to provide end-to-end traceability, enabling consumers to track the origin and journey of food products, thereby enhancing transparency and trust in the supply chain [74, 75]. Adoption of this innovative technologies leads to improving traceability and helps with the prevention of food fraud, false labelling, and tracing back to the source of origin.

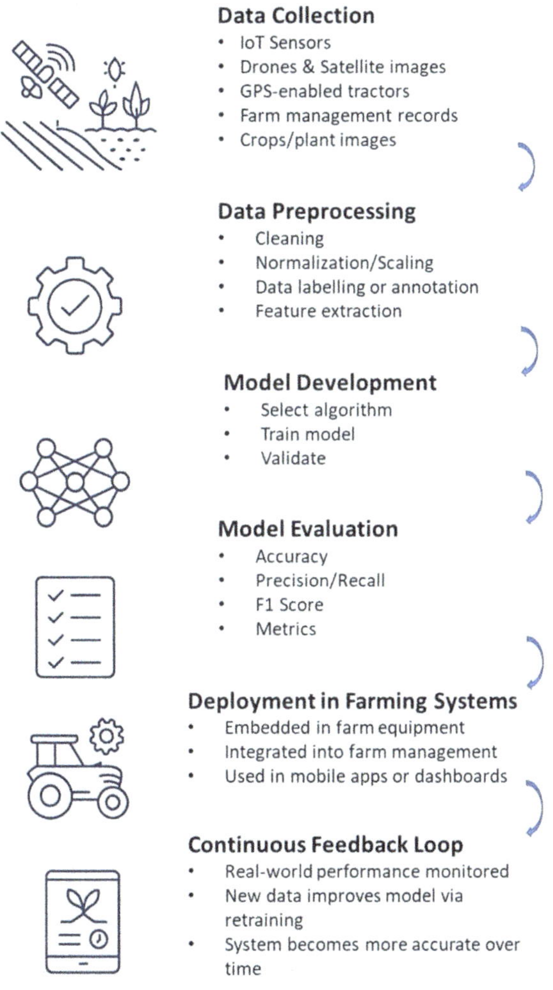

Data Collection
• IoT Sensors
• Drones & Satellite images
• GPS-enabled tractors
• Farm management records
• Crops/plant images
Data Preprocessing
• Cleaning
• Normalization/Scaling
• Data labelling or annotation
• Feature extraction
Model Development
• Select algorithm
• Train model
• Validate
Model Evaluation
• Accuracy
• Precision/Recall
• F1 Score
• Metrics
Deployment in Farming Systems
• Embedded in farm equipment
• Integrated into farm management
• Used in mobile apps or dashboards
Continuous Feedback Loop
• Real-world performance monitored
• New data improves model via retraining
• System becomes more accurate over time

Figure 11.4: General flow for the creation of machine learning algorithms and their application in smart/precision farming [61].

11.3.4 Case studies of success stories in AI-driven food system optimization

There have been several successful case studies highlighting the benefits of AI-driven food optimization. One example is the use of computer vision and AI in food processing plants to detect and remove defective products from the production line, thereby ensuring high-quality standards and reducing waste [76, 77]. The adoption of this novel technology is motivated by growing consumer concerns over food quality and safety [77]. As a result, the food industry is placing greater emphasis on the deployment of fast and reliable food evaluation systems, consequently encouraging food manufacturers and retailers to implement efficient real-time evaluations to ensure food quality and safety across the entire food manufacturing and processing stages. The adoption of computer vision is advantageous due to its nondestructive evaluation technique, as well as its ability to accurately estimate the properties of food products. This technology offers the benefits of fast, convenience of use, and minimal sample preparation.

Another example is the implementation of AI algorithms in restaurant management systems to optimize menu combinations based on customer preferences, ingredients availability, and sales data [78, 79]. AI-based technologies improved data accessibility, thereby empowering restaurateurs to collect extensive consumer information and analyze their behavior [79]. By integrating these data with information from various sources, restaurants can make a wide range of strategic and operational decisions. These AI-driven optimizations have resulted in significant value with increased customer satisfaction, improved operational efficiency, and reduced costs for businesses. In addition, AI and ML have been employed in sales forecasting for businesses such as restaurants, cafes, online meal delivery chains, hotels, and food outlets [69]. Through the analysis of past sales data, AI algorithms can generate accurate forecasts about future sales trends, enabling businesses to optimize their production and inventory management strategies. In overall, the adoption of AI and ML in the food sector has demonstrated significant promise in boosting productivity, minimizing wastage, improving food quality, and establishing a more transparent system for managing the supply chain.

11.3.5 Application of evolutionary algorithms in agri-food system optimization

Conventionally, optimization of processes is achieved by the utilization of various statistical techniques such as response surface methodology [80]. This approach is commonly employed in both industrial and laboratory settings. However, these traditional approaches either lack scalability or fail to achieve optimal solutions, sometimes resulting in prolonged execution times [81]. For example, when dealing with a large number of intricate process parameters, it becomes challenging to utilize a conventional optimization strategy [80]. Consequently, the utilization of cutting-edge technology has become prevalent in many food businesses and laboratory procedures. These are based on the deployment of AI and ML techniques in the development of several novel optimization methods. These have rapidly become popular because of their exceptional performance in optimizing processes. One such technique is the application of evolutionary algorithms (EAs). EAs are a class of computational optimization algorithms, also known as bio-inspired optimization techniques (BOTs), inspired by the process of natural selection and evolution in biology [80]. BOTs have been increasingly applied to optimize different aspects of food systems. These algorithms mimic the process of natural evolution, where candidate solutions to a problem undergo selection, recombination, and mutation to produce new generations of improved solutions [82].

The adoption of EAs in food system optimization has gained attention due to their ability to handle complex, multi-objective problems and find optimal solutions [80]. Several studies have investigated the use of EAs in food system optimization. For example, a study by Zhang et al. applied a genetic algorithm to optimize the selection and allocation of food suppliers in a supply chain network, considering factors such as cost, delivery reliability, and environmental impact [83]. They went further to note that this genetic algorithm approach was able to find optimal solutions that minimized costs and environmental impact while maximizing delivery reliability. Overall, the application of EAs in food system optimization has shown promising results in finding optimal solutions to complex problems.

11.4 Future directions for sustainable agri-food systems' optimization

Adoption of AI and ML have led to successful sustainable agri-food systems through reduction in emissions of greenhouse gases, increasing biodiversity, and improvement in livelihoods, however, there are still barriers and challenges in relation to social, economic, and political facets that need to be addressed for widespread adoption and impact [84]. Addressing these barriers include developing more comprehensive and standardized metrics for assessing sustainability, incorporating data-driven approaches, promoting interdisciplinary research and collaboration, and encouraging the adoption of innovative technologies and practices that promote sustainability [85]. Furthermore, it will be crucial to address the systemic barriers and challenges that hinder the adoption of sustainable agriculture and food systems [86]. This could require addressing issues such as access to resources and knowledge, market structures, regulations, and policy support [87]. In addition, there is great need in terms of developing holistic and quantitative methods for assessing sustainability, promoting collaboration among stakeholders, and implementing policy changes that support sustainable agri-food systems [88]. The focus of these holistic and quantitative methods in achieving sustainability of agri-food systems is to ensure practicability of these technologies and innovative practices that are not harmful to the environment, by making them affordable for all the sectors of agri-food systems in ensuring increased productivity.

In order to attain a really sustainable future for agri-food systems, it is imperative to take into account the interrelationships among biological, social, cultural, and political aspects. By using this approach, we can develop agricultural and food systems that not only tackle existing difficulties but also actively promote long-term ecological well-being, coupled with a fair distribution of resources for sustainable economic growth. One of the approaches to this is the adoption of interpretive structure referred to as “Sustainable Agri-Food Evaluation Methodology” (SAEMETH), as proposed [89]. This evaluation methodology is able to guide in the evaluation of agri-food sustainability. This study found SAEMETH as a monitoring tool based on qualitative indicators that are user-friendly and highly communicative and can be used to conduct sustainability evaluations in small-scale agri-food systems. This approach is participatory, interdisciplinary, and multi-institutional, and combines a strong theoretical foundation with a field-tested operational framework.

To further ensure successful implementation of AI and ML initiatives in the transition toward sustainable agri-food systems, there is great need to address the absence of supportive policies, restricted access to resources and knowledge, and resistance to change within the agri-food systems [90]. The limitation in conventional agricultural research approach in explaining or addressing these challenges is attributed to its reliance on a static equilibrium-centered perspective, which offers limited understanding of the dynamic nature of agri-food systems, as reviewed [90]. This viewpoint is grounded in the current framework of agri-food research and innovation, which frequently falls short in delivering sustainable results, especially when applied on a larger scale and for impoverished populations in developing nations. Hence, addressing the lack of supportive policies requires the development of food policy that aligns with the objective of establishing sustainable agri-food systems [91]. This necessitates the need for re-evaluating perspectives and fundamentally re-examining the underlying framework that agricultural and food-related policies operate upon.

Furthermore, the effectiveness of these sustainable initiatives may differ based on contextual elements such as region-specific barriers, socio-economic circumstances, and cultural norms [77]. An approach to this issue is presented in a case study undertaken by Sarabia et al. [78] in Valencia. The study proposes to achieve a sustainable agri-food system by challenging the current conventions, and adopting novel initiatives through self-organization and a transformative approach to agricultural and food policies. This relies on the implementation of local food governance and the transition toward agroecology. Their study leverages the urban transformative capacity framework (UTCF) and combines various frameworks to identify specific aspects and processes that contribute to the building of capacities for transition, before the actual development takes place. This entails the participation of multiple stakeholders at various levels, while influential social movements and their informal systems of governance play a pivotal role. Hence, it is crucial to pay close attention to these triggers during future development and implementation.

11.5 Conclusions

AI and ML possess the capacity to greatly transform the agri-food systems and tackle important issues of effectiveness, sustainability, and food security. Nevertheless, it is imperative to responsibly apply these technologies, address ethical concerns, and ensure fair access to them, to fully exploit their potential and build agri-food systems that are both resilient and sustainable for the future. Utilizing AI techniques like predictive analytics can facilitate the detection of patterns and trends in customer behavior, enabling more effective planning and allocation of resources throughout the entire supply chain. Likewise, through the integration of ML techniques in the agri-food supply chain, it becomes feasible to detect existing uses and identify areas of research that require further investigation in supply chain management. Therefore, application of AI and ML in the optimization of agri-food processing has the potential to greatly improve productivity, profitability, and sustainability in the industry, while also addressing pressing global concerns. With continuous research and development in this domain, we may anticipate the emergence of additional cutting-edge solutions that will enhance the security and resilience of the global food system.

References

[1] Tzachor, A., Richards, C. E., & Jeen, S. Transforming agrifood production systems and supply chains with digital twins. npj Science of Food 2022, 6(1), 47.
[2] Ben Ayed, R., & Hanana, M. Artificial intelligence to improve the food and agriculture sector. Journal of Food Quality 2021, 2021, 1–7.
[3] Geissdoerfer, M., et al. The Circular Economy–A new sustainability paradigm? Journal of Cleaner Production 2017, 143, 757–768.
[4] Mavani, N. R., et al. Application of artificial intelligence in food industry – A guideline. Food Engineering Reviews 2022, 14(1), 134–175.
[5] Vermeulen, S. J., Campbell, B. M., & Ingram, J. S. Climate change and food systems. Annual Review of Environment and Resources 2012, 37, 195–222.
[6] den Boer, A. C., et al. Research and innovation as a catalyst for food system transformation. Trends in Food Science & Technology 2021, 107, 150–156.
[7] Ericksen, P. J. Conceptualizing food systems for global environmental change research. Global Environmental Change 2008, 18(1), 234–245.

[8] Ingram, J. A food systems approach to researching food security and its interactions with global environmental change. Food Security 2011, 3, 417–431.
[9] Wezel, A., et al. Agroecological principles and elements and their implications for transitioning to sustainable food systems. A review. Agronomy for Sustainable Development 2020, 40, 1–13.
[10] Bandara, J. S., & Cai, Y. The impact of climate change on food crop productivity, food prices and food security in South Asia. Economic Analysis and Policy 2014, 44(4), 451–465.
[11] Anderson, C. R., et al. Agroecology Now-connecting the Dots to Enable Agroecology Transformations, Taylor & Francis, 2020, 561–565.
[12] Kerr, R. B., et al. Can agroecology improve food security and nutrition? A review. Global Food Security 2021, 29, 100540.
[13] Shift, N. Global Food System Map. 2016 [cited 2024 26 April]; Available from: https://www.slideshare.net/slideshow/global-food-system-map-57053271/57053271.
[14] Desa, G., & Jia, X. Sustainability transitions in the context of pandemic: An introduction to the focused issue on social innovation and systemic impact. In: Social Innovation and Sustainability Transition, Springer, 2022, 273–281.
[15] Kumar, L., et al. Climate change and future of agri-food production. In: Future Foods, Elsevier, 2022, 49–79.
[16] Rodriguez, J. M., et al. Barriers to adoption of sustainable agriculture practices: Change agent perspectives. Renewable Agriculture and Food Systems 2009, 24(1), 60–71.
[17] Jia, X. Agro-food innovation and sustainability transition: A conceptual synthesis. Sustainability 2021, 13(12), 6897.
[18] Melchior, I. C., & Newig, J. Governing transitions towards sustainable agriculture – Taking stock of an emerging field of research. Sustainability 2021, 13(2), 528.
[19] El Bilali, H. Research on agro-food sustainability transitions: A systematic review of research themes and an analysis of research gaps. Journal of Cleaner Production 2019, 221, 353–364.
[20] Hubeau, M., et al. A new agri-food systems sustainability approach to identify shared transformation pathways towards sustainability. Ecological Economics 2017, 131, 52–63.
[21] Campbell, B., et al. Transforming Food Systems under Climate Change through Innovation, Cambridge University Press, 2023.
[22] Kamble, S. S., Gunasekaran, A., & Gawankar, S. A. Achieving sustainable performance in a data-driven agriculture supply chain: A review for research and applications. International Journal of Production Economics 2020, 219, 179–194.
[23] Sharma, S., et al. Sustainable innovations in the food industry through artificial intelligence and big data analytics. Logistics 2021, 5(4), 66.
[24] Feng, H., et al. Applying blockchain technology to improve agri-food traceability: A review of development methods, benefits and challenges. Journal of Cleaner Production 2020, 260, 121031.
[25] Menon, S., & Jain, K. Blockchain technology for transparency in agri-food supply chain: Use cases, limitations, and future directions. IEEE Transactions on Engineering Management 2021, 71, 106–120.
[26] El Chami, D., Daccache, A., & El Moujabber, M. How can sustainable agriculture increase climate resilience? A systematic review. Sustainability 2020, 12(8), 3119.
[27] Addanki, M., Patra, P., & Kandra, P. Recent advances and applications of artificial intelligence and related technologies in the food industry. Applied Food Research 2022, 2(2), 100126.
[28] Liu, X., et al. Trends and challenges on fruit and vegetable processing: Insights into sustainable, traceable, precise, healthy, intelligent, personalized and local innovative food products. Trends in Food Science & Technology 2022, 125, 12–25.
[29] Tiwari, M., et al. Artificial intelligence in food processing. Novel Technologies in Food Science 2023, 511–550.

[30] Yaiprasert, C., & Hidayanto, A. N. AI-powered in the digital age: Ensemble innovation personalizes the food recommendations. Journal of Open Innovation: Technology, Market, and Complexity 2024, 100261.
[31] Wakchaure, Y., Patle, B., & Pawar, S. Prospects of robotics in food processing: An overview. Journal of Mechanical Engineering, Automation and Control Systems 2023, 4(1), 17–37.
[32] Hassoun, A., et al. Emerging trends in the agri-food sector: Digitalisation and shift to plant-based diets. Current Research in Food Science 2022, 5, 2261–2269.
[33] Sarker, I. H. AI-based modeling: Techniques, applications and research issues towards automation, intelligent and smart systems. SN Computer Science 2022, 3(2), 158.
[34] Taneja, A., et al. Artificial intelligence: Implications for the agri-food sector. Agronomy 2023, 13(5), 1397.
[35] Pathan, M., et al. Artificial cognition for applications in smart agriculture: A comprehensive review. Artificial Intelligence in Agriculture 2020, 4, 81–95.
[36] Shaikh, T. A., Rasool, T., & Lone, F. R. Towards leveraging the role of machine learning and artificial intelligence in precision agriculture and smart farming. Computers and Electronics in Agriculture 2022, 198, 107119.
[37] Patrício, D. I., & Rieder, R. Computer vision and artificial intelligence in precision agriculture for grain crops: A systematic review. Computers and Electronics in Agriculture 2018, 153, 69–81.
[38] Rajak, P., et al. Internet of Things and smart sensors in agriculture: Scopes and challenges. Journal of Agriculture and Food Research 2023, 14, 100776.
[39] Elbasi, E., et al. Artificial intelligence technology in the agricultural sector: A systematic literature review. IEEE Access 2022, 11, 171–202.
[40] Rai, R., et al. Machine Learning in Manufacturing and Industry 4.0 Applications, Taylor & Francis, 2021, 4773–4778.
[41] Sharma, A., et al. Machine learning applications for precision agriculture: A comprehensive review. IEEE Access 2021, 9, 4843–4873.
[42] Mamoudan, M. M., et al. Hybrid machine learning-metaheuristic model for sustainable agri-food production and supply chain planning under water scarcity. Resources, Environment and Sustainability 2023, 14, 100133.
[43] Sharma, R., et al. A systematic literature review on machine learning applications for sustainable agriculture supply chain performance. Computers & Operations Research 2020, 119, 104926.
[44] Konfo, T. R. C., et al. Recent advances in the use of digital technologies in agri-food processing: A short review. Applied Food Research 2023, 3(2), 100329.
[45] Gonesh Chandra Saha, R. M. M. S. P. S. Y. H. S. P. D. The impact of artificial intelligence on business strategy and decision-making processes. European Economic Letters (EEL) 2023, 13(3), 926–934.
[46] Abioye, E. A., et al. Precision irrigation management using machine learning and digital farming solutions. AgriEngineering 2022, 4(1), 70–103.
[47] Talaviya, T., et al. Implementation of artificial intelligence in agriculture for optimisation of irrigation and application of pesticides and herbicides. Artificial Intelligence in Agriculture 2020, 4, 58–73.
[48] Rashid, M., et al. A comprehensive review of crop yield prediction using machine learning approaches with special emphasis on palm oil yield prediction. IEEE Access 2021, 9, 63406–63439.
[49] Abideen, A. Z., et al. Food supply chain transformation through technology and future research directions – A systematic review. Logistics 2021, 5(4), 83.
[50] Plathottam, S. J., et al. A review of artificial intelligence applications in manufacturing operations. Journal of Advanced Manufacturing and Processing 2023, 5(3), e10159.
[51] Feizabadi, J. Machine learning demand forecasting and supply chain performance. International Journal of Logistics Research and Applications 2022, 25(2), 119–142.
[52] Wang, X., et al. Application of machine learning to the monitoring and prediction of food safety: A review. Comprehensive Reviews in Food Science and Food Safety 2022, 21(1), 416–434.

[53] Riahi, Y., et al. Artificial intelligence applications in supply chain: A descriptive bibliometric analysis and future research directions. Expert Systems with Applications 2021, 173, 114702.

[54] Aljohani, A. Predictive analytics and machine learning for real-time supply chain risk mitigation and agility. Sustainability 2023, 15(20), 15088.

[55] Rabaa'i, A. A., et al. The use of machine learning to predict the main factors that influence the continuous usage of mobile food delivery apps. Model Assisted Statistics and Applications 2022, 17(4), 247–258.

[56] Yaiprasert, C., & Hidayanto, A. N. AI-driven ensemble three machine learning to enhance digital marketing strategies in the food delivery business. Intelligent Systems with Applications 2023, 18, 200235.

[57] Astill, J., et al. Transparency in food supply chains: A review of enabling technology solutions. Trends in Food Science & Technology 2019, 91, 240–247.

[58] Neethirajan, S., et al. Biosensors for sustainable food engineering: Challenges and perspectives. Biosensors 2018, 8(1), 23.

[59] Ekanayake, J., & Saputhanthri, L. E-AGRO: Intelligent chat-bot. IoT and artificial intelligence to enhance farming industry. AGRIS On-line Papers in Economics and Informatics 2020, 12(1), 15–21.

[60] Lezoche, M., et al. Agri-food 4.0: A survey of the supply chains and technologies for the future agriculture. Computers in Industry 2020, 117, 103187.

[61] Araújo, S. O., et al. Machine learning applications in agriculture: Current trends, challenges, and future perspectives. Agronomy 2023, 13(12), 2976.

[62] Hansen, E. B., & Bøgh, S. Artificial intelligence and internet of things in small and medium-sized enterprises: A survey. Journal of Manufacturing Systems 2021, 58, 362–372.

[63] Uwagaba, J., Omotosho, T. D., & George, G. Exploring the barriers to artificial intelligence adoption in Sub-Saharan Africa's small and medium enterprises and the potential for increased productivity. World Wide Journal of Multidisciplinary Research and Development 2023.

[64] Cravero, A., et al. Challenges to use machine learning in agricultural big data: A systematic literature review. Agronomy 2022, 12(3), 748.

[65] Wibowo, M., Sulaiman, S., & Shamsuddin, S. M. Machine learning in data lake for combining data silos. In: Data Mining and Big Data: Second International Conference, DMBD 2017, Fukuoka, Japan, July 27–August 1, 2017, Proceedings 2, Springer, 2017.

[66] Zhou, L., et al. Machine learning on big data: Opportunities and challenges. Neurocomputing 2017, 237, 350–361.

[67] Sun, S., Zhao, J., & Zhu, J. A review of Nyström methods for large-scale machine learning. Information Fusion 2015, 26, 36–48.

[68] Ravikumar, A. The role of machine learning and computer vision in the agri-food industry. In: Artificial Intelligence Applications in Agriculture and Food Quality Improvement, IGI Global, 2022, 257–275.

[69] Ajibade, S., et al. Machine Learning (Ml) in Food Production: The Prospects and Applications, vol. 7, 2023, 325–341.

[70] Rejeb, A., et al. Examining the interplay between artificial intelligence and the agri-food industry. Artificial Intelligence in Agriculture 2022, 6, 111–128.

[71] Javaid, M., et al. Enhancing smart farming through the applications of Agriculture 4.0 technologies. International Journal of Intelligent Networks 2022, 3, 150–164.

[72] Elbasi, E., et al. Crop prediction model using machine learning algorithms. Applied Sciences 2023, 13(16), 9288.

[73] Bongarde, D., Pandit, S., & Pandit, H. Use of machine learning and artificial intelligence in food spoilage detection. International Journal of Engineering Research and Applications 2024, 14, 79–85.

[74] Galvez, J. F., Mejuto, J. C., & Simal-Gandara, J. Future challenges on the use of blockchain for food traceability analysis. TrAC Trends in Analytical Chemistry 2018, 107, 222–232.

[75] Tsoukas, V., et al. Enhancing food supply chain security through the use of blockchain and TinyML. Information 2022, 13(5), 213.
[76] Fracarolli, J. A., et al. Computer vision applied to food and agricultural products. Revista Ciencia Agronomica 2020, 51(spe), e20207749.
[77] Ma, J., et al. Applications of computer vision for assessing quality of agri-food products: A review of recent research advances. Critical Reviews in Food Science and Nutrition 2016, 56(1), 113–127.
[78] Kumar, I., et al. Opportunities of artificial intelligence and machine learning in the food industry. Journal of Food Quality 2021, 2021, 1–10.
[79] Roy, D., Spiliotopoulou, E., & de Vries, J. Restaurant analytics: Emerging practice and research opportunities. Production and Operations Management 2022, 31(10), 3687–3709.
[80] Sarkar, T., et al. Application of bio-inspired optimization algorithms in food processing. Current Research in Food Science 2022, 5, 432–450.
[81] Game, P. S., & Vaze, D. V. Bio-inspired optimization: Metaheuristic algorithms for optimization. arXiv preprint arXiv:2003.11637 2020.
[82] Binitha, S., & Sathya, S. S. A survey of bio inspired optimization algorithms. International Journal of Soft Computing and Engineering 2012, 2(2), 137–151.
[83] Zhang, S., et al. Multi-objective optimization for sustainable supply chain network design considering multiple distribution channels. Expert Systems with Applications 2016, 65, 87–99.
[84] El Bilali, H., Strassner, C., & Ben Hassen, T. Sustainable agri-food systems: Environment, economy, society, and policy. Sustainability 2021, 13(11), 6260.
[85] Bachmann, N., et al. The contribution of data-driven technologies in achieving the sustainable development goals. Sustainability 2022, 14(5), 2497.
[86] Siebrecht, N. Sustainable agriculture and its implementation gap – Overcoming obstacles to implementation. Sustainability 2020, 12(9), 3853.
[87] Nicolétis, É., et al. Agroecological and Other Innovative Approaches for Sustainable Agriculture and Food Systems that Enhance Food Security and Nutrition. A Report by the High Level Panel of Experts on Food Security and Nutrition of the Committee on World Food Security, 2019.
[88] Pretty, J. Agricultural sustainability: Concepts, principles and evidence. Philosophical Transactions of the Royal Society B: Biological Sciences 2008, 363(1491), 447–465.
[89] Peano, C., et al. Evaluating the sustainability in complex agri-food systems: The SAEMETH framework. Sustainability 2015, 7(6), 6721–6741.
[90] Thompson, J., & Scoones, I. Addressing the dynamics of agri-food systems: An emerging agenda for social science research. Environmental Science & Policy 2009, 12(4), 386–397.
[91] Galli, F., et al. How can policy processes remove barriers to sustainable food systems in Europe? Contributing to a policy framework for agri-food transitions. Food Policy 2020, 96, 101871.

Ifeoluwa Komolafe, Oluwaseun Julius, and Rodney Owusu-Darko*

Chapter 12
Food informatics: genomic data in food safety, data privacy, and security

Abstract: This book chapter explores the transformative role of genomic data within food informatics and its impact on modern food safety practices. It provides a detailed examination of how advanced genomic techniques facilitate rapid pathogen identification, outbreak tracing, and the verification of food authenticity. Through illustrative case studies – including the analysis of *Salmonella* outbreaks and instances of food fraud – the chapter demonstrates the enhanced precision, speed, and accuracy that genomic data contributes to public health responses. The integration of such data not only improves the detection and management of foodborne illnesses but also offers significant economic benefits to industry stakeholders by optimizing quality control processes and reducing contamination risks.

The chapter also addresses the challenges inherent in adopting genomic approaches. Key concerns include issues of data privacy and security, as sensitive genetic information can be susceptible to breaches that undermine consumer trust. It reviews current regulatory frameworks and ethical considerations, highlighting the delicate balance between leveraging innovative technologies and protecting individual and public interests. In response to these challenges, the text outlines promising technological solutions such as robust encryption methods, blockchain-enabled secure data management, and advanced data anonymization techniques that collectively aim to mitigate security vulnerabilities.

Looking forward, the chapter emphasizes the importance of ongoing research, international collaboration, and policy development to further refine genomic applications in food safety. By presenting both the benefits and limitations of genomic data use, the work advocates for an integrated approach that fosters innovation while ensuring stringent data protection standards. Ultimately, the chapter serves as a comprehensive resource for stakeholders seeking to understand and navigate the complex interplay between technological advancement and regulatory oversight in the realm of food safety.

Keywords: Food informatics, genomic data, food safety, data privacy, blockchain

***Corresponding author: Ifeoluwa Komolafe, Oluwaseun Julius, and Rodney Owusu-Darko**, School of Health and Life Sciences, Teesside University, Middlesbrough, UK, email: R.Owusu-Darko@tees.ac.uk

https://doi.org/10.1515/9783111441238-013

12.1 Overview of food informatics

Population rises and urbanization have significantly altered how we consume food, posing a challenge to food sustainability, safety, and security. The food industry is advancing to address this, moving beyond conventional methods that cause food waste and difficulties in traceability and sustainability. By incorporating the latest developments in the Internet of things (IoT), various aspects of food production, from processing to supply chain logistics, can be revolutionized [1, 2] The advent of big data, artificial intelligence (AI), blockchain, and food informatics has significantly transformed the food supply chain, driving improvements in efficiency, traceability, sustainability, and consumer satisfaction (Figure 12.1). Food informatics involves gathering, organizing, analyzing, and using data intelligently from agriculture, the food supply chain, processing, retail, and consumer behavior [1]. This extensive data gathering provides a holistic view of the food system, identifying inefficiencies and areas for improvement. Advanced analytical techniques, including machine learning (ML) and AI, are then applied to extract meaningful insights that can support informed decision-making [3].

Food informatics is instrumental in optimizing food production processes. By analyzing data on crop yields, weather patterns, and soil health, farmers can make data-driven decisions to enhance agricultural productivity and sustainability [3]. Data analysis can streamline operations in food processing and retail, reduce waste, and improve supply chain logistics [4]. Food informatics also plays a vital role in ensuring food security by monitoring and predicting food supply trends.

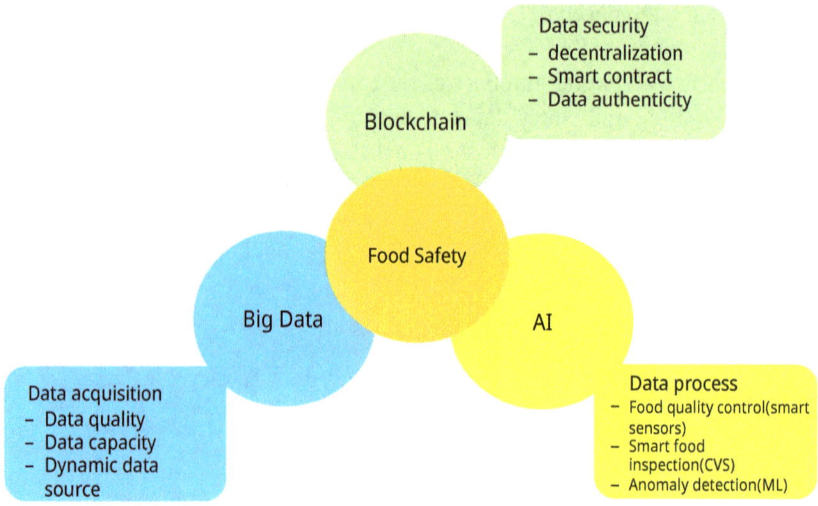

Figure 12.1: The interactions between food safety, big data, blockchain, and AI [5].

This enables policymakers to develop strategies to address potential shortages and ensure equal distribution of resources [6]. Moreover, the field contributes to consumer health by analyzing dietary patterns and food safety data, facilitating the development of healthier and safer food products [7]. Also, the emergence of next-generation sequencing (NGS) technologies has immensely improved food safety.

12.2 The emergence of next-generation sequencing (NGS) technologies

Utilizing genomic data in food safety has brought about significant advancements, providing various benefits that improve the detection, prevention, and management of foodborne illnesses. One of the primary benefits of using genomic data in food safety is the enhanced ability to detect and identify foodborne pathogens accurately. Unlike traditional methods, such as culture-based techniques, genomic technologies offer a comprehensive view of a pathogen's genetic makeup, allowing for the differentiation of closely related strains and the identification of unique genetic markers. For example, using whole genome sequencing (WGS) in regular surveillance has significantly enhanced the ability to identify *Listeria monocytogenes* strains especially in outbreak detection and surveillance [8], which enables quick intervention and containment.

In addition, genomic data significantly improves traceability and source tracking of foodborne pathogens. Scientists can precisely establish relationships and trace the origin of contamination by comparing genomic sequences from patients, food products, and environmental sources. Genomic data also plays a crucial role in monitoring foodborne pathogens' antimicrobial resistance (AMR) patterns. And with AMR emerging as a huge global issue, genomics helps identify resistance genes and track their spread among bacterial populations, which is essential for developing strategies to combat AMR [9]. Furthermore, genomic data informs regulatory policies and food safety standards by providing a scientific basis for decision-making. The detailed insights gained from genomic analyses guide the development of targeted regulations to address specific risks and vulnerabilities in the food supply chain. This includes identifying high-risk food products and production practices and setting more precise standards for pathogen levels in food products [10–12], which helps in various food safety objectives and risk analysis.

Advancements in genomic technologies have significantly impacted food safety by providing more specific and rapid techniques for identifying and characterizing foodborne pathogens. NGS technologies have become vital for genomic pathogen identification, allowing for high-throughput sequencing of whole microbial genomes directly from food samples [13, 14]. Metagenomics, which involves sequencing microbial profiles from food and environmental matrices, has proven valuable in studying the microbial community, including bacteria, viruses, fungi, and parasites. This method is particularly beneficial when the pathogen is unidentified or when multiple pathogens are suspected, as it can reveal the diversity of microbial communities and identify rare or novel pathogens that traditional methods may not detect [15]. The large volume of data produced by NGS necessitates using bioinformatics tools for analysis. Tools such as Kraken, MetaPhlAn, and MEGAHIT are employed for taxonomic classification, estimating abundance, and genome assembly. These tools use reference databases such as National Center for Biotechnology Information (NCBI), Integrated Microbial Genomes (IMG) among others, to align sequenced reads with known pathogens. Additionally, de novo assembly algorithms aid in reconstructing genomes from scratch when reference sequences are unavailable [16].

Genomic sequencing has been instrumental in tracking the source and spread of pathogens, enabling more effective interventions during foodborne outbreaks. For example, WGS was instrumental in the characterization of *Listeria monocytogenes* strains in South Africa in the 2017 outbreak, the largest listeriosis outbreak till date. The 2011 *Escherichia coli* outbreak in Germany also emphasized the use of genomic sequencing, as scientists rapidly identified the strain and its antibiotic resistance profile, thus guiding public health responses and mitigating the further spread of the outbreak. Furthermore, genomic tools enhance the traceability of foodborne pathogens throughout the food supply chain. By comparing genomic sequences from contaminated food samples with those from potential sources, investigators can effectively trace the origin of contamination [17, 18].

Genomic sequencing also shows it importance in identifying AMR genes and mutations in foodborne pathogens. It is important to understand resistance mechanisms and guide antibiotic use in food production and clinical settings. Integrating genomic technologies into public health laboratories is a significant advancement, enhancing real-time surveillance and rapid responses during foodborne outbreaks [19, 20]. Studies have shown the potential of genomic technologies to revolutionize public health practices [21–25]. However, the challenges associated with the complexity of genomic data require advanced bioinformatics expertise and computational resources. Interpreting large volumes of data, particularly in distinguishing between pathogenic and nonpathogenic strains in metagenomic samples, presents a significant challenge. Additionally, though sequencing costs have decreased, it still represents a barrier to widespread adoption, especially in low-income countries [26, 27].

12.2.1 Monitoring foodborne outbreaks

Foodborne outbreaks pose a significant challenge to public health and the economy. Outbreaks typically occur when two or more people fall ill after consuming contaminated food or drink. Tracing the source involves detecting the outbreak, formulating hypotheses, testing these hypotheses, and confirming the source.

Detection of outbreaks incorporates surveillance systems such as PulseNet with data from WGS. Public health professionals gather detailed information and conduct epidemiologic studies to identify commonalities among cases. Microbiological testing and advanced molecular techniques are used to test hypotheses and confirm the source. Controlling foodborne outbreaks involves issuing recalls and public notifications, improving food safety practices, implementing regulatory measures, and ongoing research and development to advance food safety [28, 29].

The use of genomic data has greatly improved the ability to track and control foodborne outbreaks, offering detailed insights into the origins, spread, and characteristics of pathogens. WGS has become a powerful tool in public health, allowing for precise identification and comparison of bacterial strains, which helps in timely and effective outbreak response [30, 31]. Traditional methods for tracing foodborne outbreaks, such as serotyping and pulsed-field gel electrophoresis (PFGE), have resolution and discriminatory power limitations. WGS greatly improves these methods because it provides complete genetic data of pathogens, enabling the differentiation of strains with high precision [32]. This detailed genomic information allows public health professionals and other scientists to establish phylogenetic relationships between isolates, tracing the origin of contamination with remarkable accuracy. For example, during the 2011 outbreak of *E. coli* in Europe, WGS was effective in identifying the outbreak strain and linking it to fenugreek seeds from a specific supplier [33]. These technologies are presently being used by the United Kingdom Health Security Agency (UKHSA) and its partners to investigate a Shiga toxin producing *E. coli* (STEC) outbreak in parts of the UK, which has been attributed to select sandwiches [34]. This is facilitating a swift intervention, preventing further spread and reducing the outbreak's impact. Health authorities can identify contamination routes and implement targeted interventions by analyzing the genetic sequences of isolates from patients, food sources, and the environment [35]. This level of specificity is crucial for preventing future outbreaks and protecting public health.

In a case of *Salmonella enterica*, genomic sequencing revealed patterns of antibiotic resistance and transmission pathways, guiding public health responses and regulatory policies. WGS identified a specific clone of multidrug-resistant *Salmonella* that caused widespread outbreaks linked to poultry products, leading to stricter regulations and improved food safety practices in the poultry industry [36]. Collecting and using genomic data for public health brings significant data privacy and security challenges, because genomic sequences contain highly sensitive information that can reveal the identity of individuals and their genetic predispositions to certain diseases. Protecting this information is crucial to maintaining public trust and complying with ethical and legal standards [37, 38]. Robust data privacy measures include implementing secure storage systems, encryption of genomic data, and stringent access controls to ensure that only authorized personnel can access sensitive information. Additionally, data-sharing policies must balance the need for public health surveillance with the protection of individual privacy. This involves anonymizing genomic data to prevent the reidentification of individuals [39].

12.2.2 Enhancing food authenticity and quality

Food authenticity is crucial for food safety, as it involves accurately identifying food products to prevent mislabeling and fraud. Adulteration of food products poses significant risks to consumer health and safety. However, traditional methods of species identification, such as morphological analysis and chemical assays, often need more sensitivity and specificity to detect fraudulent activities [40]. Genomic data, particularly through DNA barcoding and WGS, offers a more precise and reliable approach. DNA barcoding involves sequencing a short, standardized genome region, which serves as a unique identifier for different species [41]. This method has been effectively used to authenticate various food products, especially in the seafood industry, where mislabeling is prevalent. For instance, studies have revealed that a significant portion of seafood sold in markets is mislabeled, with cheaper species substituted for more expensive ones [42]. DNA barcoding ensures that consumers receive accurately labelled products by comparing the DNA sequences of the products to reference databases.

WGS provides even greater resolution by analyzing the entire genetic makeup of an organism. Thus, WGS has been used to verify the authenticity of high-value products such as wine, honey, and meat, ensuring that these products meet the quality standards that consumers expect [43]. Genomic tools provide powerful means to detect adulterants that might be undetectable by traditional methods. For example, using undeclared species in meat products, like the horse meat incidence in Europe, can be effectively uncovered through genomic analysis [43].

Genomic techniques can identify trace amounts of adulterants, ensuring the integrity of food products. This capability is particularly important for detecting allergens that must be declared on food labels. The presence of undeclared allergens can have severe health consequences for allergic individuals, and genomic tools can help prevent such incidents by ensuring accurate labelling. In addition to authenticity and safety, genomic data significantly enhances the overall quality of food products. In agriculture, genomic data is used to identify and select desirable traits in crops and livestock. Techniques such as marker-assisted selection and genomic selection allow breeders to accelerate the development of new varieties with improved qualities, such as disease resistance, enhanced nutritional content, and longer shelf life [44].

For instance, genomic selection in crops has led to the development of varieties that are more resistant to pests and diseases, reducing the need for chemical pesticides and increasing yield. In livestock, genomic data has improved traits such as milk production, meat quality, and reproductive efficiency [45]. These advancements enhance food quality and contribute to sustainability and food security. Moreover, genomic tools can monitor the microbiome of food production environments, such as fermentation processes in dairy and meat production. Understanding the microbial communities involved in these processes can optimize product quality, consistency, and safety [46].

Although genomic data is very helpful in food safety, security and data privacy risks should be addressed. Policies and protocols must be in place to address potential data security breaches and provide recourse for affected parties [47].

Transparency in the use of genomic data is essential to foster public trust. Clear communication about collecting, using, and protecting genomic data can alleviate public concerns and encourage cooperation with genomic initiatives. Public engagement and education are critical in addressing potential fears about data privacy and the misuse of genetic information [48].

12.2.3 Predictive microbiology

Microbial modeling plays a crucial role in assessing and predicting the behavior of microorganisms in food, aiding in food safety management and quality assurance.

It involves mathematical and computational techniques to simulate the behavior of microorganisms in food. Models predict microbial growth, survival, and inactivation under various environmental conditions, aiding in risk assessment and shelf-life estimation [49]. Microbial models assess the risk of foodborne pathogens, helping in the design and implementation of preventive controls to mitigate hazards [50]. Modeling informs the design and optimization of food processing techniques to ensure microbial safety and preservation without compromising quality [51]. The integration of food informatics and microbial modeling (Figure 12.2) offers synergistic benefits as shown below:

a. Data-driven modeling: Food informatics provides rich datasets for microbial modeling, enabling the development of more accurate and robust predictive models [52].

b. Real-time monitoring: IoT sensors and data analytics platforms facilitate real-time monitoring of microbial parameters, enhancing model accuracy and enabling proactive risk management [53].

c. Decision support systems: Integrated informatics platforms offer decision support tools that leverage microbial models to optimize food production processes, ensuring both safety and efficiency [54].

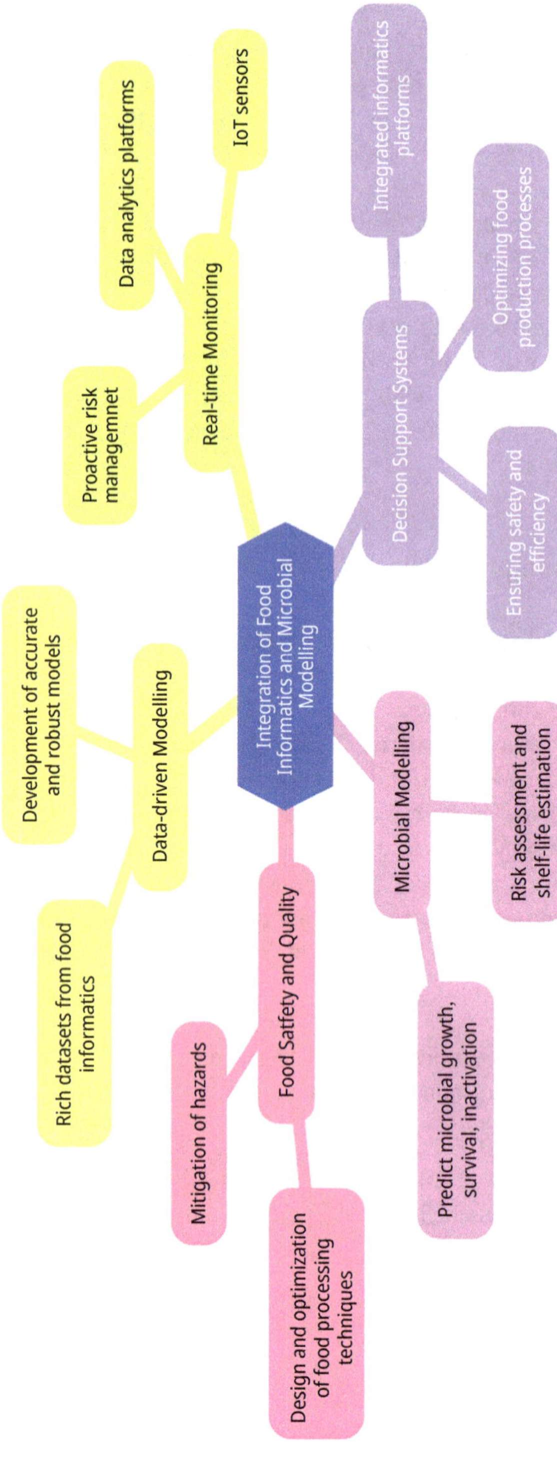

Figure 12.2: Integration of food informatics and microbial modeling.

12.2.4 Effective food safety response

Integrating genetic data with modern informatics has revolutionized established techniques of identifying, tracing, and controlling foodborne pathogens in the dynamic and complicated field of food safety. The adoption of WGS and bioinformatics tools has ushered in a new era of precision and speed in identifying microbial contaminants, thereby enhancing public health protection and reducing economic losses associated with foodborne illnesses. Timely and accurate detection of foodborne pathogens is crucial in averting outbreaks and limiting their consequences. Also, this rapid turnaround is crucial for implementing immediate control measures, such as recalls and advisories, to mitigate the spread of infection.

Precision in detecting pathogens is equally important. Genomic data enables accurate discrimination among closely related strains, which is crucial for identifying the precise origin of contamination and performing specific therapies. This high level of precision decreases the probability of false positives, thereby avoiding unnecessary product recalls and the resulting financial losses [55]. By using informatics tools, the capabilities of genomic data are significantly enhanced since they allow for the swift analysis and interpretation of extensive datasets. Bioinformatics technologies streamline data processing, minimizing the need for manual interpretation and delivering prompt actionable insights [56]. ML algorithms and AI improve predictive capabilities, allowing for the proactive identification of potential outbreaks and emerging threats.

Nevertheless, there are still obstacles to overcome in achieving widespread implementation of these modern technologies. To fully exploit their capabilities, it is essential to have standardized protocols, data-sharing agreements, and a strong bioinformatics infrastructure. Staff training and the creation of criteria for precise data interpretation are also crucial. The use of genomic data and informatics has greatly enhanced the efficiency and precision of food safety responses. These developments are essential for implementing effective intervention and prevention efforts, eventually protecting public health, and decreasing the economic impact of foodborne infections.

Rapid identification and reaction to foodborne pathogens are vital in minimizing the consequences of foodborne diseases. Failure to promptly identify the origin of contamination can result in extensive outbreaks, elevated healthcare expenses, and significant public health ramifications. Swift response times provide expedited execution of control measures, such as recalls, advisories, and sanitation interventions, thereby diminishing the propagation of infections and the intensity of outbreaks [11].

The incorporation of informatics technologies with genomic data has significantly enhanced the speed and effectiveness of food safety interventions. Bioinformatics platforms expedite the rapid analysis and comprehension of vast genomic information, converting data into promptly actionable insights with high efficiency. These platforms streamline data analysis procedures, thereby substantially decreasing the time and effort needed for manual interpretation [57]. This automation is crucial in situations where rapid decision-making is essential to contain and mitigate foodborne outbreaks.

Furthermore, ML algorithms and AI have become essential in improving the efficiency of food safety protocols. Through the identification and analysis of patterns in genetic data, these technologies can forecast possible outbreaks and detect emerging risks before they worsen. This prediction capacity enables preemptive measures, empowering health authorities to tackle contamination issues before they propagate extensively. The utilization of AI in food safety can result in expedited identification of sources of contamination and enhanced implementation of control measures, thus enhancing public health outcomes [58]. The combination of bioinformatics and AI thus represents a significant advancement in the field of food safety, ensuring that responses are both swift and precise.

The incorporation of ML and AI has also significantly bolstered the accuracy of food safety measures. Through the analysis of patterns in genomic data, these technologies can detect and forecast possible outbreaks with a high level of precision. Additionally, AI can assist in the ongoing surveillance of food safety through instantaneous analysis and prediction, enabling proactive and efficient responses [5].

Despite these advancements, challenges remain in the widespread adoption of genomic technologies and informatics tools. Standardized protocols, robust bioinformatics infrastructure, and data-sharing agreements are essential to harness the potential of these technologies fully. Additionally, continuous training for personnel and the development of guidelines for accurate data interpretation are critical for effective implementation.

12.2.5 Public health improvements

The integration of genomic data and advanced informatics tools in food safety has led to substantial improvements in public health. By enhancing the detection, tracing, and mitigation of foodborne pathogens, these technologies have significantly reduced the incidence and severity of foodborne illnesses.

The use of WGS has revolutionized pathogen detection in food safety. Traditional methods often require several days to weeks to identify pathogens, whereas WGS can provide comprehensive genetic profiles within hours. This rapid identification is crucial for implementing timely interventions and preventing widespread outbreaks [59]. Bioinformatics platforms such as the Center for Genomic Epidemiology's web-based applications enable the rapid identification of antibiotic resistance genes and multilocus sequence types. Other platforms include the Bacterial and Viral Bioinformatics Resource Center (BV-BRC), Galaxy, EnteroBase, and Rapid Annotation using Subsystem Technology (RAST). These tools allow for the precise characterization of pathogens, aiding in the development of targeted treatment and intervention strategies [60].

Identifying the origin of contamination is crucial for managing and averting foodborne outbreaks. The incorporation of genetic data into bioinformatics tools has significantly enhanced the precision and efficiency of source tracking. PulseNet, a network run by the Center for Diseases Control and Prevention (CDC) made up of public health and food regulatory agency laboratories in the United States, applies WGS to monitor the spread of foodborne diseases throughout the United States. This network has played a crucial role in pinpointing the origins of various outbreaks, resulting in expedited and more efficient public health interventions [61]. The use of genetic data in public health surveillance systems has resulted in notable enhancements in the monitoring and management of foodborne infections. The GenomeTrakr network, which is an effort by the FDA, uses WGS to construct a publicly accessible and open-source library of genomes of foodborne pathogens. This database enables the immediate monitoring and analysis of genetic data from various regions, hence improving the capacity to swiftly identify and address outbreaks [62].

This integration of genomic data and bioinformatics tools has led to substantial improvements in public health by enhancing the detection, tracing, and mitigation of foodborne pathogens. These technologies have transformed traditional food safety practices, leading to more effective and timely responses to food safety challenges. Future directions include the development of more advanced sequencing technologies and bioinformatics tools, which can further enhance the speed and accuracy of public health responses. Integrating AI and ML with genomic data analysis will improve predictive capabilities, allowing for more proactive and precise interventions.

12.2.6 Economic benefits for food industry stakeholders

The integration of genomic data and advanced informatics tools in food safety has not only improved public health outcomes but also brought significant economic benefits to the food industry. These benefits span across cost savings, enhanced market competitiveness, improved supply chain efficiency, and risk mitigation.

Integrating genetic data in food safety leads to a clear economic advantage by lowering the expenses related to outbreaks for foodborne illnesses. Conventional techniques for identifying pathogens are frequently slow and expensive, resulting in delays in pinpointing the sources of contamination and taking necessary actions to address issues. Using standard techniques like WGS allows for quick and accurate detection of infections, leading to a significant decrease in the time and resources required for outbreak investigations [63]. Through the implementation of early detection measures, organizations can effectively avert epidemics and so circumvent the significant expenses related to product recalls, legal penalties, and diminished sales. An example of the effectiveness of WGS was demonstrated during an outbreak of *L. monocytogenes* in five European countries (Austria, Denmark, Finland, Sweden, and the United Kingdom) from 2015 to 2018. The rapid identification and response made possible by WGS helped prevent prolonged disruptions and extensive recalls, resulting in significant cost savings of millions of dollars [64].

By incorporating genetic data into food safety regulations, the efficiency of the supply chain is enhanced through the improvement of traceability and accountability. Advanced informatics techniques provide the real-time surveillance and tracing of food products along the whole supply chain, from the farm to the consumer's plate. This feature guarantees the prompt detection and separation of contamination within the supply chain, hence minimizing the time and resources needed to resolve food safety concerns [65]. Moreover, improved traceability systems foster greater transparency and trust among consumers, which can translate into increased sales and customer loyalty. Retailers and producers benefit economically from the ability to assure consumers of the safety and quality of their products.

The accurate and prompt identification of foodborne pathogens using genomic technologies greatly diminishes the likelihood of extensive outbreaks, therefore reducing potential legal responsibilities and insurance expenses for the food sector. Companies that use sophisticated genetic and informatics techniques are more prepared to adhere to rigorous food safety rules, therefore avoiding penalties, sanctions, and legal consequences that may result from noncompliance. Employing state-of-the-art technologies to demonstrate strong food safety processes can improve a company's reputation with regulators and result in more efficient regulatory audits and inspections. Implementing this proactive risk management strategy not only ensures the protection of public health but also preserves the company's financial interests.

The integration of genomic data and advanced informatics tools in food safety offers substantial economic benefits for food industry stakeholders. By reducing the costs associated with outbreak management, enhancing supply chain efficiency, and mitigating risks and liabilities, these technologies create significant economic value. As the food industry continues to adopt and integrate these advanced technologies, stakeholders can expect ongoing improvements in both financial performance and public health outcomes.

12.3 Privacy concerns in genomic data collection

The utilization of genomic data in food safety has significantly advanced the detection, tracing, and mitigation of foodborne pathogens. However, alongside these advancements, the collection and use of genomic data have raised significant privacy concerns. These concerns revolve around the protection of sensitive genetic information, the potential for data misuse, and the ethical implications of genomic surveillance.

The sensitivity of genetic data in food safety is a crucial matter that involves safeguarding private information, preventing misuse, and addressing the ethical considerations associated with gathering and utilizing this data. Genomic data inherently comprises intricate and precise information regarding various organisms, including pathogens, agricultural products, and zoonotic diseases related to humans. Due to this sensitivity, it is imperative to implement stringent procedures to guarantee the privacy and security of data, while also taking ethical issues into account when using and sharing it. Genomic data pertaining to food safety frequently encompasses genetic sequences of crops and livestock employed in agriculture. These sequences may be private, indicating a substantial investment in research and development. Companies can suffer significant financial consequences if this data is accessed or used without authorization. Specifically, competitors can exploit or engage in biopiracy by utilizing comprehensive genetic data on crop varieties or livestock breeding methods, thus compromising the economic interests of the original developers [66].

Ensuring privacy is of utmost importance when it comes to safeguarding the security of genomic data. The repercussions of data breaches and unauthorized access to genomic databases can be highly significant. A breach has the potential to reveal sensitive data that could be utilized for bioterrorism or other nefarious purposes. Moreover, the unintentional disclosure of genomic data may result in breaches of privacy and the unauthorized exploitation of exclusive genetic information [67]. To reduce these dangers, it is crucial to implement strong cybersecurity safeguards. To safeguard genetic data from unauthorized access, it is imperative to incorporate encryption, access controls, and secure data storage techniques. Furthermore, it is imperative to conduct routine security audits and implement upgrades to effectively mitigate emerging threats and vulnerabilities [68].

Acquisition and utilization of genetic data in the context of food safety also give rise to noteworthy ethical and legal concerns. An important issue to consider is the possibility of discrimination based on genetic information. If genomic data indicates that certain populations have a higher vulnerability to certain medical conditions, there is a potential danger that these groups may experience stigmatization or prejudice. It is of utmost importance to ensure that genomic data is utilized in an ethical and equitable manner to avert such consequences [69]. Genomic data usage is subject to a range of national and international legislation. The Genetic Information Nondiscrimination Act (GINA) in the United States offers safeguards against discrimination that is rooted in genetic information. Nevertheless, the implementation of these rules in the realm of food safety genomics is intricate and continuously developing. Globally, legislation such as the General Data Protection Regulation (GDPR) in the European Union enforces strict rules on the gathering, retention, and use of personal data, including genetic data [70].

The notion of informed consent is essential when it comes to the gathering and utilization of genetic data. It is imperative for individuals and organizations to possess comprehensive knowledge regarding the utilization, retention, and dissemination of personal data. Obtaining informed consent can be difficult in the realm of food safety, especially when working with extensive agricultural data or environmental samples.

It is crucial to maintain ethical standards and public trust by ensuring transparency in data-gathering procedures and giving clear and accessible information to stakeholders [71]. Balancing the need for innovation in food safety with privacy concerns is a complex but essential task. While genomic data offers immense potential for improving food safety and public health, it must be collected and used responsibly.

Privacy concerns in genomic data collection are critical in the context of food safety. The sensitivity of genomic data, risks of unauthorized access, ethical and legal implications, and the need for informed consent and transparency all necessitate robust privacy protections. Balancing these concerns with the need for technological advancement requires careful consideration and the implementation of comprehensive privacy frameworks. By addressing these privacy issues, the food industry can leverage the benefits of genomic data while safeguarding the rights and interests of all stakeholders.

12.3.1 Security risks and threats in the context of food safety

The integration of genomic data and advanced informatics tools in food safety has brought substantial benefits in terms of detecting, tracing, and mitigating foodborne pathogens. However, this integration also introduces significant security risks and threats. These threats can compromise the integrity of food safety systems, expose sensitive data to unauthorized access, and potentially lead to severe public health consequences.

Data breaches are a significant security threat in the context of food safety. Given its intricate and vulnerable characteristics, genomic data is highly sought after by cybercriminals. Illegitimate entry into genomic databases can result in the pilfering of exclusive genetic data, which can be leveraged for financial advantage or employed for harmful purposes. For example, the genetic codes of exclusive plant types or animals could be unlawfully obtained and utilized to produce fake goods or provide competitors with valuable information for their breeding projects, resulting in substantial financial damages for the original creators [67]. To reduce these dangers, it is essential to establish strong cybersecurity protocols. This encompasses the use of encryption to protect data both when it is stored and when it is being transmitted, strict limits on who can access the data, and regular audits to detect and resolve any security weaknesses. In addition, implementing multifactor authentication (MFA) can offer an additional level of security, hence decreasing the probability of unauthorized entry [68].

The security of genetic data in food safety can be significantly challenged by insider threats. Individuals who have authorized access to confidential information, whether they are employees or contractors, have the potential to use their privileges for personal benefit or with evil intentions. This risk is particularly alarming since insiders generally possess extensive knowledge of the data systems and their security procedures, hence facilitating their ability to bypass protective measures [72]. Addressing insider threats necessitates a blend of technical and organizational strategies. Implementing activity monitoring and anomaly detection systems can be instrumental in identifying suspicious behavior that may indicate an insider threat. From an organizational standpoint, it is crucial to prioritize conducting comprehensive background checks, promoting a culture of security awareness, and enforcing stringent data access policies. These measures can effectively mitigate the potential risks associated with insiders.

Moreover, the possibility of bioterrorism and malevolent exploitation of genomic data poses a significant security risk. Genomic data on foodborne pathogens can be utilized to manipulate their genetic makeup, resulting in the creation of more potent strains or the production of biological weapons. These malevolent actions could result in severe public health ramifications. To mitigate this risk, it is imperative to create secure and controlled environments for the purposes of genetic research and data preservation. International cooperation and regulatory frameworks are essential for overseeing and managing the utilization of genetic data, guaranteeing that it is not exploited for detrimental intentions. Moreover, research institutes and companies are required to comply with biosafety and biosecurity policies to avoid the unauthorized access and utilization of confidential genetic data.

The food supply chain is intricate, encompassing several players from production to consumption. The intricate nature of this situation creates multiple opportunities for genomic data and food safety systems to be compromised. Targeting any part of the supply chain, cyberattacks have the potential to disrupt operations, contaminate products, and undermine the integrity of food. To improve supply chain security, it is necessary to take a comprehensive approach that involves implementing blockchain technology to provide traceability and transparency, conducting frequent security assessments, and establishing secure communication channels between all parties involved. Implementing these procedures can effectively safeguard the integrity of the food supply chain and mitigate the risks posed by cyber threats [73].

The integration of genomic data in food safety brings numerous benefits but also introduces significant security risks and threats. Data breaches, insider threats, bioterrorism, and supply chain vulnerabilities all pose substantial risks to the security and integrity of food safety systems. Addressing these risks requires a multifaceted approach, including robust cybersecurity measures, stringent access controls, continuous monitoring, and international collaboration. By implementing these measures, stakeholders can harness the benefits of genomic data while mitigating the associated security risks, ultimately enhancing the safety and security of the global food supply.

12.3.2 Consequences of data breaches on consumer trust in food safety

The integration of genomic data and advanced informatics tools in food safety has revolutionized the industry, offering enhanced capabilities for detecting, tracing, and mitigating foodborne pathogens. However, the increasing reliance on digital data and genomic information has also heightened the risks associated with data breaches. These breaches can have profound and far-reaching consequences on consumer trust, which is crucial for the sustained success and integrity of the food industry.

An immediate and very impactful outcome of data breaches in food safety is the deterioration of consumer trust. Upon discovering that their personal or sensitive information has been breached, consumers' confidence in the afflicted company and, consequently, the whole food industry can be eroded. The erosion of trust can result in less customer loyalty, lower sales figures, and a damaged business image [74].

Data breaches have financial consequences that go beyond the immediate losses resulting from fraud or theft. The loss of consumer trust can pose a significant threat to a company's long-term financial stability. Companies may encounter elevated expenses associated with customer compensation, litigation expenses, and regulatory penalties. Furthermore, it may be necessary for them to allocate substantial resources toward public relations initiatives to restore their credibility and regain the trust of consumers [75]. A breach has the potential to result in a big decline in stock values, legal actions, and hefty monetary penalties. The corporation may need to devote substantial efforts to enhance its cybersecurity infrastructure and restore confidence in clients regarding the security of their data, highlighting the significant financial consequences of such occurrences.

Data breaches in food safety can potentially result in immediate public health consequences. If the genomic data associated with foodborne infections is breached, it has the potential to be exploited for the purpose of manipulating food products or generating dangerous biological agents. These severe situations highlight the significance of safeguarding genomic material to prevent harmful exploitation [67]. Furthermore, data breaches that compromise the reliability of food safety systems might result in delays in detecting and addressing foodborne outbreaks. This can lead to extended exposure to contaminated products, increased occurrence of foodborne illnesses, and exacerbated public health hazards. Deterioration in confidence in food safety systems can also result in decreased adherence to food safety recommendations and guidelines among consumers, hence exacerbating public health challenges.

Other important outcomes of data breaches are legal and regulatory consequences. Regulatory agencies may impose severe fines and penalties on businesses that fail to protect customer data. The GDPR of the European Union sets strict guidelines for data protection and has the authority to apply heavy penalties for noncompliance

[76]. The Food and Drug Administration (FDA) and other regulatory agencies in the United States have implemented standards to safeguard genetic data in relation to food safety. Noncompliance with these principles may lead to legal repercussions and additional loss of consumer confidence. Therefore, companies must ensure strict compliance with these standards to prevent legal consequences and uphold consumer trust [69].

To minimize the negative effects of data breaches on consumer confidence, companies need to have thorough data protection procedures. These methods should incorporate strong cybersecurity protocols, frequent security assessments, and the utilization of sophisticated encryption technology to safeguard sensitive data. In addition, it is important for companies to cultivate a culture of transparency by effectively communicating their data protection processes to consumers and swiftly resolving any breaches that may occur [68]. Moreover, organizations can bolster consumer confidence by acquiring certifications and adhering to industry benchmarks for safeguarding data. Engaging in voluntary initiatives, like obtaining up-to-date accreditation for information security management, can serve as evidence of a dedication to safeguarding consumer data and restoring confidence after a security breach [77].

12.3.3 Encryption technologies, blockchain, data management, and future directions in food informatics

Encryption technologies are vital in food informatics to protect sensitive data related to food safety, quality, and supply chain management. Some key encryption technologies include Advanced Encryption Standard (AES), Public Key Infrastructure (PKI) and Homomorphic Encryption. AES, typically known as Rijndael, is widely adopted due to its strong security and efficiency. It encrypts data in blocks of 128 bits using key sizes of 128, 192, or 256 bits. Consequently, AES is used to secure data both at rest and in transit, ensuring confidentiality and integrity [78]. PKI uses digital certificates and public or private key pairs to validate entities and secure communications. It ensures that data exchanged between parties in the food supply chain remains restricted and tamperproof [79]. Lastly, Homomorphic Encryption allows computations on encrypted data without needing to decrypt it first. This ensures that sensitive data remains secure during processing and analysis, making it very effective for collaborative food informatics projects [80].

Blockchain technology is changing food informatics by improving traceability, transparency, and trust in the food supply chain. And it provides a decentralized and tamperproof ledger of all transactions. Each block in the chain contains a cryptographic hash of the previous block, a timestamp, and transaction data, ensuring immutability and transparency [81]. Blockchain also enables traceability of food products from farm to fork, thus helping to quickly address food safety issues, such as contamination or fraud, by tracking the origin and journey of food items [82]. Blockchain technology can be incorporated into smart contracts where self-executing contracts with the terms of the agreement are directly written into code. In the context of food informatics, smart contracts can automate compliance checks and prompt actions when specific conditions are met, enhancing effectiveness, and reducing human error [2]. Most importantly, secure data management is critical to protect sensitive information and ensure data integrity in food informatics. The data encryption technologies mentioned previously are essential to prevent unauthorized access and ensure data confidentiality [80]. Thus, implementing robust access control mechanisms ensures that only authorized personnel can access sensitive data.

Also, maintaining detailed audit trails helps in monitoring data access and modifications. This is essential for compliance with regulatory standards and for investigating any data violations or glitches [83]. To protect individual privacy, data anonymization techniques including, pseudonymization (GDPR), generalization [83] and perturbation [78] can be used.

Future directions in food informatics will be driven by advancements in technology and increasing demand and choice for food safety, quality, and efficiency. Increasingly, AI and ML algorithms will continue to drive advancement, providing deeper insights and predictive analytics for the entirety of the food supply chain. Also, IoT devices will become more prevalent especially in developing economies when accessibility of stable, fast, and affordable internet access becomes reality. Ultimately big data analytics would be used in understanding complex intricacies in food-related data, hopefully ensuring a much more robust, sustainable, and safe food supply. And all of this will be incorporated into innovations in smart packaging and digital twin technologies and technologies yet to be unearthed.

References

[1] Krupitzer C, Stein A. Food informatics – review of the current state-of-the-art, revised definition, and classification into the research landscape. Foods 2021 Nov 1, 10(11), 1–17.

[2] Paul PK, Aithal PS, Bhuimali A. Food informatics and its challenges and opportunities-A review [Internet]. Available from: https://ssrn.com/abstract=3042148

[3] Chen TC, Yu SY. The review of food safety inspection system based on artificial intelligence, image processing, and robotic. Food Science and Technology (Brazil) 2022, 42.

[4] Alamprese C, Amigo JM, Casiraghi E, Engelsen SB. Identification and quantification of turkey meat adulteration in fresh, frozen-thawed and cooked minced beef by FT-NIR spectroscopy and chemometrics. Meat Science 2016 Nov 1, 121, 175–181.

[5] Ding H, Tian J, Yu W, Wilson DI, Young BR, Cui X, et al. The Application of Artificial Intelligence and Big Data in the Food Industry. In: Foods, Switzerland: Multidisciplinary Digital Publishing Institute (MDPI), 2023, 12.

[6] Garcia SN, Osburn BI, Jay-Russell MT. One health for food safety, food security, and sustainable food production. In: Joshua B. Gurtler, USDA. Frontiers in Sustainable Food Systems, Switzerland: Frontiers Media S.A., 2020, 1–9, 4.

[7] Koppel N, Rekdal VM, Balskus EP. Chemical Transformation of Xenobiotics by the Human Gut Microbiota. In: Science, USA: American Association for the Advancement of Science, 2017, 356, 1246–1257.

[8] Kwong JC, Mercoulia K, Tomita T, Easton M, Li HY, Bulach DM, et al. Prospective whole-genome sequencing enhances national surveillance of listeria monocytogenes. J Clin Microbiol 2016 Feb 1, 54(2), 333–342.

[9] Struelens MJ, Ludden C, Werner G, Sintchenko V, Jokelainen P, Ip M. Real-time genomic surveillance for enhanced control of infectious diseases and antimicrobial resistance. Frontiers in Science [Internet] 2024, 2, Available from https://www.frontiersin.org/journals/science/articles/10.3389/fsci.2024.1298248.

[10] Mardis ER. DNA sequencing technologies: 2006–2016. Nature Protocols [Internet] 2017, 12(2), 213–218, Available from https://doi.org/10.1038/nprot.2016.182.

[11] Allard MW, Strain E, Melka D, Bunning K, Musser SM, Brown EW, et al. Practical value of food pathogen traceability through building a whole-genome sequencing network and database. Journal of Clinical Microbiology. American Society for Microbiology 2016, 54, 1975–1983.

[12] Abbas O, Zadravec M, Baeten V, Mikuš T, Lešić T, Vulić A, et al. Analytical methods used for the authentication of food of animal origin. Food Chemistry 2018 Apr 25, 246, 6–17.

[13] Lu H, Giordano F, Ning Z. Oxford Nanopore MinION sequencing and genome assembly. Genomics Proteomics Bioinformatics. 2016 Oct 1;14(5):265–279.

[14] Allard MW, Luo Y, Strain E, Li C, Keys CE, Son I, et al. High resolution clustering of Salmonella enterica serovar Montevideo strains using a next-generation sequencing approach. BMC Genomics 2012 Jan 19, 13(1), 13–32.

[15] Chiu CY, Miller SA. Clinical metagenomics. Nature Reviews Genetics 2019 20:6 [Internet]. 2019 Mar 27 [cited 2024 Jun 5], 20(6), 341–355, Available from https://www.nature.com/articles/s41576-019-0113-7.

[16] Breitwieser FP, Lu J, Salzberg SL. A review of methods and databases for metagenomic classification and assembly. Brief Bioinform [Internet] 2019 Mar 27 cited 2024 Jun 5, 20(4), 1125–1139, Available from https://pubmed.ncbi.nlm.nih.gov/29028872/.

[17] Grad YH, Lipsitch M, Feldgarden M, Arachchi HM, Cerqueira GC, FitzGerald M, et al. Genomic epidemiology of the Escherichia coli O104:H4 outbreaks in Europe. Proceedings of the National Academy of Sciences of the United States of America [Internet] 2011, 2012 Feb 21 [cited 2024 Jun 5] 109(8), 3065–3070, Available from https://pubmed.ncbi.nlm.nih.gov/22315421/.

[18] Mellmann A, Harmsen D, Cummings CA, Zentz EB, Leopold SR, Rico A. et al., EHEC O104:H4 in Germany 2011: Large outbreak of bloody diarrhea and haemolytic uraemic syndrome by shiga toxin–producing E. coli via contaminated food. PLoS One [Internet] 2012, [cited 2024 Jun 5] 6(7), Available from https://www.ncbi.nlm.nih.gov/books/NBK114499/.

[19] Sheppard SK, Maiden MCJ. The Evolution of Campylobacter jejuni and Campylobacter coli. Cold Spring Harbor Perspectives in Biology [Internet] 2015 Aug 1 [cited 2024 Jun 5], 7(8), Available from /pmc/articles/PMC4526750/.

[20] Köser CU, Ellington MJ, Peacock SJ. Whole-genome sequencing to control antimicrobial resistance. Trends in Genetics [Internet] 2014 [cited 2024 Jun 5], 30(9), 401, Available from /pmc/articles/PMC4156311/.

[21] Vashisht V, Vashisht A, Mondal AK, Farmaha J, Alptekin A, Singh H, et al. Genomics for Emerging Pathogen Identification and Monitoring: Prospects and Obstacles. In: BioMedInformatics, Switzerland: Multidisciplinary Digital Publishing Institute (MDPI), 2023, 1145–1177.

[22] Chen TC, Yu SY. The review of food safety inspection system based on artificial intelligence, image processing, and robotic. Food Science and Technology (Brazil) 2022, 1, 42.
[23] Hadi J, Rapp D, Dhawan S, Gupta SK, Gupta TB, Brightwell G. Molecular detection and characterization of foodborne bacteria: Recent progresses and remaining challenges. 2023 [cited 2024 Jun 5]; Available from: https://ift.onlinelibrary.wiley.com/doi/10.1111/1541-4337.13153
[24] Allard MW, Bell R, Ferreira CM, Gonzalez-Escalona N, Hoffmann M, Muruvanda T, et al. Genomics of foodborne pathogens for microbial food safety. Curr Opin Biotechnol 2018 Feb 1, 49, 224–229.
[25] Moran-Gilad J. Whole genome sequencing (WGS) for food-borne pathogen surveillance and control – Taking the pulse. Eurosurveillance [Internet] 2017 Jun 6 [cited 2024 Jun 5], 22(23), 1, Available from /pmc/articles/PMC5479979/.
[26] Wielinga PR, Hendriksen RS, Aarestrup FM, Lund O, Smits SL, Koopmans MPG, et al. Global microbial identifier. Applied Genomics of Foodborne Pathogens [Internet] 2017, [cited 2024 Jun 5] 13, Available from /pmc/articles/PMC7153444/.
[27] Quainoo S, Coolen JPM, van Hijum SAFT, Huynen MA, Melchers WJG, Van schaik W, et al. Whole-genome sequencing of bacterial pathogens: The future of Nosocomial outbreak analysis. Clinical Microbiology Reviews [Internet] 2017, Oct 1 [cited 2024 Jun 5] 30(4), 1015–1063, Available from https://pubmed.ncbi.nlm.nih.gov/28855266/.
[28] Brown E, Dessai U, Mcgarry S, Gerner-Smidt P. Use of whole-genome sequencing for food safety and public health in the united states. Foodborne Pathogens and Disease [Internet] 2019 Jul 7 cited 2024 Jun 5, 16(7), 441, Available from /pmc/articles/PMC6653787/.
[29] Van Goethem N, Descamps T, Devleesschauwer B, Roosens NHC, Boon NAM, Van Oyen H, et al. Status and potential of bacterial genomics for public health practice: A scoping review. Implementation Science 2019, 14:1 [Internet] 2019 Aug 13 [cited 2024 Jun 5] 14(1), 1–16, Available from https://implementationscience.biomedcentral.com/articles/10.1186/s13012-019-0930-2.
[30] Kovac J. Precision food safety: A paradigm shift in detection and control of foodborne pathogens. mSystems [Internet] 2019 Jun 25 [cited 2024 Jun 5], 4(3), Available from https://journals.asm.org/doi/10.1128/msystems.00164-19.
[31] Kelly S, Heaton K, Hoogewerff J. Tracing the geographical origin of food: The application of multi-element and multi-isotope analysis. Trends in Food Science & Technology [Internet] 2005 Dec [cited 2024 Jun 5], 16(12), 555–567, Available from https://www.researchgate.net/publication/222641805_Tracing_the_geographical_origin_of_food_The_application_of_multi-element_and_multi-isotope_analysis.
[32] Jackson BR, Tarr C, Strain E, Jackson KA, Conrad A, Carleton H, et al. Implementation of nationwide real-time whole-genome sequencing to enhance listeriosis outbreak detection and investigation. Clinical Infectious Diseases [Internet] 2016, Aug 1 [cited 2024 Jun 5] 63(3), 380–386, Available from https://pubmed.ncbi.nlm.nih.gov/27090985/.
[33] Mellmann A, Harmsen D, Cummings CA, Zentz EB, Leopold SR, Rico A. et al., Prospective genomic characterization of the German enterohemorrhagic Escherichia coli O104:H4 outbreak by rapid next generation sequencing technology. PLoS One [Internet] 2011, [cited 2024 Jun 5] 6(7), Available from https://pubmed.ncbi.nlm.nih.gov/21799941/.
[34] Investigation into an outbreak of Shiga toxin producing E. coli. 2024.
[35] Didelot X, Bowden R, Wilson DJ, Peto TEA, Crook DW. Transforming clinical microbiology with bacterial genome sequencing. Nature Reviews Genetics [Internet] 2012 Sep [cited 2024 Jun 5], 13(9), 601–612, Available from https://pubmed.ncbi.nlm.nih.gov/22868263/.
[36] Zhou Z, McCann A, Weill FX, Blin C, Nair S, Wain J, et al. Transient Darwinian selection in salmonella enterica serovar paratyphi a during 450 years of global spread of enteric fever. Proceedings of the National Academy of Sciences of the United States of America [Internet] 2014, Aug 19 [cited 2024 Jun 5] 111(33), 12199–12204, Available from https://pubmed.ncbi.nlm.nih.gov/25092320/.

[37] Gymrek M, McGuire AL, Golan D, Halperin E, Erlich Y. Identifying personal genomes by surname inference. Science [Internet] 2013 Jan 18 [cited 2024 Jun 5], 339(6117), 321–324, Available from https://pubmed.ncbi.nlm.nih.gov/23329047/.

[38] Xiang D, Cai W. Privacy protection and secondary use of health data: Strategies and methods. BioMed Research International. Hindawi Limited 2021, 1–11.

[39] Erlich Y, Narayanan A. Routes for breaching and protecting genetic privacy. Nature Reviews Genetics 2014 15:6 [Internet]. 2014 May 8 [cited 2024 Jun 5], 15(6), 409–421, Available from https://www.nature.com/articles/nrg3723.

[40] H-K Wong E, Hanner RH. DNA barcoding detects market substitution in North American seafood. 2008 [cited 2024 Jun 5]; Available from: http://www.fishbol.org

[41] Yogeeswaran K, Frary A, York TL, Amenta A, Lesser AH, Nasrallah JB, et al. Comparative genome analyses of Arabidopsis spp.: Inferring chromosomal rearrangement events in the evolutionary history of A. thaliana. Genome Research [Internet] 2005, Apr [cited 2024 Jun 5] 15(4), 505–515, Available from https://pubmed.ncbi.nlm.nih.gov/15805492/.

[42] Hanner R, Becker S, Ivanova N V., Steinke D. FISH-BOL and seafood identification: Geographically dispersed case studies reveal systemic market substitution across Canada. Mitochondrial DNA [Internet] 2011 Oct [cited 2024 Jun 5];22 Suppl, 1(SUPPL. 1), 106–122, Available from https://pubmed.ncbi.nlm.nih.gov/21980986/.

[43] Ballin NZ. Authentication of meat and meat products. Meat Science 2010 Nov 1, 86(3), 577–587.

[44] Allard MW, Strain E, Melka D, Bunning K, Musser SM, Brown EW, et al. Practical value of food pathogen traceability through building a whole-genome sequencing network and database. Journal of Clinical Microbiology [Internet] 2016, Aug 1 [cited 2024 Jun 5] 54(8), 1975–1983, Available from https://pubmed.ncbi.nlm.nih.gov/27008877/.

[45] Hayes BJ, Bowman PJ, Chamberlain AJ, Goddard ME. Invited review: Genomic selection in dairy cattle: Progress and challenges. Journal of Dairy Science [Internet] 2009 [cited 2024 Jun 5], 92(2), 433–443, Available from https://pubmed.ncbi.nlm.nih.gov/19164653/.

[46] Cocolin L, Alessandria V, Dolci P, Gorra R, Rantsiou K. Culture independent methods to assess the diversity and dynamics of microbiota during food fermentation. International Journal of Food Microbiology [Internet] 2013 Oct 1 [cited 2024 Jun 5], 167(1), 29–43, Available from https://pubmed.ncbi.nlm.nih.gov/23791362/.

[47] Bouzembrak Y, Steen B, Neslo R, Linge J, Mojtahed V, Marvin HJP. Development of food fraud media monitoring system based on text mining. Food Control 2018 Nov 1, 93, 283–296.

[48] The collection, linking and use of data in biomedical research and health care [Internet]. Nuffield Council on Bioethics. 2015 [cited 2024 Jun 5]. Available from: https://www.nuffieldbioethics.org/publications/biological-and-health-data

[49] Baranyi J, Roberts TA. A dynamic approach to predicting bacterial growth in food. International Journal of Food Microbiology [Internet] 1994, 23(3), 277–294, Available from https://www.sciencedirect.com/science/article/pii/0168160594901570.

[50] Ross T, Mcmeekin TA. Predictive Microbiology and Food Safety [Internet]. In: Encyclopedia of Food Microbiology, Second Edition, Elsevier, 2014, 3, 59–68, Available from http://dx.doi.org/10.1016/B978-0-12-384730-0.00256-1.

[51] Farakos SMS, Pouillot R, Spungen J, Flannery B, Doren JM Van, Dennis S. Implementing a risk-risk analysis framework to evaluate the impact of food intake shifts on risk of illness: A case study with infant cereal. Food Additives & Contaminants: Part A [Internet] 2021 Jun, 38(5), 718–730, Available from https://doi.org/10.1080/19440049.2021.1885752.

[52] Wang N, Gao J, Yuan L, Jin Y, He G. Metabolomics profiling during biofilm development of Bacillus licheniformis isolated from milk powder. International Journal of Food Microbiology 2021 Jan 16, 337, 1–12.

[53] Wei Z, Alam T, Al Sulaie S, Bouye M, Deebani W, Song M. An efficient IoT-based perspective view of food traceability supply chain using optimized classifier algorithm. Information Processing and Management 2023 May 1, 60(3), 1–13.

[54] Talari G, Cummins E, McNamara C, O'Brien J. State of the art review of big data and web-based Decision Support Systems (DSS) for food safety risk assessment with respect to climate change. Trends in Food Science and Technology [Internet] 2022, 126, 192–204, Available from https://www.sciencedirect.com/science/article/pii/S0924224421005185.

[55] Ronholm J, Nasheri N, Petronella N, Pagotto F. Navigating microbiological food safety in the era of whole-genome sequencing. Clinical Microbiology Reviews 2016 Oct 1, 29(4), 837–857.

[56] Spjuth O, Bongcam-Rudloff E, Hernández GC, Forer L, Giovacchini M, Guimera RV, et al. Experiences with Workflows for Automating Data-intensive Bioinformatics, Biology Direct. UK: BioMed Central Ltd, 2015, 10.

[57] Alkema W, Boekhorst J, Wels M, Van Hijum SAFT Microbial bioinformatics for food safety and production. Briefings in Bioinformatics 2016 Mar 1, 17(2), 283–292.

[58] Qian C, Murphy SI, Orsi RH, Wiedmann M. How can ai help improve food safety? 2022; Available from: https://doi.org/10.1146/annurev-food-060721-

[59] Brown E, Dessai U, Mcgarry S, Gerner-Smidt P. Use of whole-genome sequencing for food safety and public health in the United States. Foodborne Pathogens and Disease. Mary Ann Liebert Inc 2019, 16, 441–450.

[60] Zankari E, Hasman H, Cosentino S, Vestergaard M, Rasmussen S, Lund O, et al. Identification of acquired antimicrobial resistance genes. Journal of Antimicrobial Chemotherapy 2012, Nov 67(11), 2640–2644.

[61] Tolar B, Joseph LA, Schroeder MN, Stroika S, Ribot EM, Hise KB, et al. An overview of PulseNet USA databases. Mary Ann Liebert Inc 2019, 16 Foodborne Pathogens and Disease, 457–462.

[62] Timme RE, Sanchez Leon M, Allard MW. Utilizing the public genometrakr database for foodborne pathogen traceback. In: John M. Walker. Methods in Molecular Biology, UK: Humana Press Inc., 2019, 201–212.

[63] Kumar P, Sundermann AJ, Martin EM, Snyder GM, Marsh JW, Harrison LH, et al. Method for economic evaluation of bacterial whole genome sequencing surveillance compared to standard of care in detecting hospital outbreaks. Clinical Infectious Diseases 2021 Jul 1, 73(1), 9–18.

[64] Sarno E, Pezzutto D, Rossi M, Liebana E, Rizzi V. A review of significant European foodborne outbreaks in the last decade. Journal of Food Protection 2021 Dec 1, 84(12), 2059–2070.

[65] Membré JM, Lambert RJW. Application of predictive modelling techniques in industry: From food design up to risk assessment. Int J Food Microbiol [Internet] 2008, 128(1), 10–15, Available from http://dx.doi.org/10.1016/j.ijfoodmicro.2008.07.006.

[66] Knoppers BM. Framework for responsible sharing of genomic and health-related data. HUGO Journal 2014 Dec 1, 8(1), 1–6.

[67] Erlich Y, Narayanan A. Routes for breaching and protecting genetic privacy.In: Nature Reviews Genetics, UK: Nature Publishing Group, 2014, 15, 409–421.

[68] Kaufman DJ, Bollinger JM, Dvoskin RL, Scott JA. Risky business: Risk perception and the use of medical services among customers of DTC personal genetic testing. J Genet Couns 2012 Jun, 21(3), 413–422.

[69] Nuffield council on bioethics | 1.

[70] Phillips AM. Only a click away – DTC genetics for ancestry, health, love . . . and more: A view of the business and regulatory landscape. Applied & Translational Genomics 2016 Mar 1, 8, 16–22.

[71] McGuire AL, Achenbaum LS, Whitney SN, Slashinski MJ, Versalovic J, Keitel WA, et al. Perspectives on human microbiome research ethics. Journal of Empirical Research on Human Research Ethics 2012 Jul, 7(3), 1–14.

[72] Probst Christian W. and Hunker J GDBM. Aspects of Insider Threats. In: Probst Christian W. and Hunker J, GD and BM, editors. Insider Threats in Cyber Security [Internet], Boston, MA: Springer US, 2010, 1–15, Available from https://doi.org/10.1007/978-1-4419-7133-3_1.
[73] Tian F. An agri-food supply chain traceability system for China based on RFID & blockchain technology. 2016 13th International Conference on Service Systems and Service Management (ICSSSM) [Internet] 2016, 1–6, Available from https://api.semanticscholar.org/CorpusID:21581564.
[74] Gatzert N, Schmit J. Supporting strategic success through enterprise-wide reputation risk management. The Journal of Risk Finance [Internet] 2016 Jan 1, 17(1), 26–45, Available from https://doi.org/10.1108/JRF-09-2015-0083.
[75] Iftekhar A, Cui X. Blockchain-based traceability system that ensures food safety measures to protect consumer safety and COVID-19 free supply chains. Foods 2021 Jun 1, 10(6), 1–12.
[76] Haddara M, Salazar A, Langseth M. Exploring the impact of GDPR on big data analytics operations in the E-commerce industry. Procedia Computer Science [Internet] 2023, 219, 767–777, Available from https://www.sciencedirect.com/science/article/pii/S1877050923003599.
[77] Charlebois S, Sterling B, Haratifar S, Naing SK. Comparison of global food traceability regulations and requirements. Compr Rev Food Sci Food Saf 2014, 13(5), 1104–1123.
[78] Dworkin MJ. Advanced Encryption Standard (AES) [Internet]. 2023 May. Available from: https://nvlpubs.nist.gov/nistpubs/FIPS/NIST.FIPS.197-upd1.pdf
[79] Udugama IA, Kelton W, Bayer C. Digital twins in food processing: A conceptual approach to developing multi-layer digital models. Digital Chemical Engineering 2023 Jun 1, 7, 19.
[80] Gentry C, Halevi S. Implementing Gentry's Fully-Homomorphic Encryption Scheme, Estonia: International Association for Cryptologic Research, 2011.
[81] Mohammed A, Potdar V, Quaddus M, Hui W. Blockchain Adoption in Food Supply Chains: A Systematic Literature Review on Enablers, Benefits, and Barriers. In: Merhdad Saif. IEEE Access, Canada. Institute of Electrical and Electronics Engineers Inc., 2023, 11, 14236–14255.
[82] Tian F. An agri-food supply chain traceability system for China based on RFID & blockchain technology. In: IEEE Access, Institute of Electrical and Electronics Engineers Inc., 2016 13th International Conference on Service Systems and Service Management (ICSSSM), Kunming: 2016, 1–6.
[83] Sweeney L. L. Sweeney k-anonymity: A model for k-anonymity: A model for protecting privacy 1. International Journal on Uncertainty, Fuzziness and Knowledge-based Systems 2002, 10, 557–570.

Jokodola Temilola Tolulope, Idowu-Mogaji Grace Oluwatoyin, Abiona Stella Olusola, and Aina Adetinuke Titilade

Chapter 13 Sensory analysis and data integration using food informatics approach

Abstract: Various variables, including cost, taste, texture, appearance, and fragrance, as well as aspects related to health and safety, influence consumers' choices when choosing food products. Sensory quality is a crucial component that has a significant impact on the overall quality of food and is evaluated by the five sense organs. It is an essential tool in the food sector that aids in satisfying market expectations, producing high-quality, consumer-focused products, and achieving economic success.

Food informatics is powerful tool in tackling challenges relating to food production, processing, supply chain management, and demand estimation via the use of data generated from algorithms, computer models, and artificial intelligence (AI) to efficiently process and analyze large datasets, identify patterns and correlations, and make predictions or judgments.

Sensory analysis, which aids in assessing products using descriptive, difference, and qualitative tests, in order to create new products, minimizes cost of production and influences consumer acceptability. This engages trained sensory panelists; however in recent times, electronic sensors such as e-nose have emerged with the benefits of high sensitivity, rapid analysis, time management, and accurate data interpretation in predicting future trends for consumer acceptability of new food products.

Keywords: Internet of things (IoT), electronic nose (e-nose), sensory analysis, food informatics

13.1 Introduction

Consumers' decisions regarding food products are influenced by several factors, such as price, taste, texture, appearance, and aroma, in addition to elements pertaining to safety and health. Sensory analysis investigates sensory characteristics by the use of human

Jokodola Temilola Tolulope, Department of Food Technology, University of Ibadan, Nigeria
Idowu-Mogaji Grace Oluwatoyin, Department of Wildlife and Ecotourism Management, University of Ibadan, Nigeria
Abiona Stella Olusola, Food Science and Technology Department, Federal Polytechnic Ado Ekiti, Nigeria
Aina Adetinuke Titilade, Works and Physical Planning Department, University of Lagos, Nigeria

https://doi.org/10.1515/9783111441238-014

senses of sight, touch, smell, taste, and hearing. The overall quality of food is greatly influenced by sensory quality, which is an essential factor. For ages, food products have been accepted or rejected using such methods of analysis. Food quality evaluation in the past used to tow the path of trending technology and microbiological safety; however, significant evolution over the years have made it one of the most important approaches for innovation and application to guarantee consumer acceptance of the finished product. The importance of sensory analysis in the food industry lies in its ability to:

- Assess sensory attributes, which influence consumer acceptability, such as texture, taste, odor, and appearance.
- Evaluate sensory quality and safety, thus, enhancing quality control of product and process while also contributing to research and development.
- Offer perceptions into the qualitative aspects of food safety, assisting in the identification of variables influencing consumer perception.
- Improve product creation by learning about the tastes and expectations of the target market.
- Guarantee consumer acceptability and enhance product quality to ensure the success of the development of novel products.
- In general, sensory analysis is an essential tool in the food sector that aids in satisfying market expectations, producing high-quality, consumer-focused products, and achieving economic success.

Consumers use perceived intrinsic and extrinsic qualities to characterize the advantages of a product. Focus groups, for instance, could be organized to determine the various expectations that consumers have for new products. The success or failure of new product developments will also be greatly influenced by how accurately these variables and expectations are measured. Gaining more insight into the sensory experiences of prospective customers creates room for creativity. Due to all of these factors, researchers and the industry now view sensory analysis and consumer research as two of the most helpful instruments in various stages of the development of new products, from design to commercialization, to enhance product quality and ensure that innovations are successfully adopted by consumers [1].

13.1.1 The concept of data integration using a food informatics approach

New challenges for food production have been brought about by the growing human population, shifting food consumption patterns, and recent advancements in knowledge of food sustainability [2]. Recent research developments in the Internet of things (IoT) seem to benefit several aspects of food production, including farming, supply chain management, processing, and demand estimation. A key area in tackling these issues is food informatics, which mostly depends on data utilization. Cutting-edge technologies like the IoT, which allows connected devices to communicate via the Internet, are incorporated into this field. Because it makes it easier to create algorithms and computer models, artificial intelligence (AI) is essential in food informatics.

These set it apart from other study areas in the field by enabling machines to efficiently process and analyze large datasets, identify patterns and correlations, and make predictions or judgments based on the learned insights [2, 3]. Food informatics, then, is the gathering, processing, analysis, and intelligent application of data from retail, food processing, the food supply chain, agriculture, and smart (consumer) health to extract knowledge, perform intelligent analysis, and identify optimizations that can be applied to food production, consumption, food security, and the end of life of food products [2]. It is well known that the IoT is made up of networked sensors and computing power that often enhance commonplace items. In order to analyze and detect changes in the environment, the sensors gather data, which the IoT system can then adjust to. Procedures from AI, which hold that computers should be able to carry out tasks intelligently, and machine learning (ML), which allows machines to learn from data, can improve the analysis and system regulation process in IoT systems.

Another justification for adaptability is the measures taken to analyze and control IoT networks. The deliberate use of these techniques can enhance and optimize the current procedures. Precision agriculture, smart farming, the Internet of Food, food supply chain management, food authentication, industrial IoT (IIoT) for food production, food safety, food computing, and smart or pervasive health are some of the fields where research in this topic is spread out. These ideas frequently overlap and are not entirely separate.

Food informatics seeks to extract knowledge for intelligent analysis, managing food product end of life, optimizing food production and consumption, and guaranteeing food security.

The last phase of a food product's life cycle, which includes its disposal or use after its intended use has ended, is referred to as "end of life." In order to reduce the influence on the environment, this phase entails managing elements like waste reduction, recycling, or suitable disposal techniques. Food informatics is a strategy that establishes an information hub by using technology to digitize the food industry and address certain supply chain problems [2, 4].

13.1.2 The relevance of organoleptic evaluation and its role in sensory analysis

Food informatics can be used to increase the adaptability of modified intelligent production systems that are suited to the particular needs of the food sector. Furthermore, the use of 3D printers for additive manufacturing is one instance of how it could make it easier to incorporate cutting-edge technologies that support food product customization [5].

The general populace benefits from food informatics, but usability engineering and efficient human–computer interface are necessary. It requires knowledge of food and nutrition, database integration, web development, networking, and communication to create error-free food information systems that benefit food processors and farmers simultaneously.

An integral part of sensory analysis is organoleptic evaluation, which assesses a product's attributes using human senses. Features including color, taste, size, shape, touch, and texture are evaluated to determine quality and identity. In many industries, such as food, medicine, cosmetics, and packaging, this assessment technique is essential to meet consumer expectations and ensure product quality. The significance of organoleptic evaluation and its purposes are emphasized by the following:

1. Quality control and product development
Organoleptic evaluation is integral to quality control and product development. It ensures that products meet consistent standards and align with consumer preferences. By systematically assessing sensory characteristics, manufacturers can maintain quality and make necessary adjustments in the production process [6].

2. Consumer acceptance and market success
Understanding consumer preferences through organoleptic evaluation can predict market success. Products that meet or exceed consumers' expectations in sensory qualities are more likely to succeed in the market [7].

3. Regulatory compliance
Many industries are subject to regulatory standards that include sensory characteristics. Organoleptic evaluation helps ensure compliance with these standards, thereby avoiding legal issues and ensuring consumer safety.

4. Product differentiation
In a competitive market, products with superior sensory attributes can distinguish themselves from competitors. Organoleptic evaluation helps identify and enhance these attributes, creating a unique selling proposition.

5. Research and development
In research and development (R&D), organoleptic evaluation is used to test new formulations and modifications, ensuring that any changes do not negatively impact sensory qualities. This process is crucial for innovation and continuous improvement. By leveraging organoleptic evaluation, industries can ensure their products are not only safe and compliant but also appealing and competitive in the market.

13.2 Sensory analysis and its purpose in evaluating food products

A crucial tool for assessing food products, sensory analysis is defined as a scientific process that assesses the sensory qualities of food items, such as their appearance, aroma, taste, texture, and sound. It offers insights into the sensory qualities that influence consumer acceptance and preferences. This facilitates quality assurance, marketing, and product development.

13.2.1 Types of sensory analysis

Descriptive tests: These tests that describe a food product's sensory qualities, such as its flavor and texture, are known as descriptive tests. The flavor profile test and texture profile test are two examples.

Difference tests: These tests, which include the triangle test, duo-trio test, and paired comparison test, compare food products to find differences in sensory qualities.

Qualitative tests: Using a six-inch line scale with half-inch intensity indicators, a panel of individuals evaluates each product individually in order to measure its attributes.

13.2.2 Importance of sensory analysis

- Product development: Producers are able to create new products that satisfy consumer tastes and expectations by using sensory analysis, which offers insights into a product's sensory qualities.
- Quality control involves locating and removing sources of unintended errors, such as human subject biases or environmental influences, to assist guaranteeing that goods fulfil quality standards.
- Market success: Producers may better understand consumer preferences by using sensory analysis, which is helpful for developing launch and product marketing strategies.
- Cost-cutting measures: Sensory analysis aids in cost-cutting efforts by pointing out areas where expenses can be minimized without sacrificing quality.
- Product acceptability: By assessing the sensory characteristics that influence consumer acceptability, it assists producers in making sure that their products have been considered acceptable by consumers.

13.3 Training of panelists and sensory attributes

In sensory analysis, panelists are essential. To guarantee the impartiality, accuracy, and repeatability of their assessments, they must be carefully chosen, trained, calibrated, and validated. Accurate and consistent sensory measurements are made possible by training that improves an individual's comprehension of sensory qualities [8].

The sensory evaluation method, which tries to ascertain consumer preferences and product quality, includes evaluating food's flavor, aroma, texture, and appearance. These traits, which are emphasized below, have a significant impact on the overall acceptability and appeal of food products:

a. Taste, which describes flavors detected by the tongue's taste buds. Basic flavors like sweet, sour, salty, bitter, and savory are among them. Assessing taste aids in figuring out the food product's overall flavor profile and palatability [9].
b. Aroma, sometimes referred to as odor, adds a great deal to the entire sensory experience of food and is sensed by the olfactory system. According to Chumngoen and Tan [9], aroma evaluation aids in comprehending the complexity and smell of a food product.
c. Texture describes how food feels in the mouth, including its chewiness, crispiness, hardness, and softness. Assessing the mouthfeel and consistency of food products is crucial for determining their texture, as these factors might affect the acceptability and satisfaction of consumers.
d. A consumer's perception of food is influenced by its appearance, which is the visual component and includes color, shape, size, and overall presentation. It may inform consumers what to expect in terms of flavor and quality.

13.3.1 The role of trained panelists in conducting organoleptic analysis

The use of trained sensory panels is a crucial component of sensory analysis in food science. The primary objective of these panels is to provide an objective evaluation of the many sensory aspects of food products. Health status, culinary preferences, availability, cultural creativity, ability to focus during testing, and previous expertise with sensory evaluation are some of the factors that go into selecting the panelists [6, 10].

It may be important to undergo medical screening for certain products prior to taking part in sensory investigations [11, 12].

In order to increase internal consensus, guarantee repeatability, and boost discriminative power, sensory panels must be trained. Typically, the training procedure entails getting panelists acquainted with the terminology and methodology, enhancing their capacity for discrimination and fostering consensus among panelists. While the length and frequency of training sessions vary, it is advised to commit to a minimum of 10 h, with at least one session designated to the development of sensory terminology [13].

A number of techniques are employed to verify the panelists' performance, and panelist selection is crucial. These techniques include the sequential analysis of discrimination tests, 2-AFC tests, and triangle tests. The panelists are assessed according to their capacity to illustrate distinctions in products, their predictable scoring, their apparent approval, and their description of sensory characteristics and feelings.

The panel's performance is evaluated following testing to confirm its capabilities in terms of outcomes in addition to validation. A panel with eight to twelve people is the ideal size, and information about the panelists' nationality, gender, age, and health is recommended.

13.4 Introduction to electronic nose

The recent global remarkable increase in quantity and quality of food production and its sales have been very significant. As customers prioritize using innovative technology to preserve and improve their competitiveness, the productivity of the agri-food business faces ongoing and increasing problems. An essential component of the automation of the food business is electronics. The functions and sizes of automated food production systems vary greatly primarily based on the type of food and the manufacturer's specifications. An essential component of the automation of the food business is electronics. Depending on the type of food and the manufacturer's requirements, automated food production systems come in a wide range of sizes and functionalities. Electronic sensors are the product of the logical combination of microprocessors, computers, statistics, and advanced chemical and physical sciences. On resistive, optical, electrochemical, or piezoelectric surfaces, these comprise immobilized sensing materials [14].

E-nose is a set of gas sensors that simulates the human nose [15]. By comparing its constituent parts and examining its chemical makeup, it employs a collection of sensors to identify an odor. Table 13.1 displays the types and workings of the most prevalent e-nose sensors. They are significantly more sensitive than human noses because they have more receptor sensors with higher sensitivity. The e-nose sensors are not susceptible to fatigue or the "flu," unlike the human nose, which is sensitive, elegant, and self-healing.

The e-nose can be used to detect harmful and dangerous places that people should avoid. The e-nose tool consists of three main parts: a sample delivery mechanism, a collection of gas or chemical sensors, and a pattern recognition system. This method, which is commonly used to identify simple or complex volatile organic compounds, has emerged as one of the most useful instruments in the food industry.

In order to process complex sensory information (stimuli for the human sensory system), the e-nose, tongue, and eye have been used to characterize elements that contribute to sensory or compositional profiles, from ripening to harvest, from raw material storage to packaging and consumption. This multisensory method more accurately captures the intricacies of how people perceive various stimuli. It can become subjective for a variety of reasons; thus, this technology gives us objective data. Modeling the human olfactory system is the foundation of the e-nose concept, as Figure 13.1 illustrates why humans are able to detect smells, how they work, and their respective limitations [16]

Table 13.1: Types and mechanisms of common electronic-nose gas sensors.

Sensor type	Sensitive material	Detection principle
Acoustic sensors: quartz crystal microbalance (QMB); surface and bulk acoustic wave (SAW, BAW)	Organic or inorganic film layers	Mass change (frequency shift)
Calorimetric; catalytic bead (CB)	Pellistor	Temperature or heat change (from chemical reactions)
Catalytic field-effect sensors (MOSFET)	Catalytic metals	Electric field change
Colorimetric sensors	Organic dyes	Color changes, absorbance
Conducting polymer sensors	Modified conducting polymers	Resistance change
Electrochemical sensors	Solid or liquid electrolytes	Current or voltage change
Fluorescence sensors	Fluorescence-sensitive detector	Fluorescent-light emissions
Infrared sensors	IR-sensitive detector	Infrared-radiation absorption
Metal oxides semiconducting (MOS, Taguchi)	Doped semiconducting metal oxides (SnO2, GaO)	Resistance change
Optical sensors	Photodiode, light-sensitive	Light modulation, optical changes

Source: [14]

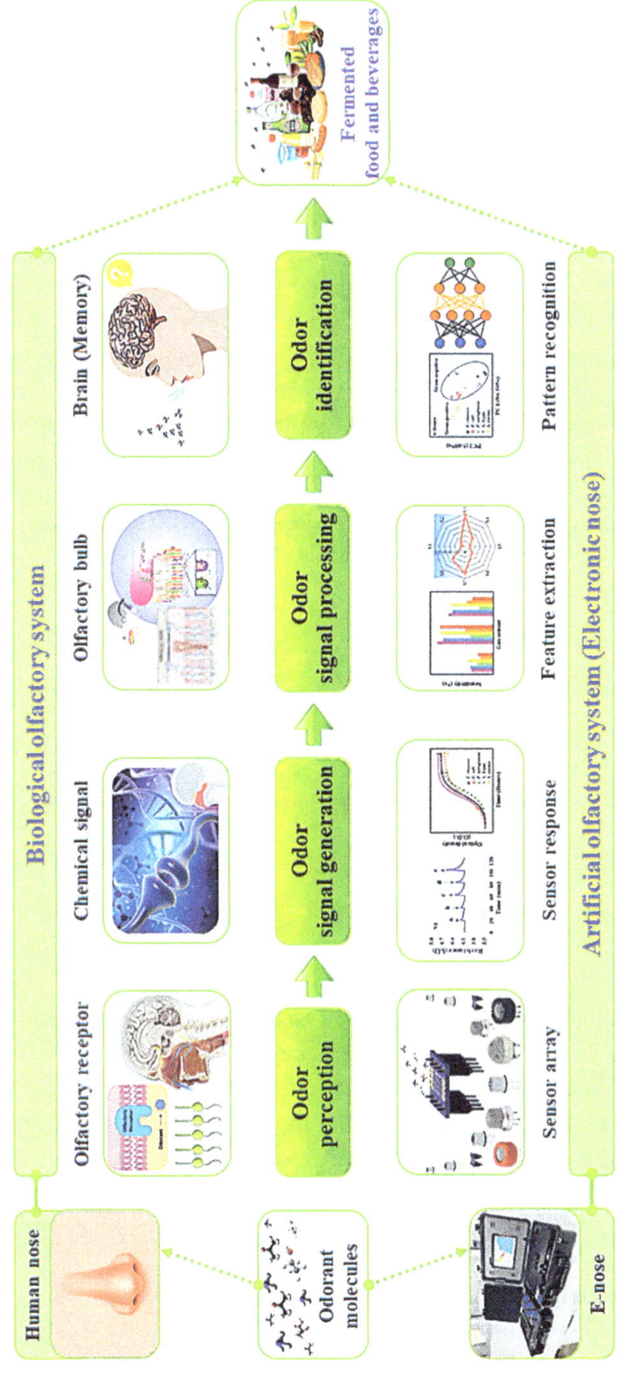

Figure 13.1: Schematic diagram of E-nose device versus biological olfactory system.
Source:[16]

13.4.1 Concept of electronic nose

The e-nose's idea is to simulate the human olfactory system (Figure 13.1). Using a sensor array (an electronic chip) and integrated pattern recognition algorithms, the portable, lightweight gadget exposes a range of plastic composite sensors to the chemical components in a vapor [17]. When the sensors come into touch with the vapor, the polymer expands like a sponge, changing the resistance of the composites.

By determining whether a pretrained chemical is present by detecting the change in resistance, a quick and accurate diagnosis can be established. The patterns are also trained to recognize and distinguish between various scents and to recognize new patterns in response to the needs of the food sector. It is crucial to remember that, in addition to the evident differences with the human olfactory system, there may be certain disadvantages to the sensors and analytical methods employed in an electronic nose. The set of gas or chemical sensors still has drawbacks, including sensitivity, calibration, and toxicity, even if these constraints have lessened in recent decades. The steps of the identification process are performed for the following reasons, much like in human olfaction:

- interaction with the scent;
- collection of samples;
- processing of data;
- matching of patterns; and
- storage and retrieval of data (Figure 13.2).

Figure 13.2: E-nose process/stages similar to human nose.
Source: [19]

Human scent can be highly subjective; e-nose design is based on more accurate scientific study and concepts. The three main parts of the human olfactory system are (I) the part that uses the brain to detect smells, which includes scent delivery systems and olfactory receptor glands; (II) the nervous system that sends signals from the brain to the rest of the body; and (III) a decision-making system that can recognize, detect, and react to smells. The system by which smell is perceived is extremely complex. People are able to discriminate between about 1 trillion olfactory stimuli, according to psychophysical tests.

Meanwhile, olfactory detection and classification are significantly impacted by human emotions and age [18]. Furthermore, two significant barriers to using the human olfactory system to detect scents are the presence of harmful substances in the sample and the duration of the test. As a result, the e-nose has emerged as a potent instrument for evaluating aroma in samples instead of using humans.

An e-nose system and the biological olfactory system are contrasted in Figure 13.3 and Table 13.2 [19]. A channel for introducing scents into the sensor chamber is provided by the e-nose system's odor delivery unit, which includes the pipes, pumps, and valves that make up the olfactory receptor section [16]. This "sensor array" of several gas sensors is the most important and central component of the olfactory receptor. A variety of sensing materials, such as conducting polymers, carbon-based nanomaterials, metal oxides, and nanocomposites, have been used to adsorb odor molecules based on both physisorption and chemisorption.

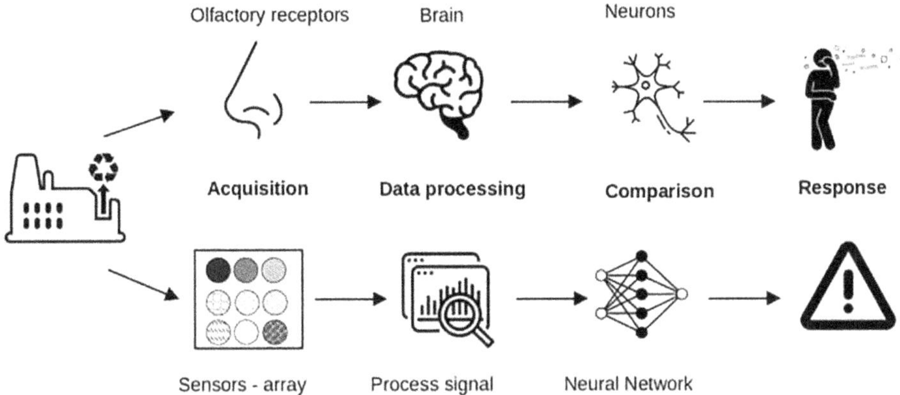

Figure 13.3: Structures of biological olfactory system and electronic nose.
Source: [16]

By triggering charge transfers, volume expansion, ion exchange, or interaction with ion species when they adsorb on the surface of the material, odor molecules can modify the electrical conductivity or resistivity of the sensing materials. Before signal processing techniques are applied, the electrical signals generated by various sensors are converted from analog to digital format using an A/D converter. The data are stored on a local computer or online platform for further study.

Because of the multivariate data from the e-nose system's gas sensor array, data analysis is typically done using supervised and unsupervised machine learning algorithms using statistical techniques like principal component analysis (PCA), hierarchical cluster analysis (CA), analysis of variance (ANOVA), linear discriminant analysis (LDA), partial least squares discriminant analysis (PLS-DA), multivariate data analysis, and artificial neural network signal amplification. To facilitate additional analysis, the data are saved locally or online.

Table 13.2: Comparison between human nose and e-nose.

Human nose	Electronic nose
$>10^8$ receptor cells	5–32 sensors
$>10^3$ types	5–32 types
Responds in a few seconds	Responds in tens of seconds to a few minutes
Sensitivity in ppb/ppt	Sensitivity in ppm/ppb
Massive neural processing in the brain	Pattern recognition, AI, artificial neural nets
Receptors regenerated every few weeks (−30 days)	Sensors replaced on a maintenance schedule (depends on application)

Source: Baum [19].

13.4.2 Types and features of aromas

Animals use their sense of smell to detect aromas, which are simple to complex mixtures of volatile chemicals that are concentrated in the air. Aromas have sometimes been called "smells" or "odors" when describing a particular connotation regarding the degree of pleasantness or unpleasantness of a fragrance. In certain cases, the fragrance may consist of a single chemical compound, while in other cases it may consist of a combination of many compounds, with only one acting as the main ingredient. However, a fragrance derived from organic materials usually contains hundreds of different compounds, each contributing to the unique characteristics and qualities of the distinctive perfume. Although odorless components are invisible to the human nose, trained panel experts can usually detect subtle changes in the relative proportions of chemical species in an aroma combination as a change in odor. The advantage of the e-nose, however, is that it can often detect some odorless compounds that the human nose cannot.

Four qualitative aspects can be used to quantify aromas in general: threshold, intensity, quality, and hedonic rating. The minimum number of aromatic components at which subjects may perceive the presence of an aroma is known as the detection threshold value. The aroma is diluted until half of the test group or human panel is unable to identify the aroma to establish detection threshold. The intensity of an aroma sense is its perceived strength, which rises with focus. The third dimension is quality, which is typically communicated through the use of descriptor types, or everyday words that characterize an aroma by connecting it to the characteristics of known substances; these terms are typically used to describe aromas generated from plants or plant components. Eight aroma groups were presented (Figure 13.4) by McGinley and McGinley, along with examples of descriptor types that were characteristic of each category:

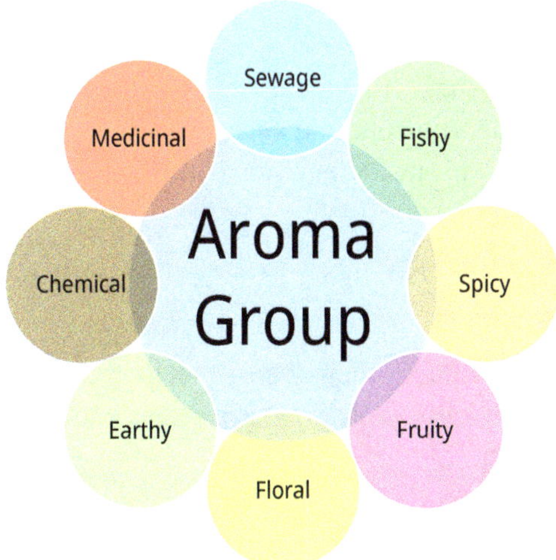

Figure 13.4: The eight food aroma groups.
Source: Adapted from [14].

i. Earthy aromas (musty, moldy, musk, stale, grassy, herbal, and woody)
ii. Floral aromas (fragrant, flowery, perfume, eucalyptus, and lavender)
iii. Fruity aromas (citrus, orange, lemon, apple, pear, pineapple, and strawberry)
iv. Spicy aromas (cinnamon, mint, peppermint, onion, dill, garlic, pepper, cloves, vanilla, almond, and pine)
v. Fishy aromas (fishy, prawns, and amine)
vi. Sewage aromas (septic, putrid, rancid, sulfurous, rotten, decayed, cadaverous, foul, sour, pungent, burnt, and swampy)
vii. Medicinal aromas (disinfectant, phenol, camphor, soapy, ammonia, alcohol, ether, anesthetic, and menthol)
viii. Chemical aromas (solvent, aromatic, varnish, turpentine, petroleum, creosote, tar, oily, and plastic) [14]

13.4.3 Application of electronic nose

E-nose systems based on a variety of gas sensor arrays are used in all major sectors:

13.4.3.1 Food industry

E-noses are indispensable instruments for quality control in the food business. They are adept at spotting deterioration, keeping an eye on freshness, and confirming the legitimacy of ingredients, all of which help to preserve the integrity of food products.

13.4.3.2 Environmental monitoring

Environmental preservation activities greatly benefit from the use of electronic noses. They are used to monitor air quality and identify contaminants in industrial and urban environments, helping to create a safer and cleaner environment.

13.4.3.3 Medical diagnosis

The use of e-noses in medical diagnosis has enormous promise. Based on their distinct features, they have the potential to distinguish between particular diseases and offer opportunities for early detection and treatment.

13.4.3.4 Pharmaceuticals

E-noses are essential for quality control in the pharmaceutical business during the manufacture of drugs. They are skilled in spotting contaminants or any changes from prescribed formulas, guaranteeing the security and effectiveness of medicinal products.

13.4.3.5 Safety and security

E-noses are essential for spotting explosive or hazardous substances in security settings. They are very useful in protecting vital infrastructure and public areas due to their great sensitivity and quick reaction times.

13.4.3.6 Wine and beverage industry

The assessment of wines and other beverages is a specialized market for the e-nose. It performs exceptionally well in evaluating their attributes and quality, which helps improve production procedures and guarantee superior products.

Others include forestry and agriculture, industrial processes, environmental toxin/pollutant analysis, space stations, medical/healthcare, identity verification pharmaceuticals and medicine, forensic science, military, toxicology/security, and food and beverage [20]. Developing mobile, reasonably priced devices for those with anosmia (loss of smell) is another growing direction in the research. E-nose systems, with an emphasis on applications in the food and beverage industry, have been used for both direct and indirect identification via odor analysis for a variety of purposes, including the inspection of product quality, studies on batch-to-batch uniformity, contamination detection, spoilage detection, adulteration detection, pathogen detection, analysis of storage conditions/shelf life, and the development of particular sensory profiles. They have been used to assess the effects of modifications to the production process and components that affect organoleptic characteristics, compare various food formulations, and analyze aromas and compare them with competitor products. Additionally, e-nose systems have demonstrated excellent performance in determining the quality of a wide range of products, such as dairy products, wine, beer, coffee, carbonated drinks, pork, beef, chicken, fish, and shrimp. E-nose systems' sensors could, nevertheless, experience a drift effect. A little bias exists in measurements taken at different time intervals because of the aging of the sensors.

13.5 The electronic nose: a revolutionary tool for food safety and quality analysis

Biochemical testing using e-noses involves the analysis of volatile organic compounds (VOCs) in biological samples, such as breath, urine, or blood. This provides valuable information about the metabolic status and biochemical processes within the body.

13.5.1 Advantages of using an electronic nose

13.5.1.1 High sensitivity

The e-nose's remarkable sensitivity is among its most noticeable features. It is more sensitive to smells than many conventional analytical techniques, even at very low concentrations.

13.5.1.2 Rapid analysis

Industries are able to make prompt judgments because to the e-nose's quick and real-time results. This agility is especially important in fields where responding quickly is essential.

13.5.1.3 Time savings and cost-effectiveness

The e-nose frequently turns out to be a more affordable option than conventional analytical techniques. This is especially noticeable when it comes to labor and time savings, which improve operational efficiency.

13.5.1.4 Nondestructive

The e-nose functions in a nondestructive manner, in contrast to several chemical analysis techniques that could change or harm samples. This keeps the samples being studied intact, which is important for many applications [19].

13.6 Data integration and flavor-profiling

A food product's quality is influenced by a variety of sensory attributes, including taste, odor, color, and texture. The goal of food innovation is to produce new experiences or improve particular sensory aspects that consumers find appealing. The conventional method of evaluating flavor perception involves using only trained sensory panels to rate sensory characteristics in the following categories: flavor, mouthfeel, aftertaste, aroma, and after-feel. Taste and smell receptors in humans combine intricately with food ingredients to produce a complex flavor that each consumer experiences differently [21]. Flavor profiling (FP) method is one of the descriptive sensory analysis methods alongside texture profile, which makes use of trained judges. FP breaks down flavor into five main components: aftertaste, character notes or attributes, intensities of those attributes, the order in which the attributes appear, amplitude, and complex phenomenon that is the overall perception of how well the analyzable and nonanalyzable flavor components are blended. The taste profile is initially graded on a five-point scale: not present, threshold, slight, moderate, and strong. However, in actuality, the traits that are absent are not denoted by zeros in the profile; rather, they are simply absent [15]. FP involves combination of sensory data with other analytical techniques to create a comprehensive FP of a food product. Data analysis proceeds with an automated learning process after preprocessing and dimensionality reduction. Three primary categories of learning strategies can be used to these processes: supervised, semi-supervised, and unsupervised learning. Numerous multivariate strategies have been developed to accomplish the data analysis, and they can be broadly divided into two basic categories [22]. Chemometric statistical approaches fall into the first group, and different types of artificial neural networks (ANN) fall into the second. Decision trees, neural networks, naïve Bayes, PLS [23], KNN, SVM, SFA, PLS-DA, DFA, SLDA, and rule learners [24] are among the frequently employed learning algorithms. Table 13.3 provides an illustration of the wide range of methodologies and applications in electronic tongues and noses.

Data interpretation can be described as understanding, organizing, and interpreting the provided data in order to draw a meaningful conclusion. Types of data interpretation include bar chart, pie chart, tables, and line graph. The step-by-step process for interpreting data includes: collect desired data, arrange the data into easy-to-read format (tables, graphs, charts, etc.), develop findings and observations, and summarize to conclude. The conclusion should be related to the data generated. Recommendations should be generated from the findings. Importance of data analysis and interpretation include the following: The management can review the data before acting to put new ideas into practice as a result of the well-structured and well-analyzed data. Predicting future trends and competitiveness is aided by it. It offers the company a number of financial advantages. Mostly, it facilitates decision-making. Interpreting data facilitates knowledge acquisition for developing a competitive strategy.

Table 13.3: Data methodologies for electronic noses for some food materials.

Application	Methodology
Determination of quality of tea from different picking periods	Adaptive pooling attention mechanism (APAM) (adaptive multiscale pooling structure + concatenation + convolutional neural network)
Prediction of tomato plants infected by fungal pathogens	PCA + discriminant functions analysis (DFA) + backpropagation neural network (BPNN) (Sun and Zheng, 2023)
Determination of Pitaya quality	Autoscaling + PCA + linear discriminant analysis
Detection and identification of subterranean termites	PCA + quality factor analysis (QFA) `
Flavor perception	This work proposes an olfactory–taste synesthesia model (OTSM) and compares the results with PCA-ELM, WF-LPP, VIP-ELM, VIP-RF, VIP-GA-SVM, BPNN, ELM, GS-SVM, and RBFNN

Source: Tibaduiza et al. [22].

13.7 Consumer preference and future implications

13.7.1 The significance of understanding consumer preferences in the food industry

The subjective preferences, which people or groups have for particular products and services are referred to as consumer preferences. A number of variables, including cultural background, socioeconomic status, individual experiences, and psychological characteristics, have an impact on these preferences. One of the key factors influencing food choice is whether or not consumers embrace and appreciate the sensory qualities of food. Consumer preferences are a major factor in how the food industry is developed. This significance can be attributed to several key factors.

13.7.1.1 Innovation and competitive advantage

Consumer preferences drive innovation as companies strive to differentiate themselves from competitors. For instance, the rise in demand for convenient, ready-to-eat meals has spurred innovations in packaging and food preservation technologies [25].

13.7.1.2 Marketing strategies

Effective marketing strategies are often built around consumer preferences. By aligning marketing messages with what consumers value – such as sustainability, health benefits, or local sourcing – companies can enhance their brand appeal and customer loyalty [26].

13.7.1.3 Supply chain and sourcing

Consumer demand influences the entire supply chain, from sourcing ingredients to production methods. For example, the increasing preference for ethically sourced and fair-trade products has encouraged companies to adopt more transparent and sustainable supply chain practices.

13.7.1.4 Regulatory and compliance trends

Consumer preferences can also impact regulatory frameworks. As consumers become more health-conscious, there is greater pressure on regulatory bodies to enforce stricter food safety standards and labeling requirements. This, in turn, affects how companies operate and comply with these regulations [26].

Consumer preferences directly affect the economic performance of food companies. Products that align well with consumer trends tend to perform better in the market, leading to higher sales and profitability. Conversely, ignoring consumer preferences can result in decreased market share and revenue [27].

13.7.1.5 Globalization and cultural influence

In a globalized market, consumer preferences vary significantly across different regions and cultures. Understanding these preferences is crucial for companies looking to expand internationally. For instance, fast food chains often tailor their menus to accommodate local tastes and dietary restrictions.

13.7.1.6 Market demand and product development

Understanding consumer preferences helps companies develop products that meet the tastes and expectations of their target market. For example, the growing preference for healthy and organic food options has led to a significant increase in the production and availability of organic produce, gluten-free products, and plant-based alternatives [28].

13.8 Sensory analysis

Over time, the function of sensory evaluation has undergone significant alteration. It aids in the creation of a successful strategy when combined with the marketing and R&D departments. Sensory testing can assist in identifying the essential sensory features that influence acceptability at an early stage of product development. Determining target customers, product rivals, and evaluating novel concepts can all be helpful. These days, information gathered from both sensory and instrumental testing are combined to determine the chemical and physical characteristics of the product that influence sensory features. Sensory analysis can determine the impact of scaling up pilot samples to large-scale manufacture. Sensory evaluation gives assurances that inferior products are not released in the market.

Since the product's sensory qualities deteriorate before its microbiological quality, sensory evaluation is typically employed to determine the shelf life of food goods. In the field of investigation, customer evaluation is often used. In order to produce new products and comprehend consumer behavior, it investigates new technologies [29].

13.8.1 Product development in the food sector

New goods have been developed as a result of increased rivalry in the food sector. Additionally, current items are continuously reevaluated, which results in enhancements like flavor or packaging. Product development can include creating an entirely new food product, coming up with concepts for a new product by creating a product profile that includes details like size and shape, or altering an existing food product by making adjustments to the original recipe.

For instance, altering a product's size or form to resemble an existing food product or adding or removing an ingredient to enhance flavor – all are examples of mimicking other well-known, branded products of like kinds. A number of intricate steps are involved in the product development process, and its success depends on the combined skills of numerous experts. Conceptualization, small-scale testing, product modification, consumer acceptability testing, final specification fixing, large-scale production, and launch are the key phases of product development. Initially, a specification is created and concepts for the new product are developed [30].

Next, a small-scale test of this concept is conducted. Numerous recipes are developed and the ingredients to be used are specified through research. Different variants are created, with different ingredients or methods. To put it another way, a professional chef or food consultant often prototypes the products. The qualified panelists assess these developed items to make sure they have the needed qualities. It might be necessary to make changes to the recipe and conduct additional testing. The product's acceptance among consumers is then assessed through extensive testing. The final product specification, which includes the precise ingredients and production techniques, is then agreed upon.

In a pilot plant, food experts collaborate to find the most efficient way to produce the product in large amounts [30]. After that, the product is mass-produced. To ensure constant product quality, this is carried out under carefully monitored circumstances. When choosing packaging, shelf-life issues are taken into account. The purpose of labeling is to comply with legal requirements. The product is finally promoted before being released. As the product is being created, sensory analysis testing is done at various points.

13.9 Food informatics

Informatics is an important field of applied science, which deals with the collection, selection, organization, processing, management as well as dissemination of information related to any system and for that, it takes the help of several techniques and technologies (Fig 13.5). It is a kind of mechanism required for the information processing and management. Based on the knowledge cluster, informatics may create a new domain; for example, the combination of informatics and medical science results in medical informatics. Similarly, integration as well as affiliation of informatics with food sciences has the potentialities of a new discipline called food informatics.

A type of system and mechanism called food informatics is in charge of creating and developing an information repository. A number of things have been made feasible by food informatics, including the regional potential food processing prospects, marketing strategy, etc. Numerous technologies, including database technology, communication and networking technology, web technology, and usability engineering, make it feasible to design and create information systems, or more simply, informatics [31].

The collection, selection, organization, processing, administration, and final distribution of information are all included in the interdisciplinary field of food informatics. The primary focus of food informatics is the processing and administration of

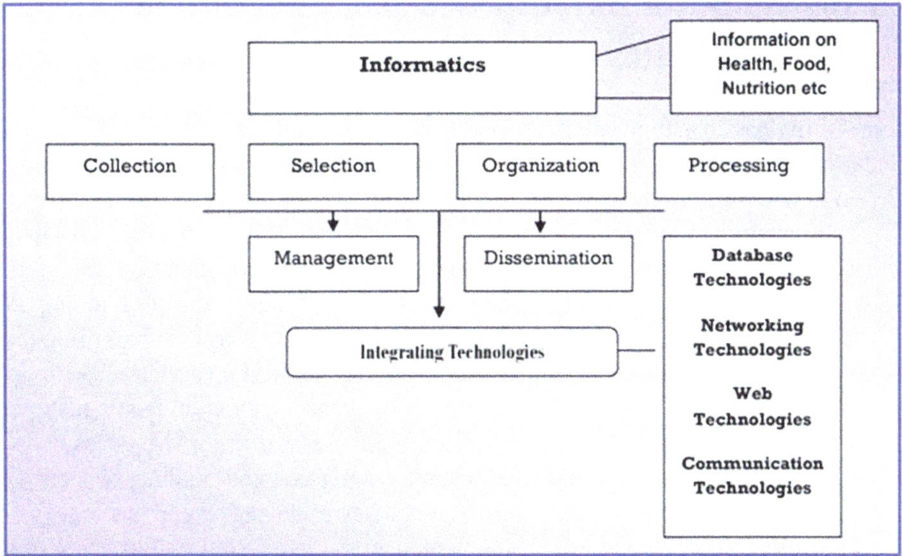

Figure 13.5: Basic dealing and dependencies of Informatics.
Source: [35].

data pertaining to food and nutrition. With the commitment to developing both systems and information infrastructure, food information systems can be viewed as a subfield of food informatics. The following are the primary tasks that the food information systems may perform:

- Gathering and choosing information on food, nutrition science, or just nutrition in general. Both the availability of food processing technologies and the handling and delivery of nutritional benefits depend on it.
- Another crucial role of a food information system is to generate and create a library on health and nutrition according to user needs [32].
- Developing food information systems using food informatics tools is another crucial issue that also relies on user expectations and demand.

Along with similar foundations developed, food informatics also works with unified information centers. A complex food information system must have users, or regular people, as well as tools, technology, and information pertaining to food. Another crucial requirement for the development of food informatics units and food information systems is food marketing, which also serves as a platform for marketing, distribution, and storage.

13.10 The potential future implication of food informatics

Some of the possible applications of food and nutrition information systems include (but are not limited to):

- Collection and selection of content and information on food and nutrition. Thus, government may get actual information and data on food. Moreover, food safety is an important matter for which food informatics may be applied [33].
- Delivery of food and nutritional information among common masses is an important example of food informatics uses, hence creating awareness regarding the foods and fruits (even information on several vegetables, nonvegetarian, etc.).
- With the applications of food informatics in general, common people may get health and medicinal benefits, including from food and fruits or vegetables.
- Food informatics is a cooperative means for designing and building of sophisticated food as well as medical informatics of a particular region and territory.
- Food informatics is also responsible for the creation of mission fulfillment of a territory by the use of healthy food information systems [34].
- Food informatics is also helpful for young academicians (including teachers, students, and researchers) in the field of food science, nutrition science, agriculture, horticulture, as well as other allied sciences.
- In clinical and medical segments among the staff members physician and allied health professionals may utilize benefits of food information systems. Here they may get nascent information on food and nutrition.
- Improving and developing public health is the core agenda of any nation and with the integration of real food informatics in community development projects it is very much easier to achieve such objectives [35].
- Collecting information on nutritional benefits of traditional foods and vegetables, etc. is helpful for nutritional product manufacturing companies [28].

13.11 Conclusion

Consumer preferences are a critical factor in the food industry, influencing everything from product development and marketing to supply chain management and regulatory compliance. By staying attuned to these preferences, food companies can not only meet the demands of their customers but also gain a competitive edge in a dynamic and ever-evolving market. Food Informatics practises nutritional information systems, which can be treated as an important tool that helps with solving adequate, optimum and proper nutrition; nutrition stages; diet control; nutritional care; malnutrition; preparation for sophisticated nutrition through adequate information delivery.

References

[1] Ruiz-Capillas, C., & Herrero, A. M. sensory analysis and consumer research in new product development. Foods 2021, Mar 10, 10(3), 582. PMID: 33802030; PMCID: PMC8001375. doi: 10.3390/foods10030582

[2] Krupitzer, C., & Stein, A. Food informatics – review of the current state-of-the-art, revised definition, and classification into the research landscape. Foods 2021, 1, 0. https://doi.org/

[3] Koenderink, J. J. P., Hulzebos, J. L., Rijgersberg, H., & Top, J. L. Food informatics: Sharing food knowledge for research and development. 2005. Retrieved from https://www.researchgate.net/publication/40117692_Food_informatics_sharing_food_knowledge_for_research_and_development?enrichId=rgreq-4843e2ed91b578163ba7540fb403490f-XXX&enrichSource=Y292ZXJQYWdlOzQwMTE3NjkyO0FTOjEwMTU5Nzc2MTc2OTQ3N0AxNDAxMjM0MjAyMzM0&el=1_x_2&_esc=publicationCoverPdf.

[4] Paul, P., Aithal, P. S., & Bhuimali, A. Food informatics and its challenges and opportunities- A review. In: Zenodo, CERN European Organization for Nuclear Research, 2017. http://doi.org/10.5281/zenodo.996933

[5] Godoi, F. C., Prakash, S., & Bhandari, B. R. 3D printing technologies applied for food design: Status and prospects. Journal of Food Engineering 2016, 179, 44–54.

[6] Lawless, H. T., & Heymann, H. Sensory Evaluation of Food – Principles and Practices, New York, NY: Springer Science+Business Media, LLC, 2010.

[7] Meilgaard, M. C., Carr, B. T., & Carr, B. T. Sensory Evaluation Techniques, 4th ed., CRC Press, 2007. https://doi.org/10.1201/b16452

[8] Mihafu, F. D., Issa, J. Y., & Kamiyango, M. W. Implication of sensory evaluation and quality assessment in food product development: A review. Current Research in Nutrition and Food Science 2020, 8(3). http://dx.doi.org/10.12944/CRNFSJ.8.3.03

[9] Chumngoen, W., & Tan, F. J. Relationships between descriptive sensory attributes and physicochemical analysis of broiler and Taiwan native chicken breast meat. Asian-Australasian Journal of Animal Sciences 2015, Jul, 28(7), 1028–1037. PMID: 26104409; PMCID: PMC4478495. doi: 10.5713/ajas.14.0275

[10] Lestringant, P., Delarue, J., & Heymann, H. 2010–2015: How have conventional descriptive analysis methods really been used? A systematic review of publications. Food Quality and Preference 2019, 71, 1–7. https://doi.org/10.1016/j.foodqual.2018.05.011

[11] Talsma, P. How much sensory panel data do we need?. Food Quality and Preference 2018, 67, 3–9. 12.005. https://doi.org/10.1016/j.foodqual.2016

[12] Djekic, I., Lorenzo, J. M., Munekata, P. E. S., Gagaoua, M., & Tomasevic, I. Review on characteristics of trained sensory panels in food science. Journal of Texture Studies 2021, Aug, 52(4), 501–509. Epub 2021 Jun 14. PMID: 34085719. doi: 10.1111/jtxs.12616

[13] Gomez-Corona, C., Pohlenz, A., Cayeux, I., & Valentin, D. Panel performance and memory in visually impaired versus sighted panels. Food Quality and Preference 2020, 80, 103807. https://doi.org/10.1016/j.foodqual.2019.103807

[14] Wilson, A. D., & Baietto, M. Applications and advances in electronic-nose technologies. Sensors 2009, 9, 5099–5148.

[15] Lee, Y.-M., Chung, S.-J., Prescott, J., & Kim, K.-O. Flavor profiling by consumers segmented according to product involvement and food neophobia. Foods 2021, 10, 598.

[16] Seesaard, T., & Wongchoosuk, C. Recent progress in electronic noses for fermented foods and beverages applications. Fermentation 2022, 8(302), 1–24.

[17] Dhar, P., Kashyap, P., Jindal, N., & Rani, R. Role of electronic nose technology in food industry. In: Conference: Emerging Sustainable Technologies in Food Processing, ESTFP-2018, 2018, 1–2.

[18] Calvi, E., Quassolo, U., Massaia, M., Scandurra, A., D'Aniello, B., & D'Amelio, P. The scent of emotions: A systematic review of human intra- and interspecific chemical communication of emotions. Brain and Behavior 2020, 10(5).

[19] Baum, H. (2021). Eletronic nose and tongue PPT. https://www.uc.edu/content/dam/refresh/cont-ed-62/olli/new-tech-sept22.pdf. Accessed June, 2024

[20] Rabehi, A., Helal, H., Zappa, D., & Comini, E. Advancements and prospects of electronic nose in various applications: A comprehensive review. Applied Sciences 2024, 14(4506), 1–46. 2024.

[21] Leygeber, S., Grossmann, J. L., Diez-Simon, C., Karu, N., Dubbelman, A.-C., Harms, A.-C., C., A., Westerhuis, J. A., Jacobs, D. M., Lindenburg, P. W., W., P., Hendriks, M. M. W. B., Ammerlaan, B. C. H., Berg, M. A. V., Rudi van Doorn, R. V., Mumm, R., Hall, R. D., D., R., Smilde, A. K., Hankemeier, T., et al. Flavor profiling using comprehensive mass spectrometry analysis of metabolites in tomato soups. Metabolites 2022, 12, 1194. https://doi.org/10.3390/me

[22] Tibaduiza, D., Anaya, M., Gómez, J., Sarmiento, J., Perez, M., Lara, C., Ruiz, J., Osorio, N., Rodriguez, K., Hernandez, I., & Sanchez, C. Electronic tongues and noses: A general overview. Biosensors 2024, 14(4), 190.

[23] Wold, S., Sjöström, M., & Eriksson, L. PLS-regression: A basic tool of chemometrics. Chemometrics and Intelligent Laboratory Systems 2001, 58, 109–130. 2001.

[24] Tibaduiza, D., Anaya, M., & Pozo, F. Pattern Recognition Applications in Engineering, Hershey, PA, USA: IGI Global, 2020, 84–102.

[25] Rogers, P. J. Food choice, mood and mental performance: Some examples and some mechanisms. In: Meiselman, H. L., & MacFie, H. J. H., editors. Food Choice, Acceptance and Consumption, 1st ed., Boston, MA, USA: Springer, 2016, 319–345. ISBN 978-1-4612-8518-2

[26] Melovic, B., Cirovic, D., Dudic, B., Vulic, T. B., & Gregus, M. The analysis of marketing factors influencing consumers' preferences and acceptance of organic food products – Recommendations for the optimization of the offer in a developing market. Foods 2020, 9, 259.

[27] Ohlhausen, P., & Langen, N. When a combination of nudges decreases sustainable food choices out-of-home – The example of food decoys and descriptive name labels. Foods 2020, 9, 557.

[28] Lawless, H. T. Laboratory Exercises for Sensory Evaluation, New York: Springer, 2013.

[29] Fuller, G. W. What Is New Food Product Development? In New Food Product Development, Boca Raton, Florida: CRC Press Taylor and Francis Group, LLC, 2011, 2011.

[30] Friede, A., Blum, H. L., & McDonald, M. Public health informatics: How information-age technology can strengthen public health. Annual Review of Public Health 2015, 16(1), 239–252.

[31] Kun, L. G. Telehealth and the global health network in the 21st century. From homecare to public health informatics. Computer Methods and Programs in Biomedicine 2011, 64(3), 155–167.

[32] Martin, S. B. Information technology, employment, and the information sector: Trends in information employment 1970–1995. Journal of the American Society for Information Science 2018, 49(12), 1053–1069.

[33] Paul, P. K., Ganguly, J., & Ghosh, M. Medical information science: Overview and a model curriculum of MSc-Information Science [Medical Information Science]. Current Trends in Biotechnology and Chemical Research 2013, 3(1), 50–54.

[34] Aarts, J., Peel, V., & Wright, G. Organizational issues in health informatics: A model approach. International Journal of Medical Informatics 2018, 52(1), 235–242.

[35] Paul, P., Aithal, P. S., & Bhimali, A. Food informatics and its challenges and opportunities – A review. International Journal on Recent Researches in Science, Engineering & Technology 2017, 5(9), 46–53. ISSN: 2347-6729., Accessed in June 2024.

Victor Ntuli*, James A. Elegbeleye, Bono Nethathe, Josphat N. Gichure, and Elna M. Buys

Chapter 14
Predictive modeling in food safety enhancement

Abstract: Predictive modeling has evolved as a necessary tool in mitigating the challenges associated with food safety. Its application is in anticipating and managing potential risks, including biological and nonbiological hazards frequently encountered throughout the food supply chain. Predictive modeling integrates engineering, microbiology, chemistry, and mathematical and computational approaches to address food safety challenges. This multidisciplinary approach is used to predict microbial responses, assess risks, and enhance the safety of food products. Recently, predictive modeling has leveraged several innovations, such as using machine learning techniques to analyze data in real-time, identify contaminated products, and forecast food recalls. This chapter highlights the various aspects of predictive modeling, its key components, types of models, practical applications in the food industry, the concept, current applications, strategies, and challenges in its implementation. The chapter also explores the latest advancements and future directions in this dynamic field, providing a comprehensive overview of how predictive modeling is transforming and enhancing food safety.

Keywords: Predictive modeling, food safety, risk assessment, machine learning, microbial growth

14.1 Introduction

With their favorable action, microorganisms have played instrumental roles in food and bioprocessing industries. Nevertheless, certain species and strains have presented complex problems for consumers and the food sector. Food safety and quality are cru-

***Corresponding author: Victor Ntuli**, Department of Food Science and Technology, Faculty of Science, Engineering and Agriculture, University of Venda, Private Bag X5050, Thohoyandou 0950, Limpopo, South Africa, e-mail: victor.ntuli@univen.ac.za
James A. Elegbeleye, Department of Consumer and Food Sciences, Faculty of Natural and Agricultural Sciences, University of Pretoria, Private Bag X20, Hatfield 0028, Pretoria, South Africa
Bono Nethathe, Department of Food Science and Technology, Faculty of Science, Engineering and Agriculture, University of Venda, Private Bag X5050, Thohoyandou 0950, Limpopo, South Africa
Josphat N. Gichure, Elna M. Buys, Department of Consumer and Food Sciences, Faculty of Natural and Agricultural Sciences, University of Pretoria, Private Bag X20, Hatfield 0028, Pretoria, South Africa

https://doi.org/10.1515/9783111441238-015

cial concerns for public health, economic stability, and food security [1, 2]. Microbial contamination, food-borne disease outbreaks, and significant economic losses due to food waste are not just existential threats facing humanity but urgent and severe challenges that must be addressed [2]. Harmful microorganisms such as *Salmonella* spp., *Staphylococcus aureus, Escherichia coli* O157, *Listeria monocytogenes, Campylobacter* spp., *Clostridium botulinum,* and viruses are frequent culprits in food contamination, leading to severe health impacts and outbreaks [3]. The economic toll of foodborne illnesses, recalls, and waste is staggering, with global food waste alone costing several billions and foodborne illnesses estimated to be US$17.6 billion annually in the United States, exacerbating food insecurity and environmental harm [4]. The gravity of these challenges necessitates a multifaceted approach, including stringent safety regulations, improved detection and control of pathogens, and innovative solutions to reduce food waste and improve supply chain efficiencies. To achieve this, understanding and anticipating the behavior of microorganisms in food ecosystems has assisted in controlling the microbial life cycle by favoring the growth of desirable microorganisms and inhibiting spoilage and pathogens [5]. One approach that has been developed to forecast how microbes will behave in food ecosystems is predictive modeling.

Predictive modeling is the use of mathematical and statistical models to anticipate the responses of pathogenic and spoilage microorganisms in food under various environmental conditions. The technique has become essential for guaranteeing food safety by utilizing the latest developments in computer science, microbiology, and data science [6]. Predictive modeling leverages past and present data to forecast potential food safety issues and enables proactive risk mitigation actions [7]. Integrating predictive models in food safety brings tangible benefits such as supply chain optimization, risk assessment, microbiological growth prediction, and shelf-life estimation [8]. Predictive modeling tools use models developed and validated using large datasets of historical food safety incidents, laboratory experiments, and real-world data from food production and distribution. Rigorous testing and validation processes ensure that the models are reliable and accurate, making them trustworthy tools used by food safety experts, regulators, and the food industry [7].

Predicting pathogen growth is one of the primary applications of predictive modeling in food safety. This is achieved through a specific type of laboratory study called a microbiological challenge test. The study involves the intentional inoculation of a food product with appropriate spoilage or pathogenic microbes to observe their activity within the product and predict its quality, including the potential for the pathogen to harm the customer. Specialized software such as the Pathogen Modelling Program (PMP) and ComBase Predictor [9] are also used to predict the behavior of microorganisms in the food using data from different challenge tests obtained from surrogates or same microbes and food matrices. These software are embedded with models that are designed to estimate the growth, survival, and inactivation of pathogens under various extrinsic and intrinsic conditions. The science of predictive modeling is invaluable in preventing foodborne disease outbreaks and also assists food manufacturers in determining the appropriate production, processing, handling, and storage of food [9].

Predicting the shelf life of food products is another crucial application of the predictive modeling technique. Here, models are used to analyze variables like temperature, humidity, and packing state to determine how long perishable goods will last on the shelf [6]. Providing more accurate expiration dates that are predicted from the models helps preserve product quality and reduce food waste. Unlike the static method of shelf-life determination, the dynamic shelf-life models utilize shelf-life predictions to provide more accurate expiry dates and versatile solutions, which are established by modifying expiration dates in response to real-time environmental data [10].

In the food processing industry, assessing the risk of microbial contamination is greatly improved through predictive modeling. Advanced models combine different sources of data, such as past contamination incidents, environmental monitoring, and production practices, to forecast the probability of contamination events [9]. These models can pinpoint potential sources of contamination and areas with high risk, allowing for targeted interventions and better allocation of resources. For instance, machine learning (ML) algorithms can analyze patterns in environmental sampling data to predict contamination risks in processing facilities [9].

In the field of supply chain optimization, predictive modeling helps tackle the challenges of food distribution networks. These models can predict potential disruptions, such as delays or spoilage, by analyzing factors like weather patterns, transportation conditions, and storage environments [10]. This predictive capability allows supply chain operations to be adjusted to ensure timely delivery and maintain food safety standards. Furthermore, combining blockchain technology with predictive models improves traceability, offering a transparent and reliable way to track food products from the farm to the consumer's plate [10].

A significant additional application of predictive modeling is in the process of conducting quantitative microbial risk assessment (QMRA). Microbial risk assessment takes into account the dynamics of pathogen proliferation, survival, and inactivation. Predictive models, which are increasingly recognized as essential steps in the risk assessment process, are the most excellent tools for evaluating the changes in food microbiological levels as a product moves through the farm-to-fork chain [11].

This chapter delves into the multifaceted realm of predictive modeling, outlining its core elements, diverse model types, and practical uses within the food industry. It examines the concept and current applications of predictive modeling, along with strategies and challenges involved in its implementation. Additionally, the chapter discusses the latest advancements and prospects in this rapidly evolving field, offering a thorough overview of how predictive modeling is revolutionizing food safety.

14.2 Current applications of predictive modeling in food safety

By facilitating proactive risk management and quality control throughout the food supply chain, predictive modeling is revolutionizing food safety [6, 12]. These advanced models predict microbial behavior and contamination risks by analyzing environmental factors, helping to handle safely and store foods such as ready-to-eat (RTE) products and dairy items [6]. Predictive models are crucial in estimating shelf life and detecting spoilage in perishable goods, optimizing inventory management, and reducing food wastes. In supply chain management, predictive models combined with blockchain technology enhance traceability and predict contamination risks, as seen in fish and other seafood supply chains [13]. Predictive models can also optimize food processing parameters, ensuring safety and quality, like adjusting meat cooking times [14]. Additionally, the concept of applying predictive modeling in the food industries aids compliance with food safety regulations and enables real-time monitoring systems for early threat detection, significantly improving response times during safety incidents [6, 12, 13]. As technology develops, predictive modeling will be increasingly important in guaranteeing safer and more effective food systems.

14.2.1 Case studies highlighting successful implementation

Predictive modeling has been successfully implemented to assess microbial risks, optimize processing parameters, and ensure food safety. The models provide valuable data for risk assessment and management to prevent foodborne illnesses. Predictive modeling has been successfully applied in a number of studies to try to offer data that are crucial for regulating food safety.

14.2.1.1 Predicting the growth *of L. monocytogenes* in fishery products

Bolívar, Costa and coworkers [15] evaluated the growth of *L. monocytogenes* in fishery products using optical density (OD) measurements in temperatures ranging from 2–20 °C under different atmosphere conditions, that is, aerobic and reduced oxygen. The Baranyi and Roberts model was used to evaluate the maximum growth rate (μ_{max}) from the obtained growth curves. Moreover, the effect of storage temperature on μ_{max} was predicted by using the Ratkowsky square root model. The study provides validated predictive models for *L. monocytogenes* growth in fishery products that can be used in microbial risk assessment and shelf-life studies.

14.2.1.2 Predictive model for *S. aureus* growth on egg products

Choi, Son [16] developed a prediction model for the growth of *S. aureus* on RTE peeled quail eggs, grilled eggs, and whole egg liquid products to improve the production of domestic food items. The growth kinetics of *S. aureus* were evaluated using the Baranyi model and the Gompertz model (primary models), while the secondary model was the square root model [16]. Moreover, root mean square errors (RMSE) were analyzed to predict the above model's sustainability. The results suggested that the models represented the actual growth of *S. aureus* in egg products. The models offer information for microbial risk assessment, which can be applied to food safety management to prevent *S. aureus*-related food poisoning in egg products [16].

14.2.1.3 Growth of the yeast *Pichia anomala* in olive fermentation

Arroyo, Quintana [17] used the modified logistic, modified Gompertz, modified Richards-Stannard, and Baranyi Roberts's models to evaluate the growth of the yeast *P. anomala* in olive fermentation. The study determined the maximum specific growth rate (μ_{max}) and lag phase period from the growth curves. Despite the good fit of all models, the most suitable models were revealed to be the modified Gompertz and Richards-Stannard models.

14.2.1.4 Microbial interaction and death of *E. coli* O157:H7 during the fermentation of green table olives

Skandamis and Nychas [18] provided a model for the microbial interactions and *E. coli* O157:H7 death during the fermentation of green table olives. The study used two distinct starter cultures, as well as varied concentrations of glucose and sucrose, to predict the survival or death of *E. coli* O157:H7 during fermentation. The investigators noted that lactic acid was significantly produced during the fermentation process. By using differential models and taking into account several factors, such as pH, the protective impact of the substrate, and protonated lactic acid, it was possible to determine if *E. coli* O157:H7 survives or dies off during the fermentation of green table olives [18].

14.2.1.5 Modeling the growth of pathogenic *E. coli* on fresh produce

Fresh produce is consumed without or with minimal processing. This implies that customers are at risk if the product is contaminated with pathogenic microorganisms. Kim, Park [19] developed a study to validate a predictive growth model of pathogenic

E. coli to ensure the safety of fresh-cut produce. The models incorporated weather conditions, irrigation methods, and post-harvest handling data. The established growth model can yield valuable information for evaluating the quantitative microbiological risk of pathogenic *E. coli* in fresh produce. Furthermore, the models can also be utilized by food businesses involved in producing, processing, distributing, and selling fresh-cut produce [19].

14.2.2 Techniques in food safety assessments that apply predictive modeling

By combining vast datasets and expertise, predictive models improve our knowledge of microbial growth, survival, and inactivation [20]. This information eventually aids in the development of efficient control strategies and safety policies. Comprehensive frameworks for evaluating risks, optimizing procedures, and guaranteeing the safety and quality of the foods are provided with essential techniques such as quantitative microbiological risk assessment (QMRA), ML, and artificial intelligence (AI), Bayesian networks, and growth and survival models [6].

14.2.2.1 Quantitative microbial risk assessment (QMRA)

QMRA is used to predict the probability and severity of adverse health consequences arising from exposure to pathogenic microbes in food [7]. It is frequently used to inform safety standards and guidelines in risk management, public health, and regulatory frameworks [7]. The process integrates four elements to assess microbial hazards in food: hazard identification, hazard characterization, dose-response assessment, and exposure assessment (Table 14.1). The application of predictive modeling greatly supports QMRA.

Table 14.1: Steps in conducting a quantitative microbial risk assessment.

Step	Procedure
Hazard identification	The process involves identifying potential health risks in food by analyzing epidemiological data, scientific literature, laboratory surveys, and outbreak reports.
Hazard characterization	Also known as the dose-response assessment. It is the qualitative and/or quantitative evaluation of the nature of the adverse health effects associated with biological agents, which may be present in food. The risk characteristics are very different for different pathogens.

Table 14.1 (continued)

Step	Procedure
Exposure assessment	It is a qualitative and/or quantitative estimation of the quantity of the hazard (the dose) at the time of consumption. Exposure assessment describes the pathway from farm to fork. When performing exposure assessment, predictive modeling is heavily utilized. It takes into consideration the following: *Consumption patterns*: Determining the quantity of contaminated food that is consumed by various demographic groupings. *Pathogen levels*: Determining or predicting the levels of pathogens in food at different phases of production, processing, distribution, and consumption. *Environmental factors*: Considering factors like cross-contamination, cooking methods, and storage conditions. *Predictive models*: The models are used to predict the growth, survival or death of pathogens in food, considering the activities from farm to fork.
Risk characterization	The process involves integrating exposure assessment and hazard characterization to estimate risk, quantify adverse health effects, and address uncertainties and variability in data.

Source: [7]

14.2.2.2 Machine learning and artificial intelligence

ML and AI enhance food safety prediction accuracy by analyzing multifaceted and complex datasets [6]. Predictive ML and AI models predict contamination risks, optimize food safety management practices, and identify critical control points. These techniques provide insights into pathogen detection, risk assessment, and spoilage prediction and also offer insights that conventional methods might overlook [6]. ML and AI in food safety use advanced algorithms models to predict and simulate outcomes based on diverse and large datasets [6]. ML and AL use models such as support vector machines (SVM), neural networks, and decision trees.

(i) Support vector machines (SVM): This is a non-probabilistic binary linear classifier that assigns new examples to categories by searching for large margin separators [6]. SVMs divide a training dataset into two sides separated by a half-space hyperplane, with larger margins associated with lower classification error [21, 22]. It works well in high-dimensional problems, but setting key parameters correctly is challenging. In summary, SVMs are managed learning models used for regression analysis and classification, effectively classifying food samples based on contamination levels, adulteration, and spoilage prediction [21, 23, 24].

(ii) Neural networks: Computer systems called neural networks are modeled after the neuronal network found in the human brain [6, 25, 26]. Three commonly used neural

networks are feedforward, convolutional, and recurrent. These are supervised models that use a directed graph model with edges and nodes to predict and optimize food safety inspection data [25, 27, 28]. Neural networks can handle incomplete datasets and are characterized by fault tolerance. An analytic hierarchy process integrated extreme learning machine (AHP-ELM) was proposed to analyze complex, discrete, high-dimensional, and nonlinear data. The model's accuracy and training time were verified using randomly selected 10% unused data. The average relative generalization error was 0.5%, nearly 20 times faster than the traditional artificial neural network (ANN) approach. Zhang et al. [28] validated an ELM model for predicting food safety risks in dairy products, achieving an accuracy level of 86%.

(iii) Decision trees: A decision tree is a tree-like structure with leaves, branches, and internal nodes representing class labels and attribution relationships [29]. It is used for identifying decision-making rules, categorized as classification trees for categorical variables and regression trees for continuous variables, for example, the absence or presence of microbial contamination and the level of bacterial growth, respectively [29]. The most common applications of decision trees in food safety include pathogen detection, spoilage prediction, risk assessment and management, adulteration detection, and process optimization. However, it may become complex when training with complex datasets.

14.2.2.3 Bayesian networks

A Bayesian network (BN) is a graphic model that consists of nodes representing variables and directed arcs connecting them [30, 31]. It can describe interactions between variables and quantify complex outcomes. BN is a supervised model for regression and classification problems that is easy to understand and capable of dealing with incomplete datasets [32, 33]. Furthermore, it is employed to simulate intricate relationships between several elements that impact food safety, including processing parameters, environmental effects, and contamination pathways. Using a directed acyclic graph (DAG), Bayesian networks are probabilistic graphical models that depict a set of variables and their conditional dependencies [32]. Table 14.2 provides a thorough overview of the Bayesian networks' functions and applications in food safety.

BNs are used in food safety for risk assessment, HACCP, and supply chain management. They help identify critical control points, predict pathogen growth and survival, and manage risks throughout the food supply chain, from farm to fork.

14.2.3 Mathematical models used in predictive microbiology

The mathematical models used in predictive modeling are grouped into kinetic, probability, empirical, and mechanistic models. The onset of the impending risk (infection

Table 14.2: Components of Bayesian networks using a directed acyclic graph (DAG).

Components	Description
Nodes	Each node represents a random variable such as hygiene practices, humidity, presence of pathogen, or temperature, either continuous or discrete.
Edges	Arrows between the nodes represent conditional dependencies between the variables. For example, when the growth of the microbial pathogen depends on temperature, the arrow will point from the temperature to the microbial pathogen growth.
Conditional probability tables (CPTs)	Every node possesses a CPT, which measures the impact of parent nodes on the node. Conditional probability distributions are used for continuous variables, for example, measuring pathogen growth given different humidity and temperature levels.

Source: Hunte et al. [34].

or intoxication-related risk) can be predicted using *kinetic models* based on the concentration levels associated with a specific microbial strain [10]. The rates of growth or death response are used to generate these models. The generation of toxins by microbes has been predicted using *probability models* [10]. In specific situations, their use has grown recently. The likelihood of bacterial growth and the toxins they produce are the only things these models predict; the rate at which the event happens is not included. Other models that have been used are grouped into empirical and mechanistic models. *Empirical models* are mathematical connections that establish a relationship between inputs and outputs without tying the structure to any physical, chemical, or other component [10]. They allow us to relate two variables through a polynomial equation. *Mechanistic models* provide significant flexibility in determining parameters and are underpinned by a comprehensive understanding of the relevant field. These models are often employed as predictive tools, significantly when conditions fluctuate. Specifically, they enable the evaluation of a theoretical model's applicability to expected experimental conditions and challenges, offering a means to anticipate and address issues before empirical testing [10]. In addition, it is essential to remember that mechanistic models only apply to scenarios within their spectrum of origin, even with a certain level of confidence. Table 14.3 presents what the mathematical models are used to predict in the food industry.

14.2.4 Growth and survival models

The growth of microorganisms under homogeneous conditions is typically described by a growth curve (Figure 14.1).

Table 14.3: Some common mathematical models applied in the food industry.

Model type	Prediction
Kinetic	– Rate of growth or death response (concentration level of microbial strain) – Chemical spoilage prediction
Probabilistic	– Production of toxins by microorganisms or sporulation
Empirical	– Relationships between inputs and outputs – Two variables' relation through a polynomial equation
Mechanistic	– Index of prediction under modified conditions – Determination of different parameters

Source: Stavropoulou et al. [10].

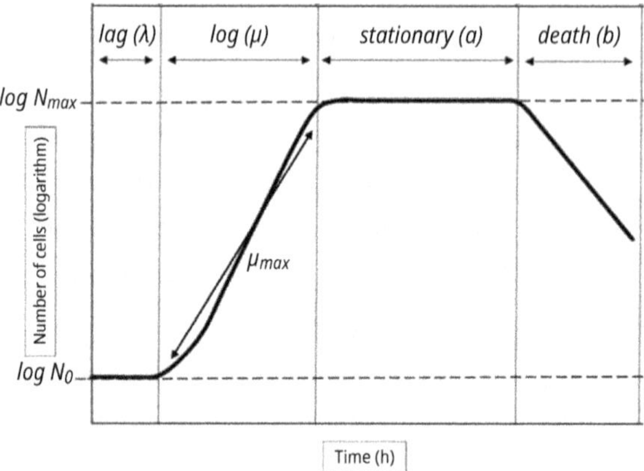

Figure 14.1: Growth curve of microorganisms under a monolithic environment.
λ, lag time (h); μ_{max}, growth rate at a specific temperature (log CFU/g h), log N_0, initial concentration at time 0 (log CFU/g); log N_{max}, maximum population density at time t (log CFU/g); a, stationary phase; b, death phase. The unit "g" can be used interchangeably with "mL" in certain contexts.

Growth and survival models are crucial for ensuring food safety because they predict how microorganisms behave under different environmental conditions that affect the food. These models aid in the comprehension of the growth, survival, and inactivation of pathogens in various food matrices and throughout food processing and storage phases [14]. Growth and survival models in predictive microbiology use statistical data and mathematical equations and are divided into three main categories: primary, secondary, and tertiary models. Table 14.4 presents the class of some of the models used in food safety predictive modeling.

Table 14.4: Predictive models used in the field of food safety.

Primary models	Secondary models	Tertiary models
Gomperts (modified)	Response surface	ComBase
Logistics	Arrhenius	Pathogen modeling program
Rosso	Square root	Growth predictor
Monod	*Z*-values	*Pseudomonas* predictor
D values of inactivation		Sym'Previus
Baranyi		

Source: McDonald and Sun [14].

14.2.4.1 Primary models

Primary models are commonly used to describe changes in microbial population over time under constant environmental conditions. The models are key in predicting microbial growth, inactivation, and survival kinetics.

(i) The most widely used growth models are the Gompertz models.

Gompertz's model: The model predicts microbial growth using the sigmoidal model. The basic equation describes exponential growth as follows:

$$\frac{dN}{dt} = \mu N \tag{14.1}$$

where N is the microbial population at time t and μ the specific growth rate.

The model can be modified to incorporate an inhibition function, which describes the stationary phase. Researchers have proposed the inclusion of the four biological parameters: λ is the lag time, $\mu_{\max}$ is the maximum specific growth rate, N_0 is inoculum, and $N_{\max}$ maximum bacterial population [35]:

$$\text{In}\left(\frac{N}{N_0}\right) = \text{In}\left(\frac{N_{\max}}{N_0}\right) \times \exp\left(-\exp\left(-\frac{\mu_{\max} e}{\text{In}\left(\frac{N_{\max}}{N_0}\right)}(\lambda - t) + 1\right)\right) \tag{14.2}$$

(ii) Inactivation models: If it is assumed that the inactivation follows a first-order kinetic, the primary model can be used to model the inactivation kinetics [36]:

$$\frac{dN}{dt} = -kN \tag{14.3}$$

After integration and transformation into decimal logarithm, the equation is:

$$\log\left(\frac{N_0}{N}\right) = \frac{t}{D} \tag{14.4}$$

where N_0 is the initial bacterial concentration, *N(t)* is the bacterial concentration at time *t*, *k* is the reaction rate, *D* is the decimal reduction time

$$(D_{\text{value}}) \text{ and } D = \frac{\text{In}(10)}{k} \tag{14.5}$$

(iii) Survival models: Bacterial survival models necessitate a thorough understanding of microbiology and offer considerable flexibility in parameter determination. When the environment changes these models are commonly employed as predictive tools. They enable researchers to evaluate a theoretical model's applicability and precision for expected experimental conditions and difficulties, assisting in the anticipation and mitigation of problems prior to conducting empirical tests [36]. In food safety, bacterial survival models play a crucial role in understanding and predicting how bacteria persist under various conditions. These models are essential for ensuring food safety and quality by allowing researchers and food safety professionals to anticipate how bacterial populations will respond to changes in the environment, such as temperature, humidity, and pH.

14.2.4.2 Secondary models

Secondary models describe the impacts of environmental variables such as temperature, pH, and water activity on the parameters of primary models:

(i) Temperature effect: To describe the effect of temperature on growth rate Ratkowsky square-root model is commonly used:

$$\sqrt{\mu_{\text{max}}} = b(T - T_{\text{min}}) \tag{14.6}$$

where μ_{max} is the maximum specific growth rate, T is the temperature, T_{min} is the minimum growth temperature, and b is an empirical constant.

(ii) pH and water activity: Secondary models commonly use polynomial or logistic equations to describe the impact of pH and water activity on the rate of microbial growth [36].

14.2.4.3 Tertiary models

Tertiary models are used to predict microbial behavior under different conditions by integrating primary and secondary models into user-friendly software tools. These software tools are known as predictive microbiology software. Table 14.5 indicates the most used software tools that provide platforms for simulating microbial growth and survival based on input variables.

Table 14.5: Predictive microbiology software used in the field of food safety.

Software	Study	Reference
ComBase	An artificial neural network model for the nonthermal inactivation of *Campylobacter jejuni* in milk and beef was developed using ComBase data, and the acceptable prediction zones approach was used to assess the model's performance and the completeness of the data.	[37]
DMFit	Software for the detection of *Listeria monocytogenes*.	[38]
GinaFit	Software to assess non-log-linear microbial survivor curves.	[39]
IPMP Global Fit	Direct data analysis tool for predictive microbiology.	[40]
MicroFit	Software for adjustment and development of mathematical models of bacterial growth.	[41]
MicroHibro	A software tool for microbial risk assessment in foods and predictive microbiology.	[42]
ValT Validation software tool	Software for predictive microbiology based on the acceptable prediction zones method.	[43]

Source: Meinert et al. [9].

14.2.5 Examples of predictive modeling in various stages of the food supply chain

Food safety benefits greatly from applying predictive modeling, which provides insights and predictions at different points in the food supply chain. Examples of predictive modeling in various stages of the food supply chain are highlighted below:

14.2.5.1 Farm-level predictive models

Farm-level geographic information system models predict the likelihood of pathogen presence in farm produce. The models use data such as temperature, rainfall, and other agricultural practices that may introduce pathogens to the farm produce [44]. The models have revealed that weather conditions significantly influence the die-off of attenuated *Salmonella enterica* serovar Typhimurium and *E. coli* on preharvest leafy greens after irrigation with contaminated water [45].

14.2.5.2 Harvesting, processing, and handling practices

Predictive microbiology research aids in designing safe food handling and processing practices, controlling contamination during slaughter and processing, and using simulation models to assess hygiene practices and equipment sanitation. Juneja, Valenzuela Melendres [46] conducted studies that evaluated the effectiveness of heat treatment in inactivating *Salmonella* in poultry processing using a first-order kinetic model. The aim was to establish critical control points and processing conditions for pathogen reduction.

14.2.5.3 Distribution and storage

The technique involves using time-temperature integrators and logistic regression models to predict spoilage risks in the transportation of perishable foods [47]. Taormina and Hardin [47] conducted a study assessing *E. coli* O157 infection risk in ground beef. The study integrated data on contamination rates, slaughter, storage, and cooking practices and developed risk-based meat inspection protocols for safety enhancement. Another study that used predictive modeling during distribution and storage explored the risk of listeriosis in RTE deli meats, utilizing data on contamination, consumer behavior, and dose-response models, and suggests implementing storage temperature and shelf-life regulations to reduce risk [48].

14.2.5.4 Retail and consumption

Predictive models are used to manage the shelf life of dairy products in retail settings by forecasting degradation based on storage conditions and microbial data [49]. Kang et al. [50] conducted a QMRA study to evaluate *Salmonella enteritidis* risk in shell eggs by combining data, consumption patterns, and dose-response models, and their study informed guidelines for egg handling and storage at retail and appropriate cooking practices to reduce the risk.

14.3 Challenges and obstacles in the application of predictive modeling in food safety

Food safety predictive modeling faces a number of obstacles, such as the requirement for complete, high-quality information and the intrinsic complexity of food systems, which include multiple interacting factors and dynamic situations. Developing models that can adapt to these ever-changing environments is technologically demanding and resource-intensive [51]. Another layer of complexity is added by complying with various regulatory frameworks, necessitating the use of reliable and understandable models [52]. The cost and resource constraints, particularly for smaller organizations, further limit the widespread adoption of these technologies. Enhancing data sharing, model transparency, and scalability alongside fostering industry collaboration is crucial to overcoming these obstacles and fully leveraging predictive modeling to ensure food safety.

14.3.1 Data quality and availability issues

Predictive food safety modeling faces substantial data availability and quality issues. The availability of reliable, pertinent, and complete datasets that precisely represent the variety of microbial activities and environmental variables is essential to the efficacy of predictive models [53]. However, obtaining such data can be daunting due to variability in data sources, inconsistent data collection methodologies, and gaps in data completeness. These issues hinder the development and validation of accurate predictive models, as models heavily rely on the quality and quantity of input data. To tackle these obstacles, stakeholders must work together to standardize data-gathering procedures, improve data integration skills, and allocate resources towards technologies that guarantee data accessibility and quality [53].

14.3.2 Model accuracy and reliability concerns

Predictive models must be accurate and dependable for food safety to benefit from their practical use. Accurate simulation of the dynamics of microbial development is crucial for predictive models, which should consider variables like temperature, pH, and interactions with the food matrix. However, complicated interactions within food systems, as well as external environmental factors, can have a significant impact on the variability of microbial behavior. It is challenging to create models that reliably anticipate microbial development in various circumstances because of this diversity [54, 55]. Thorough validation tests with various datasets and real-world scenarios are necessary to evaluate the model's performance, spot its shortcomings, and improve reliability [55]. Additionally, incorporating sensitivity analysis and uncertainty quantification techniques help quantify and manage uncertainties, improving the confidence and applicability of predictive models in food safety practices [56].

14.3.3 Regulatory and policy challenges

It is essential to navigate legal and policy frameworks for predictive modeling to be accepted and used in food safety. Predictive models must frequently undergo thorough validation and verification to be approved for use in decision-making processes. These regulations differ from one jurisdiction to the next, making it difficult to obtain global regulatory acceptability [51]. Early engagement with regulatory bodies is crucial to ensure that the model creation process is in line with regulatory standards and regulations [19]. Promoting flexible regulatory frameworks that consider advances in predictive modeling technology can help gain regulatory acceptability and encourage the application of creative solutions to improve food safety management [51].

14.3.4 Ethical considerations in predictive modeling for food safety

Predictive models for food safety are developed and applied with ethical considerations in mind. These factors include the potential socioeconomic effects of model-driven decisions on food supply chains, consumer behavior, and privacy concerns around data collecting and use. To keep the public's trust and confidence, it is imperative to ensure transparency in developing, validating, and applying predictive models [57]. Addressing the moral ramifications and promoting the proper application of predictive modeling in food safety requires active participation from stakeholders, including consumers, business representatives, and regulatory agencies. In addition to reducing ethical hazards, establishing explicit principles for data privacy, security, and fairness in decision-making processes guarantees that predictive models will improve food safety standards and practices [57, 58].

14.4 Innovations and best practices in overcoming predictive modeling in food safety

Food safety has significantly benefited from advances in data analytics and ML, which have made it possible to anticipate microbial growth, spoiling, and contamination concerns with greater accuracy. Time-series analysis, ensemble approaches, and deep learning are techniques used to manage complex and dynamic food settings. Quality assurance techniques that guarantee the dependability and transparency of these models, such as sensitivity analysis, explainable AI, and cross-validation, provide additional support. The successful integration of various technologies – such as the use of AI-driven quality control in dairy processing and blockchain for supply chain traceability – is highlighted by case studies [59]. Various innovations have increased operational efficiency, decreased waste, and improved safety. To proactively manage food safety and guarantee high-quality products for customers, these fields must continue to progress.

14.4.1 Advances in data analytics and machine learning techniques

Technological developments in data analytics and ML are transforming predictive modeling for food safety by improving models' scalability, accuracy, and efficiency. Large and complicated datasets can be analyzed using ML algorithms, such as deep learning and ensemble approaches, to derive essential insights about microbial activity and food safety concerns [60, 61]. By identifying minute patterns in data that more conventional statistical approaches would miss, these strategies enhance the prediction power of models. More stable and dependable prediction models can be created thanks to advancements in data preparation, feature selection, and model tuning. Additionally, integrating sensor technologies and real-time data streams makes it easier to continuously monitor food safety indicators, making proactive risk reduction and management techniques possible. By utilizing these innovations, stakeholders can effectively anticipate and reduce foodborne risks [60, 61].

14.4.2 Quality assurance methods for predictive models

Robust quality assurance techniques are needed at every stage of the model development life cycle to guarantee the validity and reliability of predictive models in food safety. The first step in quality assurance is to clearly define the goals and specifications for the predictive model, including the performance indicators and validation standards. Strict preparation and data validation procedures are used to address problems with data quality and guarantee consistency between datasets [62]. Validating a model entails comparing it to other datasets and real-world situations to evaluate its generalizability, accuracy, and sensitivity to input parameters. Insights into model reliability and predictive uncertainty are obtained using sensitivity analysis and uncertainty quantification approaches, which assess the model's response to changes in input data and environmental factors. Predictive models for food safety can gain regulatory acceptability and stakeholder confidence if their methods, assumptions, and limitations are documented. This promotes transparency and reproducibility of results [63].

14.5 Future prospects and emerging trends

The future of predictive modeling in food safety is set to be revolutionized by emerging trends such as the integration of advanced AI techniques and the widespread use of real-time data from IoT devices. These advancements will improve the precision and responsiveness of models, enabling real-time monitoring and quick reactions to potential food safety risks. Enhanced data integration and Explainable AI will make models more transparent and accessible, fostering greater stakeholder trust. Combining blockchain technology with predictive analytics will bolster supply chain traceability and accountability. At the same time, the focus on sustainability and reducing food waste will drive the development of more efficient and environmentally friendly food systems. As regulatory support grows and industry collaborations expand, predictive modeling will become crucial in ensuring a safe and sustainable global food supply chain.

14.5.1 Potential application of predictive modeling in future food safety initiatives

Embracing mathematical models offers a proactive approach to enhancing food safety using available datasets to simulate potential food safety risks, thereby averting potential contamination outbreaks, allergen cross-contact, or unprecedented increases in spoilage and pathogenic microorganisms. Predictive modeling has enhanced the identification of the periods and locations where the potential risk of food contamination is high from the environment or other vectors [64]. Predictive models streamline data on the entire food chain with real-time sensors used to monitor food safety and produce data, which can be processed using big data and AI technology [65]. Systematic and vigorous modeling has a central role among private and public authorities when predicting food behavior during production, distribution, and storage, thereby facilitating risk analysis and development of food safety standards and regulations. The effects of intrinsic and extrinsic factors on microbial behavior can be predicted through modeling. Shelf-life prediction models for evaluating meat quality along the cold chain have been developed by monitoring the sensory, physicochemical, and biological parameters using a real-time electronic nose [66].

14.5.2 Integration of new technologies (e.g., IoT and blockchain) with predictive modeling

Emerging cutting-edge technologies like AI), the IoT, ML, Big Data analytics, and blockchain technology, when integrated with predictive modeling, can be applied to detect and mitigate food safety hazards [54]. These qualitative approaches, which are web-based tools, have unveiled the complex interplay among factors that influence food safety. These technologies provide access to vast amounts of data, which, when analyzed using predictive and learning models, can predict patterns and anomalies in food safety risk prediction and monitoring and food safety optimization [64]. The collaborative application of emerging technologies in modeling creates opportunities to optimize efficiency, sustainability, and overall performance in food systems. Real-time monitoring of intrinsic and extrinsic quality parameters with emerging technologies will enable dynamic adjustments to processing and handling protocols based on data-driven insights [67]. Radiofrequency identification tags (RFID), wireless sensor networks (WSN), global positioning systems (GPS), and other IoT sensing devices obtain and transmit food-related information such as environmental temperature and humidity in real time, thereby facilitating information sharing by different actors in the supply chain to enhance decision support [66].

14.5.3 Research directions and areas for further development

Research direction by advancing ML algorithms to improve prediction of food safety risks has gained momentum in the recent past. Deep learning techniques such as neural networks offer an opportunity for accurate and timely detection of pathogens. For instance, deep learning, when integrated into time-lapse coherent imaging, has been used for early detection and classification of live bacteria [68]. Food microstructures have been incorporated into artificial food model systems to simulate changes in microbial ecology in different food matrices [69]. There is growing interest in innovative supply chain systems to provide intelligent real-time information on quality, shelf life, and intrinsic and extrinsic parameters to regulators, suppliers, and consumers [66].

14.6 Implications for food safety policy and practice

14.6.1 Recommendations for policymakers and regulators

Recommendations should emphasize using predictive models to advance food systems for policymakers and regulators to make more informed decisions and optimize resource allocation [70]. Decision-making and food safety objectives heavily rely on predictive models of supply chain operations to forecast food safety risks. By incorporating data on risk assessment, policymakers and regulators will be in a better position to effectively model the entire food supply chain [71]. There is a need to develop standardized protocols for data collection, sharing, and analysis to ensure consistency and reliability across the industry. With these, policymakers and regulators are encouraged to invest and encourage collaboration with research institutions and other industry partners. There is need to have accurate data on prevailing patterns based on historical occurrences and potential food safety hazards anticipated to implement preventive measures to avert potential foodborne illnesses and outbreaks [72]. Policymakers and regulators must invest in Big Data to predict food insecurity among vulnerable populations, design targeted interventions using models to increase access to safe, quality and nutritious food, and develop early warning systems to avert food crises [31].

14.6.2 Practical implications for food safety professionals and stakeholders

In the prediction of product lifespan, mathematical modeling offers a powerful tool to simulate and model shifts in microbial numbers along the food chain from farm to fork, bringing out a permanent and objective evaluation of intrinsic and extrinsic parameters [10, 73, 74]. The main application is to assess and manage food safety risks, thereby improving quality and safety by expounding information on the presence and changes in the quantities of spoilage, and pathogenic and beneficial microorganisms in a food product [75, 76]. This way, unexpected contamination and spoilage would be reduced, thereby reducing food waste, economic losses, and loss of consumer confidence. Food safety professionals and stakeholders would be in a better position to understand the logistics at all stages along the supply chain [77]. AI, big data, and blockchain have been used to model and simulate food safety issues [78]. They have been used to avert waterborne diseases globally [79].

14.6.3 Importance of ongoing research and collaboration in advancing predictive modeling for food safety

Ongoing research and collaborations are critical in advancing food safety using mathematical models. Research is necessary to advance the techniques for identifying and understanding behavior of emerging pathogens and input this information into databases. Collaborations should aim to build expertise and generate more robust predictive tools to empower policymakers, food safety authorities, and other stakeholders.

References

[1] Kamboj, S., Gupta, N., Bandral, J. D., Gandotra, G., & Anjum Njijo, C. S. Food safety and hygiene: A review. International journal of chemical studies 2020, 8(2), 358–368.

[2] Uyttendaele, M., De Boeck, E., & Jacxsens, L. Challenges in food safety as part of food security: Lessons learnt on food safety in a globalized world. Procedia food science 2016, 6, 16–22.

[3] Garcia, S. N., Osburn, B. I., & Jay-Russell, M. S. One health for food safety, food security, and sustainable food production. Frontiers in Sustainable Food Systems 2020, 4, 1.

[4] Mc Carthy, U., Uysal, I., Badia-Melis, R., Mercier, S., O'Donnell, C., Ktenioudaki, A., et al. Global food security–issues, challenges and technological solutions. Trends in Food Science & Technology 2018, 77, 11–20.

[5] Tshikantwa, T. S., Ullah, M. W., He, F., & Yang, G. Current trends and potential applications of microbial interactions for human welfare. Frontiers in microbiology 2018, 9, 372532.

[6] Wang, X., Bouzembrak, Y., Lansink, A. O., & Van Der Fels-Klerx, H. Application of machine learning to the monitoring and prediction of food safety: A review. Comprehensive Reviews in Food Science and Food Safety 2022, 21(1), 416–434.

[7] Membre, J. M., & Boue, G. Quantitative microbiological risk assessment in food industry: Theory and practical application. Food Research International 2018, 106, 1132–1139.

[8] Chen, Y., Li, H., Dou, H., Wen, H., & Dong, Y. Prediction and visual analysis of food safety risk based on TabNet-GRA. Foods 2023, 12(16), 3113.

[9] Meinert, C., Bertoli, S. L., Rebezov, M., Zhakupbekova, S., Maizhanova, A., Spanova, A., et al. Food safety and food security through predictive microbiology tools: A short review. Slovak Journal of Food Sciences 2023, 17.

[10] Stavropoulou, E., & Bezirtzoglou, E. Predictive modeling of microbial behavior in food. Foods 2019, 8(12), 654.

[11] Dogan, O. B. Implementation of Quantitative Microbial Risk Assessment and Predictive Microbiology Methods for Food Safety Assurance Applications, Nebraska, United States: The University of Nebraska-Lincoln, 2021.

[12] Aljohani, A. Predictive analytics and machine learning for real-time supply chain risk mitigation and agility. Sustainability 2023, 15(20), 15088.

[13] Ismail, S., Reza, H., Salameh, K., Kashani Zadeh, H., & Vasefi, F. Toward an intelligent blockchain IoT-enabled fish supply chain: A review and conceptual framework. Sensors (Basel) 2023, 23(11), 5136.

[14] McDonald, K., & Sun, D. W. Predictive food microbiology for the meat industry: A review. International Journal of Food Microbiology 1999, 52(1–2).

[15] Bolívar, A., Costa, J. C. C. P., Posada-Izquierdo, G. D., Valero, A., Zurera, G., & Pérez-Rodríguez, F. Modelling the growth of Listeria monocytogenes in Mediterranean fish species from aquaculture production. International Journal of Food Microbiology 2018, 270, 14–21.

[16] Choi, W.-S., Son, N., Cho, J.-I., Joo, I.-S., Han, J.-A., Kwak, H.-S., et al. Predictive model of Staphylococcus aureus growth on egg products. Food Science & Biotechnology 2019, 28, 913–922.

[17] Arroyo, F., Quintana, M. D., & Fernández, A. G. Evaluation of primary models to describe the growth of Pichia anomala and study of temperature, NaCl, and pH effects on its biological parameters by response surface methodology. Journal of Food Protection 2005, 68(3), 562–570.

[18] Skandamis, P., & Nychas, G.-J. Modeling the microbial interaction and the death of Escherichia coli O157: H7 during the fermentation of Spanish-style green table olives. Journal of Food Protection 2003, 66(7), 1166–1175.

[19] Kim, Y. J., Park, J. Y., Suh, S. H., Kim, M. G., Kwak, H. S., Kim, S. H., et al. Development and validation of a predictive model for pathogenic Escherichia coli in fresh-cut produce. Food Science & Nutrition 2021, 9(12), 6866–6872.

[20] Membre, J. M., & Lambert, R. J. Application of predictive modelling techniques in industry: From food design up to risk assessment. International Journal of Food Microbiology 2008, 128(1), 10–15.

[21] Ma, Y., Hou, Y., Liu, Y., & Xue, Y., editors. Research of food safety risk assessment methods based on big data. In: 2016 IEEE International Conference on Big Data Analysis (ICBDA), IEEE, 2016, 1–5. 10.1109/ICBDA.2016.7509812.

[22] Abe, S. Support Vector Machines for Pattern Classification, Vol. 26, Springer: London, United Kingdom, 2005.

[23] Anguita, D., Ghio, A., Greco, N., Oneto, L., & Ridella, S., editors. Model selection for support vector machines: Advantages and disadvantages of the machine learning theory. In: The 2010 International Joint Conference on Neural Networks (IJCNN), IEEE, 2010, 1–8.
[24] Pan, Y., Jiang, J., Wang, R., & Cao, H. Advantages of support vector machine in QSPR studies for predicting auto-ignition temperatures of organic compounds. Chemometrics and Intelligent Laboratory Systems 2008, 92(2), 169–178.
[25] Geng, Z., Shang, D., Han, Y., & Zhong, Y. Early warning modeling and analysis based on a deep radial basis function neural network integrating an analytic hierarchy process: A case study for food safety. Food Control 2019, 96, 329–342.
[26] Pham, D. T., Dimov, S. S., & Nguyen, C. D. Selection of K in K-means clustering. Proceedings of the Institution of Mechanical Engineers Part C Journal of Mechanical Engineering Science 2005, 219(1), 103–119.
[27] Wang, J., Yue, H., & Zhou, Z. An improved traceability system for food quality assurance and evaluation based on fuzzy classification and neural network. Food Control 2017, 79, 363–370.
[28] Zhang, R., Zhou, L., Zuo, M., Zhang, Q., Bi, M., Jin, Q., editor., et al. Prediction of dairy product quality risk based on extreme learning machine. In: 2018 2nd International Conference on Data Science and Business Analytics (ICDSBA), IEEE, 2018, 448–456.
[29] Saber, W. I., Al-Askar, A. A., & Ghoneem, K. M. Exclusive biosynthesis of pullulan using Taguchi's approach and decision tree learning algorithm by a novel endophytic Aureobasidium pullulans strain. Polymers 2023, 15(6), 1419.
[30] Marvin, H. J., Bouzembrak, Y., Janssen, E. M., van der Fels-klerx, H., Van Asselt, E. D., & Kleter, G. A. A holistic approach to food safety risks: Food fraud as an example. Food Research International 2016, 89, 463–470.
[31] Marvin, H. J., Janssen, E. M., Bouzembrak, Y., Hendriksen, P. J., & Staats, M. Big data in food safety: An overview. Critical Reviews in Food Science and Nutrition 2017, 57(11), 2286–2295.
[32] Bouzembrak, Y., Camenzuli, L., Janssen, E., & Van der Fels-klerx, H. Application of Bayesian networks in the development of herbs and spices sampling monitoring system. Food Control 2018, 83, 38–44.
[33] Sun, J., Sun, Z., & Chen, X. Fuzzy Bayesian network research on knowledge reasoning model of food safety control in China. Journal of Food, Agriculture and Environment 2013, 11, 234–243.
[34] Hunte, J. L., Neil, M., & Fenton, N. E. A causal Bayesian network approach for consumer product safety and risk assessment. Journal of Safety Research 2022, 80, 198–214.
[35] Zwietering, M., De Koos, J., Hasenack, B., De Witt, J., & Van't Riet, K. Modeling of bacterial growth as a function of temperature. Applied and Environmental Microbiology 1991, 57(4), 1094–1101.
[36] Isabelle, L., & Andre, L. Quantitative prediction of microbial behaviour during food processing using an integrated modelling approach: A review. International Journal of Refrigeration 2006, 29(6), 968–984.
[37] Boleratz, B. L., & Oscar, T. P. Use of ComBase data to develop an artificial neural network model for nonthermal inactivation of Campylobacter jejuni in milk and beef and evaluation of model performance and data completeness using the acceptable prediction zones method. Journal of Food Safety 2022, 42(4), e12983.
[38] Engelhardt, T., Ágoston, R., Belák, Á., Mohácsi-Farkas, C., & Kiskó, G. The suitability of the ISO 11290-1 method for the detection of Listeria monocytogenes. LWT-Food Science & Technology 2016, 71, 213–220.
[39] Geeraerd, A. H., Valdramidis, V. P., & Van Impe, J. F. GInaFiT, a freeware tool to assess non-log-linear microbial survivor curves. International Journal of Food Microbiology 2005, 102(1), 95–105.
[40] Huang, L. IPMP Global Fit – A one-step direct data analysis tool for predictive microbiology. International Journal of Food Microbiology 2017, 262, 38–48.

[41] Siqueira, A. A., Carvalho, P., Mendes, M. L. M., & Shiosaki, R. K. MicroFit: Um software gratuito para desenvolvimento e ajuste de modelos matemáticos de crescimento bacteriano. Brazilian Journal of Food Technology 2014, 17, 329–339.
[42] Gonzalez, S. C., Possas, A., Carrasco, E., Valero, A., Bolivar, A., Posada-Izquierdo, G. D., et al. 'MicroHibro': A software tool for predictive microbiology and microbial risk assessment in foods. International Journal of Food Microbiology 2019, 290, 226–236.
[43] Oscar, T. P. Validation software tool (ValT) for predictive microbiology based on the acceptable prediction zones method. International Journal of Food Science & Technology 2020, 55(7), 2802–2812.
[44] Dorotea, T., Riuzzi, G., Franzago, E., Posen, P., Tavornpanich, S., Di Lorenzo, A., et al. A scoping review on GIS technologies applied to farmed fish health management. Animals (Basel) 2023, 13(22), 3525.
[45] Belias, A. M., Sbodio, A., Truchado, P., Weller, D., Pinzon, J., Skots, M., et al. Effect of weather on the die-off of Escherichia coli and attenuated Salmonella enterica serovar Typhimurium on preharvest leafy greens following irrigation with contaminated water. Applied and Environmental Microbiology 2020, 86(17), e00899–20.
[46] Juneja, V. K., Valenzuela Melendres, M., Huang, L., Gumudavelli, V., Subbiah, J., & Thippareddi, H. Modeling the effect of temperature on growth of Salmonella in chicken. Food Microbiology 2007, 24(4), 328–335.
[47] Taormina, P. J., & Hardin, M. D. Food Safety and Quality-based Shelf Life of Perishable Foods, Switzerland: Springer, 2021. https://doi.org/10.1007/978-3-030-543 75–4.
[48] Tsaloumi, S., Aspridou, Z., Tsigarida, E., Gaitis, F., Garofalakis, G., Barberis, K., et al. Quantitative risk assessment of Listeria monocytogenes in ready-to-eat (RTE) cooked meat products sliced at retail stores in Greece. Food Microbiology 2021, 99, 103800.
[49] Kim, Y. J., Moon, H. J., Lee, S. K., Song, B. R., Lim, J. S., Heo, E. J., et al. Development and validation of predictive model for salmonella growth in unpasteurized liquid eggs. Korean Journal for Food Science of Animal Resource 2018, 38(3), 442–450.
[50] Kang, M. S., Park, J. H., & Kim, H. J. Predictive modeling for the growth of Salmonella spp. in liquid egg white and application of scenario-based risk estimation. Microorganisms 2021, 9(3), 486.
[51] Havelaar, A. H., Brul, S., de Jong, A., De Jonge, R., Zwietering, M. H., & Ter Kuile, B. H. Future challenges to microbial food safety. International Journal of Food Microbiology 2010, 139(Suppl 1), S79–94.
[52] Trienekens, J., & Zuurbier, P. Quality and safety standards in the food industry, developments and challenges. International Journal of Production Economics 2008, 113(1), 107–122.
[53] Messens, W., Hempen, M., & Koutsoumanis, K. Use of predictive modelling in recent work of the Panel on Biological Hazards of the European Food Safety Authority. Microbial Risk Analysis 2018, 10, 37–43.
[54] Abid, H. M. R., Khan, N., Hussain, A., Anis, Z. B., Nadeem, M., & Khalid, N. Quantitative and qualitative approach for accessing and predicting food safety using various web-based tools. Food Control 2024, 162, 110471.
[55] Membré, J.-M., & Valdramidis, V. Modeling in Food Microbiology: From Predictive Microbiology to Exposure Assessment, Oxford, United Kingdom: Elsevier, 2016.
[56] Duret, S., Guillier, L., Hoang, H. M., Flick, D., & Laguerre, O. Identification of the significant factors in food safety using global sensitivity analysis and the accept-and-reject algorithm: Application to the cold chain of ham. International Journal of Food Microbiology 2014, 180, 39–48.
[57] Sapienza, S., & Vedder, A. Principle-based recommendations for big data and machine learning in food safety: The P-SAFETY model. AI and Society 2023, 1–16.

[58] Manning, L., Baines, R., & Chadd, S. Ethical modelling of the food supply chain. British Food Journal 2006, 108(5), 358–370.
[59] Casino, F., Kanakaris, V., Dasaklis, T. K., Moschuris, S., Stachtiaris, S., Pagoni, M., et al. Blockchain-based food supply chain traceability: A case study in the dairy sector. International Journal of Production Research 2021, 59(19), 5758–5770.
[60] Deng, X., Cao, S., & Horn, A. L. Emerging applications of machine learning in food safety. Annual Reviews of Food Science and Technology 2021, 12(1), 513–538.
[61] Pandey, V. K., Srivastava, S., Dash, K. K., Singh, R., Mukarram, S. A., Kovács, B., et al. Machine learning algorithms and fundamentals as emerging safety tools in preservation of fruits and vegetables: A review. Processes 2023, 11(6), 1720.
[62] Abass, T., Itua, E. O., Bature, T., & Eruaga, M. A. Concept paper: Innovative approaches to food quality control: AI and machine learning for predictive analysis. World Journal of Advanced Research and Reviews 2024, 21(3), 823–828.
[63] Plaza-Rodriguez, C., Thoens, C., Falenski, A., Weiser, A. A., Appel, B., Kaesbohrer, A., et al. A strategy to establish food safety model repositories. International Journal of Food Microbiology 2015, 204, 81–90.
[64] Qian, C., Murphy, S., Orsi, R., & Wiedmann, M. How can AI help improve food safety?. Annual Review of Food Science and Technology 2023, 14(1), 517–538.
[65] Kumar, I., Rawat, J., Mohd, N., & Husain, S. Opportunities of artificial intelligence and machine learning in the food industry. Journal of Food Quality 2021, 2021(1), 4535567.
[66] Ren, Q.-S., Fang, K., Yang, X.-T., & Han, J.-W. Ensuring the quality of meat in cold chain logistics: A comprehensive review. Trends in Food Science and Technology 2022, 119, 133–151.
[67] Rane, N., Choudhary, S., & Rane, J. Leading-edge Artificial Intelligence (AI), Machine Learning (ML), Blockchain, and Internet of Things (IoT) technologies for enhanced wastewater treatment systems. Machine Learning (ML), Blockchain, and Internet of Things (IoT) Technologies for Enhanced Wastewater Treatment Systems 2023. (October 31, 2023).
[68] Wang, H., Ceylan Koydemir, H., Qiu, Y., Bai, B., Zhang, Y., Jin, Y., et al. Early detection and classification of live bacteria using time-lapse coherent imaging and deep learning. Light: Science and Applications 2020, 9(1), 118.
[69] Verheyen, D., & Van Impe, J. F. M. The inclusion of the food microstructural influence in predictive microbiology: State-of-the-art. Foods 2021, 10(9), 2119.
[70] Lopatkin, A. J., & Collins, J. J. Predictive biology: Modelling, understanding and harnessing microbial complexity. Nature Reviews Microbiology 2020, 18(9), 507–520.
[71] Allende, A., Bover-Cid, S., & Fernández, P. S. Challenges and opportunities related to the use of innovative modelling approaches and tools for microbiological food safety management. Current Opinion in Food Science 2022, 45, 100839.
[72] Zhang, P., Cui, W., Wang, H., Du, Y., & Zhou, Y. High-efficiency machine learning method for identifying foodborne disease outbreaks and confounding factors. Foodborne Pathogens and Disease 2021, 18(8), 590–598.
[73] Garcia, D., Ramos, A. J., Sanchis, V., & Marin, S. Modelling the effect of temperature and water activity in the growth boundaries of Aspergillus ochraceus and Aspergillus parasiticus. Food Microbiology 2011, 28(3), 406–417.
[74] Gaspar, P. D., Alves, J., & Pinto, P. Simplified approach to predict food safety through the maximum specific bacterial growth rate as function of extrinsic and intrinsic parameters. Chemical Engineering 2021, 5(2), 22.
[75] Haouet, M. N., Tommasino, M., Mercuri, M. L., Benedetti, F., Bella, S. D., Framboas, M., et al. Experimental accelerated shelf life determination of a ready-to-eat processed food. Italian Journal of Food Science 2018, 7(4), 6919.

[76] Tarlak, F. The use of predictive microbiology for the prediction of the shelf life of food products. Foods 2023, 12(24), 4461.
[77] Bruckner, S., Albrecht, A., Petersen, B., & Kreyenschmidt, J. Influence of cold chain interruptions on the shelf life of fresh pork and poultry. International Journal of Food Science & Technology 2012, 47(8), 1639–1646.
[78] Zhou, Q., Zhang, H., & Wang, S. Artificial intelligence, big data, and blockchain in food safety. International Journal of Food Engineering 2022, 18(1), 1–14.
[79] Kasprzyk-Hordern, B., Adams, B., Adewale, I. D., Agunbiade, F. O., Akinyemi, M. I., Archer, E., et al. Wastewater-based epidemiology in hazard forecasting and early-warning systems for global health risks. Environment International 2022, 161, 107143.

Vivian Chiamaka Nwokorogu, Taofeeq Garuba, and Saheed Sabiu*

Chapter 15
Gene–nutrient interactions for enhancement of health indices

Abstract: Gene–nutrient interactions represent a cutting-edge frontier in nutritional research, offering new insights into how our genetic makeup interacts with dietary components to influence health and well-being. This chapter explains the disciplines of nutrigenomics, nutrigenetics, and epigenetics, exploring how nutrients modulate gene expression and how genetic variations affect individual responses to diet. It further explains how certain genetic variants and essential nutrients, like vitamins, minerals, macro-nutrients, and phytochemicals, interact to prevent and manage chronic diseases like cancer, diabetes, and cardiovascular disorders. The potential of personalized nutrition, which tailors dietary recommendations based on genetic profiles, is highlighted as a transformative approach for optimizing health indices and mitigating disease risk. Additionally, the chapter discusses the broader public health implications, emerging technological trends, and challenges surrounding integrating gene–nutrient research into public healthcare systems. Ultimately, this chapter provides a comprehensive overview of the current state of gene–nutrient interactions and their potential to revolutionize healthcare and nutrition while presenting the major constraints and future directions. By overcoming the identified barriers and integrating these insights into public health policies, a more personalized, effective, and equitable approach to health management is achievable. The future of gene–nutrient interactions is promising, with significant opportunities for enhancing global health outcomes.

Keywords: Gene–nutrient, personalized nutrition, nutrigenomics, nutrigenetics, epigenetics, omićs

***Corresponding author: Saheed Sabiu**, Department of Biotechnology and Food Science, Faculty of Applied Science, Durban University of Technology, P.O. Box 1334, Durban 4000, South Africa, e-mail: sabius@dut.ac.za
Vivian Chiamaka Nwokorogu, Department of Biotechnology and Food Science, Faculty of Applied Science, Durban University of Technology, P.O. Box 1334, Durban 4000, South Africa
Taofeeq Garuba, Department of Plant Biology, Faculty of Life Sciences, University of Ilorin, P.M.B. 1515, Ilorin, Kwara, Nigeria

https://doi.org/10.1515/9783111441238-016

15.1 Introduction

Nutrient–gene associations, also known as nutritional genomics in recent times, have drawn significant interest, owing to current breakthroughs in genomics, revolutionizing our understanding of how nutrients interact with our genetics. Gene–nutrient interactions can be referred to as the bidirectional collaborative effects of an individual's genetic factors and dietary component intake. These interactions may have far-reaching implications for understanding disease prevention, health optimization, and the development of personalized nutrition strategies that cater to individual genetic profiles. Nutrients, as environmental components, can influence our genetic makeup. Biologically active food constituents such as polyphenols, carotenoids, and glucosinolates play crucial roles as signaling compounds. These food components transmit information and regulate gene expression through modifications in chromatin structure, DNA methylation, noncoding RNA, activation of transcription factors, and binding to nuclear receptors [1]. Studies have revealed that dietary factors can potentially influence epigenetic modifications linked to diseases, which could help prevent or reverse pathological processes [2, 3]. Gene–nutrient interaction studies encompass the fields of nutrigenetics and nutrigenomics, exploring the impact of genetic variations on the body's response to nutrients and how nutrients can modulate gene expression [4]. This emerging field of research shows great potential for advancing personalized nutrition and improving health indices, especially in disease prevention and management.

Nutrigenomics, which explores how nutrients affect gene expression, has revealed that dietary components can substantially affect our genetic activity. Diets rich in whole grains, vegetables, and fruits and low in saturated fats have shown promise in reducing obesity risk among individuals with higher genetic risk scores [5]. Also, dietary polyphenols, abundant in fruits and vegetables, have been shown to modulate gene expression involved in inflammation and oxidative stress, which are the key processes in chronic diseases like cancer, cardiovascular disease (CVD), and diabetes [6]. On the other hand, nutrigenetics focuses on how genetic variations influence individual responses to nutrients. This study area has uncovered significant genetic determinants that impact nutrient metabolism, absorption, and utilization, leading to varied health indices among different populations. Nutrigenetic studies have revealed that specific macronutrient percentages, fatty acid consumption, and ingesting nutraceuticals or dietary supplements can impact blood lipid levels [7, 8]. These findings emphasize the roles of genetic factors in developing precision diets for preventing and managing a wide range of diseases.

Since the beginning of the twenty-first century, there have been significant advances in various "omics" domains, including genomics, transcriptomics, proteomics, and metabolomics. A deeper understanding of nutrition, genetics, biochemistry, and integration of "omics" technologies would provide limitless potential for exploring nutrient–gene correlations relevant to health and diseases at molecular levels. These approaches' potential is immense in uncovering new biomarkers, creating personalized dietary recommendations, and devising enhanced nutritional and pharmaceutical approaches to tackle metabolic disease. Integration of system genetics and high throughput "omics" data has been effectively used to decode higher-order interactions and pathways involved in complex metabolic diseases [9]. Integrating gene–nutrient data with medical care offers to transform our approach to health. However, limited knowledge still exists on the roles of gene–nutrient interactions in preventing and managing diseases. Therefore, this chapter provides an overview of the current knowledge in gene–nutrient interactions, highlighting their mechanisms and key nutrients and exploring gene–nutrient interaction strategies in disease prevention, management, and personalized nutrition. Ultimately, the major challenges and future directions associated with gene–nutrient interaction strategies in enhancing health indices are emphasized, offering a roadmap for translating gene–nutrient interactions into actionable health interventions.

15.1.1 Overview of gene–nutrient interactions

Gene–nutrient interactions can be referred to as the bidirectional collaborative effects of an individual's genetic factors and dietary component intake. This interaction determines how nutrients are metabolized and utilized and how they ultimately affect one's health. Gene–nutrient interactions remain the bedrock of the field of nutrigenetics and nutrigenomics. These fields are influenced by three major factors: ethnicity-driven genome diversity; geo-cultural, economic, and taste perception differences; and gene–nutrient dose and expression [10]. The history of gene–nutrient interaction dates to the mid-twentieth century and has expanded over the years to include metabolic diseases and disorders like hypertension, obesity, phenylketonuria, stroke, type 2 diabetes, cancer, and CVDs that require dietary interventions for their management. Molecular biology advances, especially the 2003 Human Genome Project, and the current omics technologies have strengthened the field of nutritional genomics, also known as gene–nutrient interactions, which encompasses "nutrigenomics and nutrigenetics." The concept of "gene–nutrient interaction" is based on recognizing that genetic variations, known as polymorphisms, can potentially affect how an individual responds to some nutrients. SNPs (single nucleotide polymorphisms) are the most common genetic variation, occurring at about 500–2000 bp throughout the human genome, and normally found in at least 1% of the population [11]. Studies have identified single nucleotide polymorphisms (SNPs) that affect vitamin metabolism, transport, and status, particularly for vitamins D, E, C, and B-complex [12–14]. These genetic variations can consequently influence nutritional deficiencies, food intolerances, physiological processes, and the susceptibility to developing diseases such as inflammatory bowel disease (IBD) and nonalcoholic fatty liver disease (NAFLD) [15, 16]. On the contrary, certain nutrients possess the ability to regulate gene expression, which could potentially alter the risk of developing long-term health conditions [2, 3]. Although gene–nutrient interactions originated from decades of research in genetics, biochemistry, and nutrition, it currently represents a transition from perceiving nutrition in the old way of one-size-fits-all to a more individualized approach that considers individuals' distinct genetic profiles. Hence, better comprehension of gene–nutrient interactions may help explain the variability in individual responses to food and enable personalized nutritional approaches for disease management and prevention.

15.1.2 Gene–nutrient interaction and enhancement of health indices

Gene–nutrient interactions have broad impacts that go beyond personal dietary preferences. The study of gene–nutrient interactions has significant implications for preventing and managing diseases, personalized nutrition, and public health. Major gene–nutrient strategies have sought to address diverse disease predispositions and have aimed to improve human health through dietary interventions and customized nutrition. Hence, this section highlights the core importance of gene–nutrient interactions in the face of diverse disease prevalence.

15.1.2.1 Disease prevention and management

One of the fundamental aspects of gene–nutrient interactions is their crucial role in disease prevention and management. Genetic variants can influence nutrient requirements and an individual's dietary responses, while diet can modulate gene expression [17]. Gene–nutrient interactions offer a pathway to enhance health indices by enabling the identification of individuals at higher risk for certain conditions, and providing targeted nutritional interventions to mitigate these risks. Genetic polymorphisms, which are variations in DNA sequences among individuals, can significantly impact how nutrients are metabolized and utilized in the body. For disease prevention and management, carriers of the APOE ε4 allele, which is associated with an increased risk of Alzheimer's disease (AD), may benefit from diets rich in omega-3 fatty acids, which have been shown to have neuroprotective effects [18]. In type 2 diabetes mellitus, nutrient–gene interactions are central to its etiopathogenesis, particularly with polyphenols and dietary patterns that influence disease development and progression [19]. Therefore, individuals with genetic predispositions to type 2 diabetes may benefit from customized diets that consider their unique metabolic needs, leading to better glycemic control and reduced disease risk. In addition, individuals with a specific polymorphism in the Methylenetetrahydrofolate reductase (MTHFR) gene may require higher folate intake to reduce the risk of hyperhomocysteinemia and the associated CVDs [20, 21]. Understanding genetic predispositions and these gene–nutrient interactions can help recommend specific dietary adjustments to mitigate health risks. Also, identifying important disease biomarkers may assist in targeted disease prevention and management strategies.

15.1.2.2 Personalized nutrition

Personalized nutrition, individualized dieting, or precision nutrition involves the creation of regulated dietary recommendations that conform to an individual's genetic profile. This approach aims to optimize nutrient intake to improve health and prevent disease. The study of gene–nutrient interactions has greatly influenced personalized nutrition. This has led to a significant transformation in the way dietary recommendations are formulated, moving away from generic guidelines and toward customized regimens that consider an individual's genetic predispositions. The impact of genetics on disease phenotypes can be influenced by diet, while individual genotypes play a crucial role in determining specific responses to dietary interventions [17]. As a result of variations in nutrient metabolism and individual response to dietary components, personalized nutrition improves adherence to nutritional interventions and enhances the efficacy of these interventions in achieving desired health outcomes. Personalized nutrition approaches based on genetic profiles promise to reduce obesity risk, particularly for individuals with high genetic risk scores. A weighted genetic risk score from 95 obesity-related SNPs demonstrated a strong association with insulin resistance, interacting with dietary cholesterol intake and other metabolic factors [22]. Another study has shown that individuals with the FTO gene variant may respond differently to calorie restriction and physical activity in weight management programs [23]. The exploration of nutrient–gene interactions, through the integration of nutrigenetics, nutrigenomics, proteomics, and metabolomics fields, offers valuable insights into the complex link between diet, genes, and health status. This knowledge can transform personalized nutrition strategies for preventing and treating chronic diseases. Also, personalized nutrition strategies, through understanding gene–nutrient interactions, can develop functional foods and nutraceuticals designed to support the prevention and management of chronic diseases.

15.1.2.3 Enhancement of health indices

The integration of gene–nutrient interactions into nutritional science and public health systems offers the potential to significantly enhance health indices, such as body mass index (BMI), blood lipid levels, and glycemic-control-targeted interventions for populations at higher genetic risk for certain conditions. An understanding of the genetic basis of nutrient metabolism and its impact on health is crucial to design interventions that optimize health indices at both individual and population levels. Based on a public health perspective, understanding genetic variations' prevalence among different populations can provide valuable insights for making targeted nutritional policies. For instance, populations with a high prevalence of dyslipidaemia, lactose intolerance, or gluten allergy might benefit from public health policies that encourage dietary adjustments that complement these genetic predispositions, ultimately contributing to better health indices on a larger scale. The adaptation of public health policies to the specific needs of various subgroups holds the potential to reduce the occurrence of diet-related illnesses.

15.2 Mechanisms of gene–nutrient interactions

Gene–nutrient interactions represent a dynamic interplay between dietary components and genetic variables that affect individual health and well-being. It is a complex interaction of biochemical and molecular processes that may modulate gene expression and metabolic pathways and ultimately impact health. The key mechanisms by which gene–nutrient interactions exert their effects include nutrigenetics, nutrigenomics, and epigenetics, which collectively contribute to understanding how diet can modulate gene expression and impact health. An overview of the key mechanisms of gene–nutrient interaction has been depicted in Figure 15.1. Studying these mechanisms is crucial as they offer insights into personalized nutrition and disease prevention strategies tailored to an individual's genetic makeup.

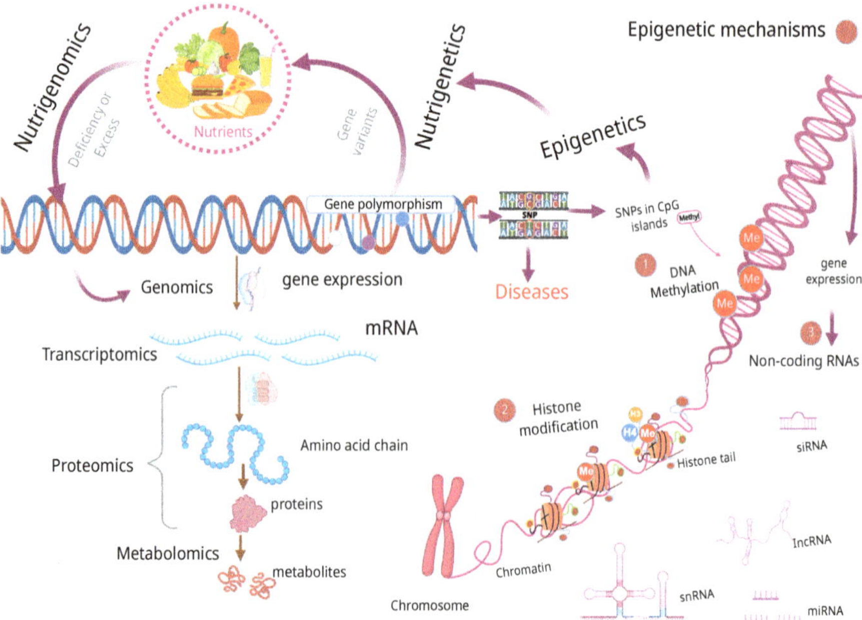

Figure 15.1: A schematic representation of the mechanisms of gene–nutrient interactions.

15.2.1 Nutrigenetics

Nutrigenetics, as a field of nutrition, explores how genetic variations or polymorphisms influence dietary components. These genetic polymorphisms, particularly SNP, can affect various aspects of nutrient metabolism, including absorption, transport, utilization, and efficacy, influencing disease risk and health outcomes. The methylenetetrahydrofolate reductase (MTHFR) gene plays a crucial role in folate metabolism, a B vitamin essential for DNA synthesis and repair. However, a common SNP in this gene, C677T, results in a thermolabile enzyme with reduced activity, leading to elevated levels of homocysteine, a risk factor for CVDs [20, 21]. Individuals with this polymorphism may require higher dietary folate intake to maintain normal homocysteine levels and reduce their risk of heart disease.

Similarly, the FTO gene, which is strongly associated with obesity, has been the subject of extensive research. Variants in the FTO gene, particularly the rs9939609 SNP, have been shown to influence BMI and predispose carriers to obesity, especially when exposed to high-calorie diets rich in fats and sugars [24, 25]. However, studies have also demonstrated that individuals with this genetic predisposition can mitigate their risk of obesity through specific dietary interventions. A nutrigenetic intervention study showed improvements in lipid abnormalities, body composition, and inflammation markers in adults with obesity and overweight, highlighting the potential of nutrigenetic strategies as coadjuvant tools for cardiometabolic disease treatment [26].

Another instance of the roles of nutrigenetics is mirrored in the aging population, where cognitive reasoning evidently depreciates, and a strong need to develop preventive strategies for those with this predisposition becomes important. The apolipoprotein E (APOE) allele is the principal cholesterol carrier protein in the brain, and genetic variation in the gene encoding the variant apolipoprotein E4 (APOE ε4) is a significant risk factor for AD. While evidence for most gene–diet interactions remains weak, strong evidence suggests that male APOE-E4 carriers exhibit significant triglyceride reductions in response to omega-3-rich fish oil [27]. These findings highlight the immense potential of nutrigenetics in providing personalized dietary recommendations that consider an individual's genetic makeup, ultimately improving human health.

15.2.2 Nutrigenomics

Nutrigenomics focuses on how nutrients influence gene expression, affecting various physiological functions and health outcomes. This field of study examines the molecular mechanisms by which dietary components modulate gene activity. Specifically, it focuses on the pathways linked to inflammation, oxidative stress, and metabolic regulation. NutriGenomeDB is a web-based platform that hosts curated gene expression data from nutrigenomics experiments and facilitates the exploration of nutrigenomics data, offering insights into the molecular effects of food compounds on human cells

[28]. Nutrigenomics is grounded in the understanding that the genome is not a static entity but is responsive to environmental factors, including diet. Biologically active food constituents play crucial roles as signaling compounds, transmitting information, and regulating gene expression through modifications in chromatin structure, DNA methylation, noncoding RNA, activation of transcription factors, and binding of nuclear receptors [1]. Through various signaling pathways, nutrients can act as ligands, enzyme cofactors, or precursors to bioactive molecules that regulate gene transcription and translation. The pathways through which nutrients like carbohydrates, fats, and amino acids modulate transcriptional activity include carbohydrate-responsive element binding protein (ChREBP), peroxisome proliferator-activated receptors (PPARs), and the activating transcription factor 4 (GCN2/ATF4) pathways [29].

One of the most well-characterized pathways in nutrigenomics is the activation of peroxisome proliferator-activated receptors (PPARs) by polyunsaturated fatty acids (PUFAs). PPARs are nuclear receptors that regulate gene expression in lipid metabolism, inflammation, and energy homeostasis. Omega-3 fatty acids, particularly those derived from marine sources, have been shown to modulate gene expression and inflammation. These PUFAs can be incorporated into cell membranes, including those of intestinal epithelial cells, where they inhibit pro-inflammatory cytokine production and induce anti-inflammatory mediators [30]. Omega-3 fatty acids significantly upregulate the expression of PPAR-γ and downregulate TNF-α and IL-1 [31]. PPARα, a nuclear receptor activated by omega-3 fatty acids, plays a crucial role in lipid metabolism and exhibits anti-inflammatory properties [32]. This gene-regulatory effect of omega-3 PUFAs is linked to their cardioprotective properties, including reducing triglyceride levels and preventing atherosclerosis. Krill oil, a source of marine omega-3 fatty acid and high-oleic sunflower oil with astaxanthin, has shown activity in altering the expression of genes involved in glucose and lipid metabolism in peripheral blood mononuclear cells [33]. This gene-regulatory effect of omega-3 PUFAs is linked to their cardioprotective properties, including reduced triglyceride levels.

Another significant area of nutrigenomics research involves the modulation of antioxidant response elements by dietary polyphenols. Polyphenols, such as those found in green tea, berries, and other plant-based foods, can enhance the expression of genes-encoding antioxidant enzymes like superoxide dismutase and glutathione peroxidase, through the activation of the nuclear factor erythroid 2-related factor 2 (Nrf2) pathway [6, 34, 35]. This upregulation of antioxidant defenses helps protect cells from oxidative damage, which is a key factor in the development of chronic diseases such as cancer, neurodegenerative disorders, and CVDs. Diet's ability to modulate gene expression through such pathways highlights the potential of nutrigenomics in contributing to disease prevention and health enhancement.

15.2.3 Epigenetics and nutrient interactions

Epigenetics is becoming increasingly important in the field of nutrient–gene interaction studies, as it has the potential to influence both current and future generations. In epigenetics, the biological processes responsible for the on/off regulation of genes are studied, while in epigenomics, changes in epigenetics at the cellular or organismal level are evaluated. Epigenetics refers to the study of heritable changes in gene function that do not involve alterations in the DNA sequence [3]. These alterations occur due to how DNA is packaged and are mediated by mechanisms such as DNA methylation, histone modifications, and noncoding RNAs. In contrast to mutation-induced changes in the genome, epigenetic alterations are modifiable by environmental, nutritional, and behavioral factors and may be reversible [36]. Diet is one of the environmental factors that influences epigenetic modifications. Dietary factors can significantly influence genetic and epigenetic processes, impacting human health and disease risk. Numerous bioactive food components, including methyl-group donors, fatty acids, and phytochemicals, can modulate epigenetic mechanisms such as DNA methylation, histone modifications, and noncoding RNAs, as shown in Figure 15.1 [2, 3]. Epigenetic alterations can affect genome stability, gene regulation, and protein activity, potentially influencing the development of metabolic syndrome, cancer, and mental health conditions [2, 6]. Glucosinolates and their hydrolysis products, like sulforaphane and phenylethyl isothiocyanate, exhibit anticancer activity by affecting DNA methylation, histone acetylation, and microRNA expression [37]. Diet–epigenomic interactions may lead to the development of preventative strategies, personalized health programs, and potential epigenetic therapies for various diseases, including cancer.

One well-documented example is the role of folate and other B vitamins in DNA methylation, a critical process for gene regulation and genome stability. DNA methylation primarily takes place at CpG islands and in repetitive regions of the human genome [10]. DNA methylation, the addition of a methyl group to the cytosine residues in DNA, is one of the most extensively studied epigenetic modifications. This process is critical for regulating gene expression, particularly in cell differentiation and genome stability. Nutrients such as folate, vitamin B_{12}, methionine, and choline are essential donors of methyl groups in one-carbon metabolism, and their availability can directly influence DNA methylation patterns. Deficiencies in these nutrients can lead to aberrant methylation patterns, which are associated with an increased risk of diseases such as cancer and neurodegenerative disorders [36]. For instance, inadequate folate intake has been shown to facilitate genomic integration of Human Papilloma Virus 16 DNA, potentially influencing carcinogenesis [38]. A large-scale epigenome-wide association study identified novel epigenetic loci associated with folate and vitamin B_{12} intake, with folate intake negatively associated with methylation at specific genomic regions [39]. However, the relationship between folate, DNA methylation, and disease outcomes remains complex, with inconsistencies in findings, possibly due to compensatory changes in interrelated metabolic pathways and the unexplored contribution of bacterial folate biosynthesis [40]. These studies highlight the connections between B vitamins, methyl-epigenetic regulation, and disease risks.

Histone modifications are a crucial aspect of epigenetic regulation that involves the process of adding or removing chemical groups to the histone proteins that encase DNA. Histones undergo posttranslational modifications with DNA and proteins, where their H3 and H4 tails can be covalently modified at multiple residues through processes such as methylation, acetylation, and phosphorylation [10]. These modifications have the potential to either augment or suppress the expression of genes, depending on the specific characteristics of the chemical modification. Dietary components have shown promising effects on cancer prevention and treatment through epigenetic modulation. Various plant-derived nutraceuticals, including curcumin, resveratrol, and sulforaphane, modulate epigenetic mechanisms such as histone modification [8]. Particularly, sulforaphane, found in cruciferous vegetables, influences histone acetylation and vitamin D-dependent gene expression in breast cancer cells [41]. Other phytochemicals, such as catechins, genistein, and quercetin, can reverse epigenetic alterations associated with oncogene activation and tumor-suppressor gene inactivation [6, 36]. These dietary compounds act as antioxidants and anticarcinogens, offering potential for cancer prevention and alternative therapeutic approaches through epigenetic modulation.

Noncoding RNAs, including microRNAs (miRNAs), also play a pivotal role in the epigenetic regulation of gene expression (Figure 15.1). Dietary factors can modulate the expression of miRNAs, which in turn can regulate the stability and translation of target mRNAs involved in key metabolic pathways. Dietary polyphenols, such as chlorogenic acid, curcumin, and resveratrol, have been shown to alter miRNA expression profiles, potentially suppressing inflammation and oxidative stress [8]. These dietary components can impact endogenous miRNA levels, influencing various metabolic processes and consequently affecting human health. Food-derived miRNAs, particularly xenotropic miRNAs, can be absorbed into the circulatory system, potentially modulating health and diseases by regulating redox homeostasis and inflammatory conditions [42]. The complex interactions between genetic predisposition, environmental factors, and dietary modifications with the human epigenome underscore the significant role of epigenetic factors in various diseases, including metabolic disorders. This growing body of evidence suggests that dietary habits are closely associated with the modulation of endogenous miRNAs, which may have implications for disease prevention and treatment.

15.3 Gene–nutrient interaction strategies in disease prevention and management

The study of genetic variant interaction with dietary components has enabled the development of more targeted strategies for preventing and managing complex and chronic illnesses. This section examines the roles of gene–nutrient interactions in

CVDs, metabolic disorders, cancer, and neurodegenerative diseases, highlighting recent advances in research and their potential for personalized medicine.

15.3.1 Cardiovascular diseases

CVDs are one of the major global health issues, contributing significantly to disease morbidity and mortality worldwide. CVDs refer to diverse abnormal conditions that affect the heart and blood vessels. These conditions may be congenital or may develop due to coronary disorders or blockages, and include common risk factors such as high blood pressure, cardiac arrest, stroke, arrhythmia, and atherosclerosis. There is a growing recognition of the importance of gene–nutrient interactions in the prevention and management of CVD, especially when it comes to lipid metabolism and inflammatory responses.

Genetic variants in genes such as apolipoprotein E (APOE), which influence lipid metabolism, have been shown to modulate responses to dietary fats. The APOE gene has three common alleles (ε2, ε3, and ε4) that influence lipid metabolism, increasing the susceptibility to CVD risk factors. Carriers of the ε2 allele generally have lower low-density lipoprotein (LDL) and total cholesterol levels, potentially reducing CVD risk [43, 44]. However, they may have higher proportions of saturated fatty acids and lower plasma omega-6 fatty acids and may be prone to hypertriglyceridemia [43]. Conversely, ε4 APOE carriers are particularly susceptible to the detrimental effects of a diet high in saturated fats and cholesterol due to higher LDL concentrations and smaller LDL particle size [44, 45]. Recent advancements in nutritional genomics, formerly known as gene–nutrient reactions, have identified specific dietary interventions to mitigate the genetic predispositions conferred by APOE polymorphisms. For example, omega-3-fatty-acids-fortified diets, particularly from fish oil, have been shown to improve lipid profiles and reduce inflammation in APOE ε4 carriers, offering a targeted approach to CVD prevention [18].

In addition, hyperhomocysteinemia (HHcy) is associated with an increased risk of CVDs in hypertensive patients [21, 46]. This disorder results from methylenetetrahydrofolate (MTHFR) C677T polymorphism, a CVD risk factor implicated with increased homocysteine levels and reduced enzyme activity. Although evidence of a strong association between the C677T polymorphism and increased CVD risk exists, others suggest it cannot reliably predict disease development [47, 48]. Through an evidence-based approach, the effect of polymorphism on homocysteine levels was shown to be influenced by sex, age, race, folic acid intake, smoking, and alcohol consumption [49]. Hence, it is crucial to diagnose and treat nutritional deficiencies and inherited disorders to correct the severe HHcy linked to CVD effects [21]. Supplementing with folic acid has been shown to effectively reduce vascular damage caused by high homocysteine levels in hypertensive rats [20]. This is achieved by suppressing immune and inflammatory responses, while also enhancing the antioxidant effects. A dose–response relationship exists between folate intake and biomarkers, with doubling folate intake resulting in a 22% increase in serum/plasma folate, a 21% increase in red blood cell folate, and a 16% decrease in plasma homocysteine [50]. Thus, further research is needed to determine optimal folate intake levels for reducing CVD risk in different populations. These findings highlight the roles of genetic factors and environmental influences in determining CVD risk.

15.3.2 Metabolic disorders

Metabolic disorders such as obesity, type 2 diabetes, and metabolic syndrome are influenced by a combination of genetic susceptibility and lifestyle factors. The interactions between genes and nutrients play essential roles in the development and progression of these diseases. Oftentimes, the reality of nutrition and dietetic practice is far more complex. The metabolism of each nutrient is influenced by several enzymes, each encoded by a gene that exists in the population with various allelic variations. Genetic variants such as FTO (fat mass and obesity-associated gene) and TCF7L2 (transcription factor 7-like 2) are well-documented in their influence on the risk of developing metabolic disorders.

The FTO gene is perhaps the most well-known genetic factor associated with obesity. Studies have shown that the FTO obesity variant is associated with increased body mass and BMI [23]. Individuals with FTO-risk alleles have a higher propensity for obesity, exhibiting preferences for high-fat foods, reduced satiety, and increased food intake [51]. These genetic variations are linked to different brain reactions that activate hunger signals and a lack of fullness. Increased polygenic risk scores, including FTO mutations, are associated with a greater risk of obesity. However, dietary interventions can mitigate these genetic risks. High-protein, calorie-restricted diets may particularly benefit individuals with high genetic risk. Plant-based and high-protein diets are associated with lower obesity risk, even in individuals with high genetic susceptibility, leading to greater reductions in inflammatory markers [52]. Furthermore, lifestyle modifications, including physical activity and dietary changes, can effectively manage weight in individuals with FTO-risk alleles [23], suggesting that modifiable factors can help overcome genetic predispositions to obesity.

In type 2 diabetes mellitus (T2DM), the TCF7L2 gene has emerged as a significant genetic determinant of disease risk. Insulin deficiency (also known as β-cell dysfunction) and insulin resistance are the key characteristics of T2DM [53]. TCF7L2 variants affect pancreatic beta-cell function, insulin secretion, and glucose metabolism through the Wnt signaling pathway [54], particularly when consuming a diet high in refined carbohydrates. Nutrients such as carbohydrates, fats, proteins, vitamins, and minerals are central to metabolic homeostasis and contribute significantly to T2DM development [55] Nutrigenomics, the study of gene–nutrient interactions, plays a crucial role in understanding and managing T2DM. Targeted dietary strategies that emphasize the consumption of whole grains, fiber, and low-glycemic index foods may improve glycemic control in individuals with TCF7L2 risk alleles, thus modulating the effects of these genetic variations.

Dietary polyphenols have emerged as promising agents for preventing and managing T2DM. These compounds can modulate gene expression related to insulin secretion, signaling, and liver gluconeogenesis [53]. Polyphenols exhibit antioxidant and anti-inflammatory properties, improving insulin sensitivity and glucose metabolism [56]. Specific polyphenols like resveratrol, quercetin, and curcumin have shown positive effects on T2DM-related genes, enhancing mitochondrial function and protecting pancreatic islets [19]. In addition, polyphenols can impact T2DM through mechanisms that are not dependent on insulin. These mechanisms include the inhibition of glucose absorption, the regulation of intestinal microbiota, and the reduction of inflammation. Different epidemiological studies have demonstrated the potential of resveratrol, catechins, curcumin, flavonoids, and anthocyanins to be antidiabetic. A better knowledge of gene–nutrient interactions and the structural activities of dietary components may lead to more effective T2DM preventive and treatment strategies.

15.3.3 Cancer

Dietary regulation of gene expression associated with carcinogenesis has drawn the curiosity of researchers interested in the function of gene–nutrient interactions in cancer prevention and treatment. Some foods have been shown to have an important role in modulating genes involved in cancer detoxification, DNA repair, and cell cycle control; this may represent an approach that can help people genetically predisposed to cancer lower their risk.

Gene–nutrient interactions play crucial roles in modulating epigenetic mechanisms, such as DNA methylation and histone modification, which are critical in the regulation of oncogenes and tumor suppressor genes. The potential of plant-derived compounds in cancer prevention and treatment is mainly through epigenetic mechanisms. Cruciferous vegetables, such as broccoli, cauliflower, and Brussels sprouts, are rich in bioactive compounds like Sulforaphane (SFN), which has shown promising anticancer effects [41, 57]. SFN acts through multiple mechanisms, including modulation of carcinogen metabolism, cell cycle arrest, and activation of Nrf2 pathways [3, 58]. These SFN mechanisms modulate the expression of phase II glutathione *S*-transferase (GST) detoxification enzymes, GSTM1 and GSTT1 genes that protect against DNA damage and carcinogenesis [57]. These enzymes are crucial in neutralizing carcinogens and protecting against DNA damage, thereby reducing cancer risk.

Additional targets for nutritional modulation in cancer prevention include dietary components that affect epigenetic processes, polyphenols, folate, vitamin B_{12}, and glucosinolates and their hydrolysis products, like phenyl ethyl isothiocyanate [6, 37].

These dietary components exhibit anticancer activity by affecting DNA methylation, histone acetylation, and microRNA expression.

Other bioactive compounds such as resveratrol, β-carotene, and quercetin can stimulate the expression of preventive factors and inhibit metastasis-related genes [6, 36]. These nutraceuticals influence gene expression by modifying histones, inhibiting DNA methyltransferases, and altering noncoding RNA expression [8, 10]. While these compounds show great potential in cancer prevention and enhancing conventional treatments, further research is needed to determine optimal dosages and potential adverse effects. Incorporating these nutraceuticals into an "epigenetic diet" could be a promising strategy for cancer prevention and management [8]. The valuable insights from different gene–nutrient interaction studies may facilitate the formulation of dietary strategies that decrease the likelihood of developing cancer and enhance the efficacy of traditional cancer treatments, offering a more holistic approach to managing the disease.

15.3.4 Neurodegenerative diseases

Neurodegenerative diseases, such as AD and Parkinson's disease, are characterized by progressive neuronal loss and cognitive decline [59], with both genetic and environmental factors contributing to disease pathogenesis. AD is a common neurodegenerative condition that accounts for a significant 60 to 70% of 55 million global dementia cases [60]. Carrying the ε4 allele of the apolipoprotein E gene (APOE4) is considered the primary genetic risk factor for late-onset AD [61]. This is due to an associated defective docosahexaenoic acid (DHA) metabolism. Gene–nutrient interactions have been identified as significant modulators of AD risks and progression, offering potential strategies for prevention and management.

The possession of the APOE ε4 allele is associated with an increased accumulation of amyloid-beta plaques and neurofibrillary tangles, leading to greater susceptibility to neuroinflammation that affects different brain regions [59]. Nevertheless, it is worth noting that not all individuals who possess the APOE4 allele necessarily develop AD [18]. This suggests that there may be certain environmental factors that play a role in influencing the risk. A precision nutrition approach for APOE ε4 carriers has been proposed, including a low-glycemic index diet with Mediterranean-style food choices and specific nutritional supplements, targeting metabolic pathways altered by APOE ε4 [62]. Nutrient–gene reaction studies suggest that diet, particularly the intake of omega-3 polyunsaturated fatty acid (PUFA) and antioxidants, can modulate this risk. Omega-3 fatty acids, abundant in fish oil, have demonstrated the ability to decrease the buildup of amyloid-beta and enhance synaptic plasticity [59]. This is especially notable for APOE ε4 gene carriers, as it presents a dietary strategy for reducing the risk of AD. While some studies indicate that APOE4 carriers might not benefit from omega-3 intake due to decreased ability to incorporate docosahexaenoic acid (DHA) into red blood cell phosphatidylcholine [63–65], others show that both carriers and noncarriers can increase omega-3 levels in brain-accessible lipid compartments through supplementation [18]. Recent animal studies demonstrated that APOE4 mice exhibit higher red blood cell DHA levels but not higher brain DHA levels, following fish oil and flavanol supplementation, than APOE3 mice [66]. However, the effects on brain DHA levels were modest and genotype-dependent. These inconsistencies in results highlight the necessity of more trials on omega-3 supplementation in individuals with the APOE4 gene, considering factors like dosage, duration, and potential co-interventions. This will help maximize the effectiveness of dementia prevention efforts.

Additionally, dietary antioxidants play a crucial role in neuroprotection by combating the "neurotoxic triad" of excitotoxicity, oxidative stress, and neuroinflammation through modulation of the Nrf2 pathway [34, 67]. Various micronutrients, such as vitamins C and E, carotenoids, and polyphenols, have been associated with reducing oxidative stress and neuroinflammation, strengthening the significance of dieting strategies in neuroprotection. These antioxidants modulate NF-κB activity differently in neurons and astrocytes, promoting neuronal survival and limiting astrocyte proliferation [68]. In Parkinson's disease, where oxidative stress contributes to dopaminergic neuron depletion, dietary antioxidants may help maintain intracellular redox homeostasis and potentially delay disease progression [69]. Hence, eating a balanced diet that includes a variety of antioxidants, along with regular exercise, and stress reduction will positively protect the brain and promote cognitive health throughout life.

In general, when it pertains to both wellness and disease management, gene–nutrient interactions are crucial. It is feasible to create individualized treatments that not only prevent disease but also enhance the management of existing conditions by recognizing and taking advantage of the interactions between nutrients and genes. Table 15.1 gives an overview of selected examples of gene–nutrient interaction studies for the improvement of health. Integrating gene–nutrient data into healthcare practices is critical, given that it holds promises of enhancing our approach to health. This can lead to more precise and effective disease prevention and management strategies.

15.4 Key nutrients and their genetic interactions

Nutrigenomics and nutrigenetics offer insights into how dietary factors interact with genetic variations to influence health indices. Nutrients can modulate gene expression through their pathways of absorption and assimilation, affecting the risk and progression of diseases. Hence, identifying important nutrient and their gene–nutrient interactions is crucial for developing personalized nutrition strategies. In this section, specific nutrients, including vitamins, minerals, macronutrients, phytochemicals, and antioxidants, and some of their genetic interactions will be discussed.

Table 15.1: An overview of selected gene–nutrient interaction studies conducted for the improvement of health and well-being.

Gene/SNPS	Number of participants (age in years)	Problem	Controlled factor or main variable(s)	Intervention or experimental factor	Key findings/outcome	References
APOE4	50 (20–80)	Alzheimer	BMI/APOE status	Omega 3 fatty acid (PUFA) supplementation/ 6 months	Consuming omega-3 fatty acids was linked to a reduced likelihood of acquiring Alzheimer's disease, particularly in persons who have the ε4 allele of apolipoprotein E (APOE4). Nevertheless, those with a higher Body Mass Index (BMI) see less significant increases in omega-3 levels than those with a lower BMI.	Balakrishnan et al. [18]
16 SNPs involving 13 Genes (ADRA2A, TCF7L2, PLIN1, NUDT3, FABP2, UCP3, MTCH2, ABCB11, LEP, NFE2LB, NPC1, AGTR2, MTHFR)	232 (45.3 ± 0.7; 48.6 ± 0.2)	Insulin resistance	Obesity (BMI:25–29.9;30–40 kg/m^2)	Nutrient intake	This study suggests that the interaction of a weighted risk score based on adiposity with nutritional and metabolic/endocrine factors affects insulin resistance phenotypes. This could help personalize nutrition recommendations for preventing and managing insulin resistance and diabetes.	Ramos-Lopez et al. [22]

ALDH2 rs671	3,495 (20–69)	Metabolic disorder/ ferritin	BMI (23 ± 3.4)	Alcohol consumption, smoking behavior, and BMI	Three novel SNPs associated with ferritin in ALDH2 were identified, with men possessing the mutated genotype of rs671 having lower serum ferritin levels. BMI was the mediation between rs671 and ferritin, with s671 genotypes strongly correlated with serum ferritin levels in drinkers.	Tao et al. [70]
MTHFR (C677T and A1298C) and CβS (844ins68pb)	34 (25 ± 3)	Folate deficiency	Serum B_{12}, red blood cell folate concentration, and homocysteine level	400 µg folate intake	The relationship between RBC, folate, B_6, B_{12} consumption, serum vitamin B_{12}, and genetic variations in 34 reproductive-age Colombian women after 3 months of 400 µg/ day folic acid supplementation.	Aris et al. [71]
MTHFR, MTR, and MTRR	54 (40–75)	Vitamin B deficiency	Homocysteine level	Methylfolate, pyridoxal-5′-phosphate, and methylcobalamin supplements	Supplementation with methyl folate, pyridoxal-5′-phosphate, and methylcobalamin reduced homocysteine and LDL-C levels in patients with specific MTHFR, MTR, and MTRR genetic polymorphisms.	Pokushalovs et al. [72]

(continued)

Table 15.1 (continued)

Gene/SNPS	Number of participants (age in years)	Problem	Controlled factor or main variable(s)	Intervention or experimental factor	Key findings/outcome	References
PPARG2 Pro12Ala (rs1801282) and IL6–174G (rs1800795)	149 (18–65 years)	Obesity	Anthropometric measurements, body composition, metabolic parameters, physical activity practice, and dietary intake	Traditional Brazilian diet and extra virgin olive oil supplementation (12 weeks)	In highly obese individuals with the Ala allele of the Pro12Ala polymorphism, extra virgin olive oil consumption may decrease % body fat and increase lean mass and fat-free mass, without weight reduction. The positive impact of this gene–diet interaction could strengthen the growing body of data supporting personalized treatments for severe obesity therapy.	Rodrigues et al. [73]
FABP2 Ala54 Thr	109 (18–65 years)	Obesity/insulin resistance	BMI ≥25 kg/m^2	Low-calorie diet intake/2 months	This moderate-fat diet improved anthropometric and biochemical characteristics in overweight or obese people. The Thr54 allele carriers responded better. This study supports the notion that genetics and environment may be utilized to treat obesity and associated issues.	Martinez-Lopez et al. [74]

MC4R V103I	1,029 (33–92)	Severe obesity	BMI	Macronutrient supplementation	High carbohydrate consumption was associated with BMI in those V103I risk allele carriers	Pichler et al. [75]
BCO1 rs6564851	189 (>15 years adolescents)	Vitamin A deficiency	BMI/pro-vitamin A carotenoid concentrations	Proposed preformed vitamin A supplementation	The low conversion efficiency of provitamin A carotenoids in individuals with the rs6564851 G allele might challenge current supplementation initiatives aimed at enhancing vitamin A levels in sub-Saharan Africa.	Graßmann et al. [76]
FTO rs9939609	2,129 (6–17 years)	Cardiovascular disease and obesity	Body composition, cardiorespiratory fitness, physical activity, inflammatory markers, and cardiovascular risk	N/A	The findings demonstrated a significant influence of the FTO genotype on body composition parameters in both the first and third year ($p < 0.05$), with those with the TT allele exhibiting lower levels of adiposity than those with the AA or AA + AT genotype. Therefore, possessing a homozygous TT gene may serve as a protection against weight gain, beginning at a young age.	Mier-Mota et al. [24]

Abbreviations; SNPs, single nucleotide polymorphism; BMI, body mass index; APOE4, apolipoprotein E4 genotype; PUFA, polyunsaturated fatty acids; ADRA2A alpha-2A adrenergic receptor; TCF7L2, transcription factor 7-like 2; PLIN1, perilipin 1; NUDT3, Nudix hydrolase 3; FABP2, fatty acid binding protein 2; UCP3, uncoupling protein 3; MTCH2, mitochondrion carrier 2; ABCB11, ATP-binding cassette, sub-family B member 11; LEP, leptin; NFE2L3, nuclear factor (erythroid 2)-like factor 3; NPC1, Niemann-Pick disease, type C1; AGTR2, angiotensin II receptor type 2; ALDH2, aldehyde dehydrogenase; MTHFR, methylenetetrahydrofolate reductase; CβS, cystathionine beta-synthase; MTRR, methionine synthase reductase; MTR, methionine synthase; LDL, low-density lipoprotein; PPARG2, peroxisome proliferator-activated receptor γ gene; 1L6-174G, interleukin-6 gene-174G; FABP2Alas54Thr, fatty acid binding protein 2 Ala54Thr; BCO1, beta-carotene oxygenase 1; FTO, fat mass and obesity-associated.

15.4.1 Vitamins and minerals

Minerals and vitamins are essential micronutrients involved in methylation, DNA repair, and synthesis, among many other biochemical activities. Genetic variations can greatly impact the effectiveness and metabolism of these micronutrients, influencing our risk of disease and its repercussions. Micronutrients such as folate (B_9), pyridoxine (B_6), cobalamin (B_{12}), and methionine are essential donors of methyl groups in one-carbon metabolism, and their availability can directly influence DNA methylation patterns through epigenetic interactions. In the context of the methylenetetrahydrofolate reductase (MTHFR) gene, which encodes an enzyme critical for folate metabolism, it has shown polymorphism, such as C677T [48]. The TT genotype has reduced enzyme activity, leading to higher homocysteine levels and an increased risk of CVDs and neural tube defects [49]. The consumption of folic acid and its beneficial gene–nutrient reaction enhancement effects are evident in the prevention and management of diseases such as CVDs [21, 47], cancer [38, 67], and hypertension [20, 46]. The ability of dietary folate to modulate these risks highlights the importance of folate and epigenetic mechanisms in personalized nutritional strategies.

Another important nutrient, vitamin D, influences immune function and bone health through interactions with genetic variations in the vitamin D receptor (VDR) gene. Polymorphisms in the VDR gene, particularly SNPs, have been associated with autoimmune diseases and bone health. The four common VDR gene SNPs extensively studied in relation to autoimmune disorders include FokI, BsmI, ApaI, and TaqI [77]. These genetic variants can influence vitamin D metabolism, potentially affecting disease risk and clinical features. Therefore, higher vitamin D intake or supplementation becomes crucial for carriers of VDR gene variants to achieve optimal bone health and immune function. Similarly, SNP variants in the SCAR-B1 gene influence vitamin E status metabolism and transport [12].

Iron, a vital mineral for oxygen transport and energy production, is also influenced by genetic factors. Hereditary hemochromatosis, primarily caused by mutations in the HFE gene, particularly C282Y, leads to excessive iron absorption and accumulation, leading to organ damage [78]. The characteristic expression of HFE-related conditions is greatly influenced by food choices. Specific diet components through epigenetic modifications directly impact iron digestion and absorption [78]. Hence, to manage iron overload in hemochromatosis patients, dietary recommendations that limit iron intake are essential for preventing disease progression and complications. Also, adopting a diverse vegetarian or semi-vegetarian diet, consuming fruits and whole grain products, and limiting alcohol intake may help lower iron absorption and lessen the need for phlebotomies in hemochromatosis management [79]. These micronutrients dietary adjustments emphasize the significance of individualized nutrition based on a genetic profile.

15.4.2 Macronutrients

Macronutrients such as carbohydrates, proteins, and fats significantly influence gene expression and metabolism, with genetic factors playing a crucial role in individual responses to these dietary components. A better comprehension of the diet–gene interactions and responses is essential for preventing and managing metabolic disorders like obesity, diabetes, and CVDs. Genetic variations in lipid regulation, carbohydrate metabolism, and energy homeostasis genes are strongly associated with obesity risk and the related metabolic syndromes [23, 25]. Specific nutrients modulate gene expression through various mechanisms, including ChREBP for sugar signaling, PPARs for fat response, and GCN2/ATF4 and mTORC1 pathways for amino acid sensing [29].

Epigenetic modifications, attributed to diet, can affect disease susceptibility and may have transgenerational effects [17]. Notably, diet–gene and epigenetic relationships modulate pathways, affecting gene expression and defects in genes such as (1) the fat mass and obesity-associated (FTO) gene; (2) lactase (LCT) gene for protein metabolism (lactase); (3) the transcription factor 7-like 2 (TCF7L2) gene with risk alleles susceptible to T2DM; (4) the FADS1 (fatty acid desaturase 1) with lipid metabolism disorder in risk alleles.

Targeted dietary strategies that focus on the intake or avoidance of certain foods and lifestyle behaviors are essential for preventing and managing disorders associated with the consumption of certain macronutrients. Carriers of TCF7L2 and FTO risk alleles are required to consume whole grains, fiber, and low glycemic index foods. Also, it is paramount for these carriers to reduce fat intake, while increasing physical activity. Recent studies have shown that a lower-fat diet can mitigate the obesity risk associated with the FTO gene [24, 25, 51]. Similarly, people with certain variations in the FADS1 gene may need to consume more preformed long-chain omega-3 fatty acids, such as DHA, found in fish oil, to maintain optimum cardiovascular and cognitive health. In addition, it is crucial to make dietary adjustments, such as consuming only lactose-free items or using lactase supplements, to effectively manage symptoms of lactose intolerance.

15.4.3 Phytochemicals and antioxidants

Phytochemicals and antioxidants derived from plants play crucial roles in modulating oxidative stress, inflammation, and gene expression related to various diseases. Phytochemicals and antioxidants are bioactive compounds that can regulate transcription factors and act as epigenetic modulators, influencing histone modifications, DNA methylation, and RNA-based mechanisms [1, 3]. Genetic variations may affect the metabolism and effectiveness of bioactive substances, impairing their ability to prevent or manage chronic diseases.

Phytochemicals like curcumin, sulforaphane, and resveratrol have been identified as epigenetic modifiers of the Nrf2 pathway, a major transcription factor regulating over 2000 genes involved in redox balance and metabolism [34, 80]. The glutathione S-transferase (GST) family of genes, which includes GSTM1, GSTT1, and GSTP1, encodes enzymes that play a critical role in detoxifying reactive oxygen species and carcinogens [57]. Mutations in these genes, especially GSTM1 and GSTT1's null variants, increase the risk of cancer, especially colon and lung malignancies, by decreasing enzyme function and increasing oxidative stress. Higher consumption of cruciferous vegetables rich in isothiocyanates reduces the risk of colorectal and gastric cancers due to their induction of detoxification enzyme expression [57].

In addition, carotenoids, including β-carotene, lutein, and zeaxanthin, are important antioxidants that contribute to eye health by protecting against oxidative stress, inflammation, and light-induced damage [81]. The β-carotene 15,15′-monooxygenase 1 (BCMO1) gene, involved in β-carotene metabolism, has polymorphisms that affect β-carotene carotenoid conversion efficiency to vitamin A. The BCMO1 gene variant rs6564851 carriers with G allele exhibit elevated plasma carotenoid concentrations, while their retinol levels were comparable to those carrying the T allele [76]. Hence, defective BCMO1 gene carriers may need to consume a higher amount of vitamin A from animal sources to ensure sufficient levels of vitamin A. This may prevent diseases associated with vitamin A deficiency, such as night blindness and immune dysfunction.

Furthermore, polyphenols are natural plant-derived phenols that exhibit various gene–nutrient interactions. Polyphenols found in fruits, vegetables, and plants have been extensively studied for their potential health benefits. These compounds can modulate gene expression, influence epigenetic mechanisms, and interact with microRNAs, potentially reducing the risk of chronic diseases such as T2DM [53, 56], and age-associated diseases [82]. Polyphenols found in juices, wines, coffee, and cocoa can activate the Nrf2 pathway, which plays a crucial role in enhancing antioxidant defenses, preventing oxidative damage [35]. Specific polyphenols like resveratrol, quercetin, and epigallocatechin gallate (EGCG) have shown promise in activating sirtuin (SIRT1), a key regulator of cellular processes in the kidney and other organs [83]. Personalized nutritional strategies that prioritize foods high in polyphenols may improve health outcomes for persons with certain SIRT gene variations, possibly decreasing their risk of diseases associated with age. Polyphenols such as quercetin, curcumin, anthocyanins, and tea polyphenols also modulate the Nrf2 pathway, offering potential treatments for diabetes and neurodegenerative diseases [19, 34]. Overall, polyphenols generally demonstrate protective effects against chronic degenerative diseases due to their antioxidant, anticarcinogenic, and anti-inflammatory properties. However, some polyphenols induce DNA damage with deleterious effects [84]. These findings highlight the complex role of dietary polyphenols in activating Nrf2-mediated antioxidant responses in human health and their potential for personalized dietary interventions. Nevertheless, additional studies are required to gain insight into the structure and functional correlation of polyphenol molecules to better comprehend the mechanisms behind their physiological effects. This will help determine whether dietary polyphenols can be used as therapeutic agents.

15.5 Personalized nutrition

The emergence of personalized nutrition signifies an important evolution in the way we approach disease prevention and management. It enables us to customize dietary recommendations based on an individual's unique genetic composition. Personalized nutrition involves customized nutritional recommendations designed to enhance health, maintain well-being, and prevent diseases [85]. The principal objective of precision nutrition is to identify personalized nutritional recommendations that are adapted to an individual's biological needs and predicted response to dietary changes using their genetic, metabolomic, and microbiome profiles. This approach acknowledges the variability in gene–nutrient interactions, where genetic differences can significantly impact nutrient metabolism and disease susceptibility. Hence, personalized nutrition leverages on genetic information to develop customized diet plans aimed at optimizing health and well-being.

15.5.1 Personalized diet plans

At the core of gene–nutrient interaction strategies are customized diet programs, which provide a targeted nutrition strategy based on an individual's genetic profile. The development of these plans involves the integration of genomic data with dietary recommendations, resulting in the harmonization of nutrient intake with genetic predispositions. This approach aims to optimize health and mitigate the risk of disease. The design of these personalized plans begins with the identification of significant genetic variations that impact nutrient metabolism and the probability of developing diseases. As an illustration, genetic testing has the potential to uncover variations in the FTO gene [86]. This FTO gene has been linked to obesity and can impact the effectiveness of various diet plans. Some individuals may experience different responses to low-fat diets, depending on their FTO variants, while others may find carbohydrate restriction more beneficial. A practical example is the case study highlighted below.

Case study

A study carried out by Mathias et al. investigated the inter-individual responses of Brazilian children and teenagers (between 9 and 13 years, 11 months, and 29 days old) to a short-term intervention involving a multi-micronutrient supplement [87]. Mathias et al. considered the fact that each person had their own unique genetic and environmental characteristics. They used a thorough technique, where they assessed the aggregated data from every participant and their individual responses, utilizing a variation of the n-of-1 trial design [87]. The *n*-of-1 design was set up to allow each participant to serve as their own control, considering their unique inter-individual variation. Mathias et al. evaluated the efficacy of a dietary intervention by comparing the changes in -omics and clinical variables measured at the beginning, 6 weeks after the intervention, and then after a 6-week washout period (with no intervention). The researchers used an elastic net regression model to assess whether several factors, including baseline clinical biochemistry, blood vitamin levels, and food intakes, could account for the variability in response to the intervention at each clinical endpoint [87]. The findings indicated that multi-micronutrients mediated the physiology of the systems associated with metabolic health. This result was determined by analyzing the effects of a 6-week intervention on total cholesterol, LDL cholesterol, mean corpuscular volume, and circulating levels of nine vitamin metabolites for 2 consecutive years [87].

This case study showed that integrating large-scale clinical data can effectively consider individual differences in response to a specific diet. This approach has the potential to shift away from the conventional method of providing general dietary recommendations. Nevertheless, the complex nature of gene–diet interactions, including concepts such as penetrance and epigenetics, hinders the straightforward use of genetic information for dietary guidance [86]. Also, the challenge of defining the most effective dosage is not just an experimental problem but also a translational one. As the area of genetic information in dietary recommendations advances, it is crucial to consider the ethical issues and practical applications associated with it. To address this, ongoing studies are evaluating the effectiveness of gene-based nutrition and lifestyle recommendations for weight management, such as the MyGeneMyDiet® study, which focuses on specific genetic variants related to weight, calorie intake, and dietary fat consumption [88].

15.5.2 Major barriers to the implementation of personalized nutrition

The complex relationship between diet, metabolism, and the microbiota in the host's cells and molecules, as well as its impact on individual health outcomes at individual levels, has not been fully explored. To fully appreciate health, it is necessary to assess how nutrition, genes, biological products, health, lifestyle patterns, gut microbiome, and the environment all interact with one another rather than only examining how nutrients or macromolecules interact with individual genes or gene products. Although personalized nutrition shows great promise, some barriers and challenges prevent its general implementation.

One of the primary obstacles is the cost associated with genetic testing and the subsequent development of customized nutritional plans. The cost of genetic testing and personalized nutrition remains a significant barrier, particularly in low- and middle-income countries [89]. While direct-to-consumer genetic testing services are growing due to decreased sequencing costs [90], the use of DNA sequencing in primary care for noncommunicable diseases is still limited by cost. The cost is further compounded due to the need for infrastructural development and specialized training of healthcare experts, such as geneticists and dietitians, to analyze genetic information and provide suitable nutritional recommendations [86]. Accessibility is a significant problem. Genetic testing and personalized dietary programs are often only available in large cities and to those with higher incomes, resulting in differences in health results.

Based on technicalities, the lack of data standardization guidelines for the implementation of personalized nutrition [90], data sparsity for comparison, and the need for improved methods further complicates its adoption. Also, the lack of expertise among healthcare professionals regarding the interpretation of genetic data, along with the variations in the formulation of dietary recommendations among practitioners, results in inconsistencies.

Furthermore, ethical considerations present serious issues in the adoption of personalized nutrition. Concerns about privacy and the potential misuse of genetic information, such as genetic discrimination by employers or insurers, have been raised. Hence, these issues relating to data privacy and genetic discrimination must be addressed [86]. Additionally, there is a need for more research to validate the efficacy of personalized nutrition in diverse populations. Most personalized nutrition studies to date have been conducted in specific ethnic groups, and it remains unclear whether the findings apply to other populations with different genetic backgrounds.

To establish credibility and efficacy in personalized nutrition approaches, guiding principles that include using validated diagnostic methods, maintaining data quality, and providing rigorous scientific evidence have been proposed [91]. Additional studies are needed to validate the efficacy of personalized nutrition across different populations. Improved standardized dietary intake measurements are crucial for result comparisons. To make personalized nutrition accessible, affordable, and effective for everyone, researchers, healthcare professionals, lawmakers, and the community need to collaborate to overcome these challenges.

15.6 Recent advances and emerging trends in gene–nutrient interactions studies

Recent years have witnessed remarkable advancements in our understanding of gene–nutrient interactions, thanks to emerging areas in the field and the incorporation of state-of-the-art omics tools. These advancements are leading to a better knowledge of the complex links between genes and nutrients, which allows for public health and nutritional strategies that are more precise and individualized.

One of the most notable trends in gene–nutrient research is the shift toward studying polygenic interactions rather than focusing solely on single-gene effects. Historically, research in this field has often centered on how variations in specific genes, such as those involved in lipid metabolism or vitamin absorption, influence individual responses to dietary components. However, recent studies highlight the importance of considering the cumulative effects of multiple genetic variants, each contributing a small impact to the overall response to nutrients. This polygenic approach provides a more comprehensive understanding of how complex traits, such as obesity or CVD, are influenced by diet in the context of an individual's genetic makeup.

Recent research in nutrigenetics highlights the shift toward studying polygenic interactions in gene–nutrient relationships. Historically, earlier studies have focused on single-gene effects on dietary responses [92]. However, recent studies highlight the importance of considering the cumulative effects of multiple genetic variants, each contributing a small effect to the overall response to nutrients. This polygenic perspective provides a more comprehensive understanding of how complex traits such as CVDs and obesity influence diet in the context of an individual's genetic makeup [86]. Studies have identified genetic variations affecting lipid regulation, carbohydrate metabolism, and energy homeostasis as key factors in obesity risk and metabolic syndromes [25, 92]. The interaction between genes and diet also influences the effectiveness of nutritional interventions. Personalized nutrition strategies based on genetic information are emerging, with potential applications in managing conditions such as dyslipidaemias, through tailored dietary recommendations [26].

Recognizing that environmental variables, including lifestyle, physical activity, and exposure to pollutants, might influence the effect of gene–nutrient interactions; another developing area is the research of gene–environment interactions. Genetic polymorphisms can affect detoxification and antioxidant functions, potentially leading to various health risks, when exposed to harmful environmental factors [93]. This broader perspective acknowledges that genetic predispositions alone do not dictate health outcomes; instead, they interact with various environmental factors to influence disease risk. For instance, studies on South Asian populations indicate that genetic predisposition may interact with obesogenic environments to influence cardiometabolic disease prevalence. This highlights the need to do more precise evaluations of lifestyle variables and include larger sets of genetic variations in future research [94].

Furthermore, there is a growing interest in the role of the gut microbiome in gene–nutrient interactions. Recent research highlights the complex interplay between diet, gut microbiome, and host metabolism, emphasizing its potential for personalized nutrition strategies. The gut microbiome, composed of diverse microorganisms, plays a crucial role in nutrient metabolism and interacts with the host genome to influence health outcomes. It is important to acknowledge that the composition of the gut microbiome can vary from person to person, which may have an impact on how individuals respond to dietary interventions [95]. This suggests that personalized nutrition approaches should consider both the individual's genetics and the composition of their gut microbiome. Researchers are employing advanced techniques such as multiple omics and modelling approaches to gain a deeper understanding of the complex link between diet, gut microbiome, and host metabolism [96]. This area of research is rapidly evolving, with potential implications for personalized nutrition strategies that consider both host genetics and microbiome composition.

15.6.1 Integration of -omics technologies in personalized nutrition

The integration of omics technologies has revolutionized research in personalized nutrition, offering a more holistic and systems-level understanding of molecular mechanisms underlying gene–nutrient interactions [97]. The current techniques employed to investigate an inter-individual response to dietary components involve utilizing omics technologies, such as genomics, metabolomics, and proteomics, which are integrated with systems biology programs. These technologies allow researchers to capture a comprehensive snapshot of biological processes at multiple levels, providing insights into individual responses to dietary interventions and disease mechanisms.

The current goal of precision nutrition is to identify personalized nutritional recommendations that are adapted to an individual's biological needs and predicted response to dietary changes using their genetic, metabolomic, and microbiome profiles. By identifying genetic variations that modulate nutritional metabolism and disease vulnerability, genomics provides the foundation for individualized nutrition. Once the variant genes are identified, transcriptomics reveals how dietary components can regulate the activity of specific genes through gene expression patterns. Bioactive diet components regulate gene expression through various mechanisms, including chromatin structure changes and activation of transcription factors [1]. For instance, using transcriptomics, polyphenols abundant in fruits and vegetables were shown to modulate gene expression to exert protective effects against chronic diseases. In addition, proteomic approaches offer the potential to uncover novel mechanisms of disease development, identify functional biomarkers, and inform targeted interventions [98]. Metabolomics is a branch of science that focuses on analyzing small molecules or metabolites. It directly assesses the biochemical processes occurring in cells and tissues, revealing how the metabolism reacts to dietary components. In diabetes research, metabolomics has identified novel biomarkers for type 2 diabetes and its complications, such as branched-chain amino acids and metabolites involved in energy and lipid metabolism [99]. The integration of metabolomics with other omics data provides a comprehensive view of metabolic networks, facilitating inferences and our understanding of biological processes [100]. Additionally, the integration of emerging technologies such as CRISPR-based gene editing, artificial intelligence, and machine learning in the analysis of gene–nutrient interactions holds promise for accelerating the discovery of novel dietary recommendations tailored to individual genetic profiles. Hence, by combining data from multi-omics areas, individual-specific responses to diet will be better comprehended, paving the way for personalized nutrition strategies that are tailored not only to genetic profiles but also to the dynamic interactions between genes, proteins, and metabolites.

15.6.2 Strategies for improving gene–nutrient interactions for public health

Recent advances in nutrigenomics and gene–nutrient interactions offer promising strategies for improving public health. To fully realize the potential of gene–nutrient interactions in public health, several strategies must be implemented to enable its adoption into practice. Establishing diverse biobanks and integrating research findings into policies are crucial for developing precision nutrition approaches [101]. This is required to collect genetic, dietary, and health data from populations across different ethnicities, ages, and socioeconomic backgrounds, fostering equity and accessibility. This approach will ensure that gene–nutrient research findings are generalizable and applicable to diverse populations, reducing result disparities [10]. Implementation of nutrition research into health policies is another essential strategy needed to bridge the gap between evidence-based interventions and their adoption in clinical and community settings [102]. Effective implementation requires collaboration between researchers, policymakers, and healthcare providers to translate scientific findings into practical recommendations that can be adapted at the population level. While it has been established that the integration of omics technologies in public health genomics can enhance disease prevention and health assessment, the issue of affordability and specialized training remains a challenge that is weakening its implementation. Education and training of healthcare professionals are crucial for the successful application of gene–nutrient strategies, including understanding ethical and legal implications [101, 103]. Due to the financial implications involved with nutritional genomics, efforts should be made to reduce the cost of genetic testing and personalized nutrition services, particularly in low- and middle-income countries [89, 103]. Overall, recent developments in the study of gene–nutrient interactions, spurred by shifts in research interests and the use of omics technology, have the potential to revolutionize public health strategies. However, to fully harness the potential of gene–nutrient research, it is crucial to make concerted efforts to improve the field. This includes translating findings into practical applications and addressing various challenges such as standardization, accessibility, equity, infrastructure development, and training.

15.7 Conclusion and future directions

The exploration of gene–nutrient interactions presents a transformative approach to understanding and enhancing health indices, reflecting the relationship between our genetic makeup and dietary intakes. As discussed in this chapter, the interplay between genes and nutrients offers a compelling framework for disease prevention, personalized nutrition, and the overall enhancement of health and well-being. The insights gleaned from nutrigenetics, nutrigenomics, and epigenetics mechanisms and their roles in cardiometabolic and chronic disease underscore the potential of these interactions to revolutionize healthcare by moving beyond a one-size-fits-all model toward more individualized and precise interventions. However, the realization of the full potential of gene–nutrient interactions is not without challenges that weaken its adoption into public health systems. The barriers to implementing personalized nutrition, including affordability, accessibility issues, standardization, training, and the issue of clear ethical guidelines, need to be addressed.

Looking forward, to make accurate, personalized nutrition recommendations and accelerate the goal of better health and well-being, the integration of omics technologies and emerging computations such as CRISPR-based gene editing, machine learning, and artificial intelligence into gene–nutrient studies is crucial. These technologies will be instrumental in refining our approaches to personalized nutrition and the development of public health strategies that are both effective and equitable. Also, future gene–nutrient interaction studies should look to investigate the influence of environmental variables, including lifestyle, physical activity, and exposure to pollutants, on personalized nutrition. Additionally, the gut microbiome, composed of diverse microorganisms, plays a crucial role in nutrient metabolism and interacts with the host genome in influencing health and disease development. There is a critical gap in the effects of an individual's microbiome on personalized nutrition and gene–nutrient reaction studies that need to be explored. Since microbiome composition varies from person to person, the integration of microbiome data may deepen our knowledge of individual responses to disease interventions. Furthermore, integrating these insights into public health policies and clinical guidelines will be crucial in realizing the full benefits of gene–nutrient interactions on a population level. Collaborative efforts across disciplines, establishing robust ethical frameworks, and promoting accessibility will be key to overcoming the current challenges and ensuring that all realize the benefits of personalized nutrition.

References

[1] Mierziak, J., Kostyn, K., Boba, A., Czemplik, M., Kulma, A., & Wojtasik, W. Influence of the bioactive diet components on the gene expression regulation. Nutrients 2021, 13(11), 3673.

[2] Park, J.-H., Kim, S.-H., Lee, M. S., & Kim, M.-S. Epigenetic modification by dietary factors: Implications in metabolic syndrome. Molecular Aspects of Medicine 2017, 54, 58–70.

[3] Pehlivan, F. E. Diet-Epigenome interactions: Epi-drugs modulating the epigenetic machinery during cancer prevention. Epigenetics to Optogenetics: A New Paradigm in the Study of Biology 2021 IntechOpen, 4, 18.

[4] Subbiah, M. R. Nutrigenetics and nutraceuticals: The next wave riding on personalized medicine. Translational Research 2007, 149(2), 55–61.

[5] Tan, P. Y., Moore, J. B., Bai, L., Tang, G., & Gong, Y. Y. In the context of the triple burden of malnutrition: A systematic review of gene-diet interactions and nutritional status. Critical Reviews in Food Science and Nutrition 2024, 64(11), 3235–3263.

[6] Borsoi, F. T., Neri-Numa, I. A., de Oliveira, W. Q., de Araújo, F. F., & Pastore, G. M. Dietary polyphenols and their relationship to the modulation of non-communicable chronic diseases and epigenetic mechanisms: A mini-review. Food Chemistry: Molecular Sciences 2023, 6, 100155.

[7] Perez-Beltran, Y. E., Rivera-Iniguez, I., Gonzalez-Becerra, K., Perez-Naitoh, N., Tovar, J., Sayago-Ayerdi, S. G., et al. Personalized dietary recommendations based on lipid-related genetic variants: A systematic review. Frontiers in Nutrition 2022, 9, 830283.

[8] Vrânceanu, M., Galimberti, D., Banc, R., Dragoş, O., Cozma-Petruţ, A., Hegheş, S.-C., et al. The anticancer potential of plant-derived nutraceuticals via the modulation of gene expression. Plants 2022, 11(19), 2524.

[9] Seldin, M., Yang, X., & Lusis, A. J. Systems genetics applications in metabolism research. Nature Metabolism 2019, 1(11), 1038–1050.

[10] Fenech, M., El-Sohemy, A., Cahill, L., Ferguson, L. R., French, T.-A. C., Tai, E. S., et al. Nutrigenetics and nutrigenomics: Viewpoints on the current status and applications in nutrition research and practice. Lifestyle Genomics 2011, 4(2), 69–89.

[11] Dhafer, A.-K., & Shaden, M. H. M. Genetic polymorphisms. In: Mahmut, Ç., Osman, E., & Gül Cevahir, Ö., editors. The Recent Topics in Genetic Polymorphisms, Rijeka: IntechOpen, 2019, p. Ch. 1.

[12] Niforou, A., Konstantinidou, V., & Naska, A. Genetic variants shaping inter-individual differences in response to dietary intakes – A narrative review of the case of vitamins. Frontiers in Nutrition 2020, 7, 558598.

[13] Krasniqi, E., Boshnjaku, A., Wagner, K.-H., & Wessner, B. Association between polymorphisms in vitamin D pathway-related genes, vitamin D status, muscle mass and function: A systematic review. Nutrients 2021, 13(9), 3109.

[14] Borel, P., & Desmarchelier, C. Bioavailability of fat-soluble vitamins and phytochemicals in humans: Effects of genetic variation. Annual Review of Nutrition 2018, 38(1), 69–96.

[15] Laing, B. B., Lim, A. G., & Ferguson, L. R. A personalised dietary approach – A way forward to manage nutrient deficiency, effects of the western diet, and food intolerances in inflammatory bowel disease. Nutrients 2019, 11(7), 1532.

[16] Meroni, M., Longo, M., Rustichelli, A., & Dongiovanni, P. Nutrition and genetics in NAFLD: The perfect binomium. International Journal of Molecular Sciences 2020, 21(8), 2986.

[17] Franzago, M., Santurbano, D., Vitacolonna, E., & Stuppia, L. Genes and diet in the prevention of chronic diseases in future generations. International Journal of Molecular Sciences 2020, 21(7), 2633.

[18] Balakrishnan, J., Husain, M. A., Vachon, A., Chouinard-Watkins, R., Léveillé, P., & Plourde, M. Omega-3 supplementation increases omega-3 fatty acids in lipid compartments that can be taken up by the

brain independent of APOE genotype status: A secondary analysis from a randomised controlled trial 1. Nutrition and Healthy Aging 2022, 7(3–4), 147–158.

[19] Felisbino, K., Granzotti, J. G., Bello-Santos, L., & Guiloski, I. C. Nutrigenomics in regulating the expression of genes related to type 2 diabetes mellitus. Frontiers in Physiology 2021, 12, 699220.

[20] Fu, L., Li, Y.N., Luo, D., Deng, S., Wu ,B., & Hu, Y.Q. Evidence on the causal link between homocysteine and hypertension from a meta-analysis of 40,173 individuals implementing Mendelian randomization. Journal of Clinical Hypertension. 2019, 21(12), 1879–1894.

[21] Levy, J., Rodriguez-Guéant, R.-M., Oussalah, A., Jeannesson, E., Wahl, D., Ziuly, S., et al. Cardiovascular manifestations of intermediate and major hyperhomocysteinemia due to vitamin B12 and folate deficiency and/or inherited disorders of one-carbon metabolism: A 3.5-year retrospective cross-sectional study of consecutive patients. The American Journal of Clinical Nutrition 2021, 113(5), 1157–1167.

[22] Ramos-Lopez, O., Riezu-Boj, J. I., Milagro, F. I., Cuervo, M., Goni, L., & Martinez, J. A. Interplay of an obesity-based genetic risk score with dietary and endocrine factors on insulin resistance. Nutrients 2019, 12(1), 33.

[23] Evans, C., Curtis, J., & Antonio, J. FTO and Anthropometrics: The role of modifiable factors. Journal of Functional Morphology and Kinesiology 2022, 7(4), 90.

[24] Mier-Mota, J., Ponce-González, J. G., Perez-Bey, A., Cabanas-Sánchez, V., Veiga-Núñez, O., Santiago-Dorrego, C., et al. Longitudinal effects of FTO gene polymorphism on body composition, cardiorespiratory fitness, physical activity, inflammatory markers, and cardiovascular risk in children and adolescents. "The UP & DOWN study". Scandinavian Journal of Medicine & Science in Sports 2023, 33(11), 2261–2272.

[25] Madrigal-Juarez, A., Martínez-López, E., Sanchez-Murguia, T., Magaña-de la Vega, L., Rodriguez-Echevarria, R., Sepulveda-Villegas, M., et al. FTO genotypes (rs9939609 T> A) are associated with increased added sugar intake in healthy young adults. Lifestyle Genomics 2023, 16(1), 214–223.

[26] Pérez-Beltrán, Y. E., González-Becerra, K., Rivera-Iñiguez, I., Martínez-López, E., Ramos-Lopez, O., Alcaraz-Mejía, M., et al. A Nutrigenetic strategy for reducing blood lipids and low-grade inflammation in adults with obesity and overweight. Nutrients 2023, 15(20), 4324.

[27] Keathley, J., Garneau, V., Marcil, V., Mutch, D. M., Robitaille, J., Rudkowska, I., et al. Nutrigenetics, omega-3 and plasma lipids/lipoproteins/apolipoproteins with evidence evaluation using the GRADE approach: A systematic review. BMJ Open 2022, 12(2), e054417.

[28] Martín-Hernández, R., Reglero, G., Ordovás, J. M., & Dávalos, A. NutriGenomeDB: A nutrigenomics exploratory and analytical platform. Database 2019, 2019, baz097.

[29] Haro, D., Marrero, P. F., & Relat, J. Nutritional regulation of gene expression: Carbohydrate-, fat-and amino acid-dependent modulation of transcriptional activity. International Journal of Molecular Sciences 2019, 20(6), 1386.

[30] Durkin, L. A., Childs, C. E., & Calder, P. C. Omega-3 polyunsaturated fatty acids and the intestinal epithelium – A review. Foods 2021, 10(1), 199.

[31] Ghasemi Darestani, N., Bahrami, A., Mozafarian, M. R., Esmalian Afyouni, N., Akhavanfar, R., Abouali, R., et al. Association of polyunsaturated fatty acid intake on inflammatory gene expression and multiple sclerosis: A systematic review and meta-analysis. Nutrients 2022, 14(21), 4627.

[32] Bougarne, N., Weyers, B., Desmet, S. J., Deckers, J., Ray, D. W., Staels, B., et al. Molecular actions of PPAR α in lipid metabolism and inflammation. Endocrine Reviews 2018, 39(5), 760–802.

[33] Rundblad, A., Holven, K. B., Bruheim, I., Myhrstad, M. C., & Ulven, S. M. Effects of fish and krill oil on gene expression in peripheral blood mononuclear cells and circulating markers of inflammation: A randomised controlled trial. Journal of Nutritional Science 2018, 7, e10.

[34] Moratilla-Rivera, I., Sánchez, M., Valdés-González, J. A., & Gómez-Serranillos, M. P. Natural products as modulators of Nrf2 signaling pathway in neuroprotection. International Journal of Molecular Sciences 2023, 24(4), 3748.

[35] Qader, M., Xu, J., Yang, Y., Liu, Y., & Cao, S. Natural Nrf2 activators from juices, wines, coffee, and cocoa. Beverages 2020, 6(4), 68.

[36] Shankar, E., Kanwal, R., Candamo, M., & Gupta, S., editors. Dietary phytochemicals as epigenetic modifiers in cancer: Promise and challenges. In: Seminars in Cancer Biology, Elsevier, 2016, 40, 82–99.

[37] Anchimowicz, J., Wyżewski, Z., & Świtlik, W. Role of the glucosinolates in cancer epigenetics. Postepy Biochemii 2023, 69(2), 96–103.

[38] Xiao, S., Tang, Y.-S., Kusumanchi, P., Stabler, S. P., Zhang, Y., & Antony, A. C. Folate deficiency facilitates genomic integration of human papillomavirus type 16 DNA in vivo in a novel mouse model for rapid oncogenic transformation of human keratinocytes. The Journal of Nutrition 2018, 148(3), 389–400.

[39] Mandaviya, P. R., Joehanes, R., Brody, J., Castillo-Fernandez, J. E., Dekkers, K. F., Do, A. N., et al. Association of dietary folate and vitamin B-12 intake with genome-wide DNA methylation in blood: A large-scale epigenome-wide association analysis in 5841 individuals. The American Journal of Clinical Nutrition 2019, 110(2), 437–450.

[40] Kok, D. E., Steegenga, W. T., & McKay, J. A. Folate and Epigenetics: Why We Should Not Forget Bacterial Biosynthesis, Taylor & Francis, Epigenomics 2018, 1147–1150.

[41] Hossain, S., Liu, Z., & Wood, R. J. Histone deacetylase activity and vitamin D-dependent gene expressions in relation to sulforaphane in human breast cancer cells. Journal of Food Biochemistry 2020, 44(2), e13114.

[42] Martino, E., D'Onofrio, N., Balestrieri, A., Colloca, A., Anastasio, C., Sardu, C., et al. Dietary epigenetic modulators: Unravelling the still-controversial benefits of miRNAs in nutrition and disease. Nutrients 2024, 16(1), 160.

[43] Leskinen, H., Tringham, M., Karjalainen, H., Iso-Touru, T., Hietaranta-Luoma, H.-L., Marnila, P., et al. APOE genotypes, lipid profiles, and associated clinical markers in a Finnish population with cardiovascular disease risk factors. Lifestyle Genomics 2022, 15(2), 45–54.

[44] Lin, Y., Yang, Q., Liu, Z., Su, B., Xu, F., Li, Y., et al. Relationship between apolipoprotein E genotype and lipoprotein profile in patients with coronary heart disease. Molecules 2022, 27(4), 1377.

[45] Thahira, A., Rasitha, C., Rajan, A., Pinchulatha, K., Harisree, P., Deepthi, S., et al. Apolipoprotein E polymorphism and it's lifestyle impact. Journal of Advanced Zoology 2024, 45(1).

[46] Zhou, L., Huang, H., Wen, X., Chen, Y., Liao, J., Chen, F., et al. Associations of serum and red blood cell folate with all-cause and cardiovascular mortality among hypertensive patients with elevated homocysteine. Frontiers in Nutrition 2022, 9, 849561.

[47] Raghubeer, S., & Matsha, T. E. Methylenetetrahydrofolate (MTHFR), the one-carbon cycle, and cardiovascular risks. Nutrients 2021, 13(12), 4562.

[48] Zarembska, E., Ślusarczyk, K., & Wrzosek, M. The implication of a polymorphism in the methylenetetrahydrofolate reductase gene in homocysteine metabolism and related civilisation diseases. International Journal of Molecular Sciences 2023, 25(1), 193.

[49] Jin, H., Cheng, H., Chen, W., Sheng, X., Levy, M. A., Brown, M. J., et al. An evidence-based approach to globally assess the covariate-dependent effect of the MTHFR single nucleotide polymorphism rs1801133 on blood homocysteine: A systematic review and meta-analysis. The American Journal of Clinical Nutrition 2018, 107(5), 817–825.

[50] Novakovic, R., Geelen, A., Ristic-Medic, D., Nikolic, M., Souverein, O. W., McNulty, H., et al. Systematic review of observational studies with dose-response meta-analysis between folate intake and status biomarkers in adults and the elderly. Annals of Nutrition and Metabolism 2018, 73(1), 30–44.

[51] Melhorn, S. J., Askren, M. K., Chung, W. K., Kratz, M., Bosch, T. A., Tyagi, V., et al. FTO genotype impacts food intake and corticolimbic activation. The American Journal of Clinical Nutrition 2018, 107(2), 145–154.

[52] Daily, J. W., & Park, S. Association of plant-based and high-protein diets with a lower obesity risk defined by fat mass in middle-aged and elderly persons with a high genetic risk of obesity. Nutrients 2023, 15(4), 1063.
[53] Kang, G. G., Francis, N., Hill, R., Waters, D., Blanchard, C., & Santhakumar, A. B. Dietary polyphenols and gene expression in molecular pathways associated with type 2 diabetes mellitus: A review. International Journal of Molecular Sciences 2019, 21(1), 140.
[54] Jan, A., Jan, H., & Ullah, Z. Transcription factor 7-like 2 (TCF7L2): A culprit gene in Type 2 Diabetes Mellitus. Сахарный Диабет 2021, 24(4), 371–376.
[55] Guo, Y., Huang, Z., Sang, D., Gao, Q., & Li, Q. The role of nutrition in the prevention and intervention of type 2 diabetes. Frontiers in Bioengineering and Biotechnology 2020, 8, 575442.
[56] Da Porto, A., Cavarape, A., Colussi, G., Casarsa, V., Catena, C., & Sechi, L. A. Polyphenols rich diets and risk of type 2 diabetes. Nutrients 2021, 13(5), 1445.
[57] Johnson, I. T. Cruciferous vegetables and risk of cancers of the gastrointestinal tract. Molecular Nutrition & Food Research 2018, 62(18), 1701000.
[58] Khan, S., Awan, K. A., & Iqbal, M. J. Sulforaphane as a potential remedy against cancer: Comprehensive mechanistic review. Journal of Food Biochemistry 2022, 46(3), e13886.
[59] Franco, R., Navarro, G., & Martínez-Pinilla, E. Lessons on differential neuronal-death-vulnerability from familial cases of Parkinson's and Alzheimer's diseases. International Journal of Molecular Sciences 2019, 20(13), 3297.
[60] WHO. Dementia. 2024.
[61] Corder, E., Saunders, A. M., Risch, N., Strittmatter, W., Schmechel, D., Gaskell, J. P., et al. Protective effect of apolipoprotein E type 2 allele for late onset Alzheimer disease. Nature Genetics 1994, 7(2), 180–184.
[62] Norwitz, N. G., Saif, N., Ariza, I. E., & Isaacson, R. S. Precision nutrition for Alzheimer's prevention in ApoE4 carriers. Nutrients 2021, 13(4), 1362.
[63] Huang, T. L., Zandi, P., Tucker, K., Fitzpatrick, A., Kuller, L., Fried, L., et al. Benefits of fatty fish on dementia risk are stronger for those without APOE ε4. Neurology 2005, 65(9), 1409–1414.
[64] Plourde, M., Vohl, M.-C., Vandal, M., Couture, P., Lemieux, S., & Cunnane, S. C. Plasma n-3 fatty acid response to an n-3 fatty acid supplement is modulated by apoE ε4 but not by the common PPAR-α L162V polymorphism in men. British Journal of Nutrition 2009, 102(8), 1121–1124.
[65] Solomon, V., Smirnova, Y., Harrington, M. G., Yassine, H. N., & Fonteh, A. N. Dietary supplementation results in a significant incorporation of DHA into RBC phosphatidylcholine of non-APOE ε4 allele but not for ε4 carriers: Human neuropathology/etiopathogenesis: Links to brain disease. Alzheimer's & Dementia 2020, 16, e038354.
[66] Martinsen, A., Saleh, R. N., Chouinard-Watkins, R., Bazinet, R., Harden, G., Dick, J., et al. The influence of APOE genotype, DHA, and flavanol intervention on brain DHA and lipidomics profile in aged transgenic mice. Nutrients 2023, 15(9), 2032.
[67] Holton, K. F. Micronutrients may be a unique weapon against the neurotoxic triad of excitotoxicity, oxidative stress and neuroinflammation: A perspective. Frontiers in Neuroscience 2021, 15, 726457.
[68] Martorana, F., Foti, M., Virtuoso, A., Gaglio, D., Aprea, F., Latronico, T., et al. Differential modulation of NF-κB in Neurons and Astrocytes Underlies Neuroprotection and Antigliosis activity of natural antioxidant molecules. Oxidative Medicine and Cellular Longevity 2019, 2019(1), 8056904.
[69] Park, H.-A., & Ellis, A. C. Dietary antioxidants and Parkinson's disease. Antioxidants 2020, 9(7), 570.
[70] Tao, Y., Huang, X., Xie, Y., Zhou, X., He, X., Tang, S., et al. Genome-wide association and gene-environment interaction study identifies variants in ALDH2 associated with serum ferritin in a Chinese population. Gene 2019, 685, 196–201.
[71] Arias, L. D., Parra, B. E., Muñoz, A. M., Cárdenas, D. L., Duque, T. G., & Manjarrés, L. M. Study exploring the effects of daily supplementation with 400 μg of folic acid on the nutritional status of folate in women of reproductive age. Birth Defects Research 2017, 109(8), 564–573.

[72] Pokushalov, E., Ponomarenko, A., Bayramova, S., Garcia, C., Pak, I., Shrainer, E., et al. Effect of Methylfolate, Pyridoxal-5′-Phosphate, and Methylcobalamin (SolowaysTM) supplementation on homocysteine and low-density lipoprotein cholesterol levels in patients with methylenetetrahydrofolate reductase, methionine synthase, and methionine synthase reductase polymorphisms: A randomized controlled trial. Nutrients 2024, 16(11), 1550.
[73] Rodrigues, A. P. S., Rosa, L. P. S., & Silveira, E. A. PPARG2 Pro12Ala polymorphism influences body composition changes in severely obese patients consuming extra virgin olive oil: A randomized clinical trial. Nutrition & Metabolism 2018, 15, 1–13.
[74] Martinez-Lopez, E., Garcia-Garcia, M. R., Gonzalez-Avalos, J. M., Maldonado-Gonzalez, M., Ruiz-Madrigal, B., Vizmanos, B., et al. Effect of Ala54Thr polymorphism of FABP2 on anthropometric and biochemical variables in response to a moderate-fat diet. Nutrition 2013, 29(1), 46–51.
[75] Pichler, M., Kollerits, B., Heid, I. M., Hunt, S. C., Adams, T. D., Hopkins, P. N., et al. Association of the melanocortin-4 receptor V103I polymorphism with dietary intake in severely obese persons. The American Journal of Clinical Nutrition 2008, 88(3), 797–800.
[76] Graßmann, S., Pivovarova-Ramich, O., Henze, A., Raila, J., Ampem Amoako, Y., King Nyamekye, R., et al. SNP rs6564851 in the BCO1 gene is associated with varying provitamin a plasma concentrations but not with retinol concentrations among adolescents from rural Ghana. Nutrients 2020, 12(6), 1786.
[77] Agliardi, C., Guerini, F. R., Bolognesi, E., Zanzottera, M., & Clerici, M. VDR gene single nucleotide polymorphisms and autoimmunity: A narrative review. Biology 2023, 12(7), 916.
[78] Rametta, R., Meroni, M., & Dongiovanni, P. From environment to genome and back: A lesson from HFE mutations. International Journal of Molecular Sciences 2020, 21(10), 3505.
[79] Milman, N. T. Managing genetic hemochromatosis: An overview of dietary measures, which may reduce intestinal iron absorption in persons with iron overload. Gastroenterology Research 2021, 14(2), 66.
[80] Bhattacharjee, S., & Dashwood, R. H. Epigenetic regulation of NRF2/KEAP1 by phytochemicals. Antioxidants 2020, 9(9), 865.
[81] Johra, F. T., Bepari, A. K., Bristy, A. T., & Reza, H. M. A mechanistic review of β-carotene, lutein, and zeaxanthin in eye health and disease. Antioxidants 2020, 9(11), 1046.
[82] Arora, I., Sharma, M., Sun, L. Y., & Tollefsbol, T. O. The epigenetic link between polyphenols, aging and age-related diseases. Genes 2020, 11(9), 1094.
[83] Tovar-Palacio, C., Noriega, L. G., & Mercado, A. Potential of polyphenols to restore SIRT1 and NAD+ metabolism in renal disease. Nutrients 2022, 14(3), 653.
[84] Qu, G., Chen, J., & Guo, X. The beneficial and deleterious role of dietary polyphenols on chronic degenerative diseases by regulating gene expression. Bioscience Trends 2018, 12(6), 526–536.
[85] Betts, J., & Gonzalez, J. Personalised Nutrition: What Makes You so Special? Nutrition Bulletin Wiley 2016, 41, 4.
[86] Mullins, V. A., Bresette, W., Johnstone, L., Hallmark, B., & Chilton, F. H. Genomics in personalized nutrition: Can you "eat for your genes"? Nutrients 2020, 12(10), 3118.
[87] Mathias, M. G., Coelho-Landell, C. D. A., Scott-Boyer, M. P., Lacroix, S., Morine, M. J., Salomao, R. G., et al. Clinical and vitamin response to a short-term multi-micronutrient intervention in Brazilian children and teens: From population data to interindividual responses. Molecular Nutrition & Food Research 2018, 62(6), 1700613.
[88] Nacis, J. S., Labrador, J. P. H., Ronquillo, D. G. D., Rodriguez, M. P., Dablo, A. M. F. D., Frane, R. D., et al. A study protocol for a pilot randomized controlled trial to evaluate the effectiveness of a gene-based nutrition and lifestyle recommendation for weight management among adults: The MyGeneMyDiet® study. Frontiers in Nutrition 2023, 10, 1238234.
[89] González-Robledo, L. M., Serván-Mori, E., Casas-López, A., Flores-Hernández, S., Bravo, M. L., Sánchez-González, G., et al. Use of DNA sequencing for noncommunicable diseases in low-income

and middle-income countries' primary care settings: A narrative synthesis. The International Journal of Health Planning and Management 2019, 34(1), e46–e71.
[90] Floris, M., Cano, A., Porru, L., Addis, R., Cambedda, A., Idda, M. L., et al. Direct-to-consumer nutrigenetics testing: An overview. Nutrients 2020, 12(2), 566.
[91] Adams, S. H., Anthony, J. C., Carvajal, R., Chae, L., Khoo, C. S. H., Latulippe, M. E., et al. Perspective: Guiding principles for the implementation of personalized nutrition approaches that benefit health and function. Advances in Nutrition 2020, 11(1), 25–34.
[92] Wang, C. Revealing obesity through diet-gene interactions. Health Science Inquiry 2020, 11(1), 158–161.
[93] Mazhaeva, T. V. Molecular and genetic aspects of health risks and their association with adverse environmental conditions and diets (systemic review). Health Risk Analysis 2022, 194, 186–97.
[94] Ahmad, S., Fatima, S. S., Rukh, G., & Smith, C. E. Gene lifestyle interactions with relation to obesity, cardiometabolic, and cardiovascular traits among South Asians. Frontiers in Endocrinology 2019, 10, 221.
[95] Song, E.-J., & Shin, J.-H. Personalized diets based on the gut microbiome as a target for health maintenance: From current evidence to future possibilities. Journal of Microbiology and Biotechnology 2022, 32(12), 1497.
[96] Jardon, K. M., Canfora, E. E., Goossens, G. H., & Blaak, E. E. Dietary macronutrients and the gut microbiome: A precision nutrition approach to improve cardiometabolic health. Gut 2022, 71(6), 1214–1226.
[97] Ramos-Lopez, O., Martinez, J. A., & Milagro, F. I. Holistic integration of omics tools for precision nutrition in health and disease. Nutrients 2022, 14(19), 4074.
[98] Tahir, U. A., & Gerszten, R. E. Omics and cardiometabolic disease risk prediction. Annual Review of Medicine 2020, 71(1), 163–175.
[99] Jin, Q., & Ma, R. C. W. Metabolomics in diabetes and diabetic complications: Insights from epidemiological studies. Cells 2021, 10(11), 2832.
[100] Wishart, D. S. Metabolomics for investigating physiological and pathophysiological processes. Physiological Reviews 2019, 99(4), 1819–1875.
[101] Dhanapal, A. C. T., Wuni, R., Ventura, E. F., Chiet, T. K., Cheah, E. S., Loganathan, A., et al. Implementation of nutrigenetics and Nutrigenomics Research and training activities for developing precision nutrition strategies in Malaysia. Nutrients 2022, 14(23), 5108.
[102] Nicastro, H. L., Vorkoper, S., Sterling, R., Korn, A. R., Brown, A. G., Maruvada, P., et al. Opportunities to advance implementation science and nutrition research: A commentary on the Strategic Plan for NIH Nutrition Research. Translational Behavioral Medicine 2023, 13(1), 1–6.
[103] Sirisena, N. D., & Dissanayake, V. H. Strategies for genomic medicine education in low and middle-income countries. Frontiers in Genetics 2019, 10, 944.

Frank Abimbola Ogundolie*, Titilayo Ibironke Ologunagba, and Christiana Eleojo Aruwa

Chapter 16 Enzyme kinetics and mechanisms in food processing

Abstract: The requirement of enzymes in the improvement of organoleptic and sensory properties of processed food as well as food quality and shelf life has positioned them as key players in the food processing industry. Application of enzymes span industries such as breweries, beverage, and cheese production, dairy, and meat industries. Industrial applications in these industries such as starch debranching, hydrolysis during fermentation, juice clarification, casein coagulation, deskinning, loaf volume and texture improvement, removal of polyphenol compounds, removal of sugar from candy scraps, and aging of wines utilize different enzymes such as glucose oxidase, amylases, phytases, pullulanase, transglutaminases, cellulases, glucose isomerases, lactases, invertases, and pectinases. These enzymes have high specificity and their activities are largely affected by factors, which include presence of metal ions, pressure, and temperature, presence of inhibitors, substrate concentration, and pH. This chapter reviews the kinetics and mechanisms of some of these enzymes in food industries.

Keywords: Enzymes, mechanisms of action, amylases, brewery, temperature, pH, organoleptic properties

16.1 Introduction

Enzymes are important biocatalysts and essential drivers of several life processes. Their high specificity positions them for their wide range of applications in cosmetics, medical, pharmaceutical/healthcare [1], beverages, feeds, textiles, and food industries

***Corresponding author: Frank Abimbola Ogundolie,** Department of Biotechnology, College of Medicine and Health Sciences, Baze University, Abuja, Nigeria, e-mails: fa.ogundolie@gmail.com, frank.ogundolie@bazeuniversity.edu.ng, 0000-0001-6112-1496
Titilayo Ibironke Ologunagba, Department of Medical Biochemistry, School of Basic Medical Sciences, The Federal University of Technology, Akure, Nigeria, e-mail: tiologunagba@futa.edu.ng
Christiana Eleojo Aruwa, Department of Biotechnology and Food Science, Faculty of Applied Sciences, Durban University of Technology, Durban 4001, South Africa, e-mail: aruwachristiana@gmail.com/ChristianaA@dut.ac.za

https://doi.org/10.1515/9783111441238-017

(Figure 16.1). The application of enzymes in food industries dates to centuries ago, and advancements in the science of these biocatalysts have significantly contributed to the refinement of products and the development of new and improved food products. In the tenderization of meat, hydrolysis of complex carbohydrate cell walls for release in breweries and clarification in beer or juice production [2], aging in fermentation processes, excessive softening of fruits in fruit industries, development of browning during ripening, preservation of vegetables and foods, biofilms for food packaging, meat tenderization, preventing sugary products from crystallization, improvement of organoleptic properties such as taste, texture, aroma development and color of some foods and meat tenderizer, and removal of free oxygen and or excess hydrogen peroxide during production [1, 3], enzymes are necessary machinery in the achievement of process optimization and product improvement.

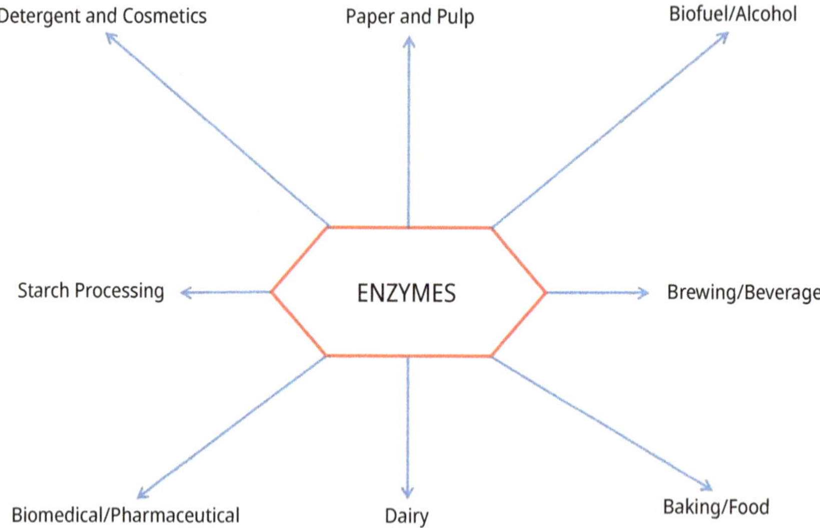

Figure 16.1: Industrial application of enzymes.

The sources of enzymes are as diverse as their multiplicity, ranging from microorganisms such as fungi, bacteria [4], algae, mushrooms, actinobacteria, and yeast to different plant parts (seeds, fruits, leaves, and flower). In plants, the quantity of the enzyme depends on the stage of plant development. Generally, it changes continually during flowering, development, maturation, and death of the plant. Enzymes can also be obtained from animal tissues, and more recently, the advent of recombinant DNA technology has produced other enzymes, improved and better adapted to relevant industrial processes [5, 6]. In general, enzymes can be classified into seven groups, which are largely based on their mode of reactions. Beyond food industries, these biocatalysts have also found applications in other industries such as biomedical and agriculture.

16.2 Classes of enzymes

Enzymes are classified and named primarily based on the reaction type they catalyze. Until recently, enzymes have been classified into seven major groups. These classes include transferases, oxidoreductases, lyases, hydrolases, isomerases, ligases [3], and the translocases. Transferases are involved in reactions such as catalyzing the transfer of functional groups, which includes the glycosyl and methyl [7, 8],; oxidoreductases, which are involved in oxidation–reduction reaction; lyases are involved in the removal/addition of atoms from/to a double bond; hydrolases are primarily involved in hydrolysis reactions; isomerases are responsible for the rearrangement of atoms during chemical reactions; ligases catalyze the condensation of molecules using ATP; and translocases catalyze the transportation of molecules, ions, and others across the membrane. These classes are according to the International Union of Biochemistry and Molecular Biology (IUBMB). The classes, mechanism of action, examples, and applications in the industries are summarized in Table 16.1.

Table 16.1: The classes of enzymes, their mechanism of action, and their applications.

Group	Mechanism of action	Examples	Application	References
Oxidoreductases	This group of enzymes is involved in oxidation-reduction.	Oxidase, catalase, oxygenase dehydrogenases, reductase, hydroxylases, peroxidases	Glucose oxidase used in food packaging. Catalase is used in cheese production.	[3, 9–11]
Transferases	Catalyze the transfer of chemical or specific functional groups, which includes methyl, glycosyl, acetyl, acyl, and amino from one substrate to the other.	Transglutaminase, glycosyltransferase, aminotransferases, acetyltransferases, phosphorylases	Transglutaminase is used in the baking and food packaging industries.	[3, 7, 8]
Lyases	This class of enzyme is involved in the removal/addition of atoms from/to a double bond.	Pectin lyase, Isocitrate lyase, pectate lyase, citrate lyase, alginate lyase, hydroxynitrile, argininosuccinate lyase, and pyruvate formate lyase	Pectate lyase and pectin lyase are involved in the preservation of vegetable foods and they are used for degumming and crushing of natural fiber.	[12-14]

Table 16.1 (continued)

Group	Mechanism of action	Examples	Application	References
Hydrolases	They are involved in hydrolysis reactions. They catalyze the hydrolytic cleavage of bonds.	Protease, lysozyme, amylase, lipase, lactase, pectinase, and invertase	Most important class of enzymes in the food industries. Used for various hydrolytic reactions such as starch hydrolysis, ripening, mist removal, deskinning.	[15-18]
Isomerases	They are involved in the rearrangement of atoms.	Triosephosphate isomerase (Oláh), glucose isomerase	Glucose isomerase for sweeteners like fructose-rich corn syrup.	[3, 15, 19]
Ligases	Combine molecules using ATP.	Ubiquitin l-igases (C–N bond), DNA ligase		[3]
Translocases	They are responsible for movement or transportation of molecules or ions across the membranes.	Ornithine translocase, carnitine-acylcarnitine translocase		[20]

16.3 Enzyme kinetics in food processing

Enzyme kinetics is the study of the rates of enzyme-catalyzed reactions and the factors that affect the reaction rates. In food processing, different unit operations are optimized to produce improved food products, with enzymes playing essential roles at different stages. This has driven scientific investigations to unravel the dynamics of different enzyme kinetics with the aim of engineering enzyme characteristics for optimal and efficient industrial applications. Depending on the desired product improvement or modification, different unit operations are deployed, consequently affecting enzyme requirements and reaction medium. Examples of such unit operations are mass transfer, mixing, size adjustment, fluid flow, heat transfer, and separation [21, 22].

According to Fellows [23], food processing is aimed at four dimensions namely: (i) providing convenience and variety in diet; (ii) supplementing nutrients required for health; (iii) increase the shelf life or period food remains wholesome (microbiologically and biochemically); and (iv) ensuring value addition. We shall briefly consider some fundamental concepts in enzyme kinetics before going further to discuss factors affecting it.

16.4 Fundamental concepts in enzyme kinetics

The primary goal of enzyme catalysis is to accelerate the rate of chemical reactions. Think of the burning of paper in air, a process sped up in the presence of air but not spontaneously initiated by air. This is due to the **energy barrier** that must be crossed to initiate the reaction. This is overcome by supplying a source of flame to the paper. This is what enzymes do. They increase chemical reaction rates without changing the reaction equilibria, thus helping to lower the energy barrier (activation energy) (Figure 16.2).

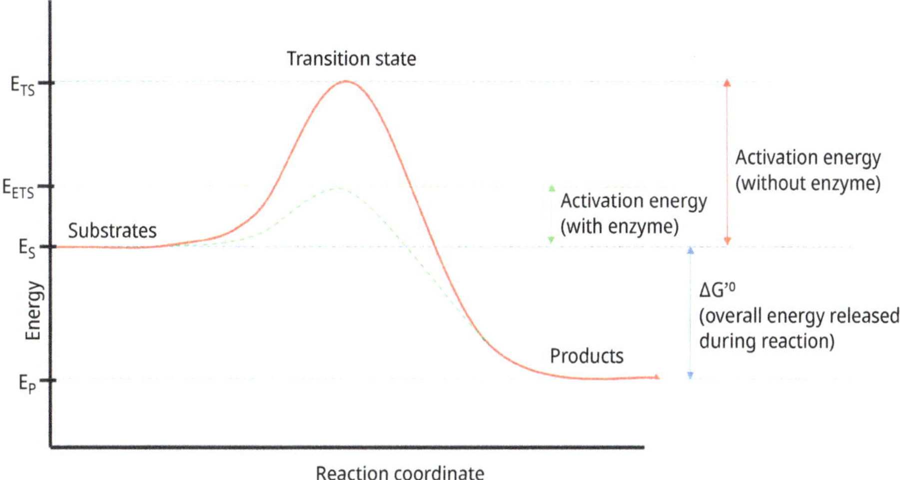

Figure 16.2: Activation energy profiles of reaction processes.

Activation energy is the energy difference between the substrate and the transition state – the energy barrier the substrate(s) is/are required to overcome before transformation to the products. Some of the physical and thermodynamic factors contributing to activation energy include entropy of motion of two molecules of substrate in solution, solvation shell of hydrogen-bonded water that surrounds and helps stabilize most biomolecules in aqueous solution, and the distortion of substrates necessary for many reactions. These factors are overcome by the binding energy of enzymatic reactions – the energy released by the formation of the enzyme–substrate (ES) complex. A simple enzymatic reaction is represented below:

$$E + S \rightleftharpoons ES \rightleftharpoons EP \rightleftharpoons E + P \tag{16.1}$$

Enzyme catalysis is facilitated by the interaction of substrates in proximity at the enzyme's active site. At this site, catalytic functional groups of the enzyme form transient covalent bonds with the substrate, forming the ES complex, thus lowering the reaction's activation energy. In addition to transient covalent bond formation, the formation of the transient ES complex also involves non-covalent interactions. The **binding energy** of an enzyme is the **free energy** released when a weak interaction is formed between the enzyme and its substrate(s). The enzyme's binding energy is used to overcome the factors contributing to activation energy described above. The binding energy serves to lower activation energy by: reducing entropy as it holds the substrates in the orientation most favorable for reaction; the formation of weak bonds between enzyme and substrates in the ES complex, which leads to elimination of the solvation shell as enzyme–substrate interactions replace hydrogen bonds in substrate-water interactions; formation of weak interactions in the reaction transition state that thermodynamically compensates for substrate electron redistribution, which is necessary for the reaction of substrates [24].

Furthermore, the Koshland induced-fit model describes an efficient mechanism of enzyme catalysis where enzymes themselves also undergo transient changes, following substrate binding at the active site, which are complementary to the transition states as the catalytic groups of the enzyme align to catalyze the formation of products [25–27]. The investigation of enzyme kinetics is categorized into three levels. The transient-state kinetics, which describes the initial stage of the enzyme–substrate interaction is characterized by quick reactions between the substrates and the enzyme [28, 29]. The techniques employed depend on quickly combining a substrate with an enzyme concentration high enough to give direct visibility to the intermediates and products generated at the active site during one enzyme cycle. Optical signals (such as absorbance, fluorescence, and light scattering) from the reaction are tracked over time to monitor reaction rates. The sequence of events that follow substrate binding can be defined by analyzing the rates and amplitudes of observable species as a function of substrate concentration [30].

The steady-state kinetics is the stage where the concentrations of reactive intermediates are constant due to the constant rates of intermediate production and breakdown. Substrate concentration is higher than enzyme concentration in this condition. This is the Michaelis–Menten principle of enzyme catalysis (Figure 16.3), and is most widely studied [28]. The substrate concentration during this phase is greater than enzyme concentration. At low substrate concentrations, the reaction velocity (V_0) increases as the substrate concentration rises; however, it eventually levels off, reaching a point where further increases in substrate concentration no longer affect the reaction rate. At this stage, the substrates occupy all binding sites on the enzyme, disallowing the formation of more ES complex intermediates.

The Michaelis–Menten equation is given by:

$$V_0 = \frac{V_{max}[S]}{K_m + [S]} \tag{16.2}$$

where V_0 is the initial reaction velocity; V_{max} is the maximum reaction velocity; $[S]$ is the substrate concentration; and K_m is the Michaelis constant.

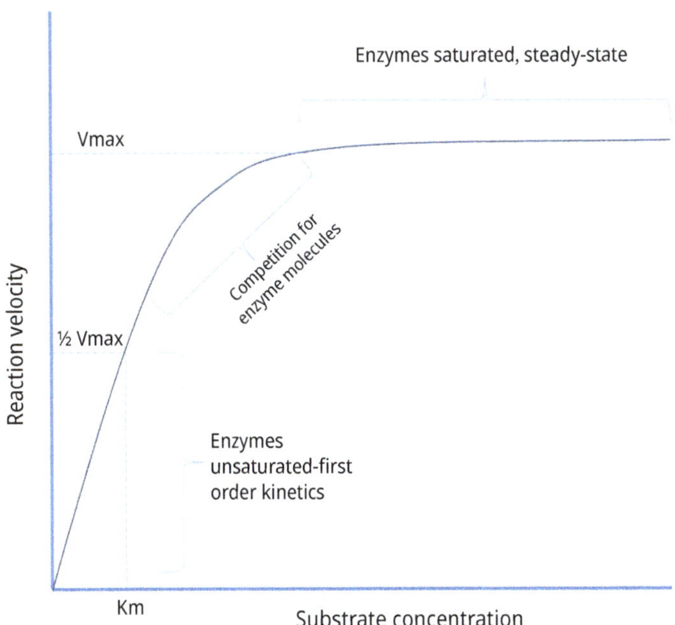

Figure 16.3: Activation energy profiles of reaction processes.

V_{max} is referred to as the maximum rate at which an enzyme can catalyze reaction and defines the capacity of the enzyme. In food processing, a higher V_{max} is beneficial for accelerating the production process and improving efficiency, but it must also be balanced with other parameters. K_m is the Michaelis constant and is the substrate concentration at which the reaction velocity is half of V_{max}. It provides insight on the enzyme's affinity for its substrate. A low K_m indicates low affinity for the substrate, meaning that the enzyme can achieve maximal activity at low concentrations of the substrate, while a high K_m indicates high affinity for the substrate meaning that a higher substrate concentration is required to bring the enzyme to maximal activity. In food processing, this is useful in the optimization of substrate concentration and enzyme concentration.

On the other hand, the k_{cat} is the rate constant of the rate-limiting step of any enzyme-catalyzed reaction. Recall from eq. (16.1) that enzyme catalysis proceeds in a number of steps ranging from the formation of ES complex, to the EP complex, and eventually to the dissociation of product from the enzyme. k_{cat} is also referred to as the turnover number and is the number of substrate molecules converted to product per enzyme molecule per unit of time when the enzyme is fully saturated with substrate. It provides a measure of the catalytic efficiency of the enzyme [24]. A high turnover number in food processing is a good thing because it means the enzyme can break down substrates quickly, which helps accelerate production cycles. High k_{cat} enzymes are frequently chosen for industrial applications where timeliness is essential [31].

Scientific efforts driven by industrial demands are yielding genetically engineered enzymes of different classifications (transglutaminases, proteases, and lipases) and origins delivering efficient alternatives to mechanical and chemical techniques with improved specificity, economic viability, and product quality [31]. With detailed understanding of the kinetic peculiarities of the enzymes, advancements in biotechnology are now being utilized to develop and improve the performance of enzymes by modifying their responses to factors that modulate their activity. Some of these factors include temperature, pH, substrate concentration, and enzyme concentration.

16.5 Factors affecting enzyme kinetics

Several factors affect enzyme kinetics in food processing. We will discuss each factor in this section.

Temperature: The degree of hotness or coldness of the reacting medium significantly impacts enzyme activity and stability. The uniqueness of enzymes as biological catalysts also makes them vulnerable to modification by temperature change. Being proteins, which are polypeptides held together by several intra- and intermolecular forces that can be altered by temperature, enzymes can be denatured by temperature and lose their native conformation and function. Two thermal parameters had previously been used to describe the effect of temperature on enzyme activity; Arrhenius activation energy, which describes the effect of temperature on catalytic rate constant, and thermal stability, which describes the effect of temperature on the thermal inactivation rate constant [32]. More recently, the introduction and validation of the equilibrium model gives a more detailed description of the effect of temperature on enzyme activity [33–35]. This model describes the response of enzyme activity to increase in temperature by an additional mechanism, wherein the active form of the enzyme (E_{act}) and the inactive form (E_{inact}) are related to the thermally denatured form (X):

$$E_{act} \rightleftharpoons E_{inact} \rightarrow X \tag{16.3}$$

Initial increase in temperature increases reaction rate due to increased molecular movement; however, further temperature increase (to T_{eq}) brings the reaction to the equilibrium between E_{act} and E_{inact}. Continuous increase in temperature leads to loss of enzyme activity, which is attributed to a shift in the E_{act}/E_{inact} equilibrium due to localized conformational changes at the enzyme's active site. This equilibrium shift then precedes the eventual irreversible denaturation of the enzyme. T_{eq} is an important consideration in technological applications and enzyme engineering approaches to improve thermal stability of enzyme. Temperature-mediated enzyme inactivation (or denaturation) is time-dependent, and this is factored into the industrial application of enzymes [32].

In optimization of food processing, the variation in optimal temperatures for different enzymes is utilized to differentially regulate enzyme activity suitable to every stage of food processing to produce the desired outcome. Enzymes involved in food processing can either be endogenous or exogenous. Endogenous enzymes are naturally present in food and may have desirable or undesirable effects, hence the need to identify and regulate them for optimal food products [36, 37]. Exogenous enzymes are added along the food processing line to improve taste, aroma, texture, shelf life, as well as nutritive value of food products [3, 38]. Hence, informed control of temperature conditions keeps the enzymes at optimal conditions and favors the production of desired products and inhibits the production of undesired products. For example, for dairy, fish, and meat products, immediate cooling post-harvest inhibits the activity of some degradation enzymes, thus preventing the development of undesirable features.

pH: Enzymes are proteins and are thus responsive to changes in the pH of their surrounding medium. Proteins are composed of amino acids, which are amphoteric, that is, they have both acidic and basic properties depending on the reaction conditions. The implication of this for enzymes is that a change in pH is capable of significantly altering the tertiary or quaternary structure of the protein and consequently leading to inactivation or denaturation. Most enzymes have a characteristic optimum pH. A change in pH alters the ionization of reactive groups at the active site of the enzyme or on the substrate, impacting their reactivity. Furthermore, if the enzyme is placed in a reaction medium that is more acidic or more basic than its optimal pH, it could lead to an alteration in the arrangement or orientation of its reactive groups such that it may be unable to attain its native conformation when the medium conditions are reversed, and its catalytic activity is not fully restored [39]. Biocatalysis can also be improved by widening the range of pH favorable to enzyme activity [40, 41].

Buffers are an essential component of the reaction medium in food processing, necessary for the maintenance of the optimal pH for enzyme activity, especially where the product tends to alter the medium pH. In addition, rapid extraction of the product essentially prevents the buildup of products' concentration to levels that cause significant pH modulation in the medium and, consequently enzyme activity [24].

Substrate concentration: Steady-state kinetics highlight the significant effect of substrate concentration on the rate of enzyme activity as a function of time. There is usually an initial increase in enzyme activity with substrate concentration until enzyme saturation. In accordance with the Michaelis–Menten principle, the reaction medium attains a plateau beyond which increase in substrate concentration does not impact reaction rate. The enzyme becomes saturated and unable to accommodate more substrate molecules. This is a key consideration in optimizing enzyme application in food processing, as the rate of reaction is controlled by systematic feeding in of the substrate to maintain substrate concentration at optimal levels for enzyme activity [24]. In addition to substrate concentration, the concentration of enzyme is also important in determining reaction rates. Higher concentrations of enzyme favor increased reaction rates; however, the substrate concentration is usually about five to six orders of magnitude greater than enzyme concentration. Higher enzyme concentrations beyond the right proportion to substrate concentration amount to wastage as the reaction is terminated when substrate is exhausted. This is not an economically viable approach to food processing; hence, the effective management of substrate concentration is central to optimization of enzyme kinetics in food processing [16].

Enzyme inhibitors: Enzyme inhibitors are molecules that interact with enzyme to inhibit its activity. Some inhibitors bind directly at the enzyme active site and compete with the substrate for the active site (competitive); some others interact with the enzyme molecule at domains different from its active site of the enzyme, inducing conformational change that impairs catalytic function (noncompetitive); while others bind to the ES complex, preventing the formation of the transition state and eventual product (uncompetitive) [24]. Industrial application of inhibitors in food processing serve to selectively control enzyme activity, moderating the individual concentration of desired constituents to attain optimal taste, aroma, texture, consistency, and nutritive value of the final food products [36, 37].

16.6 Enzyme applications in food processing

A combination of endogenous and exogenous enzymes is employed in food processing to facilitate the achievement of desired features in food products. While different techniques are being employed to optimize the activity of endogenous enzymes postharvest, several exogenous enzymes have found important applications in food processing. To avoid complex purification techniques that increase the economic burden of food processing, enzymes are being used for analytical purposes, wherein they are incorporated into biosensors for the detection of small amounts of multiple analytes [42, 43]. Enzymes are also being used as indicators; for example, catalase is used in testing the quality of milk [10].

Furthermore, enzymes are used to improve the final quality of the food products, due to their sensory properties. This is especially true of meat products where tenderness, color, and flavor are desirable features [44]. In baking industries, the use of the subclasses of amylases and proteases to enhance the production of precursors for nonenzymatic browning reactions has been employed to generally increase the acceptance of the products. Also, lipases and proteases are key enzymes employed in the development of flavor compounds in dairy products produced by various dairy industries [45].

16.7 Enzymes in starch-based industries

Starch, which is a complex carbohydrate polymer found in plants is widely regarded as an essential starting material for various derivatives and molecules used in food, breweries, and baking industries [4, 17, 46]. Due to the complexity of starch structure, which involves the presence of mono-based sugars linked together by different glycosidic bonds such as α 1–4, α 1–3, α 1–6, these bounds require the action of several biocatalysts to release monosaccharides such as fructose, glucose and galactose, disaccharides such as lactose, maltose, and sucrose, alpha-dextrins and other oligosaccharides such as maltotriose with applications in various food and baking processes.

Generally, the breakdown of starch molecules requires enzymes such as the cyclodextrinases, pullulanases, glucoamylases, and endo- and exo-amylases [17]. These enzymes are the most used class of enzyme in various food industries with high specificity for different starch- or sugar-based substrates. They can be utilized individually or in combination for more precise and desired product development (Figure 16.4).

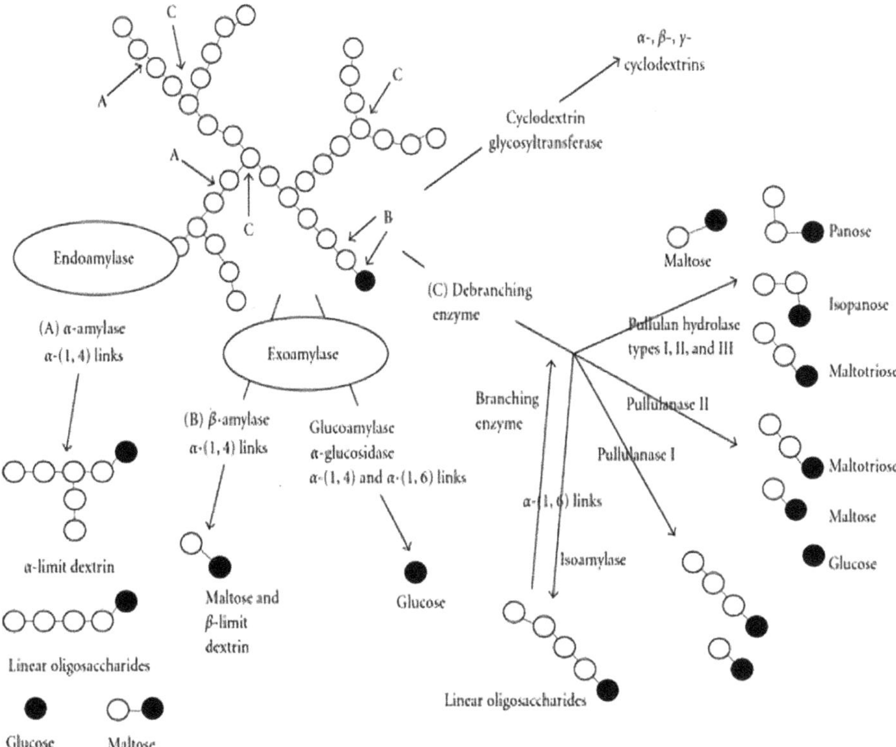

Figure 16.4: Action of hydrolyzing enzymes (starch degrading enzymes) on starch molecule for conversion into simpler sugar molecules and linear oligosaccharides, which are compounds of great importance in various starch-based industries [4, 16, 17].

16.8 Enzyme applications in food processing

The use of exo amylases such as β-amylases and the various subclasses of debranching enzyme pullulanase, which comprises of pullulanase type I, II, and III alongside glucoamylase, otherwise known as amylo-glucosidase, have been well documented [15, 16, 46]. In food industries, starch is one of the most used starting materials for producing numerous end products ranging from beers, wines, bread, and pastries to various syrups, among others. The amylolytic action of the hydrolyzing enzymes results in a rapid viscosity reduction. This process leads to saccharification and liquefaction of starch. Liquefaction of starch involves starch conversion to yield by-products that include glucose, maltose, and dextrin. Although dextrins are referred to as imperfect starch hydrolysis [47], they are good substrates in baking industries during bread production for yeast to act on; saccharification is an industrial process that involves the conversion of soluble starch to glucose. Major steps involved in starch conversion in the industry using hydrolytic enzymes and their products are shown in Figure 16.5.

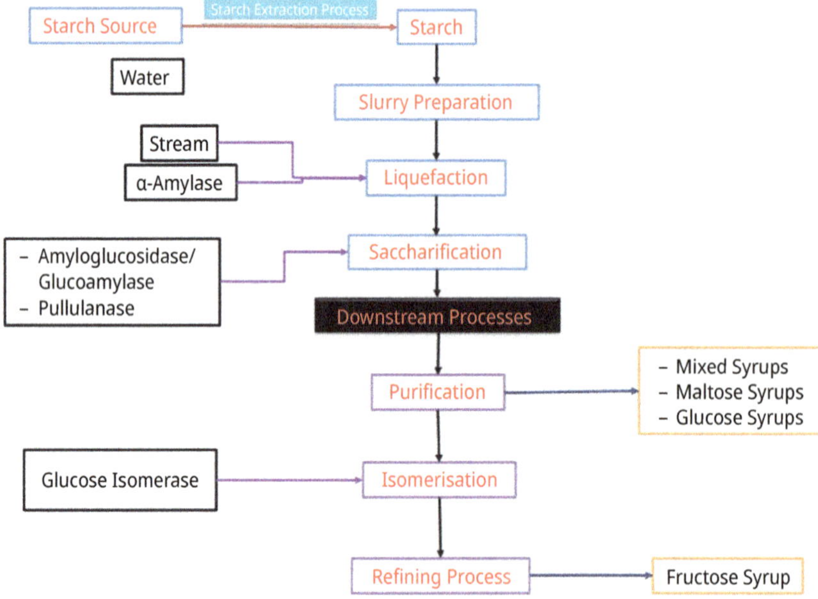

Figure 16.5: Action of hydrolyzing enzymes (starch degrading enzymes) on starch molecules for conversion into simpler sugar molecules and linear oligosaccharides, which are compounds of great importance in various starch-based industries [4, 16, 17,].

Enzymes also find application in dairy processing where milk and various milk products are subjected to enzyme treatments to produce several products. Milk is a food product that is widely consumed and, in fact, accounts for approximately $125 billion dollars trade yearly worldwide [48]. It is high in protein and contains more than 60 endogenous enzymes, which are resistant to high-temperature, short-time pasteurization [49]. Among these endogenous proteins are γ-glutamyl transpeptidase, lactoperoxidase, alkaline and acid phosphatase, lipoprotein lipase (LPL), sulfydryl oxidase and xanthine oxidoreductase, and proteases.

Rennet is a blend of pepsin and chymosin used for milk curdling in the primary phase of cheese production [50]. The milk liquid structure is preserved by casein, which prevents coagulation. Rennet proteases particularly target kappa-casein for degradation and favor the coagulation of milk necessary for cheese production. Chymosin has been bioengineered to improve productivity, contributing to about 70% cheese production, overcoming the challenge of insufficient indigenous chymosin obtained from calf [51].

Another enzyme used in dairy processing is lactase, which breaks down lactose into simpler sugars – glucose and galactose sugars – for those managing lactose intolerance. Lactase is used in frozen yoghurt production for tastier and creamier yoghurt. It is usually obtained from *Kluyveromyces* species of yeasts and *Aspergillus* species of fungi [52]. Catalase is used in place of pasteurization in Swiss cheese, where there is a higher preference for preservation of regular milk proteins and flavor enhancement. Catalase is essential in cheese making as it degrades the oxidant, hydrogen peroxide, to water and oxygen, preserving the bacterial culture used; however, it is easily denatured by pasteurization [53, 54]. Lipase is used to separate milk from fat to pave way for desirable flavors in cheese. Transglutaminase is also used in dairy products for enhancing gel consistency and reduced whey separation by enzymatic cross-linking during the production of yoghurt.

In baking industries, due to limitations in the application of chemical additives, enzymes are gaining wider acceptance in baking and fermented products as viable alternatives. The primary ingredient and source of enzyme in baking is wheat, which is a mixture of enzymes, lipids, non-starch polysaccharides, gluten, and starch. The mixing of the dough that occurs during bread making initiates a number of complex biophysical and biochemical processes catalyzed by yeast and wheat enzymes. Chemical additives may also be supplemented with enzymes added to govern the dough phase and baking phase as enzymes generate desired flavor and compensate for flour variability. Typically, during bread production, introduction of enzymes change the dough's crumb softness, gas retention, and rheology; when creating pastries and biscuits, they are used to change the product's softness; and when manufacturing bakery goods, they are added to lessen the development of acrylamide [55, 56].

Hydrolases and oxidoreductases are most widely utilized in the baking industry. The most used of the hydrolases are α-amylases [46]. α-Amylases (EC 3.2.1.1) are endoenzymes that catalyze the cleavage of α-1, 4-glycosidic bonds in the inner part of the amylose or amylopectin chain, generating oligosaccharides of varying lengths with α-configuration [57]. α-Amylases vary widely in catalytic domains and binding sites multiplicity, which significantly determines their substrate specificity, the length of the oligosaccharide fragments released after hydrolysis, and subsequently the carbohydrate profile of the final product. The profiles of heat stability differ among the many types of α-amylases [58]. Malt- or fungus-derived α-amylases are also added to flour to maximize its amylase activity, with the primary goal of raising the amounts of reducing and fermentable sugars. For flour standardization, fungal α-amylases are a better fit than malt amylases due to their poorer thermostability [3, 15].

Furthermore, in baking, to reduce the duration of mixing, curtail dough consistency, assure dough evenness, and control gluten strength and bread texture, or to prove bread flavor, proteases are essential players [59, 60]. Basically, proteases weaken the gluten network. During the production of cookies, biscuits and pastries, proteases act on wheat proteins, reducing gluten elasticity and therefore leading to the reduction observed in shrinkage of paste or dough, after molding and sheeting [55, 61].

Hemicellulases (enzymes that catalyze the hydrolysis of hemicelluloses) such as xylanase are used along with amylases, oxidoreductases, and lipases aimed at improving the organoleptic properties and rheology of dough in bakeries during bread production, loaf volume, and enhancement of sensory and textural properties of bread [62, 63]. Lipases, which hydrolyze triacylglycerols prevent the formation of undesired flavor in bread. They also play a role in improving crumb structure, staling rate, and loaf volume [64].

Oxidoreductases such as lipoxygenase serve to catalyze the oxidation of polyunsaturated fatty acids in triacylglycerols to improve mixing tolerance and dough handling properties, leading to improved loaf volume [65, 66]. Glucose oxidase, which catalyzes the oxidation of β-D-glucose to D-glucono-δ-lactone and hydrogen peroxide [67, 68], is also an enzyme used in the baking industry. Potassium bromate usage has been replaced with glucose oxidase as an alternative oxidizing agent [69, 70].

In fruit juice processing applications, a major marketing advantage in the fruit juice industry is in juice clarification [71]. Clarification is primarily associated with the removal of pectin and other fibers and carbohydrates from the juice [72, 73]. Previous methods employed involve centrifugation, membrane filtration, and the use of clarifying agents (gelatin and some other chemical additives). Given the laborious process that is necessary after these treatments, a more viable alternative is the enzymatic treatment of fruit juice [74].

To satisfy the demands of potential consumers, juice producers utilize enzymes for the breakdown of unwanted polysaccharides to produce crystal-clear juice with the characteristic appearance, taste and texture, mouth-feel, stability, and amenability to consumer taste. Generally, in fruit juice or wine production, enzymes such as amyloglucosidase, glucosidase, arabanase, xylanase, hemicellulase, amylases, cellulase, glucanase, and pectinase among others, aid in the clarification of end products [75].

Pectinases catalyze the hydrolysis of pectin, which is responsible for the cloudiness and turbidity in fruit juice. The products of pectinase enzymatic hydrolysis of pectin are pectin-protein complexes, which further flocculate, making the juice less viscous and less turbid [76, 77]. Pectinases are generally classified into pectate lyases (PAL), polygalacturonase (PG), and pectin methyl esterase (PME) [78].

Amylases employed in fruit juice processing are classified into endoamylases and exoamylases. While exoamylases work sequentially from the nonreducing end of the starch molecule to produce short-end products, endoamylases randomly cleave the center of the molecule [79]. To attain improved clarification and filterability of fruits such as unripe apple, amylases are used to inhibit retrogradation and haze formation after bottling [80]. The group of hydrolytic enzymes responsible for the catalysis and bioconversion of cellulose into soluble glucose and other simple sugars are generally called cellulases. They soften fruits, breakdown cell walls, and clarify fruit juices; these actions then promote the release of proteins, polysaccharides, enzymes, and importantly, flavors [81].

In food industries, another interesting use of enzymes involves their incorporation into foods as therapeutic agents [82]. This is related to the biological nature of enzymes compared to small molecules of conventional drugs. Enzymes are highly specific and have high affinity for their target sites and they are also biological catalysts, converting target molecules to desired products that are biodegradable, hence, no side effects. Other attractive features of enzymes for therapeutic applications are water solubility and effectiveness in a biological environment. However, there are restrictions to the therapeutic application of enzymes owing to their bulkiness, which restricts access to the intracellular domain. They are also proteins and are thus antigenic, which makes them susceptible to immunogenic clearance from the blood plasma before action [82].

Clinical proof of the application of proteolytic enzymes in cancer research has usually been acquired using an enzyme preparation that consists of chymotrypsin, papain, and trypsin in combination. Previous studies have shown that enzyme treatment can lessen the side effects of chemotherapy and radiation. Systemic enzyme treatment appears to have beneficial effects because of its anti-inflammatory properties [28]. The mode of action is yet to be fully clarified.

Some of the therapeutic enzymes already in use are PEGylated L-asparaginase in treating and management of non-Hodgkin's lymphoma, acute myeloid leukemia, and acute lymphoblastic leukemia, offering a better alternative to chemotherapy [83]. Bactericidal enzymes such as lysozyme, which is naturally produced in the human body is added in many food products. This enzyme has been discovered to show activity against urinary RNase U, and HIV, which is similar to RNase A; it selectively degrades viral RNA. It is also a promising candidate in the management of HIV infection [84, 85].

Enzymes are also used as digestive aids; for example, α-galactosidase enzyme is taken as a digestive aid by people who develop the symptoms of bloating, gas, and diarrhea upon ingesting foods like beans and *Brassica* vegetables, which include Brussels sprouts, broccoli, and cabbage [86]. Lactase supplements commercially available as lactase-fortified milk are beneficial for the breakdown of lactose to glucose and galactose, mitigating lactose intolerance symptoms such as diarrhea, gas, and bloating [86].

16.9 Conclusion

Future prospects include exploring the use of computational tools in enzyme design such as enzyme modeling and simulation, making it possible to predict the development of novel products and optimization of enzymes for industrial food processing through predictive kinetics. In addition, efforts in recombinant DNA technology and enzyme immobilization are being employed to develop other products with wider scope of application.

References

[1] Mital, S., Christie, G., & Dikicioglu, D. Recombinant expression of insoluble enzymes in *Escherichia coli*: A systematic review of experimental design and its manufacturing implications. Microbial Cell Factories 2021, 20, 1–20.

[2] Narnoliya, L. K., Jadaun, J. S., Chownk, M., & Singh, S. P. Enzymatic systems for the development of juice clarification strategies. In: Sudhir P. Singh, Ashok Pandey, Reeta Rani Singhania, Christian Larroche, and Zhi Li. Biomass, Biofuels, Biochemicals, Amsterdam, Netherlands: Elsevier, 2020, 397–412. https://doi.org/10.1016/B978-0-12-819820-9.00018-1.

[3] Motta, J. F. G., Freitas, B. C. B. D., Almeida, A. F. D., Martins, G. A. D. S., & Borges, S. V. Use of enzymes in the food industry: A review. Food Science and Technology 2023, 43, e106222 https://doi.org/10.1590/fst.106222.

[4] Ogundolie, F. A. (2015) Characterization of a purified β–Amylase from black marble vine (*Dioclea reflexa*) seeds (Masters Thesis, Federal University of Technology, Akure, Nigeria). Retrieved from http://196.220.128.81:8080/xmlui/handle/123456789/44077

[5] Restaino, O. F., Borzacchiello, M. G., Scognamiglio, I., Fedele, L., Alfano, A., Porzio, E., Manco, G., De Rosa, M., & Schiraldi, C. High yield production and purification of two recombinant thermostable phosphotriesterase-like lactonases from Sulfolobus acidocaldarius and Sulfolobus solfataricus useful as bioremediation tools and bioscavengers. BMC Biotechnology 2018, 18, 1–15.

[6] Sato, R., Minamihata, K., Ariyoshi, R., Taniguchi, H., & Kamiya, N. Recombinant production of active microbial transglutaminase in E. coli by using self-cleavable zymogen with mutated propeptide. Protein Expression and Purification 2020, 176(105730), https://doi.org/10.1016/j.pep.2020.105730.

[7] Vasić, K., Knez, Ž., & Leitgeb, M. Transglutaminase in foods and biotechnology. International Journal of Molecular Sciences 2023, 24(15), 12402, https://doi.org/10.3390/ijms241512402.

[8] Martins, I. M., Matos, M., Costa, R., Silva, F., Pascoal, A., Estevinho, L. M., & Choupina, A. B. Transglutaminases: Recent achievements and new sources. Applied Microbiology and Biotechnology 2014, 98, 6957–6964.

[9] Hanušová, K., Vápenka, L., Dobiáš, J., & Mišková, L. Development of antimicrobial packaging materials with immobilized glucose oxidase and lysozyme. Open Chemistry 2013, 11(7), 1066–1078, http://dx.doi.org/10.2478/s11532-013-0241-4.

[10] Kaushal, J., Mehandia, S., Singh, G., Raina, A., & Arya, S. K. Catalase enzyme: Application in bioremediation and food industry. Biocatalysis and Agricultural Biotechnology 2018, 16, 192–199 http://dx.doi.org/10.1016/j.bcab.2018.07.035.

[11] Li, X., Xie, X., Xing, F., Xu, L., Zhang, J., & Wang, Z. Glucose oxidase as a control agent against the fungal pathogen Botrytiscinerea in postharvest strawberry. Food Control 2019, 105, 277–284 http://dx.doi.org/10.1016/j.foodcont.2019.05.037.

[12] Kamijo, J., Sakai, K., Suzuki, H., Suzuki, K., Kunitake, E., Shimizu, M., & Kato, M. Identification and characterization of a thermostable pectate lyase from Aspergillus luchuensis var. saitoi. Food Chemistry 2019, 276, 503–510, http://dx.doi.org/10.1016/j.foodchem.2018.10.059 PMid:30409626.
[13] Dal Magro, L., Kornecki, J. F., Klein, M. P., Rodrigues, R. C., & Fernandez-Lafuente, R. Pectin lyase immobilization using the glutaraldehyde chemistry increases the enzyme operation range. Enzyme and Microbial Technology 2020, 132, e109397 http://dx.doi.org/10.1016/j.enzmictec.2019.109397.
[14] Lara-Márquez, A., Zavala-Páramo, M. G., López-Romero, E., & Camacho, H. C. Biotechnological potential of pectinolytic complexes of fungi. Biotechnology Letters 2011, 33(5), 859–868, http://dx.doi.org/10.1007/s10529-011-0520-0.
[15] Ogundolie, F. A., Ayodeji, A. O., Olajuyigbe, F. M., Kolawole, A. O., & Ajele, J. O. Biochemical Insights into the functionality of a novel thermostable β-amylase from *Dioclea reflexa*. Biocatalysis and Agricultural Biotechnology 2022, 42, 102361 https://doi.org/10.1016/j.bcab.2022.102361.
[16] Hii, S. L., Tan, J. S., Ling, T. C., & Ariff, A. B. Pullulanase: Role in starch hydrolysis and potential industrial applications. Enzyme Research 2012, 2012(1), 921362.
[17] Ogundolie, F. A. (2021) Cloning of α-amylase and pullulanase genes of *Bacillus licheniformis*FAO.CP7 from Cocoa (*Theobroma cacao* L.) Pods and biochemical characterization of the expressed enzymes. Retrieved from http://196.220.128.81:8080/xmlui/handle/123456789/4548
[18] Wongphan, P., Khowthong, M., Supatrawiporn, T., & Harnkarnsujarit, N. Novel edible starch films incorporating papain for meat tenderization. Food Packaging and Shelf Life 2022, 31, 100787 http://dx.doi.org/10.1016/j.fpsl.2021.100787.
[19] Oláh, J., Orosz, F., Keserü, G. M., Kovári, Z., Kovács, J., Hollán, S., & Ovádi, J. Triosephosphate isomerase deficiency: A neurodegenerative misfolding disease. Biochemical Society Transactions 2002, 30(2), 30–38, https://doi.org/10.1042/bst0300030.
[20] Balaji, S. The transferred translocases: An old wine in a new bottle. Biotechnology and Applied Biochemistry 2022, 69(4), 1587–1610, https://doi.org/10.1002/bab.2230.
[21] Park, S. H., Lamsal, B. P., & Balasubramaniam, V. M. Principles of food processing. In: Clark, S., Jung, S., & Lamsal., B., Edited by. Food Processing: Principles and Applications, Second Edition, © 2014 John Wiley & Sons, Ltd. New Jersey, USA: Published 2014 by John Wiley & Sons, Ltd, 2014, 1–13.
[22] Singh, R., & Singh, R. K. Role of enzymes in fruit juices clarification during processing: A review. International Journal of Biological Technology 2015, 6(1, 1–12.
[23] Fellows, P. Food Processing Technology Principles and Practice, 3rd edn, Boca Raton, FL: Woodhead Publishing, 2009.
[24] Nelson, D. L., Lehninger, A. L., & Cox, M. M. Lehninger Principles of Biochemistry, 5th ed, New York: W.H. Freeman, 2008.
[25] Csermely, P., Palotai, R., & Nussinov, R. Induced fit, conformational selection and independent dynamic segments: An extended view of binding events. Trends in Biochemical Sciences 2010, 35(10), 539–546, https://doi.org/10.1016/j.tibs.2010.04.009.
[26] Boehr, D. D., Nussinov, R., & Wright, P. E. The role of dynamic conformational ensembles in biomolecular recognition. Nature Chemical Biology 2009, 5(11), 789–796, https://doi.org/10.1038/nchembio.232.
[27] Koshland, D. E. The Key–lock Theory and the Induced Fit Theory. Angewandte Chemie International Edition 1995, 33(23–24), 2375–2378. https://doi.org/10.1002/anie.199423751
[28] Bhatia, S. Introduction to Pharmaceutical Biotechnology Volume 2, Bristol: IOP Publishing), 2018, 1–29, https://doi.org/10.1088/978-0-7503-1302-5ch1.
[29] Bagshaw, C. R. Transient-state kinetic methods. In: Roberts, G. C. K., editor. Encyclopedia of Biophysics, Berlin, 2638–2644, Heidelberg: Springer, 2013, https://doi.org/10.1007/978-3-642-16712-6_56.
[30] Begley, T. P., & Johnson, K. A. Transient state enzyme kinetics. In: Begley, T. P., editor. Wiley Encyclopedia of Chemical Biology, 2008, 1–8, Wiley and Sons, New York, USA. https://doi.org/10.1002/9780470048672.wecb609.

[31] Karkal, S. S., Venmarath, A., Velappan, S. P., & Kudre, T. G. Enzymes in meat, fish, and poultry product processing and preservation-II. In: Dutt Tripathi, A., Darani, K. K., & Srivastava, S. K., editors. Novel Food Grade Enzymes, Singapore: Springer, 2022, 193–216. https://doi.org/10.1007/978-981-19-1288-7_8.

[32] Peterson, M. E., Daniel, R. M., Danson, M. J., & Eisenthal, R. The dependence of enzyme activity on temperature: Determination and validation of parameters. The Biochemical Journal 2007, 402(2), 331–337, https://doi.org/10.1042/BJ20061143.

[33] Daniel, R. M., Danson, M. J., & Eisenthal, R. The temperature optima of enzymes: A new perspective on an old phenomenon. Trends in Biochemical Sciences 2001, 26, 223–225.

[34] Peterson, M. E., Eisenthal, R., Danson, M. J., Spence, A., & Daniel, R. M. A new, intrinsic, thermal parameter for enzymes reveals true temperature optima. Journal of Biological Chemistry 2004, 279, 20717–20722.

[35] Daniel, R. M., & Danson, M. J. Temperature and the catalytic activity of enzymes: A fresh understanding. FEBS Letters 2013, 587(17), 2738–2743, https://doi.org/10.1016/j.febslet.2013.06.027.

[36] Okafor, D. C., Ofoedu, C. E., Nwakaudu, A., & Daramola, M. O. Enzymes as additives in starch processing: A short overview. In: Kuddus, M., editor. Enzymes in Food Biotechnology, Cambridge: Academic Press, 2019, 149–168. http://dx.doi.org/10.1016/B978-0-12-813280-7.00010-4.

[37] Ordoñez, J. A., Rodríguez, M. I. C., Álvarez, L. F., Sanz, M. L. G., & Minguillón, G. Tecnologia de Alimentos: Componentes Dos Alimentos E Processos, 1ª ed., Porto Alegre: Artmed, 2005.

[38] Adetunji, C. O., Ogundolie, F. A., Ajiboye, M. D., Mathew, J. T., Inobeme, A., Dauda, W. P., Ghazanfar, S., Titilayo, O., Olaniyan, O. T., Ijabadeniyi, O. A., & Ajayi, O. O. Nano-engineered sensors for food processing. Royal Society of Chemistry 2022, 7, 151–166 https://doi.org/10.1039/9781839167966-00151.

[39] K, R. P. Enzymes: Principles and biotechnological applications. Essays in Biochemistry 2015, 59, 1–41 https://doi.org/10.1042/bse0590001.

[40] Bommarius, A. S., & Riebel, B. R. Biocatalysis: Fundamentals and Applications, Weinheim, Germany: Wiley–VCH 2004, 2004, 611 pp; ISBN 3–527-30344-8.

[41] S, W. D. W. Recent advances in enzyme development. In: Whitaker, J. R., Voragen, A. G. J., & Wong, D. W. S., editors. Handbook of Food Enzymology, New York, NY, USA: Marcel Dekker, 2003, 379–387.

[42] Wang, W., & Gunasekaran, S. Nanozymes-based biosensors for food quality and safety. Trends in Analytical Chemistry 2020, 126, 115841 http://dx.doi.org/10.1016/j.trac.2020.115841.

[43] Alvarado-Ramírez, L., Rostro-Alanis, M., Rodríguez-Rodríguez, J., Sosa-Hernández, J. E., Melchor-Martínez, E. M., Iqbal, H. M. N., & Parra-Saldívar, R. Enzyme (single and multiple) and nanozyme biosensors: Recent developments and their novel applications in the water-food-health nexus. Biosensors 2021, 11(11), 410, http://dx.doi.org/10.3390/bios11110410.

[44] Madhusankha, G. D. M. P., & Thilakarathna, R. C. N. Meat tenderization mechanism and the impact of plant exogenous proteases: A review. Arabian Journal of Chemistry 2021, 14(2), 102967, http://dx.doi.org/10.1016/j.arabjc.2020.102967.

[45] Shahani, K. M., Arnold, R. G., Kilara, A., & Dwivedi, B. K. Role of microbial enzymes in flavor development in foods. Biotechnology and Bioengineering 1976, 18(7), 891–907, https://doi.org/10.1002/bit.260180703.

[46] Ogundolie Frank Abimbola. Optimization and production of extracellular *Bacillus megaterium* (ISA08) alpha-amylase isolated from cassava dumpsite soil. Archive of Science & Technology 2022, 3(1), 92–104.

[47] Arif, A. B., Sasmitaloka, K. S., & Winarti, C. Effect of liquefaction time and enzyme addition on liquid sugar production from sweet sorghum starch by enzymatic hydrolysis. In: Susanto Budi Sulistyo, Ph. D. Jenderal Soedirman University, Purwokerto, Indonesia. IOP Conference Series: Earth and Environmental Science, IOP Publishing, 2019, March, 250, No. 1 012042, Bristol, England. doi:10.1088/1755-1315/250/1/012042

[48] Kocabaş DS, Lyne J, Ustunol Z. Hydrolytic enzymes in the dairy industry: Applications, market and future perspectives. Trends in Food Science & Technology. 2022: 119, 467–475.

[49] Fox, P. F., Cogan, T. M., & Guinee, T. P. Factors that affect the quality of cheese. In: Cheese, Elsevier, 2017, 617–641, Academic press, USA. https://doi.org/10.1016/B978-0-12-417012-4.00025-9.

[50] Merheb-Dini, C., Gomes, E., Boscolo, M., & da Silva, R. Production and characterization of a milk-clotting protease in the crude enzymatic extract from the newly isolated Thermomucor indicae-seudaticae N31: (Milk-clotting protease from the newly isolated Thermomucor indicae-seudaticae N31). Food Chemistry 2010, 120, 87–93.

[51] Khan, U., & Selamoglu, Z. Use of enzymes in dairy industry. A Review of Current Progress. Archives of Razi Institute 2020, 75(1), 131–136, https://doi.org/10.22092/ari.2019.126286.1341.

[52] Wilkinson, A. P., Gee, J. M., Dupont, M. S., Needs, P. W., Mellon, F. A., Williamson, G., et al. Hydrolysis by lactase phlorizin hydrolase is the first step in the uptake of daidzein glucosides by rat small intestine in vitro. Xenobiotica 2003, 33, 255–264.

[53] Silva, C. R. D., Delatorre, A. B., & Martins, M. L. L. Effect of the culture conditions on the production of an extracellular protease by thermophilic Bacillus sp and some properties of the enzymatic activity. Brazilian Journal of Microbiology 2007, 38, 253–258.

[54] Fox, P. F. Significance of indigenous enzymes in milk and dairy products. In: John R. Whitaker, Alphons G. J. Voragen, Dominic W.S. Wong., editors. Handbook of Food Enzymology, Florida: CRC Press, Florida, USA, 2002, 270–293.

[55] Cauvain, S., & Young, L. (2006). Ingredients and their influences. In: Baked products: science, technology and practice. Edited by Stanley P. Cauvain, Linda S. Young John Wiley & Sons 72–98. https://doi.org/10.1002/9780470995907.ch4

[56] Melim Miguel, A. S., Souza, T., Costa Figueiredo, E. V. D., Paulo Lobo, B. W., & Maria, G. Enzymes in bakery: Current and future trends. InTech 2013, doi: 10.5772/53168.

[57] Van der maarel, M. J. E. C., Van der veen, B., Uitdehaag, J. C. M., Leemhuis, H., & Dijkhuizen, L. Properties and applications of starch-converting enzymes of the α-amylase family. Journal of Biotechnology 2002, 94(2, 137–155.

[58] Synowiecki, J. The use of starch processing enzymes in the food industry. In: Polaina, J., & MacCabe, A. P., editors. Industrial Enzymes. Structure, Function and Applications, Dordrecht: Springer, 2007, 19–34.

[59] Goesaert, H., Brijs, K., Veraverbeke, W. S., Courtin, C. M., Gebruers, K., & Delcour, J. A. Wheat flour constituents: How they impact bread quality, and how to impact their functionality. Trends in Food Science and Technology 2005, 16(1–3, 12–30.

[60] Di Cagno, R., De Angelis, M., Corsettic, A., Lavermicocca, P., Arnault, P., Tossut, P., Gallo, G., & Gobbetti, M. Interactions between sourdough lactic acid bacteria and exogenous enzymes: Effects on the microbial kinetics of acidification and dough textural properties. Food Microbiology 2003, 20(1), 67–75.

[61] Kara, M., Sivri, D., & Koksel, H. Effects of high protease-activity flours and commercial proteases on cookie quality. Food Research International 2005, 2005 38(5), 479–486.

[62] Driss D, Bhiri F, Siela M, Bessess S, Chaabouni S, Ghorbel R. Retracted: Improvement of Breadmaking Quality by Xylanase GH 11 from *Penicillium occitanis* P ol6. Journal of Texture Studies. 2013 44(1), 75–84.

[63] Collins, T., Hoyoux, A., Dutron, A., Georis, J., Genot, B., Dauvrin, T., Arnaut, F., Gerday, C., & Feller, G. Use of glycoside hydrolase family 8 xylanases in baking. Journal of Cereal Science 2006, 43(1, 79–84.

[64] Stojceska, V., & Ainsworth, P. The effect of different enzymes on the quality of high-fibre enriched brewer's spent grain breads. Food Chemistry 2008, 110(4, 865–872.

[65] Feussner, I., Kuhn, H., & Wasternack, C. Lipoxygenase-dependent degradation of storage lipids. Trends in Plant Science 2001, 6(6, 268–273.

[66] Cumbee, B., Hildebrand, D. F., & Addo, K. Soybean flour lipoxygenase isozymes effects on wheat flour dough rheological and breadmaking properties. Journal of Food Science 1997, 62(2, 281–283.

[67] Bright, H. J., & Appleby, M. The pH dependence of the individual steps in the glucose oxidase reaction. The Journal of Biological Chemistry 1969, 244(13, 3625–3634.

[68] Weibel, M. K., & Bright, H. J. The glucose oxidase mechanism. The Journal of Biological Chemistry 1971, 246(9, 2734–2744.

[69] International Agency for Research on Cancer (IARC). Some naturally occurring and synthetic food components. In: Furocoumarines and Ultraviolet Radiation. IARC Monographs on Evaluation of the Carcinogenic Risk of Chemicals to Humans, Lyon, France: IARC Scientific Publication, 1986, 40, 207–220. http://monographs.iarc.fr/ENG/Monographs/vol40/volume40.pdf accessed 23 August 2012.

[70] Moore, M. M., & Chen, T. Mutagenicity of bromate: Implications for cancer risk assessment. Toxicology 2006, 221(2–3), 190–196.

[71] Tribess, T. B., & Tadini, C. C. Inactivation kinetics of pectinmethyesterase in orange juice as a function of pH and temperature-time process conditions. Journal of the Science of Food and Agriculture 2006, 86, 1328–1335.

[72] Kashyap, D. R., Vohra, P. K., Chopra, S., & Tewari, R. Applications of pectinases in the commercial sector: A review. Bioresource Technology 2001, 77, 215–227.

[73] Landbo, A. K., & Meyer, A. S. Statistically designed two step response surface optimization of enzymatic prepress treatment to increase juice yield and lower turbidity of elderberry juice. Innovative Food Science and Emerging Technologies 2007, 8, 135–142.

[74] Grzeskowiak-Przywecka, A., & Slominska, L. Saccharification of potato starch in an ultrafiltration reactor. Journal of Food Engineering 2007, 79, 539–545.

[75] Sharma, H. P., Patel, H., & Sharma, S. Enzymatic extraction and clarification of juice from various fruits–a review. Trends Post Harvest Technol 2014, 2(1, 1–14.

[76] Greice Sandri, I., Claudete Fontana, R., Menim Barfknecht, D., & Da silveira, M. M. Clarification of fruit juices by fungal pectinases. LWT – Food Science and Technology 2011, 44, 2217–2222.

[77] Kareem, S. O., & Adebowale, A. A. Clarification of orange juice by crude fungal pectinase from citrus peel. Nigerian Food Journal 2007, 25(1, 130–137.

[78] Haidar, A., Mhammad, H., & Fazaelipoor, Pectinase production in a defined medium using surface culture fermentation. International Journal of Industrial Chemistry 2010, 1(1), 5–10.

[79] Gupta, R., Gigras, P., Mohapatra, H., Goswami, V. K., & Chauhan, B. Microbial α- amylases: A biotechnological perspective. Process Biochemistry 2003, 38, 1599–1616.

[80] Saxena, S. Microbial enzymes and their industrial applications. In: Saxena, S., editor. Applied Microbiology, New Delhi: Springer, 2015, 121–154. https://doi.org/10.1007/978-81-322-2259-0_9.

[81] Pui, L., & Saleena, L. A. Enzyme-aided treatment of fruit juice: A review. Food processing: Techniques and Technology 2023, 53(1), 38–48, https://doi.org/10.21603/2074-9414-2023-1-2413.

[82] de la Fuente, M., Lombardero, L., Gómez-González, A., Solari, C., Angulo-Barturen, I., Acera, A., Vecino, E., Astigarraga, E., & Barreda-Gómez, G. Enzyme therapy: Current challenges and future perspectives. International Journal of Molecular Sciences 2021, 22(17), 9181, https://doi.org/10.3390/ijms22179181.

[83] Ensor, C. M., Holtsberg, F. W., Bomalaski, J. S., & Clark, M. A. Pegylated arginine deiminase (ADI-SS PEG20,000 mw) inhibits human melanomas and hepatocellular carcinomas in vitro and in vivo. Cancer Research 2002, 62(19, 5443–5450.

[84] Meghwanshi, G. K., Kaur, N., Verma, S., Dabi, N. K., Vashishtha, A., Charan, P. D., Purohit, P., Bhandari, H. S., Bhojak, N., & Kumar, R. Enzymes for pharmaceutical and therapeutic applications. Biotechnology and Applied Biochemistry 2020, 67, 586–601 https://doi.org/10.1002/bab.1919.

[85] Lee-Huang, S., Huang, P. L., Sun, Y., Huang, P. L., Kung, H. F., Blithe, D. L., & Chen, H. C. Lysozyme and RNases as anti-HIV components in beta-core preparations of human chorionic gonadotropin. Proceedings of the National Academy of Sciences of the United States of America 1999, 96(6), 2678–2681, https://doi.org/10.1073/pnas.96.6.2678.

[86] Lule, V. K., Garg, S., Tomar, S. K., Khedkar, C. D., & Nalage, D. Food Intolerance: Lactose Intolerance. Reference Module in Food Science-Encyclopedia of Food and Health, Elsevier, 2016, 43–48, Academic Press, USA. https://doi.org/10.1016/B978-0-12-384947-2.00312-3.

Opeyemi Titilayo Lala*, Titilayo Adenike Ajayeoba,
Oluwayomi Christianah Olaoye, Uchechi Chijioke,
Temitope Ruth Olopade, and Wuraola Olubunmi Ibitoye

Chapter 17
Emerging protein sources for future foods and security challenges

Abstract: The increasing global demand for protein, due to population growth and changing dietary preferences, demands a critical examination of alternative protein sources. This chapter discusses the potential of plant-based, insect-based, and lab-grown (cultured) proteins to address future food security challenges, while mitigating the environmental impact of traditional animal agriculture. The nutritional profiles, production methods, and sustainability aspects of these emerging protein sources are discussed, while highlighting their potential to reduce greenhouse gas emissions, land usage, and water consumption. Furthermore, we critically examined the technological, economic, regulatory, and consumer acceptance hurdles that hinders their widespread adoption. Addressing scalability and production costs, ensuring food safety, navigating regulatory landscapes, and overcoming consumer perceptions regarding sensory properties and novelty are key areas of focus. By synthesizing current research and identifying knowledge gaps, this chapter provides a comprehensive information of the emerging protein landscape, offering insights for researchers, policymakers, and industry stakeholders seeking to forge a sustainable and resilient food future. This analysis underscores continued innovation and strategic interventions to unlock the full potential of alternative proteins in ensuring global food security.

Keywords: Alternative proteins, emerging protein sources, lab-grown meat, food security, sustainability, consumer acceptance, food regulations, scalability

***Corresponding author: Opeyemi Titilayo, Lala,** Food Science and Nutrition Programme, Microbiology Department, Adeleke University, Ede, Osun, Nigeria,
e-mail: lala.opeyemi@adelekeuniversity.edu.ng

Titilayo Adenike Ajayeoba, Food Science and Nutrition Programme, Microbiology Department, Adeleke University, Ede, Osun

Oluwayomi Christianah Olaoye, Department of Animal Nutrition, College of Animal and Livestock Production, Federal University of Agriculture, Abeokuta

Uchechi Chijioke, Department of Microbiology, Adeleke University, Ede, Osun

Temitope Ruth Olopade, Department of Microbiology, College of Basic Medical and Allied Sciences, Salem University, Lokoja, Kogi

Wuraola Olubunmi Ibitoye, Food Science and Technology Programme, Bowen University, Iwo, Osun

https://doi.org/10.1515/9783111441238-018

17.1 Introduction

The demand for food, and in particular protein, is rising as the world's population rises. The United Nations projects that by 2050, there will be nearly 10 billion people on the planet, which means that food production systems will face many challenges in meeting this demand sustainably. Traditional protein sources, like fish and livestock, are limited by their resource intensity, environmental impact, and climate change susceptibility. As a result, researchers and innovators are turning to emerging sources of protein, like plant-based proteins, cultured meat, insects, and algae, to fill the gap between supply and demand [2].

There are a number of conflicting factors that make finding alternate protein sources imperative. First of all, traditional animal agriculture has a significant negative environmental impact. The production of livestock accounts for 14.5% of the world's greenhouse gas emissions and is a major cause of deforestation, biodiversity loss, and water scarcity. Furthermore, a number of fish species have been threatened by overfishing, endangering marine ecosystems and the livelihoods that depend on them [13].

Another major obstacle is the inefficiency of conventional protein synthesis techniques. To illustrate the inefficiency and resource intensity of conventional meat production, consider that it takes around 25 kilograms of feed to produce one kilogram of beef [20]. More effective methods of producing protein are clearly needed when resources like arable land and fresh water are limited.

Thirdly, the quest for more healthful protein substitutes has also been fueled by health concerns about the large consumption of red and processed meats. Consuming large amounts of these meats has been associated with a higher risk of heart problems, several types of cancer, and other illnesses [12]. Thus, increasing the variety of protein sources can improve global health outcomes.

Emerging protein sources have potential, but there are a number of issues that need to be resolved before they can be successfully included into world food systems. These difficulties are found in the fields of technology, economics, regulations, and consumer acceptability. Creating scalable and affordable processes to make substitute proteins is a major technological challenge. For instance, cultured meat, which is produced in a laboratory setting using animal cells, needs advanced bioreactors and growth media, both of which are currently costly and energy-intensive [21, 44]. Similar to this, complex technological procedures that are continuously being improved are needed to extract and process proteins from insects or algae into a form that is both pleasant and nutritionally equal to conventional proteins [58].

Another important consideration is the new protein sources' economic viability. Many alternative protein products are now more expensive than their conventional counterparts, which restricts the population's access to them. To cut costs and promote large-scale production, economic policies, subsidies, and R&D investments are required. To guarantee that these products are safe for consumption and environmentally sustainable, governments must set explicit regulations and safety requirements. Due to the extensive testing and approval processes involved, the regulatory process can be drawn out and complicated [22]. One of the biggest obstacles to the widespread use of alternative proteins may be consumer acceptance. Consumer decisions are significantly influenced by cultural preferences, dietary habits, and conceptions about what counts as "real food." For example, although eating insects is customary in some cultures, it is frequently viewed with contempt and opposition by others [59]. It takes clever marketing plans, instructional initiatives, and public awareness campaigns to change people's attitudes and promote the use of novel protein sources.

17.2 Emerging protein sources

17.2.1 Plant-based, insect, and lab-grown proteins

The global demand for protein is escalating due to population growth and increasing dietary protein consumption. This has led to a burgeoning interest in alternative protein sources that are sustainable, nutritious, and ethical. Among these, plant-based proteins, insect proteins, and lab-grown proteins stand out as promising options. Each of these sources offers unique benefits and nutritional profiles, contributing to a more sustainable and diverse food system [17].

17.2.2 Plant-based proteins

Plant-based proteins come from a variety of plants and are becoming more and more well-liked because of their favorable effects on health and the environment. Legumes, grains, nuts, seeds, and soy products are common sources. As an example, critical amino acid profiles and high protein content are well-known attributes of quinoa, lentils, and soybeans. For instance, soybeans are a complete protein source – they supply all nine essential amino acids needed by the human body – and have a protein content of roughly 36–40% [29]. Quinoa is a pseudo-cereal that provides approximately 14% protein, a significant quantity of fiber, and minerals that include iron and magnesium [57]. One of the significant advantages of plant-based proteins is their lower environmental footprint compared to animal proteins. Producing plant proteins generally requires less land and water, and emits fewer greenhouse gases [52]. Moreover, plant-based diets are associated with numerous health benefits, including reduced risks of chronic diseases such as heart disease, diabetes, and certain cancers. This is partly due to the high content of fiber, vitamins, and phytonutrients found in plants [34].

Plant-based proteins that are just starting to gain popularity include hemp and pea proteins. Pea protein, which is made from yellow peas, has an excellent balance of essential amino acids and provides around 25 grams of protein per 100 grams. It is hypoallergenic [3]. Hemp protein, which comes from hemp seeds, has 50% protein content and is high in omega-3 and omega-6 fatty acids, both of which are beneficial to cardiovascular health [28].

17.2.3 Insect proteins

Another new source that is abundant in vital nutrients and uses resources very effectively is insect proteins. Protein, good fats, vitamins, and minerals are abundant in edible insects like mealworms, locusts, and crickets. For example, crickets are high in iron and vitamin B_{12} and have roughly 65% protein by dry weight [47]. These are components that are frequently deficient in plant-based diets. Mealworms are an excellent source of fiber, omega-3 fatty acids, and a comparable protein level [58]. The farming of insects is quite sustainable. Compared to conventional livestock, insects use far less feed, water, and land, and they also produce a great deal less greenhouse gas [40]. Insects can also be raised on organic garbage, which supports efforts to reduce waste and promote a circular economy. Nutritionally speaking, insect proteins are very easily digested and include healthy fats and minerals that promote general well-being [59]. These days, businesses are creating a wide range of insect-based goods, including energy bars, protein powders, and snacks. For instance, protein bars and drinks made with cricket flour provide a sustainable substitute for traditional protein powders [58]. Insect-based bread and pasta are being innovatively incorporated into more recognizable culinary formats by startups.

17.2.4 Lab-grown proteins

A novel method of producing proteins is by lab-grown, or cultured, proteins. These proteins provide an alternative to conventional livestock farming for the production of meat because they are grown from animal cells in regulated settings. It is possible to customize lab-grown meat to have particular nutritional profiles, which could increase its health benefits. For example, it is possible to modify lab-grown beef to include less saturated fat and more beneficial fats, such as omega-3s [44]. The ability of lab-grown proteins to considerably lessen the environmental impact of meat production is one of its main advantages. Compared to normal cattle husbandry, cultured beef uses less water and land and produces less greenhouse emissions [53]. Additionally, there is less chance of contamination and foodborne illnesses due to the carefully controlled production process. Leading corporations are creating meat products using lab-grown meat. Memphis Meats and Mosa Meat, for instance, have been successful in producing lab-grown beef patties that taste and feel a lot like regular beef [49]. Finless Foods is in the process of creating lab-grown seafood, such bluefin tuna, with the goal of offering sustainable substitutes for overfished species [44].

17.3 Challenges associated with emerging food protein

The growing global population and changing dietary patterns are increasing the demand for protein, putting pressure on traditional sources like meat, dairy, and seafood. This has led to the exploration of alternative and emerging protein sources to meet the rising demand sustainably. However, these novel sources face several challenges; these challenges are shown in Figure 17.1.

17.3.1 Environmental challenges

While some novel protein sources like plant-based proteins and insects are touted as more environmentally friendly, their environmental impacts cannot be assumed and require careful evaluation across the entire life cycle [26]. Factors like land and water use, greenhouse gas emissions, and biodiversity impacts need to be considered.

17.3.2 Scalability and production costs

Scaling up production of emerging protein sources to meet global demand is a significant challenge. Many novel sources like cultivated meat, algae, and insects require the development of new value chains, infrastructure, and production processes, which can be costly and technically complex [26, 51].

17.3.3 Food safety and regulatory hurdles

Ensuring food safety and gaining regulatory approval for novel protein sources is a crucial challenge. Rigorous testing and evaluation are necessary to address potential risks, such as allergens, contaminants, and unintended effects on human health. Navigating regulatory frameworks across different regions can also be a barrier to market entry [17]

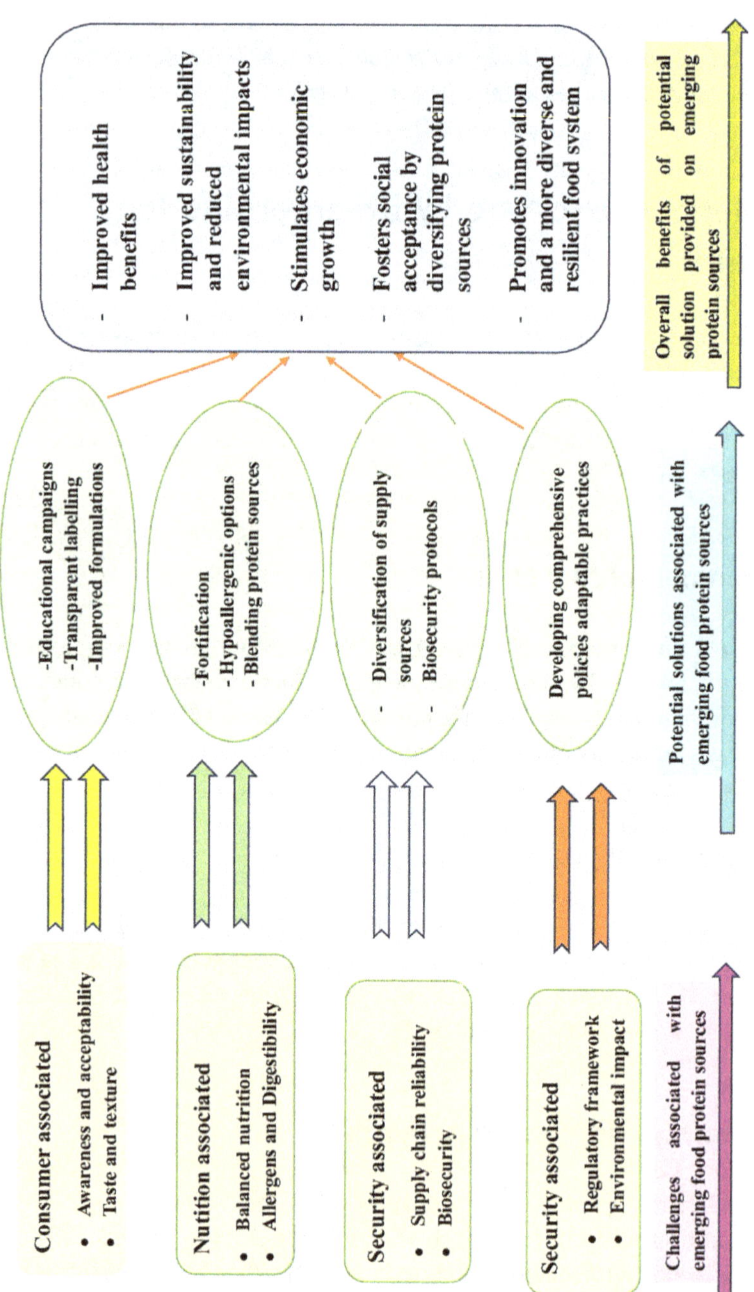

Figure 17.1: Interplay of challenges, potential solutions associated with emerging protein sources, and their benefits to the populace.

17.3.4 Consumer acceptance and sensory properties

Consumer acceptance is a critical factor for the success of emerging protein sources. Overcoming potential aversions, addressing sensory properties like taste and texture, and educating consumers about the benefits and safety of these novel foods are ongoing challenges [7, 26]. Consumer acceptance and perception pose significant challenges for the widespread adoption of alternative protein sources. The key challenges include

17.3.4.1 Sensory appeal and mimicking traditional meat

Many consumers accept that alternative proteins will have an inferior taste, texture, and sensory experience compared to traditional meat products. Replicating the sensory properties of meat is a major challenge for plant-based, cultivated, and other novel protein sources. Overcoming this perception and achieving taste and texture parity is crucial for consumer acceptance [10, 39].

Improving the taste, texture, and overall sensory experience of alternative protein sources is crucial for widespread consumer acceptance and adoption. Several studies have highlighted this as a key challenge that requires extensive research and development efforts.

Sensory appeal is a major factor influencing consumer attitudes toward alternative proteins. Many consumers expect these products to have an inferior taste and texture compared to conventional meat. Replicating the sensory properties of meat remains a significant hurdle for plant-based, cultivated, and other novel protein sources [17, 36].

Research has indicated that consumers maintain the belief that products derived from alternative proteins are worse in terms of taste, texture, and sensory experience when compared to meat-based products. In order to determine if the current trend represents a significant shift in consumer preference [62], he stressed the necessity for a study on repeated exposure and long-term consumer acceptance of the perceptual qualities of novel plant-based meat alternatives (PBMAs). Tasty food items may increase consumer receptivity to meat substitutes, which could spur innovation and technological optimization to enhance the nutritional composition and organoleptic characteristics of meat substitutes [59]. The need to optimize technological approaches to increase consumer acceptance was identified and the challenges were discussed in mimicking the sensory properties of meat, stating that "Replicating the unique sensory experience ox meat remains a major challenge for plant-based meat alternatives". They emphasized the need for research and development to improve the texture, mouth feel, and flavor of plant-based meat analogues. Extensive research and development efforts are necessary to address these sensory challenges and improve the taste, texture, and overall eating experience of alternative protein sources.

This includes optimizing existing technologies, exploring novel processing methods, and leveraging advancements in food science and ingredient development [10].

17.3.4.2 Newness and unfamiliarity

Alternative proteins are relatively new and unfamiliar to many consumers, which can create skepticism and resistance. Educating consumers about the production processes, safety, and benefits of these novel foods is an ongoing challenge. In order for consumers to accept alternative sources of food proteins, consumers must adopt framing information practices [32]

17.3.5 Price and cost competitiveness

Currently, many alternative protein products are more expensive than their conventional counterparts, which can discourage price-conscious consumers. Achieving cost parity through scalable and efficient production processes is a key challenge for widespread adoption [51].

As the global population continues to grow and the demand for protein increases, there is a pressing need to explore alternative protein sources that are sustainable, environmentally friendly, and cost-effective [55]. Plant-based, cultivated, and fermentation-derived proteins have emerged as promising alternatives to traditional animal-based proteins. However, to unlock their full potential, the alternative protein industry must overcome significant challenges related to cost and scalability.

One of the primary barriers to scaling alternative protein production is the high cost associated with sourcing, processing, and manufacturing these products. For cultivated meat, the cost of growth media and specialized facilities can be prohibitively expensive. Fetal bovine serum (FBS), a common ingredient in growth media, can cost up to $2,000 per liter, making it a major bottleneck in the production process [17]. Similarly, the overhead costs of the facilities required to produce cultivated meat at scale are significant, with much of the equipment being custom-created and requiring strict sterile conditions.

In addition to cost challenges, the alternative protein industry also faces scalability issues. Producing alternative proteins at scale requires significant investments in infrastructure, including specialized facilities and equipment. For instance, the production of cultured meat requires bioreactors; however, the bioreactors that are currently on the market were designed for medicines that have high unit costs, hence they are not appropriate for the manufacture of cultivated meat at a reasonable cost [33]. More high-volume bioreactors that are economically feasible must be created in order to meet this problem.

Governments can play a crucial role in supporting the alternative protein industry by providing subsidies or creating central facilities to help lower costs until demand grows large enough [1]. Food Innovate is a multiagency project launched by Singapore to support capacity building, resource provision, and partnership facilitation in the food production sector through innovation. By facilitating access to facilities, experts, and partners, initiatives like Food Innovate can help companies scale more easily.

To address these cost challenges, researchers are exploring alternative sources of growth media that are more sustainable and cost-effective. A team of Singaporean researchers led by Professor William Chen from the Nanyang Technological University has shown that FBS can be replaced by fermented okara, a by-product of tofu and soy milk production, which costs only $2 per liter [1]. This innovation not only reduces costs but also contributes to a more sustainable production process.

For plant-based and microorganism-based proteins, cost reduction can be achieved through improvements in sourcing, growth conditions, and processing techniques.

17.3.6 Cultural and psychological barriers

Cultural traditions, dietary preferences, and psychological factors can influence consumer attitudes toward alternative proteins. Overcoming these deeply ingrained barriers and fostering acceptance across diverse cultures and demographics remains a significant hurdle.

17.3.7 Nutritional concerns

Some consumers may have concerns about the nutritional adequacy and potential presence of anti-nutritional factors in alternative protein sources. While some novel protein sources offer promising nutritional profiles, others may lack certain essential amino acids or have anti-nutritional factors that need to be addressed through processing or breeding [7]. Ensuring a complete and balanced amino acid profile for human nutrition is crucial. Addressing these concerns through rigorous research and transparent communication is essential for building consumer trust, which requires a multifaceted approach involving product innovation, consumer education, regulatory support, and cultural shifts [10].

17.3.8 Economic and social factors

The adoption of emerging protein sources may face economic barriers, such as high production costs and limited access to capital, particularly in developing countries. Social and cultural factors, like traditional dietary preferences and perceptions, can also influence acceptance and adoption [26].

Addressing these challenges requires collaborative efforts from various stakeholders, including researchers, industry, policymakers, and consumers. Continued research, innovation, and investment are necessary to overcome technical hurdles, improve sustainability, ensure safety and nutritional quality, and foster consumer acceptance of emerging protein sources [51].

17.4 Potential deficiencies and imbalances in essential nutrients

17.4.1 Amino acid profile

Plant-based proteins present a significant nutritional issue due to their inadequate amino acid composition. One or more necessary amino acids are lacking in the majority of plant-based proteins, which might be problematic if these sources are the only ones used. Legumes and grains are two examples of plant proteins that can be combined to help reduce this problem and offer a more balanced intake of amino acids [65].

17.4.2 Micronutrient deficiencies

Despite the high nutritional content of insect proteins, it might not be able to satisfy every dietary demand. For example, despite their high protein, iron, and vitamin B_{12} content, they might not contain enough of other vital minerals, such as calcium and vitamin D [47]. Similarly, unless they are explicitly supplemented during the production process, lab-grown meats may lack some of the naturally occurring micronutrients found in traditional meats, depending on their production methods [44].

17.4.3 Digestibility and bioavailability

There can be variations in the nutrients' digestibility and bioavailability derived from these alternate sources. For instance, phytates and tannins found in some plant proteins have anti-nutritional properties that can prevent the absorption of nutrients like zinc and iron [19]. Protein digestibility may be impacted by the presence of chitin in insect proteins, contingent on the processing method used [16].

17.4.4 Ensuring nutritional adequacy for all population groups

17.4.4.1 Diverse and balanced diets

It is imperative to encourage varied and well-balanced meals in order to guarantee that newly developed protein sources fulfill nutritional needs. Nutritional imbalances may result from relying exclusively on one kind of protein source. A more comprehensive amino acid profile, for instance, can be achieved by incorporating a range of plant-based proteins, and filling any possible micronutrient deficiencies by mixing insect proteins with other dietary sources [58].

17.4.4.2 Nutritional fortification and enhancement

In order to enhance these protein sources' health benefits, the process of fortification and bioengineering can be used. For example, to address common shortages, plant-based products can be supplemented with critical vitamins and minerals like calcium, iron, and vitamin B_{12} [46]. In a similar vein, lab-grown meats can be modified to include larger concentrations of particular nutrients, thus improving their total nutritional worth [45]

17.4.4.3 Targeted nutritional guidelines

It is crucial to create dietary rules that are particular to each population group. For instance, while promoting alternative protein sources, it is important to take into account the special dietary demands of the elderly, pregnant women, and children. It ensures that these populations' particular needs for protein, vitamins, and minerals are satisfied by customizing their dietary guidelines [13].

17.4.5 Awareness and public education

In order to successfully incorporate developing protein sources into diets, it is imperative that the public be made aware of both their advantages and disadvantages. Campaigns to raise consumer awareness can assist in forming decisions about how to combine various protein sources to create a diet that is both balanced and nutrient-dense [46].

17.5 Security challenges with emerging food protein sources

17.5.1 Food safety risks

The incorporation of novel protein sources such as insects into the food system introduces new food safety concerns. Insect proteins, for example, pose risks of contamination from microbes or toxins due to their processing and storage methods, which are not as well-regulated or understood as those for traditional proteins. Furthermore, research has revealed that the protein content in insects might be overestimated, which raises questions about their nutritional value and authenticity. Ensuring the safety of these new protein sources requires rigorous testing and standardized regulatory frameworks to prevent contamination and ensure accurate nutritional labeling [54].

17.5.2 Food fraud vulnerabilities

The emerging protein market is particularly vulnerable to food fraud due to gaps in current testing methodologies and regulatory oversight. Similar to the Melamine Scandal, where harmful substances were added to food products, emerging proteins might be subject to adulteration or mislabeling, which poses significant risks to consumers' trust and products' integrity. Robust tracking systems and stringent regulatory enforcement are necessary to combat fraudulent activities and safeguard the authenticity of these protein sources [54]

17.5.3 Supply chain risks

Overreliance on a narrow range of alternative protein sources can expose the food supply chain to various risks. Disruptions in production, whether from cyber attacks, natural disasters, or the introduction of pathogens, could have severe economic and social impacts. For instance, a pathogen outbreak in a key production facility could halt production and create shortages, affecting both food accessibility and economic stability. Building resilient and diversified supply chains is crucial to mitigating these risks and ensuring a steady supply of alternative proteins [25]

17.5.4 Bioterrorism threats

While the alternative protein sector is currently less vulnerable to bioterrorism than traditional animal agriculture, its rapid expansion and fragmented market structure could attract malicious actors. The lack of consolidation within the industry means that attacks could have significant localized impacts, disrupting production and economic competitiveness. Strengthening biosecurity measures and promoting industry-wide cooperation are essential steps to safeguard against bioterrorism threats [31]

17.5.5 Technological risks

The adoption of advanced technologies in producing emerging protein sources introduces new risks related to data security, intellectual property rights, and potential technological failures. As the industry becomes increasingly reliant on specific technologies, vulnerabilities to cyber threats and operational disruptions increase. Without adequate safeguards, the industry could face significant setbacks due to data breaches or technological malfunctions. Implementing robust cybersecurity protocols and investing in resilient technological infrastructures are vital for mitigating these risks [6]

17.5.6 Market volatility

Focusing on a limited number of emerging protein sources can make the market prone to volatility. Factors such as climate change, geopolitical tensions, or sudden regulatory shifts can disrupt production and supply chains, leading to price instability and potential food insecurity. Diversifying protein sources and establishing flexible supply chains can help cushion the market against such external shocks and ensure a stable food supply.

17.5.7 Social equity issues

The concentration of production and consumption around specific emerging protein sources can exacerbate social inequalities. Small-scale producers and regions dependent on traditional protein sources may face economic disadvantages or marginalization. This shift could disrupt livelihoods and cultural practices, particularly in developing regions. Ensuring inclusive growth and equitable access to new protein markets is crucial for addressing these social equity issues [38].

17.5.8 Environmental impact

Intensive farming of a few emerging protein sources can have detrimental environmental effects. Practices such as monoculture farming can lead to deforestation, water depletion, soil degradation, and biodiversity loss. Moreover, these environmental impacts could further exacerbate climate change, undermining the sustainability of food production systems. Promoting sustainable agricultural practices and encouraging diversity in protein production are essential for minimizing environmental damage [26]

17.5.9 Nutritional imbalance

Relying heavily on a small number of emerging protein sources could result in nutritional imbalances. Different protein sources offer varying profiles of essential amino acids, vitamins, and minerals. Overreliance on any single source could lead to deficiencies or excesses in specific nutrients, adversely affecting consumer health. A balanced dietary approach, incorporating a variety of protein sources, is necessary to ensure comprehensive nutrition [8]

17.5.10 Governance challenges

The rapid growth of the alternative protein industry presents significant governance challenges. Inadequate regulation, lack of standardization, and weak enforcement mechanisms can undermine transparency, accountability, and consumer trust in the industry. Developing robust regulatory frameworks and fostering international cooperation are critical for addressing these governance issues [35].

17.6 The policy challenges with emerging food protein sources

The demand for sustainability, food security, and health considerations is causing major changes in the global food system. Alternative food protein sources have come to light as possible remedies for these issues, such as insect proteins, cultured meats, and plant-based proteins. However, there are a number of policy challenges that must be overcome before these substitutes can be incorporated into popular diets. These obstacles include things like legal frameworks, consumer acceptance, environmental impact evaluations, financial ramifications, and moral dilemmas.

The challenges include the following:

17.6.1 Lack of standardization

One of the main policy challenges is lack of standard regulations for novel proteins across various geographies. The manufacture, labeling, and marketing of alternative proteins are governed by laws unique to each nation, which results in discrepancies that restrict global trade and mislead consumers. For example, the United States and the European Union have distinct regulations on plant-based and cultured meats. While the European Food Safety Authority (EFSA) plays a central role in the approval process in the EU, the FDA and USDA jointly oversee regulation of these products in the United States [30]. Safety and approval processes – Since many newly discovered protein sources are novel, thorough safety evaluations and approval procedures are required. Regulations already in place frequently do not apply directly to these novel food products, necessitating the creation of new frameworks or their adaptation. This can lead to delay in releasing products into the market. According to [50], before being approved for commercialization, cultured meat products have to go through a rigorous safety review process. The first nation to authorize the sale of cultured meat was Singapore in 2020, underscoring the extensive and difficult regulatory procedure needed.

17.6.2 Labeling and transparency

Making well-informed choices and earning the trust of consumers depend on accurate and transparent labeling. Policy makers need to make sure that alternative protein labels include comprehensive information regarding the sources, components, and nutritional value of the alternative protein. This involves discussing controversial topics like the legitimacy of labeling plant-based goods as "meat" or "milk." The regulatory environment is made more complex in the United States by the discussions surrounding labeling rules, which have resulted in a number of state-level laws that limit the use of traditional meat and dairy words for plant-based substitutes [61].

17.6.3 Consumer acceptance

17.6.3.1 Cultural and social barriers

Food preferences are strongly impacted by social and cultural conventions. In many areas, emerging protein sources – insect-based and cultured meats in particular – face strong cultural resistance. To encourage acceptance, policymakers must address these cultural sensitivities through focused education and awareness efforts [23].

For example, although eating insects is widespread in some regions of Asia and Africa, it is frequently met with opposition in Western nations. Consumer attitudes can be changed by educational programs emphasizing the nutritional and environmental advantages of insect proteins [60].

17.6.3.2 Misinformation and perception

When it comes to new proteins' safety and potential health advantages, there is a good deal of misinformation out there. Policies that support scientific research and public education must be put in place by governments and regulatory agencies in order to combat false information and increase consumer confidence [11]. As demonstrated by the introduction of genetically modified organisms (GMOs) in the past, public outreach and education programs can play a significant role in dispelling myths and concerns regarding novel food technology [48].

17.6.4 Environmental impact Assessments

17.6.4.1 Comprehensive life cycle analyses

Even while emerging proteins are frequently promoted as being more environmentally friendly than conventional animal husbandry, thorough life cycle assessments (LCA) are required to determine their actual environmental impact. Regulations should require strict life cycle assessments (LCAs) to evaluate the effects of alternative protein production on biodiversity, water use, and carbon foot print. Incentives for sustainable practices – Studies contrasting the environmental effects of conventional and cultured meat, for example, have produced contradictory findings, highlighting the necessity of thorough and consistent environmental assessments [53].

Governments can provide businesses that implement sustainable manufacturing methods with incentives like tax cuts, subsidies, or grants. The initial high costs of adopting sustainable technology and methods in the production of alternative proteins can be somewhat compensated by these regulations. Nations like the Netherlands have positioned themselves as leaders in the sector by sponsoring and promoting research into sustainable protein sources [58].

17.6.5 Economic implications

17.6.5.1 Market competition and fairness

The conventional dairy and meat sectors may be affected by the emergence of new proteins. In order for manufacturers of traditional and alternative proteins to compete fairly, policymakers must provide level-playing fields. This could entail reevaluating the funding and assistance now provided to conventional livestock husbandry. The meat and dairy industries, for instance, benefit greatly from substantial subsidies from the US government, which may disadvantage newer protein companies that may not receive the same kind of assistance [37, 41].

17.6.5.2 Job creation and transition

There may be job losses in traditional agriculture industries as a result of the shift to new proteins. In order to maintain social justice and economic stability, policies should incorporate programs for retraining and upskilling employees, so they may move into new positions within the alternative protein sector. The negative economic effects of this transition can be lessened with the support of programs that offer education and training to workers who are being replaced by technological improvements in the food production industry [62].

17.6.6 Ethical considerations

17.6.6.1 Animal welfare

Reducing animal suffering is a major ethical motivator for developing proteins. In order to enhance animal well-being, policies should support the development of alternative proteins. This could involve introducing welfare concerns into rules governing the production of food. For example, cultured meat addresses many of the ethical issues surrounding conventional meat production by providing a method of producing meat without requiring the slaughter of animals [27].

17.6.6.2 Genetic engineering and biotechnology

Advanced biotechnological procedures are used in many of the developing protein sources, particularly in genetically modified plants and cultured meats. Transparent procedures that involve stakeholder participation and public consultation are necessary to resolve ethical concerns around genetic alteration and biotechnology. Policies that strike a balance between innovation and morality can guarantee that new technologies are created in an ethical and responsible manner [9].

17.6.6.3 Harmonizing international standards through global coordination and trade

International cooperation is necessary to enable the commerce of new proteins on a worldwide scale. It is possible to ensure that alternative proteins can be exchanged across borders without needless obstacles, by harmonizing standards and regulations via the work of organizations like the Codex Alimentarius Commission. The global market for developing proteins can be streamlined with the support of international agreements on food safety and labeling requirements, which will benefit both producers and consumers [15].

17.6.6.4 Assisting underdeveloped nations

Adopting novel proteins may provide particular difficulties for developing nations, such as restricted access to capital and technology. In order to ensure global food security and justice, international policy and aid programs should assist these nations in growing their sectors of alternative protein.

Programs that offer developing nations financial and technical support can aid them in overcoming entrance hurdles and reaping the benefits of the global transition to sustainable protein sources [42].

Finance and assistance for research and development – Innovation and progress in the field of emerging proteins rely heavily on government funding and support for research and development. Funding for pilot programs, scientific research, and public–private partnerships should be given top priority by policy makers. Significant government funding for the study of alternative proteins has sparked innovation and drawn funding from the commercial sector in nations like Singapore and Israel [4].

17.6.6.5 Collaboration and data sharing

Emerging protein development and adoption can be sped up by promoting data exchange and cooperation between academic institutions, commercial enterprises, and governmental organizations. Policies that support cooperative projects and unrestricted access to research results can spur innovation and solve shared problems. The scientific and technological obstacles related to new proteins can be surmounted with the aid of platforms that facilitate cooperation and information sharing [19].

In summary, the policy challenges with food protein sources revolve around establishing clear regulatory frameworks, addressing cultural perceptions, managing economic implications, securing government support and funding, and promoting sustainability and resilience in the food system transition toward alternative protein sources. These challenges require a comprehensive approach that considers safety, quality, economic viability, and environmental sustainability.

17.7 Future perspective of food protein sources

Proteins are essential macronutrients that play critical roles in the growth, repair, and maintenance of the human body. Traditionally, animal-based proteins have dominated the dietary landscape, but recent shifts toward sustainability, health consciousness, and ethical considerations have spurred interest in alternative protein sources. This review explores various food protein sources, including animal-based, plant-based, insect-based, and cultured proteins, highlighting their nutritional profiles, benefits, and potential challenges.

17.7.1 Animal-based proteins

Meat, dairy, eggs, and fish are examples of animal-based proteins that are regarded as complete proteins because they have all nine of the essential amino acids that the human body needs. They are effectively absorbed and utilized because they are highly bioavailable. Particularly high in vital nutrients for brain and cardiovascular health are meat and fish, which also contain iron, vitamin B_{12}, and omega-3 fatty acids [17].

However, there are ethical and environmental issues with the production of proteins derived from animals. Deforestation, water pollution, and greenhouse gas emissions are all significantly impacted by livestock husbandry [14]. Additionally, intensive farming practices raise animal welfare issues. Health concerns also arise with high consumption of red and processed meats, which are associated with increased risks of cardiovascular diseases, certain cancers, and other health issues [63].

17.7.2 Plant-based proteins

Legumes (beans, lentils, peas, etc.), grains (quinoa, brown rice, etc.), nuts, seeds, and soy products (tofu, tempeh, etc.) are sources of plant-based proteins. These sources are becoming increasingly popular because of their improved health, ethical implications, and less environmental effect. Diets high in plants have been associated with decreased risks of type 2 diabetes, heart disease, hypertension, and various types of cancer [24].

In respect to nutrition, plant proteins frequently lack one or more essential amino acids; hence, a variety of plant-based meals is needed to be combined to provide a complete amino acid profile.

For example, combining rice and beans provides a full spectrum of essential amino acids. Plant proteins are also typically lower in bioavailability compared to animal proteins, although some, like soy, are exceptions.

17.7.3 Insect-based proteins

Insects are an emerging protein source that offers a sustainable and efficient alternative to traditional animal farming. Insects that can be eaten, such as mealworms, grasshoppers, and crickets, are high in vitamins, minerals, and protein. Compared to animals, they require quite less space, water, and feed, and they emit fewer greenhouse emissions [13, 56].

Entomophagy, the custom of consuming insects, is widespread throughout the world and is especially prevalent in Asia, Africa, and Latin America.

However, acceptance in Western countries is limited due to cultural biases and lack of familiarity. There are also regulatory and safety challenges to address, such as standardizing production practices and ensuring the absence of allergens and contaminants [18].

17.7.4 Cultured proteins

Cultured or lab-grown proteins, derived from animal cells, represent a cutting-edge innovation aimed at producing meat and dairy products without the need for animal farming. This technology involves culturing animal cells in a controlled environment to grow tissue that mimics conventional meat. Cultured meat promises to reduce environmental impact, improve animal welfare, and provide a safer product, free from antibiotics and hormones [44].

While the potential benefits are substantial, cultured proteins face significant hurdles. These include high production costs, scalability challenges, and consumer acceptance. Additionally, there are regulatory barriers to navigate before cultured meat can become mainstream. Ongoing research and investment are expected to address these challenges, potentially making cultured proteins a viable alternative in the future [5].

17.7.5 Comparative analysis and future outlook

Each protein source presents unique advantages and challenges. Animal-based proteins are nutritionally complete and highly bioavailable but pose significant environmental and health risks. Plant-based proteins offer a more sustainable and health-conscious option but may require careful dietary planning to ensure complete nutrition. Insect-based proteins are highly sustainable and nutritious but face cultural and regulatory hurdles. Cultured proteins hold promise for future food security and sustainability but need technological and economic advancements to become mainstream.

17.7.6 Nutritional considerations

Balancing protein intake from various sources can optimize health benefits, while mitigating drawbacks. For example, integrating plant-based proteins into a predominantly animal-based diet can enhance nutrient diversity and reduce the environmental footprint. Similarly, supplementing traditional diets with insect-based or cultured proteins can provide additional nutritional benefits and support sustainability goals [43, 64].

17.7.7 Environmental impact

Given the effects of climate change and resource depletion, reducing the environmental impact of protein production is essential. Water, land, and greenhouse gas emissions can all be greatly reduced by switching to plant-based and alternative protein sources. Policy initiatives, consumer education, and advancements in food technology will play vital roles in driving this transition.

17.7.8 Ethical and cultural factors

Ethical considerations around animal welfare and sustainable farming practices are increasingly influencing consumer choices. Plant-based and cultured proteins offer cruelty-free alternatives that align with ethical values. Cultural acceptance and dietary habits will shape the adoption of alternative proteins, requiring targeted education and marketing strategies to overcome resistance and promote inclusivity [24].

17.8 Conclusion

In order to secure future food security and sustainability, it is critical to investigate and accept novel protein sources as the world's population continues to rise and the pressure on conventional food systems increases. This blog post explored the potential roles of various alternative protein sources, such as plant-based proteins, insect protein, and lab-grown proteins, in addressing global food security challenges. It also emphasized the need for cross-sector collaboration to accelerate the adoption and acceptance of these sources, as well as the importance of ongoing research and innovation. In conclusion, adopting new sources of protein is essential to attaining environmental sustainability and global food security. Switching to alternative proteins will reduce the environmental effect of food production, while also helping to meet the nutritional needs of an expanding population. In order to overcome these obstacles and realize the full potential of plant-based, insect-based, and lab-grown proteins, more research, funding, and legislative support are required. Through innovation and cross-sector collaboration, we can create a robust and sustainable food system for the future.

References

[1] Academy of Nutrition and Dietetics. (2016). Plant-based diets: A quick guide to getting started. https://www.eatright.org/food/nutrition/vegetarian-and-special-diets/plant-based-diets.

[2] Aiking, H., & De Boer, J. The next protein transition. Trends in Food Science & Technology 2020, Nov 1, 105, 515–522.

[3] Aiking, H. Future protein supply. Trends in Food Science & Technology 2011, Mar 1, 22(2–3), 112–120.

[4] Apostolopoulos, N., & Li, Y. Government support and innovation in the food industry: Evidence from alternative proteins. Food Policy 2020, 92, 101840.

[5] Bryant, C. J., & Barnett, J. C. Consumer acceptance of cultured meat: An updated review (2018–2020). Applied Sciences 2020, 10(15), 5201.

[6] Cheftel, J. C. Emerging risks related to food technology. In: Advances in Food Protection: Focus on Food Safety and Defense, Dordrecht: Springer Netherlands, 2011, 223–254.

[7] Caporgno, M. P., Böcker, L., Müssner, C., Stirnemann, E., Haberkorn, I., Adelmann, H., Handschin, S., Windhab, E. J., & Mathys, A. Retraction notice to "Corrigendum to "Extruded meat analogues based on yellow, heterotrophically cultivated Auxenochlorella protothecoides microalgae"" Innovative food science and emerging technologies. Innovative Food Science & Emerging Technologies 2021, 60(2020), 102319 *72*, p.102686.

[8] Colgrave, M. L., Dominik, S., Tobin, A. B., Stockmann, R., Simon, C., Howitt, C. A., Belobrajdic, D. P., Paull, C., & Vanhercke, T. Perspectives on future protein production. Journal of Agricultural and Food Chemistry 2021, Dec 9, 69(50), 15076–83.

[9] Davies, I., Macnaghten, P., & Kearnes, M. Synthetic biology and the governance of emerging technologies. Science and Public Policy 2020, 47(1), 12–21.

[10] Dekkers, B. L., Boom, R. M., & Van der Goot, A. J. (2022). https://www.tandfonline.com/doi/full/10.1080/10408398.2022.2107994

[11] Elzerman, J. E., Hoek, A. C., Van Boekel, M. A., & Luning, P. A. Consumer acceptance and appropriateness of meat substitutes in a meal context. Food Quality and Preference 2011, Apr 1, 22(3), 233–40.

[12] FAO. (2013). Edible insects: Future prospects for food and feed security. http://www.fao.org/3/i3253e/i3253e.pdf

[13] Food and Agriculture Organization (FAO). Dietary Protein Quality Evaluation in Human Nutrition, FAO, 2013, Nutrition Bulletin 38(4):42 1–8.

[14] FAO. Tackling Climate Change through Livestock: A Global Assessment of Emissions and Mitigation Opportunities, Food and Agriculture Organization of the United Nations, 2013, Pastoralism 4:1–6.

[15] FAO. (2021). Codex Alimentarius: International food standards. Retrieved from http://www.fao.org/fao-who-codexalimentarius

[16] Finke, M. D. Complete nutrient composition of commercially raised invertebrates used as food for insectivores. Zoo Biology: Published in Affiliation with the American Zoo and Aquarium Association 2002, 21(3), 269–285.

[17] Forum, F. National academies of sciences, engineering, and medicine. Alternative protein sources: Balancing food innovation. Sustainability, Nutrition, and Health: Proceedings of a Workshop, 2023. National Academics Press(US).

[18] Gahukar, R. T. Entomophagy and human food security. International Journal of Tropical Insect Science 2011, Sep, 31(3), 129–144.

[19] Gupta, R. K., Gangoliya, S. S., & Singh, N. K. Reduction of phytic acid and enhancement of bioavailable micronutrients in food grains. Journal of Food Science and Technology 2015, Feb, 52, 676–684.

[20] Godfray, H. C., Aveyard, P., Garnett, T., Hall, J. W., Key, T. J., Lorimer, J., Pierrehumbert, R. T., Scarborough, P., Springmann, M., & Jebb, S. A. Meat consumption, health, and the environment. Science 2018, Jul 20, 361(6399), eaam5324.

[21] Gravely, E., & Fraser, E. Transitions on the shopping floor: Investigating the role of Canadian supermarkets in alternative protein consumption. Appetite 2018, Nov 1, 130, 146–156.

[22] IV, G., Sturme, M., Hugenholtz, J., & Bruins, M. Review and analysis of studies on sustainability of cultured meat.

[23] Hartmann, C., & Siegrist, M. Consumer perception and behaviour regarding sustainable protein consumption: A systematic review. Trends in Food Science & Technology 2017, Mar 1, 61, 11–25.

[24] Harvard, T. H. Chan school of public health. (2020).The Nutrition Source: Protein. https://www.hsph.harvard.edu/nutritionsource/what-should-you-eat/protein/

[25] Hadi, J., & Brightwell, G. Safety of alternative proteins: Technological, environmental and regulatory aspects of cultured meat, plant-based meat, insect protein and single-cell protein. Foods 2021, May 28, 10(6), 1226.

[26] Henchion, M., Hayes, M., Mullen, A. M., Fenelon, M., & Tiwari, B. Future protein supply and demand: Strategies and factors influencing a sustainable equilibrium. Foods 2017, Jul 20, 6(7), 53.

[27] Hocquette, J. F. Is in vitro meat the solution for the future?. Meat Science 2016, Oct 1, 120, 167–176.

[28] House, J. D., Neufeld, J., & Leson, G. Evaluating the quality of protein from hemp seed (Cannabis sativa L.) products through the use of the protein digestibility-corrected amino acid score method. Journal of Agricultural and Food Chemistry 2010, Nov 24, 58(22), 11801–11807.

[29] Joshi, M., & Kumar, S. Soybean (Glycine max) seed protein analysis through two-dimensional polyacrylamide gel electrophoresis. Journal of Plant Biochemistry & Biotechnology 2015, 24(4), 450–453.

[30] Kelley, S., & Evans, J. Regulatory frameworks for novel foods: Comparison between the EU and the US. Food Policy 2020, 91, 101817.

[31] Kowalczewski, P. Ł., Pratap-Singh, A., & Kitts, D. D. Emerging protein sources for food production and human nutrition. Molecules 2023, Mar 16, 28(6), 2676.

[32] Kwasny, T., Dobernig, K., & Riefler, P. Towards reduced meat consumption: A systematic literature review of intervention effectiveness, 2001–2019. Appetite 2022, Jan 1, 168, 105739.

[33] National Institutes of Health. (2021). Protein: functions and foods. https://www.nih.gov/news-events/nih-research-matters/protein-functions-and-foods.

[34] McEvoy, C. T., Temple, N., & Woodside, J. V. Vegetarian diets, low-meat diets and health: A review. Public Health Nutrition 2012, Dec, 15(12), 2287–2294.

[35] Money, A., Srivastav, S., & Collett, K. A. The new protein economy: Policy directions. Oxford Smith School of Enterprise and the Environment 2022, Sep 14, 14, 2022–2029 14, 2022–09.

[36] Nungesser, F., & Winter, M. Meat and social change: Sociological perspectives on the consumption and production of animals. Österreichische Zeitschrift Für Soziologie 2021, Jun, 46, 109–124.

[37] OECD. Agricultural Policy Monitoring and Evaluation 2020, OECD Publishing, 2020.

[38] O'Connor, R. (2023). Alternative proteins: Addressing challenges for a sustainable future. https://www.linkedin.com/pulse/alternative-proteins-addressing-challenges-sustainable-future.

[39] Onwezen, M. C., & Van der Weele, C. N. When indifference is ambivalence: Strategic ignorance about meat consumption. Food Quality and Preference 2016, Sep 1, 52, 96–105.

[40] Oonincx, D. G., Van Itterbeeck, J., Heetkamp, M. J., Van Den Brand, H., Van Loon, J. J., & Van Huis, A. An exploration on greenhouse gas and ammonia production by insect species suitable for animal or human consumption. PloS One 2010, Dec 29, 5(12): e14445.

[41] Petrick, M., & McMahon, A. Economic aspects of alternative protein sources. International Journal of Food Science & Technology 2019, 54(4), 1000–1010.

[42] Pingali, P. Westernization of Asian diets and the transformation of food systems: Implications for research and policy. Food Policy 2007, Jun 1, 32(3), 281–298.

[43] Poore, J., & Nemecek, T. Reducing food's environmental impacts through producers and consumers. Science 2018, Jun 1, 360(6392): 987–992.

[44] Post, M. J. Cultured meat from stem cells: Challenges and prospects. Meat Science 2012, Nov 1, 92(3), 297–301.

[45] Post, M. J., Levenberg, S., Kaplan, D. L., Genovese, N., Fu, J., Bryant, C. J., Negowetti, N., Verzijden, K., & Moutsatsou, P. Scientific, sustainability and regulatory challenges of cultured meat. Nature Food 2020, Jul, 1(7), 403–415.

[46] Ritchie, H., & Roser, M. Micronutrient deficiency. Our World in Data 2017.

[47] Rumpold, B. A., & Schlüter, O. K. Nutritional composition and safety aspects of edible insects. Molecular Nutrition & Food Research 2013, May, 57(5), 802–823.

[48] Siegrist, M. Factors influencing public acceptance of innovative food technologies and products. Trends in Food Science & Technology 2008, Nov 1, 19(11), 603–608.

[49] Specht, E. A., Welch, D. R., Clayton, E. M., & Lagally, C. D. Opportunities for applying biomedical production and manufacturing methods to the development of the clean meat industry. Biochemical Engineering Journal 2018, Apr 15, 132, 161–168.

[50] Stephens, N., Di Silvio, L., Dunsford, I., Ellis, M., Glencross, A., & Sexton, A. Bringing cultured meat to market: Technical, socio-political, and regulatory challenges in cellular agriculture. Trends in Food Science & Technology 2018, Aug 1, 78, 155–166.

[51] The Good Food Institute. Challenges and breakthroughs: Contextualizing alternative protein progress 2024, April 15. https://gfi.org/blog/contextualizing-alternative-protein-progress/

[52] Tilman, D., & Clark, M. Global diets link environmental sustainability and human health. Nature 2014, Nov 27, 515(7528): 518–522.

[53] Tuomisto, H. L., & Teixeira de Mattos, M. J. Environmental impacts of cultured meat production. Environmental Science & Technology 2011, Jul 15, 45(14), 6117–6123.

[54] Traynor, A., Burns, D. T., Wu, D., Karoonuthaisiri, N., Petchkongkaew, A., & Elliott, C. T. An analysis of emerging food safety and fraud risks of novel insect proteins within complex supply chains. Npj Science of Food 2024, Jan 20, 8(1), 7.

[55] UN: World Population Prospects 2019: Highlights, Our numbers are nit our problem? United Nations Department of Economic and Social Affairs 2021, April 28(129–154).
[56] Van Huis, A. Potential of insects as food and feed in assuring food security. Annual Review of Entomology 2013, Jan 7, 58, 563–583.
[57] Vega-Gálvez, A., Miranda, M., Vergara, J., Uribe, E., Puente, L., & Martínez, E. A. Nutrition facts and functional potential of quinoa (Chenopodium quinoa willd.), an ancient Andean grain: A review. Journal of the Science of Food and Agriculture 2010, Dec, 90(15), 2541–2547.
[58] Van Huis, A., Van Itterbeeck, J., Klunder, H., Mertens, E., Halloran, A., Muir, G., & Vantomme, P. Edible insects: Future prospects for food and feed security. Food and Agriculture Organization of the United Nations 2013 FAO publication.
[59] Van Thielen, L., Vermuyten, S., Storms, B., Rumpold, B., & Van Campenhout, L. Consumer acceptance of foods containing edible insects in Belgium two years after their introduction to the market. Journal of Insects as Food and Feed 2019, Feb 15, 5(1), 35–44.
[60] Verbeke, W. Profiling consumers who are ready to adopt insects as a meat substitute in a Western society. Food Quality and Preference 2015, Jan 1, 39, 147–155.
[61] Warner, M. Contested categories: Anti-meat labeling policies and the politics of food classification. Agriculture and Human Values 2021, 38(1), 157–169.
[62] Weber, J. The future of work in food and agriculture: Perspectives from emerging protein industries. Journal of Rural Studies 2021, 82, 431–440.
[63] World Health Organization. (2015). Q&A on the carcinogenicity of the consumption of red meat and processed meat. https://www.who.int/news-room/q-a-detail/carcinogenicity of the consumption of red meat and processed meat
[64] Xiao, W., Chan, C. W., Xiao, J., Lyu, Q., Gong, N., Wong, C. L., & Chow, K. M. Managing the nutrition impact symptom cluster in patients with nasopharyngeal carcinoma using an educational intervention program: A pilot study. European Journal of Oncology Nursing 2021, Aug 1, 53, 101980.
[65] Young, V. R., & Pellett, P. L. Plant proteins in relation to human protein and amino acid nutrition. The American Journal of Clinical Nutrition 1994, May 1, 59(5), 1203S–12S.

Lamees Beekrum

Chapter 18 Food system waste remediation: waste reduction processes for a regenerative economy

Abstract: Food waste is growing at an alarming rate each year, with the highest amounts generated during the primary food production stage. Food waste has gained increasing attention in recent years due to economic, environmental, and social impacts. It is imperative that adequate steps are taken to mitigate food waste by using effective food waste management techniques. New approaches to food waste management, such as the circular economy, food waste valorization, upcycling, novel biorefinery processes, and the integration of fourth industry revolution, emphasize cutting-edge tools and procedures that improve productivity and sustainability. The circular economy is a concept that aims to reduce waste and reuse materials. Food waste valorization is a pioneering strategy that promises to turn food waste into beneficial resources. Food waste consists of a variety of useful components such as proteins, carbohydrates, lipids, and bioactives that can be valorized into new food products. Upcycling food waste using nonthermal processes has drawn scientific attention recently. Biorefinery strategies use a number of integrated processes to synergistically transform biomass feedstocks into many valuable outputs. The concept of transforming waste into bioenergy and various bio-products is a new field of study that has enormous potential to address food waste. The adoption of Fourth Industrial Revolution technologies as a collection of digital tools and real-time monitoring to assist reducing food waste is essential. Recovering and reusing as much food waste as possible is important given the rising concerns about hunger, malnutrition, and food insecurity. For sustainable food waste management, revolutionary techniques must be included within a circular economy framework.

Keywords: Biorefinery, circular economy, food waste valorization, Industry 4.0, upcycling

Lamees Beekrum, Department of Biotechnology and Food Science, Durban University of Technology, South Africa, e-mail: sharlynn@dut.ac.za

https://doi.org/10.1515/9783111441238-019

18.1 Introduction

The world population is projected to reach 9.7 billion people within the next 25 years. To ensure that this population obtains nutritious food, the global food supply is required to expand by 70% [1]. Undoubtedly, reaching this level of manufacturing is difficult. Yet, along the food supply chain from production to consumption, one-third of the food produced worldwide is lost at the postharvest and pre-consumption stages or wasted after consumption [2]. About 222 million tons of food is wasted in developed nations. According to [3], this is almost equal to sub-Saharan Africa's total net production of food. We need more diverse nutritional sources to address the nutritional issues facing current society. Food waste and its by-products are quite important in this case since they provide substantial proteins, fats, carbohydrates, micronutrients, bioactive substances, and dietary fibers [4].

Food loss is the decline in food supply or quality brought on by the choices and activities of food suppliers along the supply chain, apart from consumers, food service providers, retailers, as defined by the Food and Agriculture Organization (FAO) of the United Nations in 2019. Food waste is characterized as a reduction in food supply that results from customer, food service provider, and retailer decisions and actions [1]. Recently, there has been a growing awareness of food waste due to the reduction in the supply of fossils fuels, climate change, as well as rising economic, environmental, and social concerns [5]. Food and agricultural wastes become a huge worldwide concern as a result of this detrimental impact. One of the sustainable development goals (SDGs), specifically SDG 12 "Ensure sustainable consumption and production patterns" is to reduce food loss and waste [6].

It is vital to minimize food waste in order to protect the environment, boost the economy, and enhance society [7]. There is an opportunity to feed people while maximizing natural resources by minimizing food waste in an economical, ecologically friendly, and integrated manner. Reducing food waste is crucial for environmental sustainability as well as for reducing manufacturing expenses and enhancing the efficiency of the food system [8].

There has been an increasing amount of focus on the development of innovative, economical, and ecological waste recycling techniques. In order to accomplish this goal, waste management practices should be based on the circular economy (CE) idea. By substituting circular flows for linear flows and employing systems for closed-loop manufacture and consumption, the fundamental idea of CE is to recover valuable components from post-consumption commodities, resources, and packaging [9]. The CE is a sustainable way of transforming these wastes into valuable resources [10], and it is becoming more significant owing to several variables including an increase in wastes, especially every five years. Schroeder [11] suggest that CE business models could solve the generation of waste and diminish resource overconsumption. As a result, CE principles may help accomplish SDGs like responsible production and consumption and zero hunger.

It has been accepted that effective waste management is a necessary condition for the advancement of sustainable development, which helps achieve the global sustainability objectives (SDGs 12 and 13). Mitigation measures might involve more than simply cutting costs by optimizing manufacturing; valorization of food waste into new ingredients offers a prospect of generating additional revenue streams. According to [12], the process of converting waste, by-products, or underutilized food into valuable food material or efficiently utilizing such products is known as "food waste valorization."

The reduction in food loss and waste may be attained by practical, technological, and educational activities that apply new technology, creative approaches, and existing procedures. Additionally, the production of a food waste biorefinery can result in a variety of products that help achieve sustainable food waste management, including bioelectricity, biofuels, platform chemicals, biomaterial, biofertilizers, and animal feed [13–16]. The use of technology and digital platforms in the battle against food waste is growing [17]. A recent study by [18] highlighted the relevance of Industry 4.0 technology to advance the CE to accomplish a shift toward net zero by effective management of waste.

This chapter focusses on the impact and sources of food waste, conventional food waste management systems, the importance of the CE, and emerging sustainable technologies such as the valorization of food waste by biotechnological innovations, upcycling of food, recycling, and reusing of food and agricultural waste to commercial production of valuable bio-products and Fourth Industrial Revolution (Industry 4.0) innovations.

18.2 Impact of food waste

Food waste is an issue that has broad consequences on society, including its impact on global food security and the economic and environmental effects of food production and waste [19]. Global food waste is related to global malnutrition. According to [20], 50% of all food produced is wasted both before and after it can be consumed. Food waste affects the economy enormously. Food waste results in revenue deficits during the food distribution chain, impacting consumers, producers, and retailers. Food waste has an array of environmental effects. In addition to contributing to climate change, food waste depletes energy, soil, and water, which are essential in food production.

FAO estimates that food waste generates a carbon footprint of about 8% of all greenhouse gas emissions driven by human activity [21]. Given the consequences of climate change, especially on many developing countries and the anticipated increase in population, food production is an urgent problem. In addition to the demand for more products and energy, a growing population places more strain on farming and agricultural systems [22]. Food waste is a major contributor of pollution since it depletes commodities such as fresh water, land, fertilizers, and energy that are needed in the food value chain [21]. Moreover, food waste disposal in landfills has a significant effect on greenhouse gas emissions [23]. Sadly, according to [21], food waste is indirectly contributing to biodiversity loss.

18.3 Global scale of food waste generation

Worldwide, food loss and waste are expected to cost US$1 trillion annually [24]. Food loss and waste is a critical problem since, by 2030, approximations indicate that it would amount to 2.1 billion tons, or $1.5 trillion [25]. Food waste is predicted to amount to approximately 2.5 billion tons annually, as reported by [26], and it arises from every aspect of the food industry, including processing, production, packaging, and retail. An estimated 1,300 million tons of food are wasted on an annual basis worldwide, which is between 25% and 33% of the total amount produced [2]. From farm to fork, there is wastage of almost 88 million tons of food in Europe on an annual basis, with related costs estimated at 143 billion euros [27]. According to estimates, the agro-food industry will produce around 44% of all wastes and by-products worldwide by 2025, making it the leading producer of industrial waste [4].

The Food Waste Index Report reported an estimate of 931 million tons of food waste that was created in 2019, with households producing 61% of the waste materials, food services producing 26%, and retail producing 13% [1]. The biggest annual per capita amounts of consumption waste are found in Europe and Latin America, with 180 kilograms and 200 kilograms, respectively. There is 175 kg of consumption waste per individual per year in North Africa, West and Central Asia, North America, and Oceania. Sub-Saharan Africa (150 kg), South and Southeast Asia (110 kg), and industrialized Asia (155 kg) are the next regions in terms of annual per capita waste [28].

According to data reported by FAO in 2013, the agricultural, industrial processing, retail, and final consumption phases account for 11–23%, 17–19%, 8–17%, and more than 50% of food waste, respectively [2]. In developed nations, food waste mainly occurs in later phases of the supply chain. Ineffective waste management methods used in distribution, packaging, storage, and transportation contribute to the enormous hype around food waste [29]. Both at the consumer and supply chain levels, due to poor consumer consumption and a surplus of supply compared to demand, there is significant loss of food [30].

In developing countries, food waste is mostly observed in the beginning of the food value chain and is a result of both technical and economic issues with refrigeration, storage, and harvesting methods. Rejected food products, which fail to meet aesthetic standards result in food waste. When combined with customers' careless actions, excessive purchases and expired goods lead to significant waste production at the consumer level [2]. Food losses result from inadequate plantation fertilization, inaccurate agrochemical application, improper postharvest handling, poor climate conditions, inadequate technology, poor transportation and logistics, and ineffective packaging in food maintenance. The primary causes of food waste are expired foods, food leftovers, misuse of food, ignorance, mixing on labels, and inappropriate storage conditions [31].

In relation to food types, oil-bearing crops, roots, tubers have the greatest losses from postharvest to distribution (25%). This is followed by fruit and vegetables (22%), meat and animal products (12%), and cereals and pulses (9%). FAO reports show that roughly one-third of the global food production is lost or wasted, with the majority of this loss originating from the fruit, vegetable, and seafood industries [24].

18.4 Conventional strategies for food waste management

The many approaches of eradicating waste from the environment, such as elimination or decomposition, processing, reusing, recycling, or limiting waste output, are together referred to as waste management [32]. Figure 18.1 shows conventional strategies for the management of food waste practices.

Typically, agricultural waste is burned off or released into the environment untreated, adding to the pressure on landfills and the ecosystem. Therefore, instead of merely disposing of these wastes, they need to be considered as possible raw materials to prevent contamination and the discharge of harmful substances into the ecosystem. Improved approaches to agricultural waste management, incentives, and technology use will be vital for this [33]. Composting, incineration, landfilling, and utilizing waste as animal feed are among the traditional agricultural and food waste management practices that are currently utilized globally [34].

Food waste that contains a lot of water that is incinerated, in addition to energy loss from water evaporating during the process and dioxin emission, can have a negative environmental impact. Furthermore, by changing their chemical makeup, incinerated substrates make it more difficult to extract valuable molecules and recycle nutrients, which affect their economic worth [35]. According to [36], the incineration of food waste cannot be considered an environmentally favorable procedure because of its significant energy consumption, release of greenhouse gas emissions, and generation of incineration ashes.

Nowadays, a substantial proportion of food and agricultural waste inevitably finds its way into landfills, especially in underdeveloped countries, with more than 90% used for food and loss waste treatment [37]. Food wastes serve as ideal breeding grounds for a variety of microorganisms resulting in bacterial contaminations and infectious diseases. The disposal of food waste in landfills has resulted in the direct or indirect emission of greenhouse gases including carbon dioxide and methane. This

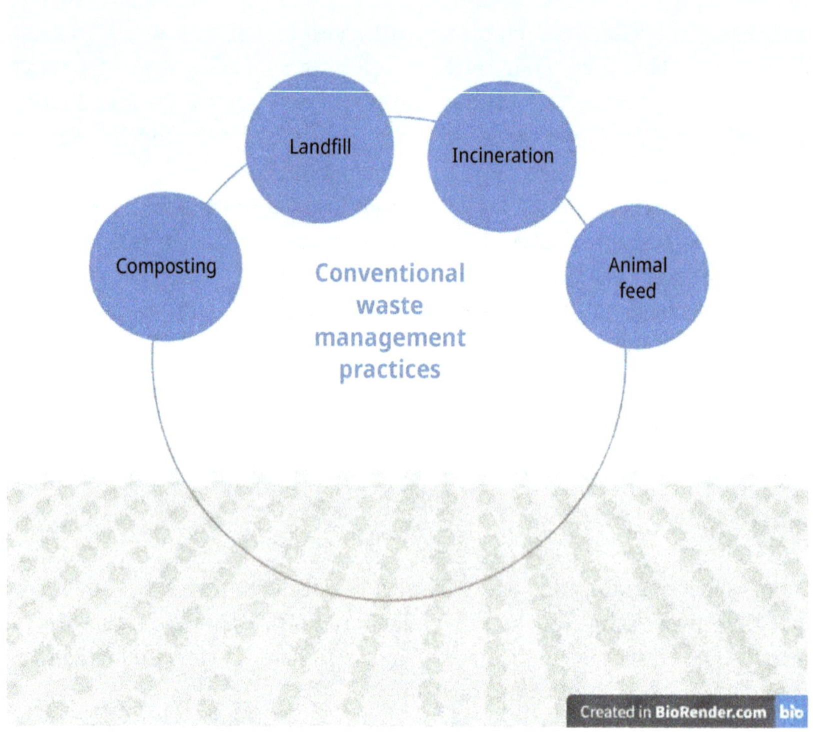

Figure 18.1: Conventional methods of food waste management activities (created with BioRender.com).

poses a serious threat to air quality [38], and accounts for approximately 3% of the total global greenhouse gases emissions [39]. Furthermore, the accumulation in greenhouse gases would result in global warming and climate change, both of which are detrimental to living species [40].

Composting, sometimes referred to as land spread, is a more efficient method as it may lower the amount of food waste that is disposed in landfills and eliminates the need to employ biofertilizers from farmers. However, the use of composting is limited since it is comparatively more expensive than landfills and incineration. In the long term, treating biowastes with food and agricultural wastes as animal feed is thought to be the most economical method; however, because of the variety of unknown compounds included in the biowastes, this application is highly regulated. Furthermore, the disintegration of biowaste during the conversion process at times becomes uncontrolled and produces undesirable by-products, making the management of waste as a whole even more complex [40]. Food waste that is treated conventionally causes issues with the economy, environment, and society. This results in a number of profitable and sustainable valorization techniques that manage food waste offering alternatives to generate value-added products.

18.5 Emerging technologies toward sustainable circular economy

The CE model minimizes waste and protects the value of commodities until it is viable [41]. Materials and components are continually recycled and reused, in contrast to linear models where they are discarded as waste after usage [42]. By replacing linear models with circular models and employing closed-loop systems for production and consumption, the fundamental idea of CE is to recover valuable post-consumption goods, packaging, and resources [9]. Figure 18.2 offers an outline of the CE concept with the application of food waste valorization.

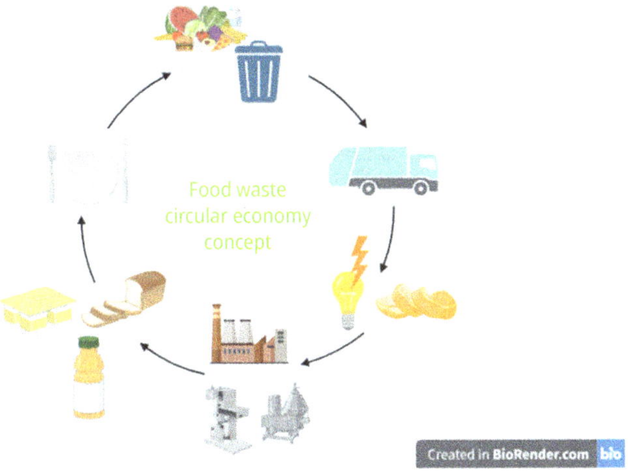

Figure 18.2: Food waste valorization and the circular economy (created with BioRender.com).

The CE model describes a development paradigm whereby a company's waste is utilized as the starting point for another. According to Borrello et al. [43], food products that have expired and been returned to retail stores can be used to create new animal-derived products through the regeneration system of the CE. Tan and Lamers [44] emphasized the significance of the circular economy idea in minimizing climate change, pollution, and microbial contamination resulting from landfills and water degradation. The shift from a linear to a circular economy is imperative [45]. Through the development and closure of material cycles, the economic cycle seeks to achieve two major objectives: improved management of natural resources and increased avoidance of waste generation [46, 47].

The success of CE initiatives in the food industry depends greatly on factors like increasing consumer awareness, attempting to recycle by-products and products with imperfect appearances, and encouraging consumers to refrain from producing food waste [48].

18.6 Innovative strategies for food waste valorization

Food waste reduction or prevention is the ideal solution; however in situations when these are not achievable, the valorization of food waste is regarded as one of the finest strategies. Valorization is the process of transforming food waste that has been rejected into food and feed commodities. In addition, considerations need to be made for the composition, quality, and durability of the waste. Food waste conversion necessitates assessing market circumstances in terms of economic viability, technological feasibility, legal reportability, and environmental sustainability and utility [7].

The goal of food waste valorization approaches is to transform food waste into valuable commodities that may be included into the food supply chain. The composition of the produced food waste material is the cornerstone of every technique for food waste valorization [49]. Food chain operations are diversified through the valorization of by-products, creating new employment possibilities and revenue sources for the population that can help communities in need. This is crucial since reaching the SDGs for hunger and malnutrition requires tackling poverty [50]. Despite having varying chemical composition, food waste from various sources may serve as great viable raw materials for various bioprocesses that produce biofuels and other commercially valuable bioproducts. Figure 18.3 shows the emerging strategies in food waste management.

Figure 18.3: Emerging techniques in food waste management (created with BioRender.com).

Due to the economic potential of transforming food processing and agricultural waste into bio-based products rather than the problems associated with food decomposition from landfills, there have been numerous studies to find novel and sustainable bioprocesses to transform food waste into valuable bio-products. According to [51], methods of valorization are thus regarded as extremely beneficial in the waste management system. Waste valorization has not yet been extensively explored because of the high costs associated with transporting waste materials to processing facilities, retaining residues to prevent breakdown, and upstream processing technologies [52]. Waste valorization techniques should be developed to both extract valuable compounds from biomass feedstocks and make efficient use of the residual by-products, so that their sustainability and cost-effectiveness can be guaranteed. According to [52], this will decrease the usage of natural resources and the overall outcomes of new methods of valorization on the economy and environment.

18.7 Biorefinery approach

A biorefinery approach converts biomass feedstocks into several valuable products in a synergistic manner using a variety of integrated processes. These processes include thermochemical, physical, biological, and chemical processes [53]. A fruit waste biorefinery may substantially boost the amount of fruit waste biomass that is converted into commercially viable goods, optimizing the value produced from fruit waste [54, 55].

The biorefinery strategy can reduce the greenhouse gas emissions linked to such product supply chains by employing fruit wastes as a renewable resource to create chemicals that are typically manufactured using conventional feedstocks [52]. However, most traditional high-efficiency extraction techniques rely on the incorporation of an organic solvent, which frequently results in an unstable waste stream and necessitates an energy-intensive evaporation step afterwards. Although novel extraction techniques reduce the requirement for organic solvents, they come with a significant initial investment and necessitate continuous modification and verification, particularly when used in industrial settings [56].

18.8 Production of valuable bio-products in microbial fermentation

Fermentation using microorganisms has been reported to be a cutting-edge method for transforming food waste into beneficial products. Food wastes are considered to have an appreciable amount of fatty acids, proteins, minerals, and phenolic compounds, making them an attractive and sustainable crude feedstock for microbial fermentation that produces a variety of bioactive products. Food wastes can be bio transformed into valuable bio-products with improved functionality, which can lower production costs [57].

According to [58], agricultural wastes may sometimes be employed as low-cost sources of carbon and nitrogen for the growth of microbes that can ferment bioactive substances. Alternatively, microbial fermentation may employ food waste as a basic substrate, which can create a variety of bio-products such as proteins, enzymes, antioxidants, and pigments [59]. Among the most frequent by-products of food waste fermentation are lactic acid and ethanol [60, 61].

18.9 Bioethanol fermentation

By substituting petroleum-based energy sources with bioethanol, a renewable biofuel, the harmful environmental consequences of petroleum-based energy sources can be minimized. Demand for and price of bioethanol are both rising, in part because the cost of commercial gasoline is rising. Corn and sugar crops have been investigated extensively as feedstocks for the manufacture of bioethanol in many different nations. This strategy is not ecologically or economically sustainable for a number of clear reasons. Therefore, in order to produce bioethanol in a way that is both environmentally and economically sustainable, it is imperative to develop alternative, sustainable, and green feedstocks. In this case, it has been shown that food waste that is high in carbohydrates is a great feedstock for the aerobic or anaerobic fermentation processes used to produce bioethanol. [62] showed that bioethanol could be produced sustainably from solid-state fermentation.

One viable option to alleviate the rising worldwide problem of food waste and replenishing energy reserves is through the waste bioconversion into ethanol. Food waste can be a perfect substitute substrate for creating a decentralized bioprocess based on its substantial carbohydrate and nitrogen content. In order to make the process economically viable, bottleneck problems related to substrate collection and transportation, pretreatment, fermentative organisms, and product separation can be resolved by optimizing the process [63].

18.10 Upcycling food waste

Food products made from resources or food materials that could be considered as waste or by-products are known as “upcycled” foods. This novel approach, which transforms excess materials into food products with increased value, presents a potential way to reduce food waste [64]. Thus, foods that have been upcycled enhance value, prevent waste, and restore the food to their intended human consumption usage [65].

Upcycled foods are made from materials that are frequently not meant for human consumption, such as residues, by-products, underutilized or damaged foods, and waste from food processing. Consequently, upcycling transforms waste materials and by-products into useful and nutritious components that may be employed in the food distribution network as new food commodities. Recycled foods and ingredients are frequently referred to as value-added surplus goods or waste-to-value food products [66].

The processing of food waste requires a minimum of one process for extraction, isolation, pretreatment, purification, and separation to create final goods with high added value [67]. Organic acids, oils, lipids, and protein are the macromolecular food waste components that are usually made up of polysaccharides like cellulose, hemicellulose, lignin, and starch. Food waste also contains functional ingredients such as valuable colorants, mineral elements, vitamins, phenolics, and other bioactives that might be used in the food, pharmaceutical, and packaging sectors. The approach of producing value-added functional products by upcycling of food waste into functional ingredients involves the application of established techniques. These methods include solvent extraction, gasification, pyrolysis, and precipitation processes, among others. However, there may be limitations due to the high cost of applying conventional processes, their increased energy consumption, and their potential to impair target compound stability and performance due to heat application. Furthermore, conventional solvents that deplete natural resources and produce greenhouse gas emissions are frequently used in traditional procedures [67, 68].

Conversely, emerging methods use nonthermal conditions and sustainable, greener solvents to increase safety of products, save processing costs and time, and maintain ingredient stability. The nonthermal technologies for upcycling food waste that have attracted scientific interest are microwave-assisted processes, ultrasound-assisted techniques, and pulsed electric field (PEF) approaches. Their use may help create useful compounds like flavorings, colorants, and preservatives as well as turn waste into biofuel, which will advance the CE by encouraging the effective use of resources [67].

18.10.1 Pulsed electric field

PEF applies high-strength electric field pulses lasting microseconds to cause electroporation in cell membranes [68]. PEF has been applied in upcycling to boost bioactive antioxidant molecule recovery yield. Numerous product waste components have been the subject of extensive research on the recovery of antioxidant chemicals using PEF [68, 69] to remove color pigments, proteins, and sugars from fruit and vegetable waste [67].

18.10.2 Ultrasound-assisted extraction

By increasing the permeability of plant tissues via the application of high-intensity sound waves, the ultrasonic technique improves the transformation of food waste into appropriate final products [70]. Its foundation is the cavitation effect brought on by ultrasonic radiation, which is a sound wave with oscillations at wavelengths higher than 16 kHz [71]. Ultrasound-assisted pretreatment methods were tested for the production of biofuels from cellulose and hemicellulose substances, and lignin as well as for enhancing the hydrolysis of food waste [72].

18.10.3 Microwave-assisted extraction

Microwave-assisted extraction, or MAE, is another recently developed process that has been studied for food waste valorization. It makes use of electromagnetic radiation with a microwave wavelength range. By heating the water inside the plant cells, MAE increases pressure, which in turn causes the cell structure to be disrupted. The procedure shortens the time, energy, expense, and quantity of raw materials needed for extraction while also increasing the rate at which intracellular molecules are released. Microwave irradiation is a useful pretreatment method for boosting the production of biofuel and extracting new food components including pectin, protein, and phenolic chemicals [67].

18.10.4 Three-dimensional food printing

Three-dimensional (3D)-printed food is a rapidly evolving breakthrough technique in food waste reduction and sustainability. According to [73], 3D food printing is becoming a viable method for transforming food manufacturing procedures in order to minimize food waste and promote sustainability. It facilitates the creation of meals with personalized textures and patterns using less complicated procedures. This innovative method is thought to be able to convert multistage procedures into a single phase, which will revolutionize the food industry's manufacturing process. Layer by layer, food is added to create dishes using 3D food printing; the most widely used method among the many forms of 3D printing is extrusion printing [74].

18.11 Extraction of valuable bioactives from food processing wastes

Among the value-added products made from food waste are antioxidants, bioactives, biopolymers, biopeptides, bio nanocomposites, fine chemicals, industrial enzymes, nutraceuticals, single-cell proteins, polysaccharides, activated carbon adsorbent, chitosan, corrosion inhibitors, organic acids, pigments, sugars, wax esters, and xanthan gum [75]. These value-added goods, which are mostly made up of sugars, pectin, proteins, lipids, polysaccharides, fibers, polyphenols, vitamins A and E, vital minerals, fatty acids, volatiles, anthocyanins, and pigments, are produced by separating fruit and vegetable wastes or by-products [8].

In the current global context, there may be numerous possibilities for revenue generation through the sustainable use of agricultural food waste and/or by-products to develop high-value goods with possible uses in the food, pharmaceutical, or cosmetic sectors. According to [8], food waste derived from various agro-food companies, such as fruits, vegetables, sugar, drinks, meat, fish, and marine goods, might be a cost-effective and attractive source of compounds with possible functional or bioactive properties. These substances can be separated and utilized for a variety of purposes, including food additives, medications, cosmetics, and nutraceuticals [76]. Furthermore, by value addition of agricultural wastes and by-products, local food security is preserved, thus supporting sustainable food production [77].

New techniques for valorization of fruit waste retain nutrients and important components, making it possible to extract them and use them directly or modify them into more complex, value-added compounds. Unlike conventional waste management methods, these methods recover valuable substances from fruit waste and reintegrate them into other product distribution networks. This process reduces the amount of waste that is subsequently disposed of, thus creating products that are more valuable than the waste [51, 52].

The food industry has made substantial advances in the extraction and use of functional food components by utilizing cutting-edge techniques. These developments will improve the preservation of sensory, nutritional, and functional properties of food, as well as the extraction of key ingredients. Moreover, large quantities of protein, fat, and carbohydrates have been found in food processing waste, suggesting potential for value addition [34]. New concepts that facilitate the effective application of agricultural food wastes and by-products to yield value-added products have surfaced with the introduction of new technology and the concepts of green chemistry [77].

18.12 Advances in Fourth Industrial Revolution (Industry 4.0)

Global food sustainability as well as the blue economy may be improved by utilizing new technologies, such as digitalization, digital platforms, and Industry 4.0 advancements like artificial intelligence, Big Data, smart sensors, Internet of Things, and cloud computing to decrease and valorize waste to promote connection and information exchange [78, 79]. The foundation of Industry 4.0 describes a broad spectrum of technologies employed in the manufacturing sector, comprising supply chain management and product design [80].

It offers real-time information on production, component flow, and machinery, assimilating this data to assist managers in decision-making, performance monitoring, and tracking parts and products [81]. This includes the Internet of Things as well as smart factories and products [82, 83]. Food waste might be reduced by using these technologies to improve efficiency and value addition of food supply networks. For instance, smart sensors and digital technologies might be used to offer real-time expiry dates in place of conventional labels with "use-by" or "best-before" dates [82, 84]. With an emphasis on toxins, humidity, pH, freshness, temperature, contaminants, and pathogens, food sensors technology has advanced remarkably. These sensors are essential for preserving food quality, ensuring food safety, and maintaining packaging requirements [85]. Monitoring plant physiology, health, and disease on-site may greatly increase agricultural output and productivity, which has major financial advantages [86]. This reduces food loss during the production stage. Therefore, by enabling real-time monitoring, management, and communication amongst supply chain layers, utilizing Industry 4.0 technology reduces food loss and waste. As digital platform technologies have advanced, so too have smart food sharing technologies gained popularity as a way to minimize food waste.

18.13 Regulatory framework and policy considerations

According to an estimate from the USDA in 2010, 31%, or 133 billion pounds, of the food supply that was available in the US was discarded at the retail and consumer levels [87]. The vast quantity of food waste offers an unexplored opportunity to recycle it into something that consumers would find acceptable. For such a shift, a strong legal and regulatory framework is essential. As upcycled food is a relatively new industry, extensive regulations have not yet been created [67].

Foods or ingredients that have been upcycled in the US must adhere to the regulations set out by the FDA, USDA, and other food safety authorities. Given the emerging popularity of upcycled products, governments have to consider implementing measures that would encourage major stakeholders in the agricultural food sector to invest in innovative technology and waste-reduction strategies. Establishing standards that would facilitate the streamlining of the production and supply chain components of food waste processing should be the primary objective of the regulatory policies. In order to support distributors, farmers, processors, and suppliers in their waste reduction initiatives, legislation should also provide incentives for the adoption of best practices [67]

18.14 Conclusion and future directions

Food prevention is at the top of the food waste chain, which is seen to be the most environmentally friendly management strategy. The first step in minimizing food waste is to prevent waste from being created. Globally, there are several initiatives underway to reduce food waste in order to enhance sustainability and food security. With a population of almost nine billion people, the world would need 70% more food in 2050, or at least 2 billion tons, if the required methods for decreasing food waste are not implemented and food production and consumption are not planned. By turning food waste into raw materials in a closed, regenerative cycle, the CE model seems to provide a solution to the problem of food waste. A circular economy would improve community sustainability, mitigate the effects of global warming, and benefit the environment by replacing the current food system. Value advancements in food supply chains and the alleviation of severe hunger in underdeveloped nations may both benefit from the valorization and management of food waste.

By using biotransformation techniques and converting food wastes into biofuels and other beneficial bio-products, food waste production should be reduced. Promoting the utilization of food and other renewable waste in integrated waste-based biorefineries would lower the cost of the bioprocess while also reducing environmental pollution. The private sector, including the food industry can optimize food processing methods, streamline supply chains, and connect farmers to markets to effectively reduce food waste. Food waste management may find an innovative and sustainable solution in the complexity of sensors made possible by sophisticated manufacturing such as nanomanufacturing and their integration with Internet of Things. Thus, in order to achieve significant environmental sustainability and a regenerative economy, it is essential that present methods for managing food waste undergo a fundamental change.

References

[1] Nations, U. World population projected to reach 9.8 billion in 2050, and 11.2 billion in 2100. Department of Economic and Social Affairs 2017 (1).

[2] Gustavsson, J., Cederberg, C., Sonesson, U., Van Otterdijk, R., & Meybeck, A. Global Food Losses and Food Waste, FAO Italy, 2011.

[3] Boon, E. K., & Anuga, S. W. Circular economy and its relevance for improving food and nutrition security in Sub-Saharan Africa: The case of Ghana. Materials Circular Economy 2020, 2, 1–14.

[4] Singh, K., Kumar, T., Prince, V. K., Sharma, S., & Rani, J. A review on conversion of food wastes and by-products into value added products. International Journal of Chemical Studies 2019, 7(2), 2068–2073.

[5] Bilal, M., Qamar, S. A., Ashraf, S. S., Rodríguez-Couto, S., & Iqbal, H. M. Robust nanocarriers to engineer nanobiocatalysts for bioprocessing applications. Advances in Colloid and Interface Science 2021, 293, 102438.

[6] Cf, O. Transforming Our World: The 2030 Agenda for Sustainable Development, New York, NY, USA: United Nations, 2015.

[7] Capanoglu, E., Nemli, E., & Tomas-Barberan, F. Novel approaches in the valorization of agricultural wastes and their applications. Journal of Agricultural and Food Chemistry 2022, 70(23), 6787–6804.

[8] Socas-Rodríguez, B., Álvarez-Rivera, G., Valdés, A., Ibáñez, E., & Cifuentes, A. Food by-products and food wastes: Are they safe enough for their valorization?. Trends in Food Science & Technology 2021, 114, 133–147.

[9] De Sousa Jabbour Ab, L., FCdO, F., Santibanez Gonzalez, E. D., & Chiappetta Jabbour, C. J. Are food supply chains taking advantage of the circular economy? A research agenda on tackling food waste based on Industry 4.0 technologies. Production Planning & Control 2023, 34(10), 967–983.

[10] Dahiya, S., Katakojwala, R., Ramakrishna, S., & Mohan, S. V. Biobased products and life cycle assessment in the context of circular economy and sustainability. Materials Circular Economy 2020, 2, 1–28.

[11] Schroeder, P., Anggraeni, K., & Weber, U. The relevance of circular economy practices to the sustainable development goals. Journal of Industrial Ecology 2019, 23(1), 77–95.

[12] Roy, P., Mohanty, A. K., Dick, P., & Misra, M. A review on the challenges and choices for food waste valorization: Environmental and economic impacts. ACS Environmental Au 2023, 3(2), 58–75.

[13] Dahiya, S., Kumar, A. N., Sravan, J. S., Chatterjee, S., Sarkar, O., & Mohan, S. V. Food waste biorefinery: Sustainable strategy for circular bioeconomy. Bioresource Technology 2018, 248, 2–12.

[14] Maina, S., Kachrimanidou, V., & Koutinas, A. A roadmap towards a circular and sustainable bioeconomy through waste valorization. Current Opinion in Green and Sustainable Chemistry 2017, 8, 18–23.

[15] Mak, T. M., Xiong, X., Tsang, D. C., Iris, K., & Poon, C. S. Sustainable food waste management towards circular bioeconomy: Policy review, limitations and opportunities. Bioresource Technology 2020, 297, 122497.

[16] Zabaniotou, A., & Kamaterou, P. Food waste valorization advocating Circular Bioeconomy-A critical review of potentialities and perspectives of spent coffee grounds biorefinery. Journal of Cleaner Production 2019, 211, 1553–1566.

[17] Benyam, A. A., Soma, T., & Fraser, E. Digital agricultural technologies for food loss and waste prevention and reduction: Global trends, adoption opportunities and barriers. Journal of Cleaner Production 2021, 323, 129099.

[18] Kurniawan, T. A., Meidiana, C., Othman, M. H. D., Goh, H. H., & Chew, K. W. Strengthening waste recycling industry in Malang (Indonesia): Lessons from waste management in the era of Industry 4.0. Journal of Cleaner Production 2023, 382, 135296.

[19] Kibler, K. M., Reinhart, D., Hawkins, C., Motlagh, A. M., & Wright, J. Food waste and the food-energy-water nexus: A review of food waste management alternatives. Waste Management 2018, 74, 52–62.
[20] Parfitt, J., Barthel, M., & Macnaughton, S. Food waste within food supply chains: Quantification and potential for change to 2050. Philosophical Transactions of the Royal Society B: Biological Sciences 2010, 365(1554), 3065–3081.
[21] Footprint, F. W. Food Wastage Footprint: Impacts on Natural Resources: Summary Report, Québec City, QC, Canada: Food & Agriculture Org, 2013.
[22] Foley, J. A., Ramankutty, N., Brauman, K. A., Cassidy, E. S., Gerber, J. S., Johnston, M., et al. Solutions for a cultivated planet. Nature 2011, 478(7369), 337–342.
[23] Garnett, T. Where are the best opportunities for reducing greenhouse gas emissions in the food system (including the food chain)?. Food Policy 2011, 36, S23–S32.
[24] FAO. The State of Food and Agriculture 2019: Moving Forward on Food Loss and Waste Reduction, United Nations, Rome, 2019.
[25] Hegnsholt, E., Unnikrishnan, S., Pollmann-Larsen, M., Askelsdottir, B., & Gerard, M. Tackling the 1.6-billion-ton Food Loss and Waste Crisis, The Boston Consulting Group, Food Nation, State of Green, Boston, Massachusetts. 2018.
[26] Safdie, S. Global food waste in 2023. Greenly 2023, March, 24.
[27] Corrado, A., & Palumbo, L. Essential farmworkers and the pandemic crisis: Migrant labour conditions, and legal and political responses in Italy and Spain. Migration and Pandemics: Spaces of Solidarity and Spaces of Exception 2022, 145–166.
[28] Kannah, R. Y., Merrylin, J., Devi, T. P., Kavitha, S., Sivashanmugam, P., Kumar, G., et al. Food waste valorization: Biofuels and value added product recovery. Bioresource Technology Reports 2020, 11, 100524.
[29] Asgher, M., Qamar, S. A., Bilal, M., & Iqbal, H. M. Bio-based active food packaging materials: Sustainable alternative to conventional petrochemical-based packaging materials. Food Research International 2020, 137, 109625.
[30] Vilariño, M. V., Franco, C., & Quarrington, C. Food loss and waste reduction as an integral part of a circular economy. Frontiers in Environmental Science 2017, 5, 257971.
[31] De Moraes, C. C., De Oliveira Costa, F. H., Pereira, C. R., Da Silva, A. L., & Delai, I. Retail food waste: Mapping causes and reduction practices. Journal of Cleaner Production 2020, 256, 120124.
[32] Sufficiency, E., Qamar, S. A., Ferreira, L. F. R., Franco, M., Iqbal, H. M., & Bilal, M. Emerging biotechnological strategies for food waste management: A green leap towards achieving high-value products and environmental abatement. Energy Nexus 2022, 6, 100077.
[33] Rao, P., & Rathod, V. Valorization of food and agricultural waste: A step towards greener future. The Chemical Record 2019, 19(9), 1858–1871.
[34] Bilal, M., & Iqbal, H. M. Sustainable bioconversion of food waste into high-value products by immobilized enzymes to meet bio-economy challenges and opportunities–A review. Food Research International 2019, 123, 226–240.
[35] Ma, H., Wang, Q., Qian, D., Gong, L., & Zhang, W. The utilization of acid-tolerant bacteria on ethanol production from kitchen garbage. Renewable Energy 2009, 34(6), 1466–1470.
[36] Ma, Y., & Liu, Y. Turning food waste to energy and resources towards a great environmental and economic sustainability: An innovative integrated biological approach. Biotechnology Advances 2019, 37(7), 107414.
[37] Thi, N. B. D., Lin, C.-Y., & Kumar, G. Waste-to-wealth for valorization of food waste to hydrogen and methane towards creating a sustainable ideal source of bioenergy. Journal of Cleaner Production 2016, 122, 29–41.
[38] Karthikeyan, O. P., Mehariya, S., & Wong, J. W. C. Bio-refining of food waste for fuel and value products. Energy Procedia 2017, 136, 14–21.

[39] Papargyropoulou, E., Lozano, R., Steinberger, J. K., Wright, N., & Bin Ujang, Z. The food waste hierarchy as a framework for the management of food surplus and food waste. Journal of Cleaner Production 2014, 76, 106–115.
[40] Ng, H. S., Kee, P. E., Yim, H. S., Chen, P.-T., Wei, Y.-H., & Lan, J. C.-W. Recent advances on the sustainable approaches for conversion and reutilization of food wastes to valuable bioproducts. Bioresource Technology 2020, 302, 122889.
[41] Scarpellini, S., Portillo-Tarragona, P., Aranda-Usón, A., & Llena-Macarulla, F. Definition and measurement of the circular economy's regional impact. Journal of Environmental Planning and Management 2019, 62(13), 2211–2237.
[42] Walzberg, J., Lonca, G., Hanes, R. J., Eberle, A. L., Carpenter, A., & Heath, G. A. Do we need a new sustainability assessment method for the circular economy? A critical literature review. Frontiers in Sustainability 2021, 1, 620047.
[43] Borrello, M., Caracciolo, F., Lombardi, A., Pascucci, S., & Cembalo, L. Consumers' perspective on circular economy strategy for reducing food waste. Sustainability 2017, 9(1), 141.
[44] Tan, E. C., & Lamers, P. Circular bioeconomy concepts – A perspective. Frontiers in Sustainability 2021, 2, 701509.
[45] Mu'azu, N. D., Blaisi, N. I., Naji, A. A., Abdel-Magid, I. M., & AlQahtany, A. Food waste management current practices and sustainable future approaches: A Saudi Arabian perspectives. Journal of Material Cycles and Waste Management 2019, 21, 678–690.
[46] Ingrao, C., Faccilongo, N., Di Gioia, L., & Messineo, A. Food waste recovery into energy in a circular economy perspective: A comprehensive review of aspects related to plant operation and environmental assessment. Journal of Cleaner Production 2018, 184, 869–892.
[47] Zeller, V., Towa, E., Degrez, M., & Achten, W. M. Urban waste flows and their potential for a circular economy model at city-region level. Waste Management 2019, 83, 83–94.
[48] Esposito, M., Tse, T., & Soufani, K. The circular economy takes on food waste. Stanford Social Innovation Review 2016, 14(2).
[49] Carmona-Cabello, M., García, I., Sáez-Bastante, J., Pinzi, S., Koutinas, A., & Dorado, M. Food waste from restaurant sector–Characterization for biorefinery approach. Bioresource Technology 2020, 301, 122779.
[50] Torres-León, C., Ramírez-Guzman, N., Londoño-Hernandez, L., Martinez-Medina, G. A., Díaz-Herrera, R., Navarro-Macias, V., et al. Food waste and byproducts: An opportunity to minimize malnutrition and hunger in developing countries. Frontiers in Sustainable Food Systems 2018, 2, 52.
[51] Eriksson, M., & Spångberg, J. Carbon footprint and energy use of food waste management options for fresh fruit and vegetables from supermarkets. Waste Management 2017, 60, 786–799.
[52] Esparza, I., Jiménez-Moreno, N., Bimbela, F., Ancín-Azpilicueta, C., & Gandía, L. M. Fruit and vegetable waste management: Conventional and emerging approaches. Journal of Environmental Management 2020, 265, 110510.
[53] Moreno, A. D., Ballesteros, M., & Negro, M. J. Biorefineries for the valorization of food processing waste. The Interaction of Food Industry and Environment: Elsevier 2020, 155–190.
[54] Leong, Y. K., & Chang, J.-S. Valorization of fruit wastes for circular bioeconomy: Current advances, challenges, and opportunities. Bioresource Technology 2022, 359, 127459.
[55] Lee, A., Lan, J. C.-W., Jambrak, A. R., Chang, J.-S., Lim, J. W., & Khoo, K. S. Upcycling fruit waste into microalgae biotechnology: Perspective views and way forward. Food Chemistry: Molecular Sciences 2024, 8, 100203.
[56] Kandemir, K., Piskin, E., Xiao, J., Tomas, M., & Capanoglu, E. Fruit juice industry wastes as a source of bioactives. Journal of Agricultural and Food Chemistry 2022, 70(23), 6805–6832.
[57] Dursun, D., & Dalgıç, A. C. Optimization of astaxanthin pigment bioprocessing by four different yeast species using wheat wastes. Biocatalysis and Agricultural Biotechnology 2016, 7, 1–6.

[58] Javed, U., Ansari, A., Aman, A., & Qader, S. A. U. Fermentation and saccharification of agro-industrial wastes: A cost-effective approach for dual use of plant biomass wastes for xylose production. Biocatalysis and Agricultural Biotechnology 2019, 21, 101341.
[59] Sadh, P. K., Kumar, S., Chawla, P., & Duhan, J. S. Fermentation: A boon for production of bioactive compounds by processing of food industries wastes (by-products). Molecules 2018, 23(10), 2560.
[60] Tang, J., Wang, X., Hu, Y., Zhang, Y., & Li, Y. Lactic acid fermentation from food waste with indigenous microbiota: Effects of pH, temperature and high OLR. Waste Management 2016, 52, 278–285.
[61] Waqas, M., Rehan, M., Khan, M. D., & Nizami, A.-S. Conversion of food waste to fermentation products. Encyclopedia of Food Security and Sustainability 2019, 1, 501–509.
[62] Passadis, K., Christianides, D., Malamis, D., Barampouti, E., & Mai, S. Valorisation of source-separated food waste to bioethanol: Pilot-scale demonstration. Biomass Conversion and Biorefinery 2022, 12(10), 4599–4609.
[63] Bibra, M., Samanta, D., Sharma, N. K., Singh, G., Johnson, G. R., & Sani, R. K. Food waste to Bioethanol: Opportunities and challenges. Fermentation 2023, 9(1), 8.
[64] Bhatt, S., Ye, H., Deutsch, J., Jeong, H., Zhang, J., & Suri, R. Food waste and upcycled foods: Can a logo increase acceptance of upcycled foods?. Journal of Food Products Marketing 2021, 27(4), 188–203.
[65] Aschemann-Witzel, J., Bizzo, H. R., Doria Chaves, A. C. S., Faria-Machado, A. F., Gomes Soares, A., De Oliveira Fonseca, M. J., et al. Sustainable use of tropical fruits? Challenges and opportunities of applying the waste-to-value concept to international value chains. Critical Reviews in Food Science and Nutrition 2023, 63(10), 1339–1351.
[66] Moshtaghian, H., Bolton, K., & Rousta, K. Challenges for upcycled foods: Definition, inclusion in the food waste management hierarchy and public acceptability. Foods 2021, 10(11), 2874.
[67] Bangar, S. P., Chaudhary, V., Kajla, P., Balakrishnan, G., & Phimolsiripol, Y. Strategies for upcycling food waste in the food production and supply chain. Trends in Food Science & Technology 2024, 143, 104314.
[68] Peiró, S., Luengo, E., Segovia, F., Raso, J., & Almajano, M. P. Improving polyphenol extraction from lemon residues by pulsed electric fields. Waste and Biomass Valorization 2019, 10, 889–897.
[69] Franco, D., Munekata, P. E., Agregán, R., Bermúdez, R., López-Pedrouso, M., Pateiro, M., et al. Application of pulsed electric fields for obtaining antioxidant extracts from fish residues. Antioxidants 2020, 9(2), 90.
[70] Kaveh, S., Gholamhosseinpour, A., Hashemi, S. M. B., Jafarpour, D., Castagnini, J. M., Phimolsiripol, Y., et al. Recent advances in ultrasound application in fermented and non-fermented dairy products: Antibacterial and bioactive properties. International Journal of Food Science & Technology 2023, 58(7), 3591–3607.
[71] Phimolsiripol, Y., Buadoktoom, S., Leelapornpisid, P., Jantanasakulwong, K., Seesuriyachan, P., Chaiyaso, T., et al. Shelf life extension of chilled pork by optimal ultrasonicated Ceylon Spinach (Basella alba) extracts: Physicochemical and microbial properties. Foods 2021, 10(6), 1241.
[72] Flores, E. M., Cravotto, G., Bizzi, C. A., Santos, D., & Iop, G. D. Ultrasound-assisted biomass valorization to industrial interesting products: State-of-the-art, perspectives and challenges. Ultrasonics Sonochemistry 2021, 72, 105455.
[73] Jagadiswaran, B., Alagarasan, V., Palanivelu, P., Theagarajan, R., Moses, J., & Anandharamakrishnan, C. Valorization of food industry waste and by-products using 3D printing: A study on the development of value-added functional cookies. Future Foods 2021, 4, 100036.
[74] Pant, A., Lee, A. Y., Karyappa, R., Lee, C. P., An, J., Hashimoto, M., et al. 3D food printing of fresh vegetables using food hydrocolloids for dysphagic patients. Food Hydrocolloids 2021, 114, 106546.
[75] Banu, R., Kumar, G., Gunasekaran, M., & Kavitha, S. Food Waste to Valuable Resources: Applications and Management, Academic Press, London, 2020.

[76] Nayak, A., & Bhushan, B. An overview of the recent trends on the waste valorization techniques for food wastes. Journal of Environmental Management 2019, 233, 352–370.

[77] Ben-Othman, S., Jõudu, I., & Bhat, R. Bioactives from agri-food wastes: Present insights and future challenges. Molecules 2020, 25(3), 510.

[78] Abbate, S., & Centobelli, P. Unveiling the impact of digitized supply chains on mitigating food loss and waste: A multiple case study analysis. Available at SSRN 4846307.

[79] Anbarasu, V., Karthikeyan, P., & Anandaraj, S., editors. Turning human and food waste into reusable energy in a multilevel apartment using IoT. In: 6th International Conference on Advanced Computing and Communication Systems (ICACCS), IEEE, Coimbatore, India, 2020.

[80] Singh, A., Madaan, G., Hr, S., & Kumar, A. Smart manufacturing systems: A futuristic roadmap towards application of industry 4.0 technologies. International Journal of Computer Integrated Manufacturing 2023, 36(3), 411–428.

[81] Lu, Y. Industry 4.0: A survey on technologies, applications and open research issues. Journal of Industrial Information Integration 2017, 6, 1–10.

[82] Stock, T., & Seliger, G. Opportunities of sustainable manufacturing in industry 4.0. Procedia CIRP 2016, 40, 536–541.

[83] Lasi, H., Fettke, P., Kemper, H.-G., Feld, T., & Hoffmann, M. Industry 4.0. Business & Information Systems Engineering 2014, 6, 239–242.

[84] Tichoniuk, M., Biegańska, M., & Cierpiszewski, R. Intelligent packaging: Sustainable food processing and engineering challenges. In: Sustainable Food Processing and Engineering Challenges, Elsevier, Cambridge, 2021, 279–313.

[85] Mustafa, F., & Andreescu, S. Chemical and biological sensors for food-quality monitoring and smart packaging. Foods 2018, 7(10), 168.

[86] Roper, J. M., Garcia, J. F., & Tsutsui, H. Emerging technologies for monitoring plant health in vivo. ACS Omega 2021, 6(8), 5101–5107.

[87] Buzby, J. C., Farah-Wells, H., & Hyman, J. The estimated amount, value, and calories of postharvest food losses at the retail and consumer levels in the United States. USDA-ERS Economic Information Bulletin, 2014(121), 1–39.

Ayodeji Amobonye, Branly Nguena-Dongue, Naveen Kumar, Buka Magwaza, Grace Abel, Hassan T. Abdulameed, Chinmay Hazare, and Santhosh Pillai*

Chapter 19 Food packaging innovations: current and future applications in smart food development

Abstract: Food packaging as a critical component of food processing has evolved tremendously from the groundwork done by Nicholas Appert in the nineteenth century. This development has mainly been necessitated by the notoriety of conventional plastic packaging in the environment, the demand for functional products, as well as increased consumer awareness. Over the last century, plastic polymers, particularly polyethylene and polypropylene, have dominated the industry; however, the worldwide menace of plastic pollution has reinforced the push for eco-friendly packaging, especially in the food industry. Hence, various innovations have since emerged in food packaging to address these problems, aligning with several UN SDGs (sustainable development goals), including cleaner production, climate action, and zero hunger. These innovations include active-, edible-, intelligent-, and nano packaging. While active packaging incorporates various active compounds to extend food shelf life, edible packaging can be consumed and also biodegrades as efficiently as the food it protects. Similarly, intelligent packaging facilitates the safety and transport of packaged food using advanced technologies, while nanopackaging utilizes nanoscale materials to enhance numerous functionalities in food packaging. We present a comprehensive overview of these highlighted food packaging innovations, thus presenting a theoretical framework for future works and the evolving smart food industry.

*__Corresponding author: Santhosh Pillai,__ Department of Biotechnology and Food Science, Faculty of Applied Sciences, Durban University of Technology, PO Box 1334, Durban 4000, South Africa, e-mail: santhoshk@dut.ac.za

*__Corresponding author: Ayodeji Amobonye,__ Department of Biotechnology and Food Science, Faculty of Applied Sciences, Durban University of Technology, PO Box 1334, Durban 4000, South Africa; Department of Polymer Chemistry and Technology, Kaunas University of Technology, Radvilenu Rd. 19, 50254 Kaunas, Lithuania, email: ayodeji.amobonye@ktu.lt

Branly Nguena-Dongue, Naveen Kumar, Buka Magwaza, Grace Abel, Chinmay Hazare, Department of Biotechnology and Food Science, Faculty of Applied Sciences, Durban University of Technology, PO Box 1334, Durban 4000, South Africa

Hassan T. Abdulameed, Department of Toxicology, Advanced Medical and Dental Institute, Universiti Sains Malaysia, Kepala Batas, 13200, Penang, Malaysia; Department of Biochemistry, Kwara State University, Malete, Nigeria

https://doi.org/10.1515/9783111441238-020

Keywords: Bio-based packaging, edible films, food storage, functional packaging, innovative packaging, smart food industry

19.1 Introduction

Industrial food packaging (FP) can be traced to Nicholas Appert, a French confectioner and food scientist, who is considered by many as the "father of canning" [1]. Records from the Napoleonic era show that Nicholas Appert invented corked glass bottles and metal cans for the preservation of the French army rations [1]. However, the emergence of plastic polymers as well as the affordability and versatility of the materials have facilitated their wide adoption as FP materials in the last century. Plastic polymers have gained wide acceptability in food processing as they possess a broad range of physicochemical characteristics, which are essential in the wide spectrum of FP operations and objectives. According to Verma et al. [2], food packages serve several functions, which include but are not limited to physical protection, barrier protection, transportation, storage, information transmission, marketing, as well as security (anti-tampering). Among the various plastic polymers, polyethylene and polypropylene have been mostly utilized; polyethylene-based packaging is the gold standard in beverages, milk-based products, water, grocery, retail, and disposal bags, as well as pastries and frozen food bags as they facilitate longer storage with reduced compromise in organoleptic properties. However, these plastic materials have raised huge global concerns due to their accumulative tendencies and their detrimental effects on the environment [3]. For example, estimates at the beginning of this decade showed that a large proportion of the ~8 billion newly manufactured plastic items are have ended up discarded in various environments just after single use.

Thus, the evolution of innovative FP approaches has been one of the core components of the recent clamor for a more sustainable planet. In addition to ameliorating the damaging effect of plastic pollution on the environment, it is believed that innovative FP has the potential to promote food security and reduce household poverty. An analysis of the current UN-SDGs (United Nations sustainable development goals) shows that increased innovation in FP will remarkably enhance the attainment of SDG 2 (zero hunger), SDG 3 (good Health and well-being), SDG 9 (industry, innovation, and infrastructure), SDG 12 (responsible consumption and production), and SDG 13 (climate action). Furthermore, the cumulative effects of such innovation would also ultimately affect life below water and on land, which are the core mandates of SDGs 14 and 15 respectively. Furthermore, the increasing consumer preferences for environmentally friendly products have also accelerated developments of various technologies including FP technologies. However, due to the wide range of food product categories and significant differences in consumer expectations, it has been observed that there exists no universal strategy with respect to innovative food packaging [4]. Hence, needs arise for the parallel advancement of various innovative approaches in

FP to fill in these gaps; and it is quite remarkable that a lot of success has been achieved in the past three decades in this regard [5–7].

The market of innovative FP was globally estimated to be US$5 billion in 2023, a value that was also forecasted to increase significantly to US$8 billion between 2024 and 2032 (Transparency Market Research, 2023). As earlier highlighted, there are parallel developmental approaches in FP; however, it was observed that the most remarkable scientific and industrial successes have been recorded with active-, edible-, intelligent- (smart-), and nanopackaging. While active FP involves the functionalization of packaging materials with active antioxidant or antimicrobial compounds, which are systemically released to delay food spoilage [8, 9], edible FP is designed to be consumed safely along with the food products they enclose [10, 11]. While intelligent FP incorporates indicators, sensors, and other advanced technologies to promote the functionality, safety, and communication of the products [12, 13]; food nanopackaging is centered around the application of nanomaterial substances that are ≤100 nm, to improve the functionality and properties of FP materials [14, 15]. Hence, this chapter presents an in-depth review and discussion of the current innovations in FP with special emphasis on active-, edible-, intelligent-, and nanopackaging. It also highlights the future direction in the packaging aspects of smart food development with the aim of creating a strong theoretical framework for prospective research and development.

19.2 Active packaging in food industry

Active packaging (AP) is based on the addition of bioactive components into the matrix of materials meant for packaging purposes or within the package headspace [16]. These active components thus interact with the food or the food environment to improve overall performance of the system, hence, maintaining the organoleptic properties, ameliorating microbial and chemical contamination, and consequently increasing the enclosed food's shelf life [17]. According to Firouz, Mohi-Alden, and Omid [18], preserving foods during storage and monitoring their sensorial quality has become very crucial; hence, APs that might have been enhanced by flavors, antimicrobials, preservatives, and/or antioxidants, are considered one of the most notable approaches in the FP industry. Hence, various AP agents have been developed to enhance food stability [19] (Table 19.1). The presence of oxygen in food decreases food quality and shelf life by increasing oxidation and promoting aerobic microorganism growth; hence, the incorporation of oxygen absorbents removes oxygen from packaged food during storage, thus prolonging the desired condition of perishable food products [20]. Studies have since highlighted various oxygen absorbents in food packaging such as oxygen scavengers [21, 22], palladium catalytic system [23], iron carbonate [24], and aloe emodin [25]. Recently, Cheng et al. [26] demonstrated that ethylene absorbent-incorporated sachets and 1-methylcyclopropene-incorporated paper cards remarkably reduced ethylene release and respiration rate while maintaining the texture and taste of peaches.

Moisture absorbers in FP reduce mold and bacteria growth thus retarding spoilage; and the commonly used agents are calcium oxide and silica gel [27]. Similarly, organic molecules like fructose, cellulose, and carboxymethyl cellulose are popular compounds used in active packaging but due to the lower absorbing capacity and stability bigger sachets are required for silica gel desiccants [28]. Bovi et al. [29] developed a product termed Fruitpad, which was based on fructose (30%) for the packaging of fresh strawberries; it exhibited a significantly high amount of moisture absorption while maintaining the weight of fruit (~0.9% loss), much lower than the acceptable limit of 6%. Numerous antibacterials and antifungals have also been applied in FP, especially, essential oils and hydrophilic extracts from plants. Depending on the volatility of the antimicrobials incorporated in food systems, they can be grouped into nonvolatile or volatile compounds [30]. Recently, huge interest has risen in essential oil utilization in FP due to their broad as well as significant antimicrobial and antioxidant characteristics [31–33]. In one instance, Remya et al. [34] developed a multilayer film of ethylene–vinyl alcohol supplemented with ginger essential oil and O_2 scavenger; the bioactive package was demonstrated to possess significant antimicrobial activity and reduce the release of total volatile base nitrogen and lipid oxidation while preventing fish spoilage. Films containing essential oils from flaxseed, grapeseed, rose, and ginger also showed significant radical scavenging ability while prolonging fresh meat shelf life by more than 20% [35].

The utilization of natural polymers in packaging has long gained popularity due to their eco-friendliness, biocompatibility, biodegradability, and/or bioactivity [36]. These materials have demonstrated the capacity to maintain food quality by controlling and maintaining the internal condition of the product and acting as antimicrobial agents [37, 38]. Starch, cellulose, collagen, chitosan, and polylactic acid are currently among the main natural biopolymers used for food packaging materials [39]. Recently, Nilsuwan et al. [40] developed gelatin film fortified with epigallocatechin gallate, which significantly retarded lipid oxidation and has immense applicability in high-lipid foods packaging. A chitosan film supplemented with glycerol (40%) was also demonstrated to reduce water solubility as well as exhibit remarkable UV-blocking, and antimicrobial potentials as well as attenuate malondialdehyde levels [25]. Additionally, Zheng et al. [41] developed a functional film by grafting gallic, ferulic, and caffeic acid to chitosan and blending with collagen.

Table 19.1: Recent advancements in active food packaging.

Active agent	Matrix	Food	Benefits/ improvements	References
Nitrogen and oxygen scavenger	Polypropylene-based material with nitrogen and oxygen scavenger	Potato chips	Retention of 80% of the β-carotene; maintained sensory quality for more than 200 days of storage of chips at 25 °C and 75% RH	[22]
Catalytic system with palladium	O_2 scavenging film based on palladium catalytic system	Linseed oil	Maintained of linseed oil quality for over 6 months at 45 °C	[23]
Iron carbonate ($FeCO_3$)	O_2-absorbing sachets	Soybean sprouts	Increased oxygen absorption and durability of sprouts up to 12 days at 5 °C	[24]
Ethylene absorber	Low-density polyethylene supplemented with ethylene scavengers	Banana and kiwi	Reduced ethylene concentration (0.45 µl/L bananas, 0.50 µl/L kiwi fruit); maintained the physicochemical and sensory properties of bananas and kiwi fruit, respectively, for 9 and 30 days	[42]
Ethylene absorbent and 1-methylcyclopropene	1-Methylcyclopropene paper cards and ethylene absorbent sachets	Peaches	Reduced respiration rate, delayed ethylene release, preserved firmness, and volatile compounds, extended shelf life	[26]
Active O_2 scavenging element (mixture of iron filings and sodium polyacrylate)	Polylactic acid tray and film	Lulo fruit	The enhanced tray removed O_2 and CO_2 up to the respective values of 16.8% and 5% and increased the shelf life of lulo fruits stored at 12 ± 1 °C and 75 ± 2% RH to more than 25 days	[43]
Carvacrol	Polyethylene packaging	Crisps	Anti-*Salmonella enterica* activity (30% carvacrol) and reduced oxidation rate without loss of organoleptic properties	[44]
Tetraethyl citrate	Bio-based film made of polylactic acid	Pork fillets	Antioxidant activity, and antimicrobial activity; reduced lipid oxidation	[45]

Table 19.1 (continued)

Active agent	Matrix	Food	Benefits/ improvements	References
Avocado ethyl acetate extract	Cellulose-reinforced papers	Pear slices	Antimicrobial barrier to *E. coli* and *B. cereus*, thus protecting pear slices stored at 4 °C for 5 days.	[46]
Clove essential oil and rosehip seed oil	Functionalized bleached Kraft paper	Fresh cheese and beef	Antioxidant and antibacterial activities	[47]
Pumice and potassium permanganate (KMnO4)	Low-density polyethylene-based multilayer	Avocado	Water absorption and ethylene scavenging capacity; reduced loss in firmness and mass	[48]
Cinnamic acid	Film made of sodium alginate and pectin	Beef	Improved color and microbiological quality of beef samples	[49]
Carvacrol: eugenol mix (80:20)	Active paper sheets containing EOs – βCD complex	Lemons	Increased activities of antioxidant enzymes and lowered decay incidence	[50]
Thyme essential oil	Zein ultrafine fiber membranes	Meat	Antioxidant activity and inhibition against *E. coli* and *Staphylococcus* spp.	[51]
Black cumin oil	Multiple-layer active film made of polyester combined with chitosan and alginate layer	Chicken	Prolonged shelf life during refrigeration	[52]
Chlorogenic acid (CGA)	Sweet whey/starch blend films	Bananas	Acted as a barrier to oxygen, antioxidant effects, inhibited browning, and extended postharvest shelf life	[53]
Bioactive compounds (*trans*-cinnamaldehyde and citral)	Copolymer film fortified with β-cyclodextrin inclusion complex	Fresh beef	Extended shelf life.	[54]
Sodium carbonate and calcium hydroxide supplemented with polyacrylate	Active packaging CO_2- and moisture-absorbing pad	Mushrooms	Modified the atmosphere of the package to the required conditions of the shiitake mushrooms (10.7–11.6% O_2 and 7.1–8.9% CO_2) by absorbing the CO_2 and water vapor	[55]

Table 19.1 (continued)

Active agent	Matrix	Food	Benefits/ improvements	References
Carvacrol and thymol	Low-density polyethylene films	Hummus spread	Prolonged inhibitory activity against bacteria	[56]
Ginger essential oil and O_2 scavenger	Chitosan-based film	Chilled fish	Increased the shelf life	[34]
Carvacrol	Polyamide films	Tomatoes, lychee, cherries. and grapes	Antifungal properties, reduced deterioration, and extended shelf life	[57]

19.3 Edible food packaging

Edible packaging is a class of FP produced from components that can be eaten together with the enclosed food products; they are typically designed to be harmless to humans on consumption, environmentally friendly, as well as biodegradable. This approach offers a consumable, innovative, and sustainable alternative to conventional packaging, combining functionality with environmental benefits and hygiene. While the use of edible packaging is currently seen by many as a novel sustainable option to rising fossil-based plastic packaging the concept of using edible substances as food coatings has long history. Records revealed that the Sumerians preserved meats by stuffing them in animal intestines around 3000 BC; the Japanese also used boiled milk skins to cover food in the fifteenth century [58]. Notably, the major characteristics of edible packaging are edibility, biodegradability, reduced environmental impact, antimicrobial, as well as safety [59, 60]. Furthermore, these materials provide the added benefit of selectively exchanging respiratory gases while reducing moisture and aromas, and keeping ingredient loss to the minimum [61]. Edible packaging – which is mainly based on naturally sourced materials and polymers – has wide use in the food industry especially in fruits and vegetables, meat, fish, dairy products, pastries products, sweets, etc. Biopolymers that are used in edible packaging may be grouped into lipids, polysaccharides, proteins, as well as composites. The polysaccharide biopolymers include pectin, carrageenan, alginate, starch, and xanthan gum [62]. On the other hand, the protein biopolymers include casein, whey protein, gelatine, gluten, zein, etc. [63]; it is believed that the unique fibrous structure of proteins such as casein and gelatin confer better mechanical structure when compared to polysaccharides [62]. Lipid-based biopolymers are glossy, moisture-repellent, and cost-effective; they reduce the complexity of packaging and include beeswax, carnauba, and shellac [64]. However, a combination of more than one biopolymer known as composites can enhance desirable FP properties. It has been noted that the films produced by mixing lipids, proteins, and polysaccharides possess higher mechanical and barrier resistance nonetheless would be less moisture-deterrent as compared to pure lipids.

The many desirable characteristics of edible packaging can further be promoted by incorporating food-grade compounds possessing properties that would extend food shelf life of products. For instance, the use of antimicrobials, which may be naturally sourced antibacterial [65], antifungal [66], and antiviral compounds [67], has been shown to prevent microbial spoilage of packaged foods. Essential and cold pressed oils, for example, are known to selectively inhibit food spoilage microbes once added in edible films [68]. Similarly, organic acids such as propionic, malic, acetic, benzoic, citric, and tartaric acid; proteins such as lactoferrin, lysozyme, peroxidase, and nisin are some of the well-known antimicrobial substances with potential in this regard. Recently, the use of antimicrobial peptides along with biopolymers has proven to be an effective additive in packaging films, showing antimicrobial properties, anti-biofilm properties, and even antioxidant properties [69]. Similarly the use of natural antioxidants in edible FP was demonstrated to retard oxidation of the food, thus preventing spoilage [70]. Moreover, edible packaging can serve as an additional source of nutrition and can be fortified with various supplements and nutraceuticals to enhance their value and functionality. Studies have also established that the incorporation of probiotics and prebiotics in such packaging can promote gut health by delivering beneficial lactic acid bacteria and the fibers that nourish the organisms [71, 72].

These additional supplements/fortificants can be incorporated into the edible biopolymeric films using techniques like encapsulation, which ensures stability and controlled release, or spray-coating, which involves applying nutrient-rich solutions to the packaging surface. Edible packaging that aligns with ethical and cultural values is more likely to gain consumer trust and acceptance, hence, the need to consider dietary, cultural, and religious restrictions such as vegetarian, vegan, sattvic, kosher, and halal requirements. For instance, using collagen derived from fish is unsuitable as a component of edible packaging material, which may be preferred by vegetarians. Similarly, dairy-based proteins such as casein would be considered inappropriate by vegans, and pork-based products do not meet halal or sattvic dietary standards. Additionally, medical dietary restrictions, such as allergies or intolerances to specific ingredients, also need to be considered; for example, individuals with casein intolerance cannot consume packaging containing casein [73]. Therefore, understanding and integrating these considerations is essential for the development and widespread acceptance of edible packaging solutions. Despite the many advantages of edible food packaging, several other challenges need to be addressed. For instance, maintaining the stability and bioavailability of nutrients throughout the packaging's shelf life is crucial. Ensuring adherence with regulations and preserving sensory qualities of FP materials are critical for consumer acceptance. Additionally, achieving cost-effectiveness and scalability for commercial production remains a significant hurdle.

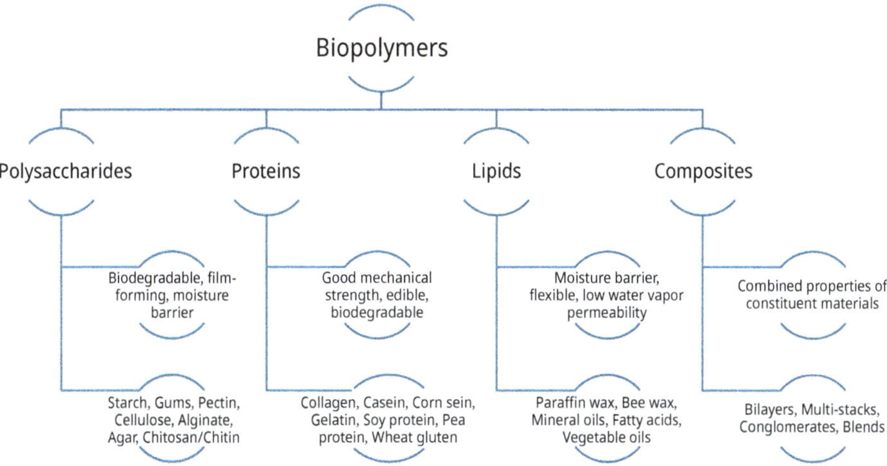

Figure 19.1: Overview of materials used in edible packaging.

19.4 Nanopackaging in food

Nanomaterials have recently become more popular due to their miniaturized sizes, which facilitate their versatility in various industrial applications as well as additional properties such as antimicrobial, ethylene-scavenging, UV scavenging, and gas barrier properties [15, 74]. Nanomaterials have been demonstrated severally to improve the shelf life and quality of food/drinks by protecting the enclosed products against microbiological, chemical, and physical contaminations [75–77]. Nanopackaging in the food value chain mainly involves the application of antimicrobial nanoparticles, and nanocomposites/materials in food packaging. Packaging with antimicrobial nanoparticles (NPs) including gold, silver, copper, zinc oxide, graphene, etc., have shown potential in extending food shelf life via growth inhibition of food pathogens such as *Salmonella* [75], *Listeria monocytogens* [78], *E. coli* [79], as well yeasts (*Candida albicans*) and molds (*Aspergillus niger*) [80]. For instance, Li et al. [81] demonstrated the effective functionality of copper NPs in food packaging as they hindered *Staphylococcus aureus* and *E. coli* proliferation in packaged meats. Similarly, the shelf life of minced meat was significantly prolonged by the inclusion of TiO_2 NPs in the packaging as the NPs were observed to suppress the growth of Enterobacteriaceae, *S. aureus*, as well as psychotropic bacteria [82]. Similarly, Pandey et al. [83], observed that Ag NPs increased meat longevity by hindering the growth of bacterial.

Different matrices – including lignocellulosic fibers, alginate, starch pectin, and carboxymethyl cellulose – have been used together with nanoparticles to form nanocomposites with enhanced functional properties in food packaging [84, 85]. Most of these polymeric matrices are natural polymers and they also include cellulose, chitosan, guar gum, starch, and zein [77]. Synthetic polymers, especially polyethylene, have also been demonstrated to be efficient for the same purpose [86, 87]; however, there are concerns about their impact on environmental sustainability. Recently, a nanocomposite with remarkable antioxidant and antimicrobial properties was developed for cheese packaging by incorporating ZnO NPs into chitosan-guar gum-Rossell calyx matrix [88]. Kumar et al. [89], also incorporated ZnO NPs into chitosan-gelatin matrix for the same purpose. In both cases, the developed films showed improved thermal stability, elongation at break, and compactness, in addition to their significant antimicrobial activities. Despite the remarkable successes recorded with nano-based packaging material in various food types, there are still concerns about potential, undesirable, and unhealthy drift of these miniaturized materials into the food matrices [90–92]. Some authors have highlighted the migration of silver and titanium NPs into blood circulation; and these may be detrimental in animal organs due to their insolubility and accumulation [93, 94]. In addition, the accumulation of Zn NPs and AgO NPs (100 µg/ml) was linked to a genotoxic effect in human epidermal cells [95]. However, it has been shown in some studies that NPs that migrated from nanocomposite materials into food products fall below the standard limit of 5 mg/kg by the European Commission (EC) [92, 96–98]. For instance, minute quantities of Zn (2.44 mg/kg) and Ag (<0.01 mg/ kg) were reported to migrate from nanocomposites (LDPE-Ag-ZnO) used in poultry product packaging [95]. However, regulators such as the US Food Drug Administration, European Food Safety Authority, as well as Environmental Protection Agency have also promulgated necessary standards to guide nanomaterial utilization and their migration.

Table 19.2: Nanomaterials in food packaging.

Nanomaterial	Packaging matrix	Food	Functionality	References
Silver NPs	Polyethylene/ polypropylene	Button mushroom	Improved antioxidant activity and delayed aging at 4 °C	[99]
Chitosan NPs	Potato starch	Cheese	Inhibition of lipid oxidation during storage and extension shelf of cheese	[100]
Chitosan NPs loaded with *Citrus aurantium* essential oil	Direct fruit coating	Button mushroom	Inhibition of the brown patches via inhibition of yeasts, mesophilic, and psychrophilic bacteria	[101]
Cinnamaldehyde	Chitosan	Trout fillet	Improved thermal properties; activity against *L. monocytogenes*, and *Salmonella entiritidis* and *E.coli*	[102]

Table 19.2 (continued)

Nanomaterial	Packaging matrix	Food	Functionality	References
Copper NPs	Polyethylene	Sweet dairy product	Inhibition against *E.coli* and *S. aureus*, decreased water vapor permeability	[103]
Copper-Titanium oxide NPs	Carboxymethyl cellulose	Banana	Improved water vapor barrier properties, delayed browning of banana, antibacterial activity	[104]
Titanium oxide NPs	Agar-κ carrageenan–anthocyanin	Pork	Improved color stability and UV-Vis barrier properties	[105]
Zinc oxide NPs	Gum arabic/chitosan	Banana	Effective barrier against gas and moisture	[106]
Zinc oxide NPs	Semolina flour	Mozzarella cheese	Exhibited strong antibacterial Activity against *E.coli* and *A. niger*, and extended the shelf life of cheese during storage at low-temperature	[80]
Zinc oxide NPs	Agar-nisin cinnamon essential oil	Mince fish	Inhibition against *E.coli*, *L. monocytogenes*, and *S. aureus*	[107]
Zinc oxide NPs	Mahua oil-based polyurethane and chitosan	Carrot	Shelf life extension of carrots, improved barrier properties, and inhibition against *S. aureus* and *E. coli*	[108]

19.5 Intelligent packaging

The European Commission (2004) states that intelligent packaging (IP) materials are "materials and articles that monitor the condition of packaged food or the environment surrounding the food" [109]. Hence, the word "intelligent" simply indicates a product's ability to switch "ON" or "OFF" on its packaging in reaction to shifting internal or external stimuli, thereby informing customers or end users of the product's condition [110]. Unlike active food packaging technologies, which work actively to increase enhance the shelf life of food, the IP's goal is to provide food quality information to the various stakeholders, vis-a-vis, consumers, suppliers, manufacturers, and retailers. Generally, IP systems are based on external, discrete components such as two-dimensional films or three-dimensional objects, and they are classified based on their functionality into indicators, data carriers, and sensors [111].

19.5.1 Indicators

Indicators in intelligent food packaging are designed to increase consumer convenience and/or provide information about food quality; for instance, information about the concentration of a particular substance, the interaction extent between various entities, or the presence or absence of a material may be passed via these indicators [112]. The fundamental mechanism of these indicators relies on the distinct reactions that occur between different materials, which could be biological, mechanical, or chemical, leading to the irreversible discoloration of the indicator as a clear reaction [18]. Indicators are further classified as freshness, gas, and temperature indicators. Temperature indicators, which come either as simple- or time–temperature integrators (TTIs), draw consumer attention to the possibility of the presence of unwanted microbes and/or the denaturation of proteins during procedures such as refrigeration or thawing [113]. Consumers easily understand the information provided by TTI devices as they provide information directly related to the product quality at a specific measure of temperature. Freshness indicators allow the tracking of food product quality during transit and storage, based on chemical or microbiological changes; for example, the volatile amines or total volatile basic nitrogen content are measures for freshness meant for seafood [114]. These compounds are produced as food deteriorates and they are typically quantifiable via conductometric analysis and pH changes. The sensitivity to additional microbial metabolites, such as CO_2, alcohols, aldehydes, esters and diacetyl, is the basis for other freshness indicators [115]. RipeSense^{TM}, a commercially available freshness indicator, measures the freshness of pears, kiwi, mango, as well as other fruits and vegetables by using a monitoring some aromatic and ethylene generated [116]. Gas indicators track variations in the internal environment brought on by chemical or enzymatic changes in the food matrix, microbial metabolism, and penetration events across the packaging material; also they can find leaks or evaluate how well active packaging elements (such as CO_2 and O_2 scavengers) are working [13, 117]. Redox dyes (such as toluidine blue, 2,6-dichloroindophenol, and methylene blue), reducing compounds (such as some monosaccharides and disaccharides), and bases (such as NaOH and KOH) have also been noted to be quite functional as indicators in this regard [118].

19.5.2 Data carriers

Data carriers, also commonly known as automatic identification devices, enhance the movement of information within the food supply chain, and consequently food safety and quality; they are designed to facilitate automation, traceability, as well as prevent counterfeiting and theft[110]. The barcode labels and radio frequency identification (RFID) tags are notable data carrier devices in FP. Barcodes, which include one-and two- dimensional barcodes, are quite prevalent in large-scale retail establishments because of their affordability and user-friendliness, which helps with inventory control, product replenishment, and checkout. On the other hand, RFIDs are the most advanced types of data carrier device with two unique features, *viz*., they can store numerous different codes on a tag and can transmit information over long distances, which improve automated identification and traceability of the products [119, 120]. Although RFID has long been recognized, its market penetration has lagged behind that of barcodes, mostly because of cost considerations. RFID tags have been used to control the freshness of meat [121]. Also, RFID tags based on temperature and gas sensor, together with a reader, were applied in the tracking system for pork freshness[122]. Similarly, RFID along with CO_2 and oxygen sensors was also demonstrated in the monitoring of vegetables as well as the real-time evaluation of the packaged milk quality [123, 124].

19.5.3 Sensors

A cutting-edge technology for intelligent food packaging systems is sensors [18]. The majority of sensors consist of four main parts: a receptor, the transduction element, the signal processor; and the signal display component [125]. Several types of sensors, including electrochemical and luminescence sensors, have been created recently. These sensors fall into two primary categories: biosensors and chemical sensors. The recognition layers in biosensors and chemical sensors are biological materials (antibodies, enzymes, antigens, nucleic acids, etc.) and chemical compounds (carbon, platinum, silicon, etc.) respectively [126]. For example, the Food Sentinel System (SIRA Technologies, USA), alerts retailers and product consumers on the exposure of the product to extreme conditions that may significantly impact its safety. This Food Sentinel System is based on a biosensor carrying pathogen-specific antibodies, modelled into membranes that are attached to barcodes. Similarly, flexible biosensors were created by Flex Alert (Canada) to identify toxins in packaged meals at any stage of the supply chain; these biosensors are specifically designed to detect aflatoxins, *Salmonella* spp., *Listeria* spp., and *E. coli* O157 [127]. Additionally, NIMA Company (San Francisco, CA, USA) created an antibody-based sensor to identify peanut allergens in chocolate candy bars, including archin (Ara h1) [128].

19.6 Current market for innovative food packaging

According to Verma et al. [129], the need for environmental conservation, effective resource utilization, and dynamic consumer lifestyles/demands are the major driving forces behind the growth of the various smart innovations in food packaging. With the rise in consumer taste and knowledge, IP is envisaged to lead in innovative FP technology due to the following reasons: the level of safety and traceability of products it offers consumers, the need to meet the rising demand for export products, as well as the role of regulatory bodies. As the largest consumer of active-, and intelligent packaging solutions, the food and beverage sector utilizes innovative food packaging materials for various commodities, which include bakery products (bread, cakes), meat products, beverages (fruits and vegetables), and even ready-to-eat foods/prepared foods. It is believed that the demand is higher in packaging for fruits and vegetables and meat products where maintaining freshness and quality is premium. However, the high cost of IP materials and equipment has been identified as a limiting factor to the growth of the market [130]. Various market research reports have highlighted the increasing economic value of innovative food packaging materials and their significant contribution to the overall packaging and food industries. For example, reports by Future market insights indicated that the global market for antimicrobial nanocoating will experience a steady growth rate from an estimated US$930.5 million in 2022 to US$5,893.7 million in 2032. Some leading companies involved in the innovative food packaging industries have been identified, which includes Exopack and Tetra Pak (Table 19.1).

Table 19.3: Leading companies in the innovative food packaging industries.

Company	Brand name for packaging products and application	Market value (USD)
Exopack LLC, USA	Halo®, ClearShield®, and Maraflex® in cheese and red, meat performance and packaging	82 million
Trinseo, USA	Ligos A 9200 used in dry lamination of flexible structures in contact with food	100 million
Tetra Pak, South Africa	Tetra Recart® as sustainable food packaging for canned foods	*12.7 billion
Amcor-JD Farm	NFC (near-field communication)-enabled packaging for effective and interactive consumer engagement	*13.89 billion
Varcode Ltd. and Media Sourcery, Inc., Israel	Smart Tag™ for cold chain logistics	9.8 million
Notpla, UK	Ooho, clear transparent packaging sachets	120 million
Celluforce, Canada	Cellulose nanocrystals	10.7 million
American Process, USA	Cellulose nanocrystals	NR
FiloCell (Kruger company), South Africa	Nanofibrillated cellulose	NR

NR, not reported; *– values in Euros.

A critical review of the patent landscape for innovative food packaging also showed that interest continues to grow, notwithstanding challenges such as high production costs, as well as strict safety and hygiene regulatory standards. For example, the applications for nano-food related patents were on the rise between 2011 and 2015 as revealed by Google Patent analysis [131]. This can be as a result of higher investments in research and development, more awareness of food safety, which is influencing favorable consumer perception, and opening the market potential of innovative food packaging. A review on sustainable FP in 2002–2022 revealed the exploration of innovative food packaging from biodegradable, renewable agro-wastes [132]. An example is the utilization of nanocellulose as an alternative to plastic, which has a growing market value majorly in industries the developed nations, with a market that is projected to rise to above US$60 billion in the next three years [133]. Moreover, the rise in the number of publications and patent development indicates that both the scientific and commercial audience are making efforts to expand the markets for innovative FP and the food industry as a whole. The amount of publications generated on sustainable FP rose by about 20 times from 31 in 2012 to 431 publications in 2022 [134].

19.7 Conclusion

Recent progress in the food industry has seen the introduction of active-, nano-, and smart packaging technologies. Implementing these innovations has significantly benefited the industry by providing quick, simple, and cost-effective methods to maintain quality and/or oversee environmental conditions across the food chain. Future developments in food packaging should focus more on ecological sustainability, focusing on reusable, user-friendly materials that effectively communicate with consumers, while addressing the pertinent issues of food waste, quality management, and foodborne diseases. For instance, biogenic IP is a promising area with immense potential to reduce carbon footprint and ensure food safety. Moreover, recent innovations leveraging agro-industrial by-products have also revealed that a circular bioeconomy could be created by extracting bioactive compounds from agro-residues and embedding them into food packaging materials. Efforts should also be directed at integrating multiple IP devices into a unified system to maximize the advantages of this approach as well as the utilization of artificial intelligence and machine learning. In addition, it is also posited that the combination of active, smart, and green packaging technologies may facilitate development of versatile food packing systems with little or no adverse interactions between their components. However, extensive research is still required in many aspects of innovative food packaging, especially with regards to the migration of NPs into food products, the absorption of these materials into the human system, and their impacts on both consumer health and various environments. These important investigations will be faster achieved with collaborative efforts between researchers, industries, regulatory agencies, NGOs, and other relevant stakeholders. As highlighted in this chapter, various pilot- and bench-scale studies have demonstrated the potential of different groundbreaking innovations in smart food development. However, these innovations are yet to be fully deployed in the common markets mainly as a result of their current high manufacture cost, and gaps in relation to their regulation and standardization, among many other limitations. It is believed that these challenges will be circumvented in the near future as deeper insights are being gained into these relatively new food packaging innovations.

References

[1] Risch, S. J. Food packaging history and innovations. Journal of Agricultural and Food Chemistry 2009, 57(18), 8089–8092.

[2] Verma, M. K., Shakya, S., Kumar, P., Madhavi, J., Murugaiyan, J., & Rao, M. V. R. Trends in packaging material for food products: Historical background, current scenario, and future prospects. Journal of Food Science and Technology 2021, 58(11), 4069–4082.

[3] Amobonye, A. E., Bhagwat, P., Singh, S., & Pillai, S. Chapter 10 – Biodegradability of Polyvinyl chloride. In: Sarkar, A., Sharma, B., & Shekhar, S., editors. Biodegradability of Conventional Plastics, Cham: Elsevier, 2023, 201–220.

[4] Macht, J., Klink-Lehmann, J., & Venghaus, S. Eco-friendly alternatives to food packed in plastics: German consumers' purchase intentions for different bio-based packaging strategies. Food Quality and Preference 2023, 109, 104884.

[5] Thirupathi Vasuki, M., Kadirvel, V., & Pejavara Narayana, G. Smart packaging – An overview of concepts and applications in various food industries. Food Bioengineering 2023, 2(1), 25–41.

[6] Vanderroost, M., Ragaert, P., Devlieghere, F., & De Meulenaer, B. Intelligent food packaging: The next generation. Trends in Food Science and Technology 2014, 39(1), 47–62.

[7] LaCoste, A., Schaich, K. M., Zumbrunnen, D., & Yam, K. L. Advancing controlled release packaging through smart blending. Packaging Technology and Science: An International Journal 2005, 18(2), 77–87.

[8] Westlake, J. R., Tran, M. W., Jiang, Y., Zhang, X., Burrows, A. D., & Xie, M. Biodegradable active packaging with controlled release: Principles, progress, and prospects. ACS Food Science and Technology 2022, 2(8), 1166–1183.

[9] Zhou, M., Han, Y., McClements, D. J., Cheng, C., & Chen, S. Co-encapsulation of anthocyanin and cinnamaldehyde in nanoparticle-filled carrageenan films: Fabrication, characterization, and active packaging applications. Food Hydrocolloids 2024, 149, 109609.

[10] Bhuimbar, M. V., Bhagwat, P. K., & Dandge, P. B. Extraction and characterization of acid soluble collagen from fish waste: Development of collagen-chitosan blend as food packaging film. Journal of Environmental Chemical Engineering 2019, 7(2), 102983.

[11] Li, L., Luo, X., Liu, Y., Teng, M., Liu, X., Zhang, X., & Liu, X. Edible packaging revolution: Enhanced functionality with natural collagen aggregates. Food Hydrocolloids 2024, 156, 110331.

[12] Sani, M. A., Zhang, W., Abedini, A., Khezerlou, A., Shariatifar, N., Assadpour, E., et al. Intelligent packaging systems for the quality and safety monitoring of meat products: From lab scale to industrialization. Food Control 2024, 160, 110359.

[13] Khodaei, S. M., Gholami-Ahangaran, M., Karimi Sani, I., Esfandiari, Z., & Eghbaljoo, H. Application of intelligent packaging for meat products: A systematic review. Veterinary Medicine and Science 2023, 9(1), 481–493.

[14] Ahmad, A., Qurashi, A., & Sheehan, D. Nano packaging–Progress and future perspectives for food safety, and sustainability. Food Packaging and Shelf Life 2023, 35, 100997.

[15] Ahmad, K., Li, Y., Tu, C., Guo, Y., Yang, X., Xia, C., & Hou, H. Nanotechnology in food packaging with implications for sustainable outlook and safety concerns. Food Bioscience 2024, 58, 103625.

[16] Yildirim, S., Röcker, B., Pettersen, M. K., Nilsen-Nygaard, J., Ayhan, Z., Rutkaite, R., et al. Active packaging applications for food. Comprehensive Reviews in Food Science and Food Safety 2018, 17(1), 165–199.

[17] Janjarasskul, T., & Suppakul, P. Active and intelligent packaging: The indication of quality and safety. Critical Reviews in Food Science and Nutrition 2018, 58(5), 808–831.

[18] Firouz, M. S., Mohi-Alden, K., & Omid, M. A critical review on intelligent and active packaging in the food industry: Research and development. Food Research International 2021, 141, 110113.

[19] Rai, P., Mehrotra, S., & Sharma, S. K. Challenges in assessing the quality of fruit juices: Intervening role of biosensors. Food Chemistry 2022, 386, 132825.

[20] Mexis, S., & Kontominas, M. Effect of oxygen absorber, nitrogen flushing, packaging material oxygen transmission rate and storage conditions on quality retention of raw whole unpeeled almond kernels (*Prunus dulcis*). LWT-Food Science and Technology 2010, 43(1), 1–11.

[21] Dombre, C., Guillard, V., & Chalier, P. Protection of methionol against oxidation by oxygen scavenger: An experimental and modelling approach in wine model solution. Food Packaging and Shelf Life 2015, 3, 76–87.

[22] Júnior, L. M., Ito, D., Ribeiro, S. M. L., da Silva, M. G., & Alves, R. M. V. Stability of β-carotene rich sweet potato chips packed in different packaging systems. LWT 2018, 92, 442–450.

[23] Faas, N., Röcker, B., Smrke, S., Yeretzian, C., & Yildirim, S. Prevention of lipid oxidation in linseed oil using a palladium-based oxygen scavenging film. Food Packaging and Shelf Life 2020, 24, 100488.

[24] Cantelli, K. C., Carrão-Panizzi, M. C., Moreira, F. K. V., Steffens, J., Zeni, J., & Steffens, C. Evaluation of packaging systems with O_2-absorbers on quality of minimally processed soybean sprouts. Food Science and Technology International 2023, 29(4), 310–317.

[25] Yang, W., Ning, Y., Ren, Z., Xu, S., Li, J., & Wang, L. Fabricating anti-ultraviolet and antibacterial films with lower water solubility from chitosan, glycerol and aloe emodin for pork preservation. Food Hydrocolloids 2024, 153, 110030.

[26] Cheng, C., Liang, X., Wei, W., Zhang, N., Yao, G., & Yan, R. Enhanced shelf life quality of peaches (*Prunus persica* L.) using ethylene manipulating active packaging in e-commerce logistics. Scientia Horticulturae 2024, 326, 112701.

[27] Guo, X., Liu, L., Hu, Y., & Wu, Y. Water vapor sorption properties of TEMPO oxidized and sulfuric acid treated cellulose nanocrystal films. Carbohydrate Polymers 2018, 197, 524–530.

[28] Gaikwad, K. K., Singh, S., & Ajji, A. Moisture absorbers for food packaging applications. Environmental Chemistry Letters 2019, 17(2), 609–628.

[29] Bovi, G. G., Caleb, O. J., Klaus, E., Tintchev, F., Rauh, C., & Mahajan, P. V. Moisture absorption kinetics of FruitPad for packaging of fresh strawberry. Journal of Food Engineering 2018, 223, 248–254.

[30] Drago, E., Campardelli, R., Pettinato, M., & Perego, P. Innovations in smart packaging concepts for food: An extensive review. Foods 2020, 9(11), 1628.

[31] Mukurumbira, A., Shellie, R., Keast, R., Palombo, E., Muir, B., & Jadhav, S. The antimicrobial efficacy of native Australian essential oils in liquid and vapour phase against foodborne pathogens and spoilage microorganisms. Food Control 2023, 151, 109774.

[32] Wu, J., Sun, X., Guo, X., Ge, S., & Zhang, Q. Physicochemical properties, antimicrobial activity and oil release of fish gelatin films incorporated with cinnamon essential oil. Aquaculture and Fisheries 2017, 2(4), 185–192.

[33] Wińska, K., Mączka, W., Łyczko, J., Grabarczyk, M., Czubaszek, A., & Szumny, A. Essential oils as antimicrobial agents – Myth or real alternative?. Molecules 2019, 24(11), 2130.

[34] Remya, S., Mohan, C. O., Venkateshwarlu, G., Sivaraman, G. K., & Ravishankar, C. N. Combined effect of O_2 scavenger and antimicrobial film on shelf life of fresh cobia (*Rachycentron canadum*) fish steaks stored at 2 °C. Food Control 2017, 71, 71–78.

[35] Wrona, M., Silva, F., Salafranca, J., Nerín, C., Alfonso, M. J., & Má, C. Design of new natural antioxidant active packaging: Screening flowsheet from pure essential oils and vegetable oils to *ex vivo* testing in meat samples. Food Control 2021, 120, 107536.

[36] Horst, C., Pagno, C. H., Flores, S. H., & Costa, T. M. H. Hybrid starch/silica films with improved mechanical properties. Journal of Sol-Gel Science and Technology 2020, 95, 52–65.

[37] Mostafavi, F. S., & Zaeim, D. Agar-based edible films for food packaging applications-A review. International Journal of Biological Macromolecules 2020, 159, 1165–1176.

[38] Hou, X., Xue, Z., Xia, Y., Qin, Y., Zhang, G., Liu, H., & Li, K. Effect of SiO_2 nanoparticle on the physical and chemical properties of eco-friendly agar/sodium alginate nanocomposite film. International Journal of Biological Macromolecules 2019, 125, 1289–1298.

[39] Perera, K. Y., Jaiswal, A. K., & Jaiswal, S. Biopolymer-based sustainable food packaging materials: Challenges, solutions, and applications. Foods 2023, 12(12), 2422.

[40] Nilsuwan, K., Guerrero, P., De la Caba, K., Benjakul, S., & Prodpran, T. Fish gelatin films laminated with emulsified gelatin film or poly (lactic) acid film: Properties and their use as bags for storage of fried salmon skin. Food Hydrocolloids 2021, 111, 106199.

[41] Zheng, T., Tang, P., & Li, G. Development of composite film based on collagen and phenolic acid-grafted chitosan for food packaging. International Journal of Biological Macromolecules 2023, 241, 124494.

[42] Öztürk, M., & Ayhan, Z. Combined effects of ethylene scavenging-active packaging system and modified atmosphere to reduce postharvest losses of ethylene sensitive produce: Banana and kiwifruit. Packaging Technology and Science 2023, 36(11), 951–967.

[43] Escobar, H. J., Garavito, J., & Castellanos, D. A. Development of an active packaging with an oxygen scavenger and moisture adsorbent for fresh lulo (*Solanum quitoense*). Journal of Food Engineering 2023, 349, 111484.

[44] Wrona, M., Manso, S., Silva, F., Cardoso, L., Salafranca, J., Nerín, C., et al. New active packaging based on encapsulated carvacrol, with emphasis on its odour masking strategies. Food Packaging and Shelf Life 2023, 40, 101177.

[45] Karabagias, V. K., Giannakas, A. E., Andritsos, N. D., Moschovas, D., Karydis-Messinis, A., Leontiou, A., et al. Novel Polylactic Acid/Tetraethyl Citrate Self-Healable active packaging films applied to pork fillets' shelf-life extension. Polymers 2024, 16(8), 1130.

[46] Acquavia, M. A., Benítez, J. J., Bianco, G., Crescenzi, M. A., Hierrezuelo, J., Grifé-Ruiz, M., et al. Incorporation of bioactive compounds from avocado by-products to ethyl cellulose-reinforced paper for food packaging applications. Food Chemistry 2023, 429, 136906.

[47] Irimia, A., Grigoraș, V. C., & Popescu, C.-M. Active cellulose-based food packaging and its use on foodstuff. Polymers 2024, 16(3), 389.

[48] Wang, C., & Ajji, A. Development and application of low-density polyethylene-based multilayer film incorporating potassium permanganate and pumice for avocado preservation. Food Chemistry 2023, 401, 134162.

[49] Tong, W. Y., Rafiee, A. R. A., Leong, C. R., Tan, W.-N., Dailin, D. J., Almarhoon, Z. M., et al. Development of sodium alginate-pectin biodegradable active food packaging film containing cinnamic acid. Chemosphere 2023, 336, 139212.

[50] López-Gómez, A., Navarro-Martínez, A., & Martínez-Hernández, G. B. Effects of essential oils released from active packaging on the antioxidant system and quality of lemons during cold storage and commercialization. Scientia Horticulturae 2023, 312, 111855.

[51] Peixoto, E. C., Fonseca, L. M., Da Rosa Zavareze, E., & Gandra, E. A. Antimicrobial active packaging for meat using thyme essential oil (*Thymus vulgaris*) encapsulated on zein ultrafine fibers membranes. Biocatalysis and Agricultural Biotechnology 2023, 51, 102778.

[52] Takma, D. K., & Korel, F. Active packaging films as a carrier of black cumin essential oil: Development and effect on quality and shelf-life of chicken breast meat. Food Packaging and Shelf Life 2019, 19, 210–217.
[53] Dai, X., Dong, F., Dong, Z., Bai, Z., & Mao, L. Enhanced antibacterial and antioxidant activities of chlorogenic acid loaded sweet whey/starch active films for edible food packaging. LWT 2024, 199, 116118.
[54] Chen, H., Li, L., Ma, Y., Mcdonald, T. P., & Wang, Y. Development of active packaging film containing bioactive components encapsulated in β-cyclodextrin and its application. Food Hydrocolloids 2019, 90, 360–366.
[55] Wang, H. J., An, D. S., Rhim, J.-W., & Lee, D. S. Shiitake mushroom packages tuned in active CO_2 and moisture absorption requirements. Food Packaging and Shelf Life 2017, 11, 10–15.
[56] Krepker, M., Shemesh, R., Poleg, Y. D., Kashi, Y., Vaxman, A., & Segal, E. Active food packaging films with synergistic antimicrobial activity. Food Control 2017, 76, 117–126.
[57] Shemesh, R., Krepker, M., Nitzan, N., Vaxman, A., & Segal, E. Active packaging containing encapsulated carvacrol for control of postharvest decay. Postharvest Biology and Technology 2016, 118, 175–182.
[58] Teixeira-Costa, B. E., & Andrade, C. T. Natural polymers used in edible food packaging – History, function and application trends as a sustainable alternative to synthetic plastic. Polysaccharides 2021, 3(1), 32–58.
[59] Petkoska, A. T., Daniloski, D., D'Cunha, N. M., Naumovski, N., & Broach, A. T. Edible packaging: Sustainable solutions and novel trends in food packaging. Food Research International 2021, 140, 109981.
[60] Moeini, A., Pedram, P., Fattahi, E., Cerruti, P., & Santagata, G. Edible polymers and secondary bioactive compounds for food packaging applications: Antimicrobial, mechanical, and gas barrier properties. Polymers 2022, 14(12), 2395.
[61] Shahidi, F., & Hossain, A. Preservation of aquatic food using edible films and coatings containing essential oils: A review. Critical Reviews in Food Science and Nutrition 2022, 62(1), 66–105.
[62] Mohamed, S. A., El-Sakhawy, M., & El-Sakhawy, -M. A.-M. Polysaccharides, protein and lipid-based natural edible films in food packaging: A review. Carbohydrate Polymers 2020, 238, 116178.
[63] Milani, J. M., & Tirgarian, B. An overview of edible protein-based packaging: Main sources, advantages, drawbacks, recent progressions and food applications. Journal of Packaging Technology and Research 2020, 4(1), 103–115.
[64] Milani, J. M., & Nemati, A. Lipid-based edible films and coatings: A review of recent advances and applications. Journal of Packaging Technology and Research 2022, 6(1), 11–22.
[65] Manzoor, A., Yousuf, B., Pandith, J. A., & Ahmad, S. Plant-derived active substances incorporated as antioxidant, antibacterial or antifungal components in coatings/films for food packaging applications. Food Bioscience 2023, 53, 102717.
[66] Silva-Weiss, A., Ihl, M., PdA, S., Gómez-Guillén, M., & Bifani, V. Natural additives in bioactive edible films and coatings: Functionality and applications in foods. Food Engineering Reviews 2013, 5, 200–216.
[67] Priyadarshi, R., Purohit, S. D., Roy, S., Ghosh, T., Rhim, J.-W., & Han, S. S. Antiviral biodegradable food packaging and edible coating materials in the COVID-19 era: A mini-review. Coatings 2022, 12(5), 577.
[68] Wang, M., Wei, Z., & Zhang, Z. Antimicrobial edible films for food preservation: Recent advances and future trends. Food and Bioprocess Technology 2024, 17(6), 1391–1411.
[69] Luo, X., Peng, Y., Qin, Z., Tang, W., Duns, G. J., Dessie, W., et al. Chitosan-based packaging films with an integrated antimicrobial peptide: Characterization, in vitro release and application to fresh pork preservation. International Journal of Biological Macromolecules 2023, 231, 123209.

[70] Rodríguez, G. M., Sibaja, J. C., Espitia, P. J., & Otoni, C. G. Antioxidant active packaging based on papaya edible films incorporated with *Moringa oleifera* and ascorbic acid for food preservation. Food Hydrocolloids 2020, 103, 105630.

[71] Hua, Q., Wong, C. H., & Li, D. Postbiotics enhance the functionality of a probiotic edible coating for salmon fillets and the probiotic stability during simulated digestion. Food Packaging and Shelf Life 2022, 34, 100954.

[72] Ceylan, H. G., & Atasoy, A. F. New bioactive edible packing systems: Synbiotic edible films/coatings as carries of probiotics and prebiotics. Food and Bioprocess Technology 2023, 16(7), 1413–1428.

[73] He, M., Sun, J., Jiang, Z. Q., & Yang, Y. X. Effects of cow's milk beta-casein variants on symptoms of milk intolerance in Chinese adults: A multicentre, randomised controlled study. Nutrition Journal 2017, 16, 1–12.

[74] Cheng, H., Chen, L., McClements, D. J., Xu, H., Long, J., Zhao, J., et al. Recent advances in the application of nanotechnology to create antioxidant active food packaging materials. Critical Reviews in Food Science and Nutrition 2024, 64(10), 2890–2905.

[75] Shankar, V. S., Thulasiram, R., Priyankka, A., Nithyasree, S., & Sharma, A. A. Applications of nanomaterials on a food packaging system – A review. Engineering Proceedings 2024, 61(1), 4.

[76] Prakash, S., Kumari, M., & Chauhan, A. K. The intervention of nanotechnology in food packaging: A review. Journal of Materials Science 2024, 59, 2585–2601.

[77] Peerzada, J. G., Ojha, N., Jaabir, M. M., Lakshmi, B., Hannah, S., Chidambaram, R., et al. Advancements in eco-friendly food packaging through nanocomposites: A review. Polymer Bulletin 2024, 81(7), 5753–5792.

[78] Jampílek, J., & Kráľová, K. Benefits of chitosan-based and cellulose-based nanocomposites in food protection and food packaging. In: Sarma, H., Joshi, S.J., Prasad, R., Jampilek, J. (eds) Biobased Nanotechnology for Green Applications. Nanotechnology in the Life Sciences. Springer, Cham, 121–160 Biobased Nanotechnology for Green Applications 2021, 121–160.

[79] Wahab, Y. A., Al-Ani, L. A., Khalil, I., Schmidt, S., Tran, N. N., Escribà-Gelonch, M., et al. Nanomaterials: A critical review of impact on food quality control and packaging. Food Control 2024, 163, 110466.

[80] Jafarzadeh, S., Rhim, J. W., Alias, A. K., Ariffin, F., & Mahmud, S. Application of antimicrobial active packaging film made of semolina flour, nano zinc oxide and nano-kaolin to maintain the quality of low-moisture mozzarella cheese during low-temperature storage. Journal of the Science of Food and Agriculture 2019, 99(6), 2716–2725.

[81] Li, F., Liu, Y., Cao, Y., Zhang, Y., Zhe, T., Guo, Z., et al. Copper sulfide nanoparticle-carrageenan films for packaging application. Food Hydrocolloids 2020, 109, 106094.

[82] Hosseinzadeh, S., Partovi, R., Talebi, F., & Babaei, A. Chitosan/TiO_2 nanoparticle/*Cymbopogon citratus* essential oil film as food packaging material: Physico-mechanical properties and its effects on microbial, chemical, and organoleptic quality of minced meat during refrigeration. Journal of Food Processing and Preservation 2020, 44(7), e14536.

[83] Pandey, V. K., Upadhyay, S. N., Niranjan, K., & Mishra, P. K. Antimicrobial biodegradable chitosan-based composite Nano-layers for food packaging. International Journal of Biological Macromolecules 2020, 157, 212–219.

[84] Rabee, M., Elmogy, S. A., Morsy, M., Lawandy, S., Zahran, M. A. H., & Moustafa, H. Biosynthesis of MgO nanoparticles and their impact on the properties of the PVA/gelatin nanocomposites for smart food packaging applications. ACS Applied Bio Materials 2023, 6(11), 5037–5051.

[85] Momtaz, M., Momtaz, E., Mehrgardi, M. A., Momtaz, F., Narimani, T., & Poursina, F. Preparation and characterization of gelatin/chitosan nanocomposite reinforced by NiO nanoparticles as an active food packaging. Scientific Reports 2024, 14(1), 519.

[86] Youssef, A. M., El-Aziz ME, A., & Morsi, S. M. M. Development and evaluation of antimicrobial LDPE/TiO_2 nanocomposites for food packaging applications. Polymer Bulletin 2023, 80(5), 5417–5431.

[87] Barros, C., Miranda, S., Castro, O., Carneiro, O., & Machado, A. LDPE-Nanoclay films for food packaging with improved barrier properties. Journal of Plastic Film and Sheeting 2023, 39(3), 304–320.
[88] El-Sayed, S. M., & Youssef, A. M. Eco-friendly biodegradable nanocomposite materials and their recent use in food packaging applications: A review. Sustainable Food Technology 2023, 1(2), 215–227.
[89] Kumar, S., Mudai, A., Roy, B., Basumatary, I. B., Mukherjee, A., & Dutta, J. Biodegradable hybrid nanocomposite of chitosan/gelatin and green synthesized zinc oxide nanoparticles for food packaging. Foods 2020, 9(9), 1143.
[90] Adeyeye, S. A. O., & Ashaolu, T. J. Applications of nano-materials in food packaging: A review. Journal of Food Process Engineering 2021, 44(7) e13708.
[91] Alexander, S., Panikar, D. R., & Sreenivasan, A. T. Synthesis and characterization of modified graphene oxide based molecularly imprinted polymer for the detection of amoxicillin. Journal of Molecular Structure 2023, 1285, 135528.
[92] Perera, K. Y., Hopkins, M., Jaiswal, A. K., & Jaiswal, S. Nanoclays-containing bio-based packaging materials: Properties, applications, safety, and regulatory issues. Journal of Nanostructure in Chemistry 2024, 14(1), 71–93.
[93] Anirudhan, T., Athira, V., & Sekhar, V. C. Electrochemical sensing and nano molar level detection of bisphenol-a with molecularly imprinted polymer tailored on multiwalled carbon nanotubes. Polymer 2018, 146, 312–320.
[94] Rovera, C., Ghaani, M., & Farris, S. Nano-inspired oxygen barrier coatings for food packaging applications: An overview. Trends in Food Science and Technology 2020, 97, 210–220.
[95] Garcia, C. V., Shin, G. H., & Kim, J. T. Metal oxide-based nanocomposites in food packaging: Applications, migration, and regulations. Trends in Food Science and Technology 2018, 82, 21–31.
[96] Connolly, M., Zhang, Y., Brown, D. M., Ortuño, N., Jordá-Beneyto, M., Stone, V., et al. Novel polylactic acid (PLA)-organoclay nanocomposite bio-packaging for the cosmetic industry; migration studies and in vitro assessment of the dermal toxicity of migration extracts. Polymer Degradation and Stability 2019, 168, 108938.
[97] Velásquez, E., Espinoza, S., Valenzuela, X., Garrido, L., Galotto, M. J., Guarda, A., & López de Dicastillo, C. Effect of organic modifier types on the physical–mechanical properties and overall migration of post-consumer polypropylene/clay nanocomposites for food packaging. Polymers 2021, 13(9), 1502.
[98] Manikantan, M., Pandiselvam, R., Arumuganathan, T., Indurani, C., & Varadharaju, N. Low-density polyethylene based nanocomposite packaging films for the preservation of sugarcane juice. Journal of Food Science and Technology 2022, 59, 1629–1636, 1–8.
[99] Wang, T., Yun, J., Zhang, Y., Bi, Y., Zhao, F., & Niu, Y. Effects of ozone fumigation combined with nano-film packaging on the postharvest storage quality and antioxidant capacity of button mushrooms (*Agaricus bisporus*). Postharvest Biology and Technology 2021, 176, 111501.
[100] Ma, Y., Zhao, H., Ma, Q., Cheng, D., Zhang, Y., Wang, W., et al. Development of chitosan/potato peel polyphenols nanoparticles driven extended-release antioxidant films based on potato starch. Food Packaging and Shelf Life 2022, 31, 100793.
[101] Karimirad, R., Behnamian, M., Dezhsetan, S., & Sonnenberg, A. Chitosan nanoparticles-loaded *Citrus aurantium* essential oil: A novel delivery system for preserving the postharvest quality of *Agaricus bisporus*. Journal of the Science of Food and Agriculture 2018, 98(13), 5112–5119.
[102] Hosseini, S. F., Ghaderi, J., & Gómez-Guillén, M. C. Tailoring physico-mechanical and antimicrobial/antioxidant properties of biopolymeric films by cinnamaldehyde-loaded chitosan nanoparticles and their application in packaging of fresh rainbow trout fillets. Food Hydrocolloids 2022, 124, 107249.
[103] Lomate, G. B., Dandi, B., & Mishra, S. Development of antimicrobial LDPE/Cu nanocomposite food packaging film for extended shelf life of peda. Food Packaging and Shelf Life 2018, 16, 211–219.

[104] Ezati, P., Riahi, Z., & Rhim, J.-W. CMC-based functional film incorporated with copper-doped TiO_2 to prevent banana browning. Food Hydrocolloids 2022, 122, 107104.

[105] Perera, K. Y., Jaiswal, S., & Jaiswal, A. K. A review on nanomaterials and nanohybrids based bio-nanocomposites for food packaging. Food Chemistry 2022, 376, 131912.

[106] La, D. D., Nguyen-Tri, P., Le, K. H., Nguyen, P. T., Nguyen, M. D.-B., Vo, A. T., et al. Effects of antibacterial ZnO nanoparticles on the performance of a chitosan/gum arabic edible coating for post-harvest banana preservation. Progress in Organic Coatings 2021, 151, 106057.

[107] Abdollahzadeh, E., Mahmoodzadeh Hosseini, H., & Imani Fooladi, A. A. Antibacterial activity of agar-based films containing nisin, cinnamon EO, and ZnO nanoparticles. Journal of Food Safety 2018, 38 (3) e12440.

[108] Indumathi, M., & Rajarajeswari, G. Mahua oil-based polyurethane/chitosan/nano ZnO composite films for biodegradable food packaging applications. International Journal of Biological Macromolecules 2019, 124, 163–174.

[109] Sarmah, J. K., Ali, A. A., Saikia, R., Dey, R. R., & Dutta, R. R. Functionalized carbon nanostructures for smart packaging. In Barhoum, A., Deshmukh, K. (eds): Handbook of Functionalized Carbon Nanostructures: From Synthesis Methods to Applications, Cham: Springer, 2023, 1–31.

[110] Ghaani, M., Cozzolino, C. A., Castelli, G., & Farris, S. An overview of the intelligent packaging technologies in the food sector. Trends in Food Science and Technology 2016, 51, 1–11.

[111] Fernandez, C. M., Alves, J., Gaspar, P. D., Lima, T. M., & Silva, P. D. Innovative processes in smart packaging. A systematic review. Journal of the Science of Food and Agriculture 2023, 103(3), 986–1003.

[112] Shao, P., Liu, L., Yu, J., Lin, Y., Gao, H., Chen, H., & Sun, P. An overview of intelligent freshness indicator packaging for food quality and safety monitoring. Trends in Food Science and Technology 2021, 118, 285–296.

[113] Wang, Y., Liu, K., Zhang, M., Xu, T., Du, H., Pang, B., & Si, C. Sustainable polysaccharide-based materials for intelligent packaging. Carbohydrate Polymers 2023, 313, 120851.

[114] Liu, R., Ning, Y., Zhu, Q., Yang, D., Li, J., & Wang, L. A colorimetric total volatile base nitrogen-responsive film with antibacterial and leaching-resistant properties for accurate tracking fish quality. LWT 2023, 181, 114684.

[115] Pereira, P. F., De Sousa Picciani, P. H., Calado, V., & Tonon, R. V. Electrical gas sensors for meat freshness assessment and quality monitoring: A review. Trends in Food Science and Technology 2021, 118, 36–44.

[116] Fuertes, G., Soto, I., Carrasco, R., Vargas, M., Sabattin, J., & Lagos, C. Intelligent packaging systems: Sensors and nanosensors to monitor food quality and safety. Journal of Sensors 2016, 2016(1), 4046061.

[117] Kumar, J., Akhila, K., & Gaikwad, K. K. Recent developments in intelligent packaging systems for food processing industry: A review. Journal of Food Processing and Technology 2021, 12, 895.

[118] Mohammadian, E., Alizadeh-Sani, M., & Jafari, S. M. Smart monitoring of gas/temperature changes within food packaging based on natural colorants. Comprehensive Reviews in Food Science and Food Safety 2020, 19(6), 2885–2931.

[119] Frith, J. A Billion Little Pieces: RFID and Infrastructures of Identification, Cambridge: MIT Press, 2024.

[120] Nikola, G., Schulz, K., & Ivanochko, I. Use of RFID Technology in retail supply chain. Developments in Information and Knowledge Management for Business Applications 2021, 1, 555–587.

[121] Eom, K.-H., Hyun, K.-H., Lin, S., & Kim, J.-W. The meat freshness monitoring system using the smart RFID tag. International Journal of Distributed Sensor Networks 2014, 10(7), 591812.

[122] Sen, L., Hyun, K. H., Kim, J. W., Shin, J. W., & Eom, K. H. The design of smart RFID system with gas sensor for meat freshness monitoring. Advanced Science and Technology Letters 2013, 41, 17–20.

[123] Eom, K. H., Kim, M. C., Lee, S., & Lee, C. W. The vegetable freshness monitoring system using RFID with oxygen and carbon dioxide sensor. International Journal of Distributed Sensor Networks 2012, 8(6), 472986.

[124] Potyrailo, R. A., Nagraj, N., Tang, Z., Mondello, F. J., Surman, C., & Morris, W. Battery-free radio frequency identification (RFID) sensors for food quality and safety. Journal of Agricultural and Food Chemistry 2012, 60(35), 8535–8543.

[125] Osmólska, E., Stoma, M., & Starek-Wójcicka, A. Application of biosensors, sensors, and tags in intelligent packaging used for food products: A review. Sensors 2022, 22(24), 9956.

[126] Costa, S. P., Nogueira, C. L., Cunha, A. P., Lisac, A., & Carvalho, C. M. Potential of bacteriophage proteins as recognition molecules for pathogen detection. Critical Reviews in Biotechnology 2023, 43(5), 787–804.

[127] George, J., Kumar, R., Aaliya, B., & Sunooj, K. V. 2003 Packaging solutions for monitoring food quality and safety. In: Hebbar, H.U., Sharma, R., Chaurasiya, R.S., Ranjan, S., Raghavarao, K. (eds). Engineering Aspects of Food Quality and Safety, Cham: Springer, 2023, 411–442.

[128] Pollet, J., Delport, F., Janssen, K., Tran, D., Wouters, J., Verbiest, T., & Lammertyn, J. Fast and accurate peanut allergen detection with nanobead enhanced optical fiber SPR biosensor. Talanta 2011, 83(5), 1436–1441.

[129] Verma, S. K., Prasad, A., & Katiyar, V. State of art review on sustainable biodegradable polymers with a market overview for sustainability packaging. Materials Today Sustainability 2024, 26, 100776.

[130] Gomes, H., Navio, F., Gaspar, P. D., Vngj, S., & Jmlp, C. Radio-frequency identification traceability system implementation in the packaging section of an industrial company. Applied Sciences 2023, 13(23), 12943.

[131] Yata, V. K., Tiwari, B. C., & Ahmad, I. Nanoscience in food and agriculture: Research, industries and patents. Environmental Chemistry Letters 2018, 16(1), 79–84.

[132] Cristofoli, N. L., Lima, A. R., Tchonkouang, R. D. N., Quintino, A. C., & Vieira, M. C. Advances in the food packaging production from agri-food waste and by-products: Market trends for a sustainable development. Sustainability 2023, 15(7), 6153.

[133] Klein, H. S., & Luna, F. V. Cellulose. In: Klein, H. S., & Luna, F. V., editors. Brazilian Crops in the Global Market: The Emergence of Brazil as a World Agribusiness Exporter since 1950, Cham: Springer Nature Switzerland, 2023, 269–293.

[134] Cunha, K. C. T., & Mazieri, M. R. Intelligent packaging and value generating: Technological development opportunities based on patent analysis. World Patent Information 2024, 76, 102258.

Shivute, Fimanekeni Ndaitavela, Natascha Cheikhyoussef, and Ahmad Cheikhyoussef*

Chapter 20 Ethical and regulatory perspectives on food biotechnology

Abstract: Agricultural biotechnology is a sustainable technology that will bolster the humanitarian efforts toward food security and improve agricultural productivity around the world in a sustainable and environmentally safe manner. Achieving food security is the second Sustainable Development Goal (SDG), aimed at ending hunger, improving nutrition, and promoting sustainable agriculture by 2030. There is a rapid development of new plant breeding techniques (NPBTs) for producing plants with increased level of nutrients, medicinal properties, and enhanced organoleptic properties. Interestingly, there are several ethical debates around NPBTs and their anticipated advantages and disadvantages toward enhanced land productivity, food nutritional value, human well-being, and environmental impact. Agricultural biotechnology has experienced several advances in the development and application of genetic modification techniques in plants, also referred to as "gene editing" or "genome editing." However, several debates and diverging views are ongoing in terms of how these new technologies should be regulated to efficiently utilize their capability toward ensuring food security.

Keywords: Ethics, food biotechnology, genetically modified organisms, genome editing, regulations, transgenic plants

20.1 Introduction

In a recent review on the importance of agricultural biotechnology for sustainable food security, agricultural biotechnology is defined as "a range of tools, including traditional and new breeding techniques, which alter living organisms, or parts of organisms, to make or modify products, improve plants or animals, or develop microorganisms for specific agricultural uses" [1]. Food biotechnology has a great role and

***Corresponding author: Ahmad Cheikhyoussef,** Multidisciplinary Research Services, Centre for Research Services, University of Namibia, Windhoek 10026, Namibia, e-mail: acheikhyoussef@unam.na
Shivute, Fimanekeni Ndaitavela, Multidisciplinary Research Services, Centre for Research Services, University of Namibia,Windhoek, 10026, Namibia
Natascha Cheikhyoussef, Ministry of Higher Education, Technology, and Innovation (MHETI), Windhoek, Namibia.

https://doi.org/10.1515/9783111441238-021

promising applications that can lead to the development of crops with elevated nutritional value. However, due to public sentiments, there should be a control over the wide use of such technologies, to ensure responsible and equal distribution of their benefits across different communities and regions [2]. Genetic modification of crops has been shown to improve agricultural outputs by increasing crop yield, enhancing nutritional value, reducing the environmental impact, and contributing toward global food security in terms of crops quantity, quality, and safety [3]. However, there are some disadvantages for genetic modification of crops, such as overproduction, unnatural biological diversity, health concerns, and domination of multinational agribusiness corporations. Significant developments in genetic editing techniques enabled the full mechanization of hybrid breeding of important economic crops [4–6]. New plant breeding techniques (NPBTs) are used to develop new variants with improved traits such as early maturing and more precise varieties than conventional breeding techniques, via DNA modifications in the newly developed seeds and cells. They have been developed to overcome several limitations in conventional breeding. Examples of NPBTs include zinc-finger nucleases, transcription activator-like effector nucleases (TALENs), and the mega-nuclease and clustered regularly interspaced short palindromic repeats (CRISPR) systems [1]. NPBTs are used in the food and agriculture sectors for genetic improvement of plant varieties, characterization and conservation of genetic resources, and diagnosis of plant diseases [7, 8]. New genomic techniques (NGTs), particularly genome editing (GE), are potential methods that can develop improved crops that contribute to addressing global demand for increasing food and energy. As such, the ethical and sociocultural concerns surrounding these promising methods must be addressed as a priority to ensure trust subsists between the public, food manufacturer, and the government or policy maker [9]. Harfouche et al. [9] argues that to gain public trust, ethically responsible solutions must be developed that are socially sound and responsive, as well as relevant to the people from different cultural backgrounds. Communication methods with the public are also key in building trust. The aim of this chapter is to give insights into NPBTs and their roles in food biotechnology, as well as to highlight the debates and the ongoing diverging scenarios on the regulations of NPBTs, their ethical guidelines, the strategies to ensure how such legislation can be harmonized to certify safety, and the public acceptance of the developed products.

20.2 Genome editing techniques

There is a need to differentiate between genome-edited (GE) foods and genetically modified (GM) foods because these terminologies are sometimes misinterpreted. GE in plants is a key technology in sustainable bioeconomy, which allows targeted insertion or deletion, or replacement of a gene at a specific site without the addition of foreign DNA [10–12]. Examples of GE technology is the famous clustered regularly interspaced short palindromic repeats (CRISPR), which is known to be cheaper in terms of cost and time of development [13, 14]. Meanwhile, GM foods or crops are a well-known term that has been used in the literature to describe crops that have been subjected to a genetic material alteration (most likely the DNA) in a way that they don't exist naturally by natural recombination or mating. Several studies indicated that GE technology has gained preference over GM technology as many of the critics are alluding to its same history of GM organisms (GMOs) or considering GE as an extension or a new generation of GMOs [15–18]. Li et al. [19] summarized the latest targeted genome-modification techniques and their evolution from randomly generated mutations to precision substitutions, followed by one or a combination of the following processes of small DNA fragments insertions, substitutions, and deletions of, and finally, precision manipulation of large DNA segments to be achieved. Furthermore, the latest developments in base editing [20–22], prime editing [23–25], and other CRISPR-associated systems [26–29] enabled the scientists to achieve great technological foundations in plant basic research and precision molecular breeding success in several plants such as cotton [30], wheat [31, 32], rice [33, 34], potato [35–37], cabbage [38], maize [39], tomato [40], rapeseed [41], banana [42], chestnut [43], and watermelon [44].

In a recent review, Enfissi et al. [45] indicated that there is an opportunity in the accelerated development and the evolving pathways of NPBTs to advance our current understanding of metabolomic approaches, especially regarding industrial research development and safety perspectives. Furthermore, although the metabolomic analysis may not provide nonspecific identifying constituents of gene editing [46], it produces a precise comparative analysis to the available natural variations and outputs from NPBTs [45]. They recommended the use of pan-genome approaches [22] to extrapolate the metabolomes, facilitating the possible identification of biochemical variation resulting from NPBT [45].

Several NPBTs such as Sequence Specific Points (SSP) in the DNA have been developed (Table 20.1) to achieve gene editing in plants, including engineered sequence-specific nucleases (SSNs) for the provision of double stranded breaks (DSBs) [47]. Other NPBTs are the clustered regularly interspaced short palindromic repeat (CRISPR)-associated (Cas) protein SSNs [47–49], Transcription activator-like effector nucleases (TALENS) [50, 51, 53], pan-genome approaches [22, 52], the zinc-finger nucleases (ZFN) [53–55], and meganucleases [53, 56].

20.3 Impact of genome editing

GE allows scientists to precisely alter DNA sequences within an organism [57], and has the potential to significantly impact some important fields, including medicine, agriculture [58, 59] as well as research, by enabling the correction of genetic defects,

Table 20.1: New plant breeding techniques (NPBTs) – principles and applications.

Technique	Principle	Applications	References
CRISPR-Cas	The CRISPR-Cas9 system comprises two fundamental components: the Cas9 protein and the guide RNA (gRNA). Functioning as a molecular scissor, Cas9 cleaves the DNA at predetermined target sites, guided by the gRNA, which aligns with the target DNA sequence.	Medical research, human gene therapy, plant science, and crop improvement	[47–49]
pan-genome approaches	They represent a comprehensive overview of the genomic variation in a population or species to identify genomic regions associated with diverse traits of interest.	Crop improvement	[22, 52]
TAL effector nucleases (TALENS)	TALENs act as molecular scissors by introducing double-strand breaks (DSBs) to the DNA at a given location. The DSBs are subsequently repaired by the cell itself using different repair pathways such as nonhomologous end joining (NHEJ) or homologous recombination (HR).	Genome editing	[50, 51, 53]
Zinc-finger nucleases (ZFN)	ZFNs are able to introduce DNA double-strand breaks in target genes that can lead to stimulation of the cell's endogenous homologous recombination machinery.	Improve the efficiency of gene targeting	[53–55]
Meganucleases	They are naturally occurring endonucleases from different organisms such as bacteria, fungi, and algae. They have a long DNA recognition site with variable size differing between 12 and 30 bp, and can be modified by introducing changes to the amino acids found in the recognition site of the protein to adapt for restriction of a specifically chosen sequence.	Gene therapy, crops editing	[53, 56]

development of disease-resistant crops, and better understanding of gene function [60]. However, ethical concerns regarding unintended consequences [61], potential for misuse [62], and societal differences need to be scrupulous, alongside its benefits.

Proper implementation of GE will not only support continental policies on agri-food agendas, such as the European Union's farm-to-fork strategy and green deal ambitions [63] but, also help to realize several sustainable development goals (SDGs) such as SDG1, SDG3, and SDG13 [64–66]. The socioeconomic factors (economic conditions and regulation) that are affecting the emergence and impact of new genomic techniques (NGTs) in agriculture have been studied and they emphasize that the emergence and impact of NGTs depend on three factors, including consumer preferences, supply chain, and intellectual property rights (IPR) strategies [63].

In a recent review by Tuncel et al. [26] on genome-edited foods, it has been reported that GE has powerful applications and promising potential in improving and enhancing crop and animal productivity, leading to the realization of food security [67]; specifically, CRISPR-Cas-based technologies can be applied to several food products to increase the nutritional value of crops via macronutrient engineering (protein) [68], and crops biofortification (vitamins and minerals) [69–72], or antinutrients' reduction (phytase and oxalate). The availability of responsible regulations is an essential step toward supporting GE and public acceptance for extra valorization and industrial deployment as well as the commercialization of GE foods [26].

20.4 Ethical guidelines and agricultural biotechnology

There are mixed views on the theological acceptance of agricultural biotechnology [3]. Abushal et al. [3] indicated in their report that scholars and leaders of monotheistic religions had no consensus on the acceptability of GM organisms (GMOs), considering there needs to be prudent ethical and scientific compliance measures and regulatory controls in place. Furthermore, the monotheistic religions support has enormous consequences to build public trust and their acceptances to engage several policy makers and government entities embarking on new policies to be developed and adopted [3].

A recent study highlighted the improvements of plants, and the way researchers are leveraging the advancements in plant breeding and genomic analyses techniques such as site-directed nucleases for the enhancement of food crop traits via the revolutionary and evolving agricultural biotechnology techniques [73]. Furthermore, Steinwand and Ronald argued that deployment of gene-edited products from the laboratory to the field remains stalled/hampered by the presence of biological and regulatory bottlenecks. These bottlenecks require further revisions and development to meet the increasing demand for food crops to achieve sustainable food security [73]. There is a heterogeneity in regulations and public opinions regarding GE despite its wide applications to improve traits diversity from consumers' perspectives of increasing the content of food antioxidants, non-browning properties, and allergen-free characteristics of foods [10, 74, 75], and farmers perspectives of increasing the agronomic performance, disease and pest resistance, and herbicide tolerance of crops [76].

Considering the historical evolution of the two compared technologies, the development of regulations and implementation of GE are mostly based on the existing GMO frameworks [77] in several countries and regions, such as in the European Union (EU) [78] and New Zealand [79]. However, within the EU member states, there are different views on GE technology [80], whereby, only last year on 5th of July, a proposal was published by the European Commission (EC) on exempting specific new genomic techniques (NGT) plants from GMO regulations [81]. The latest regulations in EU 2023 exempted the Category-1 NGT plants, which defined "NGTs that could occur naturally or through conventional breeding"; meanwhile, Category-2 NGT plants, which apply to all other NGT plants that are not included under Category-1 NGT, are to be regulated as GMOs [10]. Furthermore, the latest regulations on new genomic techniques for plants are anticipated to increase the agriculture-based innovation and sustainable development in the region and benefit European farmers, plants breeders, seed companies, and plant scientists [82], promoting international trade [83].

One important aspect to consider is the potential impact on society and the environment [84, 85]. Improving biotechnology has great potential to solve socioeconomic issues as far as food security is concerned, but it could also create new ones. Gene manipulation of different crops to produce hybrids is one of the most common problems in most biotech experiments, where other organisms are made to sacrifice [86]. Therefore, due to such harsh sacrifices, numerous plant and animal species have been mutated regularly. Food ethics is important because the food we choose must minimize the harm that would be caused to the environment, consumers, farmers, and others. Food consumption ethics include having a basic awareness of the wages of farmers, sustainable food production, plant-based alternatives, and carbon footprint, among others. The application of plant biotechnology offers a critical edge to enhance the balance between preserving the environment, practicing sustainable agriculture, and maintaining global food security [67, 87]. It is a promising breeding skill for developing resilient crops, resource utilization reduction, and safeguarding biodiversity, and finally to foster sustainable and secure food production systems.

20.5 Ethics and gene editing

Gene editing is one of the genetic engineering protocols whereby the DNA is inserted, deleted, modified, or replaced in the genome of a living organism [88, 89]. The process enables plant breeders to achieve very precise changes to DNA [90]. GE was introduced to bring changes to a plant or other organisms by targeting a specific location in a gene within the DNA [91–93]. Plant genetic transformation is an important pathway to improve plant yield, quality, and tolerance to abiotic and biotic stressors [94, 95]. During the last decade, gene editing of different organisms witnessed several achievements in a short time, therefore it attracts several researchers to comprehensively review the latest progress and offer opinions on future perspectives [91, 96, 97]. New gene-editing technologies such as (GETs) (e.g., CRISPR) offer the opportunity to accelerate sustainable agri-food production but for realizing it though, more needs to be addressed in terms of social acceptability of these technologies as well as the regulatory and economic issues [98, 99]. One major challenge in gene editing is the potential for off-target effects, where unintended mutations occur in nontargeted regions of the genome [100, 101]. To overcome such a case, advanced CRISPR variants with unique features such as high specificity and accuracy, like Cas12b (C2c1), have been developed to minimize off-target effects [102–104].

Key ethical questions in gene therapy and GE include how to distinguish between beneficial and harmful applications of these technologies [105]. Who decides which traits are normal and which constitute a disability or disorder [106]? Will the high costs of gene therapy make it available only to the wealthy [107, 108]? Environment ethical issues, especially the release of GMOs into the environment, can cause serious ecological damage [109, 110]. The common frightening impact is the crossbreed of GM crops with any wild relatives, leading to the spread of modified genes and the creation of invasive species [111, 112]. Potential health risks to humans include the possible exposure to new allergens and new cases of food allergy in the GM foods [113], as well as the transfer of antibiotic-resistant genes to gut microflora [114]. The applications of biotechnology in the production of food products in the food processing sector make use of microorganisms for the preservation of food and to produce a range of value-added products such as enzymes, flavor compounds, vitamins, microbial cultures, and food ingredients. Biotechnology uses microorganisms and enzymes to ferment foods through processes like brewing and bread-making [115]. Therefore, modern biotechnology genetically modifies crops and livestock for improved traits like disease resistance or nutrition. Biotech crops can increase farming profitability by enhancing crop quality and may, in some cases, increase yields. Crossbreeding, selective breeding, and mutation breeding are some of the conventional types of biotechnology methods used to make genetic changes [116]. However, these breeding methods require mixing all genes from two different sources and then, to be used to create common crops like modern corn varieties [117], and seedless watermelons [118]. Therefore, in the future, genetic therapies may be used to prevent, treat, or cure certain inherited disorders. The highest profile to protect against GMOs has been Genetic Use Restriction Technologies (GURTs) or "terminator seed" technology [119].

20.6 Food biotechnology regulations

Regulatory approaches in food biotechnology are put in place to ensure the safety and use of GMOs in food. These approaches vary considerably among different countries. The Cartagena Protocol on Biosafety (CPB) of the Convention on Biological Diversity

(CBD) is an international agreement that aims to safeguard the handling, transport, and use of living modified organisms (LMOs) resulting from modern biotechnology that may cause adverse effects on biological diversity and bring risks to human health [120]. The CPB was adopted in 2000 by member states and entered into force in the year 2003. The CPB encouraged countries to put in place national legislation in the regulation of GMOs. The Nagoya-Kuala Lumpur Supplementary Protocol is a treaty intended to supplement the Cartagena Protocol on Biosafety by providing international rules and procedures on liability and provide reparation to the damage on biodiversity resulting from the LMOs [121]. The protocol was adopted in the year 2010. The widespread rejection of genetic engineering generated several challenges and controversies, especially when the GETs were applied to agricultural crops and food products through the crop's cultivation and food production [122]. The main dispute and debates on the GETs are focusing on the insufficiency or the doubt of current regulations due to the great array of approaches in the European Union and USA [122]. For the African scenarios [123], although, there was a positive attitude from African farmers and consumers toward GM crops (South Africa, Burkina Faso, and Sudan), due to political constraints, the GM crops' farming is very limited to some African countries such as Kenya, Nigeria, Uganda, and Tanzania [3, 124]. However, several African countries such as Nigeria, Kenya, Uganda, and Ethiopia, have released guidelines for using GE tools on plants and animals [125, 126]. The USDA is responsible for administering the food labeling regulations in the USA, including the National Organic Program (NOP) and the National Bioengineered Food Disclosure Standard (NBFDS) [127]. Under the Code of Federal Regulations (CFR) Title 7 Part 205, the prohibition of transgenic (inserting a foreign gene) or non-transgenic gene editing methods are included [128]. Furthermore, the application of GE technology in food production with organic label is prohibited, although the existence of GE technology will not prevent a farm's organic certification process [127]. Kendig et al. [129] reported on the implementation of the National Bioengineered Food Disclosure Law (NBFDL) in the USA, starting in January 2022, which requires that all US-based food manufacturers reveal whether their products contain bioengineered ingredients [130]. A debate has been created on the argument that the law hinders rather than ensures transparency when it comes to biotechnology use in food [131].

The latest approved EC's proposal on NGTs in February 2024 was considered as new era for the GE technology in EU after the recent vote in the European Parliament [132], with important amendments mainly to emphasize listing of all NGT1 plants and being made available online to public, prohibiting NGT plants-based patents, and organic agriculture affiliation in the EU, and that all Category-1 NGT plants, planting materials, and their products must be labeled as "New Genomic Techniques" products [133]. Table 20.2 presents some of the regulatory approaches for GEd plants under food biotechnology regulations around the world. The scenarios in UK and China [134, 135] are different, whereby the GMO regulation is applied to transgenic plants and do not include the GEd organisms [136]; however, in England, they exempted the GEd crops from GMO regulations if the genetic alterations are confirmed to be like the ones done via traditional breeding methods [137, 138]. In other countries, GE refers to a different technology that requires its own regulations, autonomous of the local GMO regulatory framework, either with prior confirmation (e.g. Argentina, Japan, and Canada) or without prior confirmation (e.g., the USA and Australia) by the government before introducing the GE products on the market [77]. The previously mentioned scenarios confirm the heterogeneity of landscape for GE governmental regulations and strategies [99], which can lead to a disruption of international trade [83]. Therefore, GEd foods labeling is an important regulation that is advocated for by several consumers and organizations to increase consumers' trust [138], and to assure the safety and ethical handling of GEd foods in the USA [139] and Japan [140].

Table 20.2: Regulatory approaches for genome editing (GEd) plants around the world.

Country	GEd regulation approach	Statutory situation	References
Argentina	Are not regulated as GMOs	Light regulatory regime	[135]
Australia	GEd crops without introduced SDN-1 to guide homology-directed repair are not regulated as GMOs; those classified as SDN-2, SDN-3, and organisms modified using ODM, are regulated as GMOs	Moderate middle-ground regulatory model	[135]
Brazil	Are not regulated as GMOs	Light regulatory regime	[135]
Canada	Aligned with regulatory approach for GMOs	No GEd-specific regulations	[135]
Chile	Aligned with conventionally bred crops	Light regulatory regime	[135]
China	GEd crop processing, production, marketing, import, and export are still regulated under GMO legislations	Moderate middle-ground regulatory model	[134, 135]
Colombia	Implemented regulatory frameworks to regulate GMOs	Light regulatory model	[144]
European Union	European Union (EU) subjects GEd organisms to GMO regulations	Strict regulatory framework	[10]
India	GEd plants falling under the categories of SDN-1 and SDN-2, are exempt from GMO regulations	Light regulatory regime	[135]
Israel	GEd crops without insertion and/or incorporation of foreign DNA are not regulated as GMOs	Light regulatory regime	[135]

Table 20.2 (continued)

Country	GEd regulation approach	Statutory situation	References
Japan	GEd crops without foreign genes are not considered an LMO and are exempted from regulation; meanwhile, those created using DNA sequence templates (SDN-2 and SDN-3) are subject to regulation	Light regulatory regime	[134, 135]
Kenya	GEd crops without foreign genetic material or have inserted foreign genetic material that is undetectable are not regulated as GMOs	Light regulatory regime	[135]
Namibia	Regulated, if it falls within the scope of the Biosafety Act No. 7. of 2006); no separate regulations are yet in existence regarding GEd	No GEd-specific regulations	[145]
New Zealand	GEd crops are currently regulated as GMOs	Strict regulatory framework	[135]
Nigeria	GEd plants with or without rDNA rDNA that have been removed in the final product, are not regulated as GMO	Light regulatory regime	[135]
Norway	GEd crops are current regulated as GMOs	Strict regulatory framework	[135]
Russia	GEd legal status and regulation are still under discussion	Strict regulatory framework	[135]
South Africa	GEd crops are currently regulated as GMOs	Strict regulatory framework	
United Kingdom	GEd plants with genetic changes that could have occurred in nature or via traditional breeding methods, are exempt from GMO regulations	The Genetic Technology (Precision Breeding) Act 2023; Light regulatory regime	[137, 138]
Uruguay	Emerging capacities in Uruguay for research and development of GED crops	No specific regulations for gene-edited crops.	[146]
USA	GEd crops devoid of foreign DNA and posing no risk to other plants or exhibiting food safety attributes distinct from conventionally bred crops are exempted from GMO regulations	Light regulatory regime	[130, 131, 139]

Caradus [141] presented an extensive review on GM and GE plants' regulations, and suggested several processes to regulate GM and GE plants. Furthermore, he recommended that for an effective regulatory system to be established as transparent, science-based evidence, and allowing public participation [131, 142], it must fulfil the following requirements: (1) pre-market mandatory approval; (2) established safety standards (3) transparent system; (4) public participatory approach; (5) use of external scientists' expertise and scientific advice; (6) independent agency decisions; (7) post-approval activities; and (8) availability of enforcement authority and resources. An additional recommendation was to emphasize the essential role of scientists to understand and engage with the frameworks of technological determinism and Responsible Research and Innovation (RRI) [143].

20.7 Conclusions and future perspectives

The process of maintaining genome integrity is essential in agricultural biotechnology, especially in ensuring food integrity. There are several efforts to develop plant breeding programs aimed at enhancing crop productivity, considering the environmental conditions in each country. There must be governing legislation for GM products' labeling to enable their international acceptance. The future of biotechnology in food may find a niche in alternative food production because, with the increasing global population and concerns about the environment, biotechnology is being used to develop alternative foods, such as lab-grown meat and plant-based meat substitutes, offering a sustainable alternative to traditional meat production. With time, food biotechnology will ensure improved safety and quality of food, and create innovative solutions to address global food challenges. There should be a multidisciplinary approach for NPBTs to establish strong cooperation between all parties to address all challenges and concerns on the administrative, economic, and legislative aspects and allow the full potential to explore all anticipated benefits of NPBTs. The ethics of biotechnology are complex and multifaceted. Concerns over consumer awareness of biotechnology have increased over the last three decades, yet most consumers remain confused over science, especially if the communication strategy is ambiguous and not well-formulated, considering the public, cultural, and societal diversity.

References

[1] Tyczewska, A., Twardowski, T., & Woźniak-Gientka, E. Agricultural biotechnology for sustainable food security. Trends in Biotechnology 2023, Mar 1, 41(3), 331–341.

[2] Das, S., Ray, M. K., Panday, D., & Mishra, P. K. Role of biotechnology in creating sustainable agriculture. PLOS Sustainability and Transformation 2023, Jul 13, 2(7), e0000069.

[3] Abushal, L. T., Salama, M., Essa, M. M., & Qoronfleh, M. W. Agricultural biotechnology: Revealing insights about ethical concerns. Journal of Biosciences 2021, Sep, 46(3), 81.

[4] Chen, G., Zhou, Y., Kishchenko, O., Stepanenko, A., Jatayev, S., Zhang, D., & Borisjuk, N. Gene editing to facilitate hybrid crop production. Biotechnology Advances 2021, Jan 1, 46, 107676.

[5] Huang, K., Wang, Y., Li, Y., Zhang, B., Zhang, L., Duan, P., Xu, R., Wang, D., Liu, L., Zhang, G., & Zhang, H. Modulation of histone acetylation enables fully mechanized hybrid rice breeding. Nature Plants 2024, Jun, 3, 1–7.

[6] Xuedan, L., Fan, L., Yunhua, X., Feng, W., Guilian, Z., Huabing, D., & Wenbang, T. Grain shape genes: shaping the future of rice breeding. Rice Science 2023, Sep 1, 30(5), 379–404.

[7] Dong, O. X., & Ronald, P. C. Genetic engineering for disease resistance in plants: Recent progress and future perspectives. Plant Physiology 2019, May 1, 180(1), 26–38.

[8] Van Esse, H. P., Reuber, T. L., & Van der Does, D. Genetic modification to improve disease resistance in crops. New Phytologist 2020, Jan, 225(1), 70–86.

[9] Harfouche, A. L., Petousi, V., Meilan, R., Sweet, J., Twardowski, T., & Altman, A. Promoting ethically responsible use of agricultural biotechnology. Trends in Plant Science 2021, Jun 1, 26(6), 546–559.

[10] Atimango, A. O., Wesana, J., Kalule, S. W., Verbeke, W., & De Steur, H. Genome editing in food and agriculture: From regulations to consumer perspectives. Current Opinion in Biotechnology 2024, Jun 1, 87, 103127.

[11] Karavolias, N. G., Horner, W., Abugu, M. N., & Evanega, S. N. Application of gene editing for climate change in agriculture. Frontiers in Sustainable Food Systems 2021, Sep 7, 5, 685801.

[12] Woźniak-Gientka, E., & Tyczewska, A. Genome editing in plants as a key technology in sustainable bioeconomy. EFB Bioeconomy Journal 2023, Nov 1, 3, 100057 29.

[13] Bullock, D. W., Wilson, W. W., & Neadeau, J. Gene editing versus genetic modification in the research and development of new crop traits: An economic comparison. American Journal of Agricultural Economics 2021, Oct, 103(5), 1700–1719.

[14] Martin-Laffon, J., Kuntz, M., & Ricroch, A. E. Worldwide CRISPR patent landscape shows strong geographical biases. Nature Biotechnology 2019, Jun, 37(6), 613–620.

[15] Borrello, M., Cembalo, L., & Vecchio, R. Role of information in consumers' preferences for eco-sustainable genetic improvements in plant breeding. PLoS One 2021, Jul 29, 16(7), e0255130.

[16] Cummings, C., & Peters, D. J. Who trusts in gene-edited foods? Analysis of a representative survey study predicting willingness to eat-and purposeful avoidance of gene edited foods in the United States. Frontiers in Food Science and Technology 2022, Jun 1, 2, 858277.

[17] Son, E., & Lim, S. S. Consumer acceptance of gene-edited versus genetically modified foods in Korea. International Journal of Environmental Research and Public Health 2021, Apr 6, 18(7), 3805.

[18] Strobbe, S., Wesana, J., Van Der Straeten, D., & De Steur, H. Public acceptance and stakeholder views of gene edited foods: A global overview. Trends in Biotechnology 2023, Jun 1, 41(6), 736–740.

[19] Li, B., Sun, C., Li, J., & Gao, C. Targeted genome-modification tools and their advanced applications in crop breeding. Nature Reviews Genetics 2024, Apr, 24, 1–20.

[20] Rees, H. A., & Liu, D. R. Base editing: Precision chemistry on the genome and transcriptome of living cells. Nature Reviews Genetics 2018, Dec, 19(12), 770–788.

[21] Tong, H., Wang, X., Liu, Y., Liu, N., Li, Y., Luo, J., Ma, Q., Wu, D., Li, J., Xu, C., & Yang, H. Programmable A-to-Y base editing by fusing an adenine base editor with an N-methylpurine DNA glycosylase. Nature Biotechnology 2023, Aug, 41(8), 1080–1084.

[22] Zhao, Q., Feng, Q., Lu, H., Li, Y., Wang, A., Tian, Q., Zhan, Q., Lu, Y., Zhang, L., Huang, T., & Wang, Y. Pan-genome analysis highlights the extent of genomic variation in cultivated and wild rice. Nature Genetics 2018, Feb, 50(2), 278–284.

[23] Anzalone, A. V., Randolph, P. B., Davis, J. R., Sousa, A. A., Koblan, L. W., Levy, J. M., Chen, P. J., Wilson, C., Newby, G. A., Raguram, A., & Liu, D. R. Search-and-replace genome editing without double-strand breaks or donor DNA. Nature 2019, Dec, 576(7785):149–157.

[24] Yu, G., Kim, H. K., Park, J., Kwak, H., Cheong, Y., Kim, D., Kim, J., Kim, J., & Kim, H. H. Prediction of efficiencies for diverse prime editing systems in multiple cell types. Cell 2023, May 11, 186(10), 2256–2272.

[25] Zong, Y., Liu, Y., Xue, C., Li, B., Li, X., Wang, Y., Li, J., Liu, G., Huang, X., Cao, X., & Gao, C. An engineered prime editor with enhanced editing efficiency in plants. Nature Biotechnology 2022, Sep, 40(9), 1394–1402.

[26] Tuncel, A., Pan, C., Sprink, T., Wilhelm, R., Barrangou, R., Li, L., Shih, P. M., Varshney, R. K., Tripathi, L., Van Eck, J., & Mandadi, K. Genome-edited foods. Nature Reviews Bioengineering 2023, Nov, 1(11), 799–816.

[27] Ahmar, S., Hensel, G., & Gruszka, D. CRISPR/Cas9-mediated genome editing techniques and new breeding strategies in cereals–current status, improvements, and perspectives. Biotechnology Advances 2023, Sep, 2, 108248.

[28] Ghoshal, B. A birds-eye-view on CRISPR-Cas system in agriculture. The Nucleus 2024, Apr, 67(1), 89–96.

[29] Zaman, Q. U., Raza, A., Lozano-Juste, J., Chao, L., Jones, M. G., Wang, H. F., & Varshney, R. K. Engineering plants using diverse CRISPR-associated proteins and deregulation of genome-edited crops. Trends in Biotechnology 2024 May 1;42(5):560–574.

[30] Qin, L., Li, J., Wang, Q., Xu, Z., Sun, L., Alariqi, M., Manghwar, H., Wang, G., Li, B., Ding, X., & Rui, H. High-efficient and precise base editing of C• G to T• A in the allotetraploid cotton (*Gossypium hirsutum*) genome using a modified CRISPR/Cas9 system. Plant Biotechnology Journal 2020, Jan, 18(1), 45–56.

[31] Ni, P., Zhao, Y., Zhou, X., Liu, Z., Huang, Z., Ni, Z., Sun, Q., & Zong, Y. Efficient and versatile multiplex prime editing in hexaploid wheat. Genome Biology 2023, Jun 29, 24(1), 156.

[32] Singh, M., Kumar, M., Albertsen, M. C., Young, J. K., & Cigan, A. M. Concurrent modifications in the three homeologs of Ms45 gene with CRISPR-Cas9 lead to rapid generation of male sterile bread wheat (*Triticum aestivum* L.). Plant Molecular Biology 2018, Jul, 97, 371–383.

[33] Jiang, Y., Chai, Y., Qiao, D., Wang, J., Xin, C., Sun, W., Cao, Z., Zhang, Y., Zhou, Y., Wang, X. C., & Chen, Q. J. Optimized prime editing efficiently generates glyphosate-resistant rice plants carrying homozygous TAP-IVS mutation in EPSPS. Molecular Plant 2022, Nov 7, 15(11), 1646–1649.

[34] Xu, R., Liu, X., Li, J., Qin, R., & Wei, P. Identification of herbicide resistance OsACC1 mutations via in planta prime-editing-library screening in rice. Nature Plants 2021, Jul, 7(7), 888–892.

[35] Mali, S., Dutta, M., & Zinta, G. Genome editing advancements in potato (*Solanum tuberosum* L.): Operational challenges and solutions. Journal of Plant Biochemistry and Biotechnology 2023, Dec, 32(4), 730–742.

[36] Kumari, N., Kumar, A., Sharma, S., Thakur, P., Chadha, S., & Dhiman, A. CRISPR/Cas system for the traits enhancement in potato (Solanum tuberosum L.): Present status and future prospectives. Journal of Plant Biochemistry and Biotechnology 2024, Mar, 5, 1–21.

[37] Tiwari, J. K., Buckseth, T., Challam, C., Zinta, R., Bhatia, N., Dalamu, D., Naik, S., Poonia, A. K., Singh, R. K., Luthra, S. K., & Kumar, V. CRISPR/Cas genome editing in potato: Current status and future perspectives. Frontiers in Genetics 2022, Feb 2, 13, 827808.

[38] Ma, C., Zhu, C., Zheng, M., Liu, M., Zhang, D., Liu, B., Li, Q., Si, J., Ren, X., & Song, H. CRISPR/Cas9-mediated multiple gene editing in Brassica oleracea var. capitata using the endogenous tRNA-processing system. Horticulture Research 2019, Dec 1, 6:20.

[39] Liu, X., Zhang, S., Jiang, Y., Yan, T., Fang, C., Hou, Q., Wu, S., Xie, K., An, X., & Wan, X. Use of CRISPR/Cas9-based gene editing to simultaneously mutate multiple homologous genes required for pollen development and male fertility in maize. Cells 2022, Jan 27, 11(3), 439.

[40] Zsögön, A., Čermák, T., Naves, E. R., Notini, M. M., Edel, K. H., Weinl, S., Freschi, L., Voytas, D. F., Kudla, J., & Peres, L. E. De novo domestication of wild tomato using genome editing. Nature Biotechnology 2018, Dec, 36(12), 1211–1216.

[41] Cheng, H., Hao, M., Sang, S., Wen, Y., Cai, Y., Wang, H., Wang, W., Mei, D., & Hu, Q. Establishment of new convenient two-line system for hybrid production by targeting mutation of OPR3 in allopolyploid *Brassica napus*. Horticulture Research 2023, Dec, 10(12), 1–9.

[42] Zhang, S., Wu, S., Hu, C., Yang, Q., Dong, T., Sheng, O., Deng, G., He, W., Dou, T., Li, C., & Sun, C. Increased mutation efficiency of CRISPR/Cas9 genome editing in banana by optimized construct. PeerJ 2022, Jan 5, 10, e12664.

[43] Pavese, V., Moglia, A., Corredoira, E., Martínez, M. T., Torello Marinoni, D., & Botta, R. First report of CRISPR/Cas9 gene editing in castanea sativa mill. Frontiers in Plant Science 2021, Aug 25, 12, 728516.

[44] Tian, S., Jiang, L., Cui, X., Zhang, J., Guo, S., Li, M., Zhang, H., Ren, Y., Gong, G., Zong, M., & Liu, F. Engineering herbicide-resistant watermelon variety through CRISPR/Cas9-mediated base-editing. Plant Cell Reports 2018, Sep, 37, 1353–1356.

[45] Enfissi, E. M., Drapal, M., Perez-Fons, L., Nogueira, M., Berry, H. M., Almeida, J., & Fraser, P. D. New plant breeding techniques and their regulatory implications: An opportunity to advance metabolomics approaches. Journal of Plant Physiology 2021, Mar 1, 258, 153378.

[46] Fedorova, M., & Herman, R. A. Obligatory metabolomic profiling of gene-edited crops is risk disproportionate. The Plant Journal 2020, Sep, 103(6), 1985–1988.

[47] Gaj, T., Gersbach, C. A., & ZFN, B. C. F. TALEN, and CRISPR/Cas-based methods for genome engineering. Trends in Biotechnology 2013, Jul 1, 31(7), 397–405.

[48] Anzalone, A. V., Koblan, L. W., & Liu, D. R. Genome editing with CRISPR–Cas nucleases, base editors, transposases and prime editors. Nature Biotechnology 2020, Jul 1, 38(7), 824–844.

[49] Song, P., Zhang, Q., Xu, Z., Shi, Y., Jing, R., & Luo, D. CRISPR/Cas-based CAR-T cells: Production and application. Biomarker Research 2024, May 31, 12(1), 54.

[50] Li, T., & Yang, B. TAL effector nuclease (TALEN) engineering. Enzyme Engineering: Methods and Protocols 2013, 978, 63–72.

[51] Lin, J. Y., Liu, Y. C., Tseng, Y. H., Chan, M. T., & Chang, C. C. TALE-based organellar genome editing and gene expression in plants. Plant Cell Reports 2024, Mar, 43(3), 61.

[52] Li, W., Liu, J., Zhang, H., Liu, Z., Wang, Y., Xing, L., He, Q., & Du, H. Plant pan-genomics: Recent advances, new challenges, and roads ahead. Journal of Genetics and Genomics 2022, Sep 1, 49(9), 833–846.

[53] Sprink, T., Metje, J., & Hartung, F. Plant genome editing by novel tools: TALEN and other sequence specific nucleases. Current Opinion in Biotechnology 2015, 32, 47–53.

[54] Mukhametzyanova, L., Schmitt, L. T., Torres-Rivera, J., Rojo-Romanos, T., Lansing, F., Paszkowski-Rogacz, M., Hollak, H., Brux, M., Augsburg, M., Schneider, P. M., & Buchholz, F. Activation of recombinases at specific DNA loci by zinc-finger domain insertions. Nature Biotechnology 2024, Jan, 31, 1–1.

[55] Porteus, M. H., & Carroll, D. Gene targeting using zinc finger nucleases. Nature Biotechnology 2005, Aug 1, 23(8), 967–973.

[56] May, D., Paldi, K., & Altpeter, F. Targeted mutagenesis with sequence-specific nucleases for accelerated improvement of polyploid crops: Progress, challenges, and prospects. The Plant Genome 2023, Jun, 16(2):e20298.

[57] Doudna, J. A. The promise and challenge of therapeutic genome editing. Nature 2020, Feb 13, 578(7794), 229–236.

[58] Gao, C. Genome engineering for crop improvement and future agriculture. Cell 2021, Mar 18, 184(6), 1621–1635.

[59] Gordon, D. R., Jaffe, G., Doane, M., Glaser, A., Gremillion, T. M., & Ho, M. D. Responsible governance of gene editing in agriculture and the environment. Nature Biotechnology 2021, Sep, 39(9), 1055–1057.

[60] Manzoor, S., Nabi, S. U., Rather, T. R., Gani, G., Mir, Z. A., Wani, A. W., Ali, S., Tyagi, A., & Manzar, N. Advancing crop disease resistance through genome editing: A promising approach for enhancing agricultural production. Frontiers in Genome Editing 2024, Jun 26, 6, 1399051.
[61] Piergentili, R., Del Rio, A., Signore, F., Umani Ronchi, F., Marinelli, E., & Zaami, S. CRISPR-Cas and its wide-ranging applications: From human genome editing to environmental implications, technical limitations, hazards and bioethical issues. Cells 2021, Apr 21, 10(5), 969.
[62] Paris, K. Genome Editing and Biological Weapons: Assessing the Risk of Misuse, Cham: Springer, 2023.
[63] Lemarié, S., & Marette, S. The socio-economic factors affecting the emergence and impacts of new genomic techniques in agriculture: A scoping review. Trends in Food Science & Technology 2022, Nov 1, 129, 38–48.
[64] Jenkins, D., Dobert, R., Atanassova, A., & Pavely, C. Impacts of the regulatory environment for gene editing on delivering beneficial products. Vitro Cellular & Developmental Biology-Plant 2021, Aug, 57(4), 609–626.
[65] Lemke, S. L. Gene editing in plants: A Nutrition professional's guide to the science, regulatory, and social considerations. Nutrition Today 2022, Mar 1, 57(2), 57–63.
[66] Smyth, S. J. Contributions of genome editing technologies towards improved nutrition, environmental sustainability and poverty reduction. Frontiers in Genome Editing 2022, Mar 17, 4, 863193.
[67] Zaidi, S. S., Vanderschuren, H., Qaim, M., Mahfouz, M. M., Kohli, A., Mansoor, S., & Tester, M. New plant breeding technologies for food security. Science 2019, Mar 29, 363(6434), 1390–1391.
[68] Falk, M. C., Chassy, B. M., Harlander, S. K., Hoban, I. V. T. J., McGloughlin, M. N., & Akhlaghi, A. R. Food biotechnology: Benefits and concerns. The Journal of Nutrition 2002, Jun 1, 132(6), 1384–1390.
[69] Bouis, H. E., Hotz, C., McClafferty, B., Meenakshi, J. V., & Pfeiffer, W. H. Biofortification: A new tool to reduce micronutrient malnutrition. Food and Nutrition Bulletin 2011, Mar, 32(1_suppl1), S31–40.
[70] Li, J., Martin, C., & Fernie, A. Biofortification's contribution to mitigating micronutrient deficiencies. Nature Food 2024, Jan, 5(1), 19–27.
[71] Morelli, L., & Rodriguez-Concepcion, M. Open avenues for carotenoid biofortification of plant tissues. Plant Communications 2023, Jan 9, 4(1) 100466.
[72] Watkins, J. L., & Pogson, B. J. Prospects for carotenoid biofortification targeting retention and catabolism. Trends in Plant Science 2020, May 1, 25(5), 501–512.
[73] Steinwand, M. A., & Ronald, P. C. Crop biotechnology and the future of food. Nature Food 2020, May, 1(5), 273–283.
[74] Chaudhry, A., Hassan, A. U., Khan, S. H., Abbasi, A., Hina, A., Khan, M. T., & Abdelsalam, N. R. The changing landscape of agriculture: Role of precision breeding in developing smart crops. Functional & Integrative Genomics 2023, Jun, 23(2), 167.
[75] Wang, T., Zhang, C., Zhang, H., & Zhu, H. CRISPR/Cas9-mediated gene editing revolutionizes the improvement of horticulture food crops. Journal of Agricultural and Food Chemistry 2021, Mar 18, 69(45), 13260–13269.
[76] Ricroch, A. E., Martin-Laffon, J., Rault, B., Pallares, V. C., & Kuntz, M. Next biotechnological plants for addressing global challenges: The contribution of transgenesis and new breeding techniques. New Biotechnology 2022, Jan 25, 66, 25–35.
[77] Tachikawa, M., & Matsuo, M. Divergence and convergence in international regulatory policies regarding genome-edited food: How to find a middle ground. Frontiers in Plant Science 2023, Jan 30, 14, 1105426.
[78] Bruetschy, C. The EU regulatory framework on genetically modified organisms (GMOs). Transgenic Research 2019, Aug 1, 28(Suppl 2), 169–174.

[79] Caradus, J. Impacts of growing and utilising genetically modified crops and forages–a New Zealand perspective. New Zealand Journal of Agricultural Research 2023, Sep 3, 66(5), 389–418.
[80] Meyer, M., & Heimstädt, C. The divergent governance of gene editing in agriculture: A comparison of institutional reports from seven EU member states. Plant Biotechnology Reports 2019, Oct, 13, 473–482.
[81] European Commission: Proposal for a REGULATION of the EUROPEAN PARLIAMENT and of the COUNCIL on Plants Obtained by Certain New Genomic Techniques and Their Food and Feed, and Amending Regulation (EU) 2017/625; 2023
[82] Dima, O., Custers, R., De Veirman, L., & Inzé, D. EU legal proposal for genome-edited crops hints at a science-based approach. Trends in Plant Science 2023, Dec 1, 28(12), 1350–1353.
[83] Purnhagen, K., Ambrogio, Y., Bartsch, D., Eriksson, D., Jorasch, P., Kahrmann, J., Kardung, M., Molitorisová, A., Monaco, A., Nanda, A. K., & Romeis, J. Options for regulating new genomic techniques for plants in the European Union. Nature Plants 2023, Dec, 9(12), 1958–1961.
[84] Asveld, L., Osseweijer, P., & Posada, J. A. Societal and ethical issues in industrial biotechnology. Sustainability and Life Cycle Assessment in Industrial Biotechnology 2019 Jul 22:121–41.
[85] Trump, B., Cummings, C., Klasa, K., Galaitsi, S., & Linkov, I. Governing biotechnology to provide safety and security and address ethical, legal, and social implications. Frontiers in Genetics 2023, Jan 11, 13, 1052371.
[86] Nicholl, D. S. An Introduction to Genetic Engineering, Cambridge University Press, 2023, Mar 2.
[87] Hafeez, U., Ali, M., Hassan, S. M., Akram, M. A., & Zafar, A. Advances in breeding and engineering climate-resilient crops: A comprehensive review. International Journal of Research and Advances in Agricultural Sciences 2023, Jun 25, 2(2), 85–99.
[88] Chen, Y. C. Introductory Chapter: Gene Editing Technologies and Applications. London, UK: IntechOpen, 2019, May 14.
[89] Saha, P. U., Ali, A., & Khan, J. Genetically engineered microorganisms. In: Bhat, T.A., & Al-Khayri, J.M. (Eds.). Genetic Engineering, Apple Academic Press, 2023, Sep 15, 125–157.
[90] Nerkar, G., Devarumath, S., Purankar, M., Kumar, A., Valarmathi, R., Devarumath, R., & Appunu, C. Advances in crop breeding through precision genome editing. Frontiers in Genetics 2022, Jul 14, 13, 880195.
[91] Khalil, A. M. The genome editing revolution. Journal of Genetic Engineering and Biotechnology 2020, Dec 1, 18(1), 68.
[92] Van Eck, J. Applying gene editing to tailor precise genetic modifications in plants. Journal of Biological Chemistry 2020, Sep 18, 295(38), 13267–13276.
[93] Wada, N., Ueta, R., Osakabe, Y., & Osakabe, K. Precision genome editing in plants: State-of-the-art in CRISPR/Cas9-based genome engineering. BMC Plant Biology 2020, Dec, 20, 1–2.
[94] Morelli, L., & Rodriguez-Concepcion, M. Open avenues for carotenoid biofortification of plant tissues. Plant Communications 2023, Jan 9, 4(1) 100466.
[95] Su, W., Xu, M., Radani, Y., & Yang, L. Technological development and application of plant genetic transformation. International Journal of Molecular Sciences 2023, Jun 26, 24(13), 10646.
[96] Chuang, C. K., & Lin, W. M. Points of View on the tools for Genome/Gene editing. International Journal of Molecular Sciences 2021, Sep 13, 22(18), 9872.
[97] Sharma, A., Ali, A., & Khan, J. Tools used in genetic engineering. In: Bhat, T.A., & Al-Khayri, J.M. (Eds.). Genetic Engineering, Apple Academic Press, 2023, Sep 15, 45–76.
[98] Zhang, D., Hussain, A., Manghwar, H., Xie, K., Xie, S., Zhao, S., Larkin, R. M., Qing, P., Jin, S., & Ding, F. Genome editing with the CRISPR-Cas system: An art, ethics and global regulatory perspective. Plant Biotechnology Journal 2020, Aug, 18(8), 1651–1669.
[99] Turnbull, C., Lillemo, M., & Hvoslef-Eide, T. A. Global regulation of genetically modified crops amid the gene edited crop boom–a review. Frontiers in Plant Science 2021, Feb, 24, 12, 630396.

[100] Wang, D. C., & Wang, X. Off-target genome editing: A new discipline of gene science and a new class of medicine. Cell Biology and Toxicology 2019, Jun, 35(3), 179–183.

[101] Han, H. A., Pang, J. K., & Soh, B. S. Mitigating off-target effects in CRISPR/Cas9-mediated in vivo gene editing. Journal of Molecular Medicine 2020, May, 98(5), 615–632.

[102] Wang, Q., Alariqi, M., Wang, F., Li, B., Ding, X., Rui, H., Li, Y., Xu, Z., Qin, L., Sun, L., & Li, J. The application of a heat-inducible CRISPR/Cas12b (C2c1) genome editing system in tetraploid cotton (*G. hirsutum*) plants. Plant Biotechnology Journal 2020, Dec, 18(12), 2436–2443.

[103] Javaid, N., & Choi, S. CRISPR/Cas system and factors affecting its precision and efficiency. Frontiers in Cell and Developmental Biology 2021, Nov 24, 9, 761709.

[104] Movahedi, A., Aghaei-Dargiri, S., Li, H., Zhuge, Q., & Sun, W. CRISPR variants for gene editing in plants: biosafety risks and future directions. International Journal of Molecular Sciences 2023, Nov 13, 24(22), 16241.

[105] Delhove, J., Osenk, I., Prichard, I., & Donnelley, M. Public acceptability of gene therapy and gene editing for human use: A systematic review. Human Gene Therapy 2020, Jan 1, 31(1–2), 20–46.

[106] Almeida, M., & Ranisch, R. Beyond safety: Mapping the ethical debate on heritable genome editing interventions. Humanities and Social Sciences Communications 2022, Apr 20, 9(1), 1–4.

[107] Alhakamy, N. A., Curiel, D. T., & Berkland, C. J. The era of gene therapy: From preclinical development to clinical application. Drug Discovery Today 2021, Jul 1, 26(7), 1602–1619.

[108] Ayanoğlu, F. B., Elçin, A. E., & Elçin, Y. M. Bioethical issues in genome editing by CRISPR-Cas9 technology. Turkish Journal of Biology 2020,44(2), 110–120.

[109] Tait, J. Environmental risks and the regulation of biotechnology. In: Lowe, P., Marsden, T., & Whatmore, S. (Eds.) Technological Change and the Rural Environment, Routledge, 2023, Jun 8, 168–202.

[110] Dolezel, M., Lang, A., Greiter, A., Miklau, M., Eckerstorfer, M., Heissenberger, A., Willée, E., & Züghart, W. Challenges for the post-market environmental monitoring in the European Union imposed by novel applications of genetically modified and genome-edited organisms. BioTech 2024, May 15, 13(2), 14.

[111] Rout, G. R., Swain, R., & Sahoo, D. P. Journey of genetically modified crops: Status and prospects. Magna Scientia Advanced Research and Reviews 2023,7(1), 103–128.

[112] Sakib, S. N. The role of innovation in driving the bioeconomy: The challenges and opportunities. Handbook of Research on Bioeconomy and Economic Ecosystems 2023, 288–311.

[113] Mahdizade Ari, M., Dadgar, L., Elahi, Z., Ghanavati, R., & Taheri, B. Genetically engineered microorganisms and their impact on human health. International Journal of Clinical Practice 2024,2024(1), 6638269.

[114] Wu-Wu, J. W., Guadamuz-Mayorga, C., Oviedo-Cerdas, D., & Zamora, W. J. Antibiotic resistance and food safety: Perspectives on new technologies and molecules for microbial control in the food industry. Antibiotics 2023, Mar 10, 12(3), 550.

[115] Bhowmik, S. N., & Patil, R. T. Application of microbial biotechnology in food processing. In: Prasad, R., Gill, S. S., & Tuteja, N (Eds.) Crop Improvement Through Microbial Biotechnology, Elsevier, 2018, Jan 1, 73–106.

[116] Harikrishna, S. P., Gajghate, R., Devate, N. B., Shiv, A., Mehta, B. K., Sunilkumar, V. P., Rathan, N. D., Mottaleb, K. A., Sukumaran, S., & Jain, N. Breaking the yield barriers to enhance genetic gains in wheat. In: Kashyap, P.L., et al. (Eds.) New Horizons in Wheat and Barley Research: Global Trends, Breeding and Quality Enhancement, Singapore: Springer Singapore, 2022, Mar 27, 179–226.

[117] Andorf, C., Beavis, W. D., Hufford, M., Smith, S., Suza, W. P., Wang, K., Woodhouse, M., Yu, J., & Lübberstedt, T. Technological advances in maize breeding: Past, present and future. Theoretical and Applied Genetics 2019, Mar 1, 132, 817–849.

[118] Krishnan, R. R., Sivakumar, B., Jagannath, N., Rao, N. R., Suresh, N., Nagella, P., Al-Khayri, J. M., & Jain, S. M. CRISPR-Cas9 genome editing of crops: Food and nutritional security. In: Abd-Elsalam, K.

A., Ahmad, A., & Zhang, B. H. (Eds.) CRISPRized Horticulture Crops, Academic Press, 2024, Jan 1, 161–190.

[119] George, D. R., Danciu, M., Davenport, P. W., Lakin, M. R., Chappell, J., & Frow, E. K. A bumpy road ahead for genetic biocontainment. Nature Communications 2024, Jan 20, 15(1), 650.

[120] Convention on Biological Diversity. Cartagena Protocol on Biodiversity to the Convention on Biological Diversity, Secretariat of the Convention of Biological Diversity. 2000 ISBN: 92–807-1924-6.

[121] Convention on Biological Diversity (2023). The Nagoya – Kuala Lumpur supplementary protocol on liability and redress to the Cartagena protocol on biosafety. https://bch.cbd.int/protocol/supplementary. Accessed on 1 July 2024.

[122] Bartkowski, B., & Baum, C. M. Dealing with rejection: An application of the exit–voice framework to genome-edited food. Frontiers in Bioengineering and Biotechnology 2019, Mar 22, 7, 57.

[123] Gbadegesin, L. A., Ayeni, E. A., Tettey, C. K., Uyanga, V. A., Aluko, O. O., Ahiakpa, J. K., Okoye, C. O., Mbadianya, J. I., Adekoya, M. A., Aminu, R. O., & Oyawole, F. P. GMOs in Africa: Status, adoption and public acceptance. Food Control 2022, Nov 1, 141, 109193.

[124] Sanou, E. I., Gheysen, G., Koulibaly, B., Roelofs, C., & Speelman, S. Farmers' knowledge and opinions towards bollgard II® implementation in cotton production in western Burkina Faso. New Biotechnology 2018, May 25, 42, 33–41.

[125] ISAAA. (2019). Ugandan president ready to sign GMO bill into law. https://www.isaaa.org/kc/cropbiotechupdate/article/default.asp?ID=18030 (Accessed 1 July 2024).

[126] ISAAA. (2021). Genome editing in Africa's agriculture 2021: An early take-off. https://africenter.isaaa.org/wp-content/uploads/2021/04/GENOME-EDITING-IN-AFRICA-FINAL.pdf (Accessed 1 July 2024).

[127] Jones, M. S., & Brown, Z. S. Food for thought: Assessing the consumer welfare impacts of deploying irreversible, landscape-scale biotechnologies. Food Policy 2023, Nov 1, 121, 102529.

[128] ECFR 2024. https://www.ecfr.gov/current/title-7/subtitle-B/chapter-I/subchapter-M/part-205 (Accessed 27 June 2024).

[129] Kendig, C., Selfa, T., & Thompson, P. B. Biotechnology ethics for food and agriculture. Science 2022, Jun 17, 376(6599), 1279–1280.

[130] USDA, Agricultural marketing service, BE disclosure (2022); https://www.ams.usda.gov/rules-regulations/be Accessed on 1 July 2024.

[131] Jaffe, G., & Kuzma, J. New bioengineered (aka GM) food disclosure law: Useful information or consumer confusion?. FDLI Update 2021, 17.

[132] European parliament: New genomic techniques: MEPs back rules to support green transition of farmers. EP press release;2024. Available on https://www.europarl.europa.eu/news/en/press-room/20240202IPR17320/new-genomic-techniques-meps-back-rules-to-support-green-transition-of-farmers Accessed on 30 June 2024.

[133] European parliament: Plants obtained by certain new genomic techniques and their food and feed; 2024. Available on https://www.europarl.europa.eu/doceo/document/TA-9-2024-0067_EN.html Accessed on 30 June 2024.

[134] Mallapaty, S. China's approval of gene-edited crops energizes researchers. Nature 2022, Feb 24, 602(7898), 559–560.

[135] Yang, F., Zheng, K., & Yao, Y. China's regulatory change toward genome-edited crops. Trends in Biotechnology 2024, Jul 1, 42(7), 801–806.

[136] Sprink, T., Wilhelm, R., & Hartung, F. Genome editing around the globe: An update on policies and perceptions. Plant Physiology 2022, Nov 1, 190(3), 1579–1587.

[137] Vora, Z., Pandya, J., Sangh, C., & Vaikuntapu, P. R. The evolving landscape of global regulations on genome-edited crops. Journal of Plant Biochemistry and Biotechnology 2023, Dec, 32(4), 831–845.

[138] Cummings, C., Selfa, T., Lindberg, S., & Bain, C. Identifying public trust building priorities of gene editing in agriculture and food. Agriculture and Human Values 2024, Mar, 41(1), 47–60.

[139] Lindberg, S. A., Peters, D. J., & Cummings, C. L. Gene-edited food adoption intentions and institutional trust in the United States: benefits, acceptance, and labeling. Rural Sociology 2023, Jun, 88(2), 392–425.

[140] Tabei, Y., Shimura, S., Kwon, Y., Itaka, S., & Fukino, N. Analyzing twitter conversation on genome-edited foods and their labeling in Japan. Frontiers in Plant Science 2020, Oct 22, 11, 535764.

[141] Caradus, J. R. Processes for regulating genetically modified and gene edited plants. GM Crops & Food 2023, Dec 31, 14(1), 1–41.

[142] Anderson, J. A., Ellsworth, P. C., Faria, J. C., Head, G. P., Owen, M. D., Pilcher, C. D., Shelton, A. M., & Meissle, M. Genetically engineered crops: Importance of diversified integrated pest management for agricultural sustainability. Frontiers in Bioengineering and Biotechnology 2019, Feb 20, 7, 24.

[143] Ruder, S. L., & Kandlikar, M. Governing gene-edited crops: Risks, regulations, and responsibilities as perceived by agricultural genomics experts in Canada. Journal of Responsible Innovation 2023, Jan 2, 10(1), 2167572.

[144] Fernandes, P. M., Favaratto, L., Merchán-Gaitán, J. B., Pagliarini, R. F., Zerbini, F. M., & Nepomuceno, A. L. Regulation of CRISPR-edited plants in Latin America. In: Global Regulatory Outlook for CRISPRized Plants, Academic Press, 2024, 197–212.

[145] Namibia biosafety act no. 7 of 2006, Accessed on 12 February 2025. https://www.lac.org.na/laws/annoSTAT/Biosafety%20Act%207%20of%202006.pdf

[146] Segretin, M. E., Soto, G. C., & Lorenzo, C. D. Latin America: A hub for agro-biotechnological innovations. Annals of Botany 2025 Apr 1;135(4):629–42.

Oluwatosin Ademola Ijabadeniyi, Christiana Eleojo Aruwa, and Titilayo Adenike Ajayeoba

Chapter 21 Conclusion and outlook

As a synchronized one-stop collation on key topics surrounding current and future trends in food biotechnology (FB), this book portrays the works and opinions of experts, researchers, and professionals across varied cross-disciplinary FB fields and subfields, as they relate to academia, industry, and global economies. Products from such stakeholders and interdisciplinary collaborations are targeted at reducing food waste and emerging intercontinental hunger trends, encouraging innovative derivations of high nutritive foods via application of new food developing tools, as well as the added advantage of guiding consumer choices, and enhancing consumer rights, protection, and safety, among others. Food and nutrition trends impact health and both factors are key drivers in the use of biotechnological tools for innovative food advancements [1]. Given the increase in consumer awareness and demand for "safe," "natural," "organic," and innovative superfoods for management or treatment of diseases, the agrifood industry must awake to bridge the supply gap [2]. Furthermore, regulatory inspections should be coupled with comprehensive training programs to effectively promote a strong food safety culture. This should entail all stakeholders working to ensure adequate food systems surveillance from farm to fork, and consumer protection in the face of rising global demands.

It is estimated that one of every nine individuals do not have access to a life that is active and healthy, and malnutrition remains a major global health risk relative to infections such as tuberculosis, HIV/AIDS, and malaria [3]. Food insecurity, malnutrition, hunger, low per-capita income are also most prominent in the African continent relative to other countries around the globe [4]. As such, combating the malnutrition and nutrition scourge from a multidimensional perspective cannot be overemphasized to steer the teeming world population from both current and emerging disease trends [5]. This is where FB tools play a major role in shaping the future of innovative foods and foods products. The demand for functional foods, nutraceuticals, or personalized healthy food products continue to show increasing market trends. In line with the derivation of such high-value foods, there is also an increased awareness with optimizing food products production processes in a way that significantly reduces food waste for environmental protection. FB workflows may hold the key to bridging

Oluwatosin Ademola Ijabadeniyi, Department of Biotechnology and Food Science, Durban University of Technology, Steve Biko Road, Berea 16, Durban, 4001, South Africa, e-mail: oluwatosini@dut.ac.za
Christiana Eleojo Aruwa, Department of Biotechnology and Food Science, Faculty of Applied Sciences, Durban University of Technology, Durban 4001, South Africa, e-mail: christianaa@dut.ac.za
Titilayo Adenike Ajayeoba, Food Science and Nutrition Programme, Microbiology Department, Adeleke University, Ede, Osun

https://doi.org/10.1515/9783111441238-022

emerging market demand gaps for new raw material [6] and food products, and could extend its application towards the generating of innovative food packaging or modified atmosphere environments. Food product raw materials or the extrinsic or intrinsic features of foods may be engineered to produce highly functional, nutrient-rich products that enhance negative health indices and mitigate adverse outcomes. Achieving this cannot be extricated from FB advancements that will continue to impact food production and biotransformation processes from start to finish whether solely biological, physicochemical, biochemical in nature, or a combination of these. These have birthed the emergence of market-savvy terminologies such as nutraceuticals, functional foods within the growing field of nutrigenomics. Again, the upgrade of conventional methods or introduction of new biotechnological advancements contributes to significantly enhance process efficiency, thus improving profitability for all stakeholders across the food production chain and market [7].

The FB market involves product, technology, and application segments, and is estimated to grow from its current value of about $20 billion to over $50 billon in value by the year 2032, a trend spurred on by ever-increasing food demands [7]. These demands are largely driven by tools incorporated within FB strategies that are dependent on genetic insights into manufacturing processes and organisms' growth, and the engineering of microbes (metabolic and genetic) to suit specific needs [8], all in a bid to enhance quality and yield of advanced products. These technologies further encourage new capabilities development and the acquisition of patent products from innovative agrifood-based processes which improve the ability of small enterprises to stay afloat in evolving, and competitive global market trends. Asides increased food demands, an additional factor driving the FB market include evolving DNA-based tools (hybridization, genetic engineering) which increase the derivation of bio-derived synthetics from biocatalytic [9], fermentative, and genetic engineered workflows, and transgenic crops and food products (pests-, disease-, and drought-resistant varieties, etc.). In contrast, restraints around the FB market include length of research time to bring innovative products to market, and issues with consumer acceptance of genetically "modified" products. Nevertheless, opportunities such as advancements in technologies, increased demand for healthy foods, attributed to the FB market cannot be overemphasized. Other FB market segments include gene editing, ribonucleic acid interference (RNAi), and biochip technologies. In recent a development, the conventional, closed, batched fermentation process has been repurposed for the generation of transformed natural colors (Givaudan's acid-stable blue phycocyanin produced by the extremophilic *Galdieria* microalga) with benefits for various culinary applications in the food industry. Such example demonstrates the immense potential which could accrue from meeting points between "old" tools and innovation [7, 10]. In other advances mushroom mycelia have been transformed into eco-friendly, organic packaging materials, as well as mushroom with bacon-like features, laboratory cultivated hamburger meats have also been developed [11].

Again, advances in "omics" and bioinformatic workflows have led to the discovery of new sets of extracellular biological molecules and factors [*Penicillium scleroti-*

genum and *Aspergillus niger*-derived carbohydrate-active enzymes (CAZymes)] from food residues (sweet potato starch), further enhancing the concept of "waste" to wealth [12]. Loss or wastage is experienced across all points in the food chain [13], and reduction of such wastes is key to ensuring global food security [14]. While correlated problems such as climate change, globalization, and urbanization are key disruptors of the food chain, they also exacerbate food insecurity issues [15]. Hence, food industry stakeholders; institutions, businesses, individuals, policymakers, governments, and agencies need to work collaboratively to ensure sustainability of innovations, research, planning, and leadership through co-funded projects and use of realistic and implementable national and international frameworks and strategies [16].

FB has also ushered in a new age of safer, smarter foods, and food products, which are traceable with digital tools [Internet of things (IoT), blockchain, artificial intelligence (AI), and machine learning (ML)] [17]. These digitalization tools have made the processing of "big data" more cost-effective while mitigating food waste and enhancing real time monitoring and transparency across a "smarter" food chain/system. Blockchain also suffices for improving business intelligence models by assisting with making better business decisions and waste reduction [18], and reducing production cost [19], hence its enhanced adoption in the food industry. Likewise, the emergence of innovative virtual reality platforms and programs could aid in the training and capacity building of FB industry stakeholders for more innovative outcomes in the near and far future [20].

FB has an immense capacity to enhance food quality, processes, and safety [21], and production, and within sustainable food models that ensure optimization of food wastes. It achieves this through the engineering of microbial, animal and plant genes, foods shelf-life extension using innovative packaging tools [11], new food additives, raw materials derivation via upgraded microbial fermentation processes [22], enzymes [6], integration of AI and blockchain in food production and surveillance, as well as novel ways of manufacturing foods using three-dimensional printing technologies [23]. Thus, it is expected that advancements across FB research interdisciplinary fields and industry will continue to generate novel raw materials, products, and enhanced production workflows. Innovative, cutting-edge research is essential for addressing challenges related to food quality, safety, and shortages. Example of such research is the use of CRISPR-Cas9 technology in developing climate resilient crops with the added advantage of achieving food security [24]. The application of CRISPR in food production and processing should, however, include compliance with regulatory standards and the execution of thorough risk assessments [25]. It is also crucial to consider existing research and development on genetical modified foods while addressing concerns related to safety and also conducting thorough risk assessments. This textbook therefore portrays an outlook and in-depth expatiation of current, emerging, and futuristic FB processes. It also showcased study gaps that could guide future biotechnological trends shaping the food industry, especially as they relate to the "nutrient" content of current and future foods. The book's chapters covered a broad range of FB topics, but which are not limited to those itemized as follows:

i. microbial engineering tools for fermentation process enhancement;
ii. food waste optimization;
iii. novel food packaging development;
iv. functional food and nutraceutical development;
v. emerging novel food challenges (hypo-allergenicity, acceptance, etc.);
vi. impact of blockchain and AI in food systems advancements;
vii. bioengineering of foods key constituents (macro- and micronutrients);
viii. engineering, derivation, and application of enzymes and enzyme kinetics in food processes; and
ix. food industry waste management and optimization.

This book has achieved the coverage of state-of-the-art topics in food biotechnology in a bid to ensure enhanced global health outcomes and steer future food research in the right direction, while piquing the interest of all agrifood industry stakeholders and policymakers on their roles in maintaining stable and sustainable food biotechnology models and workflows.

References

[1] Liu, Y., Gong, J. S., Marshall, G., Su, C., Hall, M., Li, H., Xu, G. Q., Shi, J. S., & Xu, Z. H. Protein engineering of NADH pyrophosphatase for efficient biocatalytic production of reduced nicotinamide mononucleotide. Frontiers in Bioengineering and Biotechnology 2023, 11, 1159965.

[2] Kieliszek, M., Piwowarek, K., Kot, A. M., Wojtczuk, M., Roszko, M., Bryła, M., & Trajkovska, P. A. Recent advances and opportunities related to the use of bee products in food processing. Food Science & Nutrition 2023, 11, 4372–4377.

[3] Food and Agriculture Organization (FAO). The state of food security and nutrition in the world, 2019. (Accessed September 20, 2024, at http://www.fao.org/state-offood-security-nutrition).

[4] Jimoh, B. Africa still most food insecure continent – FAO, 2014. (Accessed August 30, 2024, at https://http://www.vanguardngr.com/2014/03/africa-still-food-insecure-continentfao/?utm_source=dlvr.it&utm_medium=twitter).

[5] Lancet. A future direction for tackling malnutrition, 2019. (Accessed August 30, 2024, at https://www.thelancet.com/journals/lancet/article/PIIS0140-6736(19)33099-5/fulltext).

[6] Naik, B., Kumar, V., Goyal, S. K., Dutt Tripathi, A., Mishra, S., Joakim Saris, P. E., Kumar, A., Rizwanuddin, S., Kumar, V., & Rustagi, S. Pullulanase: Unleashing the power of enzyme with a promising future in the food industry. Frontiers in Bioengineering and Biotechnology 2023, 11, 1139611.

[7] Straits research report (SRFB584DR). Food biotechnology market, 2024. (Accessed September 22, 2024, at https://straitsresearch.com/report/food-biotechnology-market).

[8] Zhang, C., Ottenheim, C., Weingarten, M., & Ji, L. Microbial utilization of next-generation feedstocks for the biomanufacturing of value-added chemicals and food ingredients. Frontiers in Bioengineering and Biotechnology 2022, 10, 874612.

[9] Mesbah, N. M. Industrial biotechnology based on enzymes from extreme environments. Frontiers in Bioengineering and Biotechnology 2022, 10, 870083.

[10] Food Business News. Biotechnology: Growing options for natural colors production, 2024. (Accessed September 17, 2024, at https://www.foodbusinessnews.net/articles/25745-biotechnology-growing-options-for-natural-colors-production).
[11] Verdone, M. Emerging biotechnology trends in the food industry. University of Wisconsin (UW) online collaboratives, 2020. (Accessed September 12, 2024, at https://uwex.wisconsin.edu/stories-news/the-affinity-column-biotechnology-developments-in-the-food-industry/).
[12] Barrett, K., Zhao, H., Hao, P., Bacic, A., Lange, L., Holck, J., & Meyer, A. S. Discovery of novel secretome CAZymes from *Penicillium sclerotigenum* by bioinformatics and explorative proteomics analyses during sweet potato pectin digestion. Frontiers in Bioengineering and Biotechnology 2022, 10, 950259.
[13] Natural Resources Defense Council (NRDC) Report. Tackling food waste in cities: A policy and program toolkit. The Rockefeller Foundation 2019, 79, 1–53.
[14] Food and Agriculture Organization (FAO). Do good. Save food, 2016. (Accessed August 22, 2024, at https://www.fao.org/save-food).
[15] Rodin, J. The Resilience Dividend, Managing Disruption, Avoiding Disaster, and Growing Stronger in an Unpredictable World, London: Judith Rodin, Profile Books Ltd, 2015, 398.
[16] Tesfahun, W. Climate change mitigation and adaptation through biotechnology approaches: A review. International Journal of Agriculture, Forestry and Life Sciences 2018, 2, 62–74.
[17] United States Food and Drug Administration (FDA). Statement from acting FDA Commissioner Ned Sharpless, M.D., and Deputy Commissioner Frank Yiannas on Steps to Usher the U.S. into a New Era of smarter food safety, 2019. (Accessed September 17, 2024, at https://www.fda.gov/news-events/press-announcements/statement-acting-fdacommissioner-ned-sharpless-md-and-deputy-commissioner-frank-yiannas-steps-usher.)
[18] The Grocer. How blockchain is turning back the tide on food waste, 2019. (Accessed September 18, 2024, at https://www.thegrocer.co.uk/promotional-features/how-blockchain-is-turning-back-the-tide-on-food-waste/590970.article).
[19] Frangoul, A. Blockchain to save food industry $31 billion, 2019. (Accessed September 16, 2024, at https://www.cnbc.com/2019/11/27/blockchain-to-save-food-industry-31-billion-new-research-says.html).
[20] Hyde, R. How virtual reality training could reduce your chance of food poisoning, 2019. (Accessed September 21, 2024, at https://theconversation-com.cdn.ampproject.org/v/s/theconversation.com/amp/howvirtual-reality-training-could-reduce-your-chance-of-food-poisoning-95195?fbclid=IwAR0eAYYs6DNYGk65NOPFWWCbnSJNXE2lgZpVEtIhzySbYaZZW2D1hxlGpcU&_js_v=0.1.)
[21] Lahiri, D., Nag, M., Dutta, B., Sarkar, T., Pati, S., Basu, D., Abdul Kari, Z., Wei, L. S., Smaoui, S., Wen Goh, K., & Ray, R. R. Bacteriocin: A natural approach for food safety and food security. Frontiers in Bioengineering and Biotechnology 2022, 24, 1005918.
[22] Huang, X., Li, H., Han, T., Wang, J., Ma, Z., & Yu, X. Isolation and identification of protease-producing *Bacillus amyloliquefaciens* LX-6 and its application in the solid fermentation of soybean meal. Frontiers in Bioengineering and Biotechnology 2023, 11, 1226988.
[23] Cabrera-Barjas, G., Banerjee, A., Valdes, O., Moncada, M., Sirajunnisa, A. R., Surendhiran, D., Ramakrishnan, G., Rani, N. S., Hamidi, M., Kozani, P. S., & Kozani, P. S. Food biotechnology: Innovations and challenges. In: Bhat, R., editor. Future Foods – Global Trends, Opportunities, and Sustainability Challenges, Cambridge, MA, USA: Academic Press, 2022, 697–719.
[24] Ndudzo, A., Makuvise, A. S., Moyo, S., & Bobo, E. D. CRISPR-Cas9 genome editing in crop breeding for climate change resilience: Implications for smallholder farmers in Africa. Journal of Agriculture and Food Research 2024, 28, 101132.
[25] Brookes, G., & Smyth, S. J. Risk-appropriate regulations for gene-editing technologies. GM Crops & Food 2024, 15, 1–14.

Index

https://doi.org/10.1515/9783111441238-023

www.ingramcontent.com/pod-product-compliance
Lightning Source LLC
Jackson TN
JSHW061107110725
87393JS00021B/81

* 9 7 8 3 1 1 1 4 4 1 2 1 4 *